Concepts in
MODERN
BIOLOGY

DAVID KRAUS, B.S., M.S. in Ed., Chairman Science Department, Far Rockaway High School, New York City. Member of the New York State Biology Syllabus Committee and President of the Biology Chairmen's Association of New York City. Member of the Committee on Social Implications of Biology Teaching of the National Association of Biology Teachers. Winner of NSTA Teaching Award. Served on curriculum project at Sloan-Kettering Institute and did research on *Hydra* at Adelphi University and at Boyce Thompson Institute.

CONTRIBUTING EDITORS

OTHO E. PERKINS, Supervisor of Science, Columbus Public Schools, Columbus, Ohio. Teacher of science, elementary, high school and university. Taught science on TV; author of several science books.

ROBERT L. LEHRMAN, Chairman Science Dept., Roslyn High School, New York. Science teacher for 25 years; author *Race, Evolution, and Mankind,* several other science texts and other popularized scientific books.

LOUIS P. BANCHERI, Chairman Science Dept., H. Frank Carey Jr.-Sr. High School, New York. Member of Molloy College staff; consultant to Diocese of Rockville Centre. NSF Institutes at Waldemar Medical Research Foundation, Univ. of Calif. Berkeley and Polytechnic Institute of Brooklyn.

JACK ROBBINS, District Supervisor of Science, Long Beach, N.Y. City School District. Author of several science texts. Educational consultant, N.Y.S. Dept. of Education, Bureau of Elementary Curriculum. Adjunct Professor of Chemistry and Physical Science, Pace College.

Concepts
in
MODERN
BIOLOGY

DAVID KRAUS

OTHO E. PERKINS

CAMBRIDGE BOOK COMPANY, INC.
A subsidiary of Cowles Communications, Inc.
Bronxville, New York

Library of Congress Catalog Number: 72-84784

ACKNOWLEDGMENTS

Artwork   Versatron Corporation

Cover photo courtesy CCM, GENERAL BIOLOGI-CAL, Inc. Diatom and radula. A radula is a tongue-like organ with rows of horny protuberances which functions as a rasp to macerate ingested food. It is found in all mollusks except the oyster and mussel group.

CONTENTS

vi

UNIT FOUR: REPRODUCTION AND DEVELOPMENT

UNIT FIVE: GENETICS

UNIT SIX: PLANTS AND ANIMALS IN THEIR ENVIRONMENT

UNIT SEVEN: EVOLUTION

APPENDIX

PREFACE

Achieving clear explanations for the average student has been one of the major goals in writing this book. If the teacher can rely upon the textbook to present the major concepts and content of the course, he can set aside time for his students to learn additional aspects of science "by practicing the activities of the scientist."

Today's biology teacher is rightly concerned with inquiry and the processes of science. He wants to utilize challenging inquiries in the laboratory and field, to introduce enrichment through guided independent study, to use stimulating new visual aids, to promote provocative classroom discussions, and to guide pupils in projects and research investigations. But it is still essential for the student to learn the basic concepts of biology as a foundation for further study and as a basis for his contributions as a citizen and potential scientist. The allowance of time for the activities aspect of biology teaching can be achieved through use of a book which is appropriate for the students and for the objectives of the course.

With the aid of my colleagues and editors, I have attempted to introduce into this book the following characteristics:

(1) *Adjusted to the ability of the average pupil.* Recent years have witnessed an enormous increase in the depth and sophistication of the biologist's understanding of life, but there has been little concomitant evolution of pupil ability to cope with these understandings. How deep can we go with our students? The author and his colleagues believe that they have made a realistic appraisal of the proper depth of treatment. This appraisal is based on study of the syllabi of various states and on classroom experience gained in teaching the concepts of modern biology to average students. *Concepts in Modern Biology* contains the kinds of explanations which have proven successful in the classroom. Each topic is explained in detail instead of being merely touched upon to give a semblance of completeness. Throughout, there has been an effort to keep the writing simple and direct.

(2) *Conceptual approach.* The subject matter has been selected and organized to help the student develop an understanding of the major concepts of modern biology. The concept of *organic evolution* is introduced early and remains as the pervading theme for the book. *Homeostasis* and the *complementarity of structure and function* receive repeated emphasis. The concept of *energy* utilization provides a unifying background in biochemistry and ecology. The *adaptation of the organism to its environment*

is a concept which threads the sections on plant and animal maintenance, ecology, and evolution.

(3) *Up-to-date*. The subject matter reflects the modernization introduced by the yellow, blue, and green versions of the Biological Sciences Curriculum Study. *Concepts in Modern Biology* can be used as a basic text, as a supplementary text, or as a review for courses which are based upon BSCS innovations in course content. The student will find up-to-date explanations concerning *biochemistry, cellular respiration and ATP, photosynthesis, cellular structure, biochemical genetics and DNA,* the *heterotroph hypothesis,* and *evolution.*

(4) *Balanced treatment*. Appropriate attention is given to the molecular, the cellular, and the organismal aspects of modern biology. Biochemistry, introduced early, serves as a basis for understanding life processes both within the cell and in the multicellular organism. But the total living organism is not neglected and receives extensive consideration, particularly in the unit on ecology.

(5) *Organized for flexibility*. To meet the needs of varying groups of students, many courses of study now provide for a *basic core* of content and for semi-optional *extended areas* which take up the same topics in greater depth. In consonance with this approach, each of the major topics of this book has been written in such manner that the basic core material appears first and is then followed by in-depth treatment.

The chapters are sub-divided into A, B, and C sections to aid the teacher in assigning end-of-chapter exercise material. This carries the corresponding A, B, and C letters. Section letters enclosed by a circle indicate the more difficult or enrichment material, usually toward the end of the chapters. Since the various sections are self-contained, the teacher can easily omit a circled-letter section without disturbing the continuity of the teaching program. This organization of content, with parallel end-of-chapter materials, permits considerable flexibility in the use of the book.

(6) *Type-animal study. Paramecium, Hydra,* the earthworm, the grasshopper, and man are utilized as type animals to illustrate the evolution of adaptive mechanisms. For courses which examine man in greater depth, there are separate enrichment sections on human structure and function.

(7) *Illustrations.* Many of the 350 illustrations were formulated for direct use in the classroom. These should help the student to summarize and visualize difficult ideas that have been explained in the text. The book also contains many charts that are useful for teaching and for providing a broad overview.

(8) *End-of-chapter materials.* Approximately 2,000 questions are provided for purposes of study and review. The numerous completion and multiple-choice questions are arranged in the same sequence as the ideas in the chapter. Additional chapter tests contain questions in an arbitrary sequence. Reasoning exercises, at the end of each section, provide challenging ideas for brief essays or classroom discussion.

The textbook is only one aspect of a course in biology. Good teachers have always supplemented the text with numerous additional readings and activities, both inside and outside of the classroom. I hope that this book, in the hands of the pupils, will permit my colleagues to devote greater time and effort for their unique personal contribution — providing experiences which foster pupil inquiry, creativity, and imagination.

I am deeply indebted to my colleagues for their painstaking care in reading the manuscript, for their numerous suggestions, and for their valued advice. I am also deeply indebted to my wife, Beatrice, for her unfailing assistance, encouragement and patience.

—— DAVID KRAUS

THE STUDY OF LIFE

Living things consist of matter, but they have characteristics that seem to go beyond matter. Although a stone hugs the earth, a tree seems to be freer as it grows upward toward the sky. Moreover, an animal can move from place to place and a man can send his thoughts around the globe in a matter of moments, indicating even more freedom and complexity. We can observe these things, but we should also observe that living things exist in an environment and are dependent upon matter and energy as well as on other living things.

Scientists have already stated many of the interrelationships of the living and nonliving. In your study of biology, learn what has already been discovered and is now held to be true. But go beyond this; use your own mind to come to grips with nature and the ideas of others. Observe, think, conjecture, and check your conjectures by observation and experimentation. In this way, you will be acting as a scientist. Whatever your future occupation, the scientist's way of thinking will have carry-overs into your life as a thoughtful and constructive citizen.

CHAPTER 1 / The Nature of Life

Biologists cannot yet give an all-inclusive, satisfactory definition of life itself. At present, life is best defined in terms of the things that living things can do, or their *life processes.* Any organization of substances which can carry on the life processes is said to be "alive." *Biology* is the study of living things (*bios,* life + *logia,* study).

A | THE CONCEPT OF LIFE

Some of the processes that are characteristic of living things are familiar to everyone. We all know that living things can grow. A crystal or an icicle can increase in size, but these nonliv-

ing structures get larger simply by adding more material, of a kind identical to themselves, to their outer surfaces. A living thing takes in raw materials and then transforms them chemically into more of the living thing. In most plants, the raw materials are simple chemicals taken from air, water, and soil.

Animals (and many plants) use other living things — food — as their raw material. The food is broken down chemically into simpler substances and then rebuilt into new parts of the living organism. Every living thing starts its life from another organism of the same kind, so that life passes on from one generation to the next through the process of *reproduction.*

Most living things are able to do various kinds of work. Many animals can move from place to place; they can fly, run, swim, or dig. They can chew and swallow their food, and their hearts pump blood. Even plants have the ability to move, though most are rooted in place. The roots of a tree can split a boulder. To do this work, all organisms must consume fuel, just as a steam engine can do work only by using coal or oil. The fuel used by living things is the food they take in or manufacture; for example, animals and nongreen plants take in food whereas green plants manufacture it. Although a steam engine burns its fuel rapidly and produces a great deal of heat, living things oxidize food slowly and produce much lower temperatures.

The sum total of all the life processes of any organism is known as its *metabolism.* All the processes of metabolism require food, which is either used to produce more living substance (*assimilation*), or is oxidized to provide the energy that is needed to do the work of living (*respiration*). In many creatures, respiration requires the organism to absorb oxygen from the air or water in which it lives. All metabolic processes produce waste materials, which must be discharged into the surroundings.

Living things adapt to their environment. Factors in the environment of living things include temperature, light, moisture, oxygen, air currents, soil conditions, and the presence of other living things. As environmental factors change, living things must change to meet the new situation or they will not survive. A change in a living thing that results in survival in a new environment is called an *adaptation.*

THE LIFE PROCESSES

Process	Definition
Nutrition	Those activities of an organism by which it takes raw materials from its environment and makes them usable.
Transport	The intake (absorption) and distribution (circulation) of usable materials throughout an organism.
Respiration	The oxidation of food in an organism, resulting in the release of energy and waste products.
Excretion	The process by which an organism gets rid of the wastes of metabolism. (Excretion should not be confused with **egestion,** which is the removal of undigested food. Egestion may take the form of elimination or defecation.)
Synthesis	The process by which simple materials are united to form more complex materials.
Regulation	The coordinated response of an organism to a changing environment so as to maintain its stability.
Growth	The increase in size of an organism that results from the synthesis and organization of materials into new materials and structures.
Reproduction	The unique characteristic of living organisms by which they produce more of their own kind. This process is essential to the species, not to the individual organism.

THE CELL — BASIC UNIT OF LIFE

Three hundred years ago scientists would have doubted that living things are made up of tiny chambers invisible to the naked eye. Today, the concept that plants and animals are composed of microscopic structures called *cells* is universally accepted.

The *cell* is the unit of structure of living things, somewhat as a brick is the unit of structure of a brick wall. There is a crucial difference, however. Whereas all the bricks are alike, there are many kinds of cells in a typical animal or plant, each having its own function to perform in the life of the organism, and each having its special structure that enables it to perform that function.

Development of the Microscope

Developments in all fields of science have led to improvements in scientific instruments. In turn, more accurate machines and instruments have helped to introduce new discoveries. The mass of new information that resulted from the development of the microscope is an example of how new instruments help to advance man's knowledge.

About 1590, Zacharias Janssen, a Dutch lens maker, built the first simple microscope. (A *simple microscope* is a microscope with only one set of lenses.)

In the 1670's, Anton van Leeuwenhoek, a Dutch naturalist and lens maker, constructed a simple microscope that could magnify objects as much as 270 times. With his simple microscope Leeuwenhoek saw objects that had never been seen before because they were invisible to the naked eye. Among his discoveries were bacteria, protozoa (one-celled animals), sperm cells, red blood cells, and yeast cells.

Robert Hooke, an English physicist, combined two sets of lenses and produced a *compound microscope*. In 1665, while examining a thin piece of cork with his microscope, Hooke discovered that it was not solid as had been believed, but was made up of walled compartments that looked like the honeycomb of a beehive. Hooke called these hollow boxes "cells." We know now that Hooke saw the walls of empty dead plant cells. He concluded that the cell was an empty box.

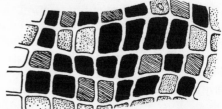

Fig. 1–1. The microscopic structure of a piece of cork.

In 1831, Robert Brown, a Scottish botanist, described a small, spherical structure in the center of living plant cells. He called this central structure the *nucleus* (plural: *nuclei*).

In 1838, Matthias Schleiden, a German botanist, concluded that all plants are composed of cells. About the same time, Theodor Schwann, a German zoologist, was studying animal tissues under the microscope. Because animal cells do not have cell walls as plant cells do, they are more difficult to see. However, by concentrating his search on the nuclei that Robert Brown had earlier described, Schwann concluded that all animals are also composed of cells. Thus, Schleiden and Schwann strongly established the concept that all living things are composed of cells. This was not a new concept: in 1824 Henri Dutrochet, a French physiologist, had expressed the same idea.

In 1839, Johannes Purkinje, a Czech physiologist, used the word "protoplasm" when he referred to the living material within the cell. At the same time, in Germany, Hugo von Mohl and Max Schultze were studying the functions of protoplasm. Schultze called protoplasm "the physical basis of life" and the idea developed that living things carry on their life activities because of the activity of their cells. This idea was stated as: *The cell is the unit of function of living things.*

The German scientist Rudolf Virchow is called the "father of pathology" (the study of disease). In 1858, Virchow explained that an organism becomes sick because its cells do not function properly. Through his study of microorganisms, Virchow contributed the important concept that *cells arise only from previously existing cells.*

By the end of the 19th century the cell theory was widely accepted. This resulted from the development of better stains for observation of cells, from improvement in lenses, from the development of methods for slicing sections of tissue thin enough for observation under the microscope, and from the spread of information by scientific societies through their meetings and journals.

Summary of the Cell Theory

1. *The cell is the unit of structure of plants and animals.* Plants and animals are composed of cells and of products made by cells.

2. *The cell is the unit of function of plants and animals.* Living things carry on their activities because of the activity of their component cells.

3. *All cells arise from pre-existing living cells.*

A *theory* is a general statement that unifies many isolated facts into a broad idea. It is a good, useful theory if it leads to the discovery of new facts and more advanced theories. The cell theory has been most effective in leading to biological discoveries, and has now become so universally accepted that it is referred to as the "cell doctrine."

Exceptions to the Cell Doctrine

To be useful, it is not necessary that a doctrine be universal, and modern investigations have revealed several exceptions to the cell doctrine as it was stated in the nineteenth century.

1. Although the cell is considered the basic "unit" of life, there are smaller structures *within* the cell which might be considered to be more fundamental units because they are capable of reproducing more structures of the same kind. Two of these fundamental units capable of reproducing themselves are the mitochondria and the chloroplasts.

2. Some structures and organisms are not built of separate cells each consisting of a bit of cytoplasm, a nucleus, and a surrounding membrane. For example, the long strands of striated muscle tissue contain many nuclei that are not separated from one another by cell membranes. Similarly, at one stage in its life, the slime mold is a structure several inches in length which contains thousands of nuclei not separated by cell membranes. (Fig. 1–2.)

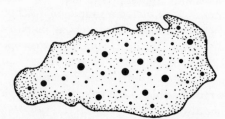

Fig. 1–2. One stage of a slime mold showing multinucleate structure.

3. Since viruses can multiply, they seem to be living. Nevertheless, their relatively simple structure (they have no cytoplasm) does not coincide with our concept of a cell. Moreover, certain viruses have been obtained in the form of a dry crystalline chemical which may start to "live" again. How can a nonliving chemical (such as a crystal) be thought of as a cell?

REASONING EXERCISES

1. How does the biologist determine whether a thing is living or nonliving?
2. Demonstrate as effectively as you can that a crystal is not a living thing.
3. Which of the life processes is more important to the survival of the species than to the survival of the individual?
4. Why is nutrition necessary to metabolism? (Make sure that you can define nutrition and metabolism before you attempt to answer the question.)
5. What is a cell? Name three structures that you would expect to find in a typical cell.
6. Distinguish between the simple microscope and the compound microscope.
7. What is the cell doctrine? What are some exceptions to the cell doctrine?

B | INSTRUMENTS USED IN THE STUDY OF CELLS

As you know, most cells are too small to be seen with the naked eye. With the aid of different types of microscopes, scientists have developed detailed concepts of cell structure and function.

The Compound Microscope

A student compound microscope is shown in Fig. 1–3. As you read this section, refer to the diagram to identify its parts. The compound microscope contains a combination of two lens systems. The *eyepiece* or (*ocular*) is a system of lenses at the top of the body tube. The *objective* is the system of lenses at the lower end of the tube. (Note: There are two objectives shown in Fig. 1–3: *high power* and *low power*.) The objectives produce a magnified image of the specimen, and this image is further magnified by the eyepiece.

The compound microscope commonly used in the high school laboratory has two interchangeable objectives. The shorter objective is low power in its magnification, and the longer one is high power. The magnifying power is always marked on the objective. If the two objectives are $10\times$ and $43\times$ and the eyepiece has a magnifying power of $10\times$, the specimen can be magnified to $100\times$ (low power objective $\times$ eyepiece) and $430\times$ (high power objective $\times$ eyepiece). The objectives are fastened to the *nosepiece*, which may be rotated to bring either of the two objectives into position below the eyepiece.

The specimen to be observed is usually placed on a glass microscope slide over the opening in the *stage*. By means of an adjustable *mirror* below the stage, light is directed through the opening in the stage and is passed through the

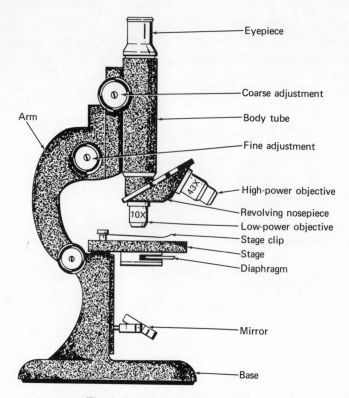

Fig. 1–3. A compound microscope.

specimen on the slide to reach the objective. In some microscopes, an electric light bulb, called a *sub-stage illuminator,* is attached below the stage and is used instead of the mirror. Illumination of the specimen is improved in some microscopes by a *condenser* (lens) placed in the opening in the stage, or below the stage.

A *diaphragm,* mounted below the stage, regulates the amount of light (actually the cone of light) reaching the objective lens. One type of diaphragm, the *disc diaphragm,* has a series of holes of various sizes in a flat disc. The disc is rotated to select the proper size opening. Another type of diaphragm is the *iris diaphragm.* Here the size of a single opening is varied, as with the iris of the eye. This regulation

is accomplished by a circle of flat, thin strips of metal.

The heavy *base* supports the microscope and is attached to the *arm* which holds the body tube and acts as a carrying handle.

A *coarse adjustment* knob and a *fine adjustment* knob can be turned to vary the distance between the objective and the specimen to produce a sharp image. The sharpest image of the specimen is formed when the proper distance between it and the objective is achieved. This occurs when the objective is at the focal point for that set of lenses. For the object to be in focus under high power, the objective lens must be closer to the object than under low power.

At higher magnifications (for example, 970×), more light is necessary.

Therefore, when a higher power objective having an oil immersion lens (usually 97×) is used, steps must be taken to avoid the loss of light in the air space between the microscope slide and the objective lens. This is done by placing a drop of a special oil on the slide and very carefully lowering the objective lens into it. Then focus away from the slide, as with every other objective. This will prevent the objective lens and slide from breaking. They should not come in contact with each other. An objective designed to be used with oil is known as an *oil-immersion objective*.

A microscope reverses the image of an object seen under it. Therefore, a microorganism viewed under the microscope may appear to move in the opposite direction from which it is moving.

Magnification and Resolving Power

We have noted that a microscope can produce a magnified image. Another point to be considered is the amount of resolution, or the resolving power of a microscope. The *resolving power* of an optical system is its ability to distinguish clearly and in detail between objects which lie very close to each other. Two microscopes may have the same magnifying power but may differ in resolving power.

Remember, the best image is not always the largest, but is the clearest. The most important factor determining the resolving power of a compound microscope is the objective lens; the eyepiece can magnify but does not increase the resolution of the microscope.

The Stereomicroscope

The stereomicroscope is sometimes called the binocular microscope because it has two oculars (eyepieces). However, the important aspect of this instrument is that it also has *two objectives* (one for each eye) and thus reveals the object under observation in three dimensions. Since a stereomicroscope does not have high power; it is used to observe relatively large objects such as minute insects and small crystals. It may be used with light which is either reflected from the object or transmitted through it, if the object permits.

The Phase-Contrast Microscope

Because living cells contain a large amount of water, they usually appear transparent under the light of the compound microscope. In order to contrast transparent cell structures under the compound microscope, the specimen must generally be stained with special dyes. Staining results in the absorption of different wave lengths of light by different structures in the cells of the specimen, and the structures are discerned by their different colors. However, staining may kill the contents of the cell and may also distort it.

The phase-contrast microscope, on the other hand, makes it possible for us to examine structures in the *living cell*.

Some parts of the cell are either thicker or denser than others. Because of these variations, the light travels faster through some cell materials than through others. The phase-contrast microscope converts these differences in the velocity of light into differences

The different structures in a cell have different effects upon the light that passes through them. As a result the light waves passing through one structure may be out of phase with the light passing through a nearby structure or substance. This means that the crests and troughs at the waves are out of step.

The phase-contrast microscope converts phase differences into differences in brightness. Thus, the contents of the living cell can be distinguished by contrasts in varying shades of gray.

The Electron Microscope

Instead of using visible light, the electron microscope, invented in 1937, uses a stream of high-speed, short wave electrons to give greater resolution. The source of electrons is a hot cathode filament which is enclosed in a vacuum chamber in order to prevent the electrons from colliding with air molecules which would scatter them. This filament acts as a gun to shoot out a narrow beam of electrons. This beam is collected and focused on the specimen by electromagnetic condensers which act as the "lens" to focus the electron beam. After passing through the specimen, the electrons are collected and pass through another series of electromagnetic "lenses" which results in an increase in the magnification.

The electron microscope can produce images magnified 200,000 times. It can be used for viewing large viruses and even certain large molecules. One disadvantage of the electron microscope is that it cannot be used to study living cells. The preparation of the cells and the action of the electron stream kills the cells. Only dead cells can be studied with the ordinary electron microscope.

High-Speed Centrifuge

Cytologists (biologists who study cells) try not to distort the cells or parts of cells being studied. Sometimes, however, they may purposely disrupt cells by using a blender or a special mortar and pestle (for grinding) and thus release the cell contents in a liquid. When the liquid containing the cell fragments

is placed in a tube and then whirled at high speed in a centrifuge, cell components sort out in different layers according to their density. Batches of mitochondria, ribosomes, and nuclei, separated by this method, can then be studied with the electron microscope and by biochemical methods.

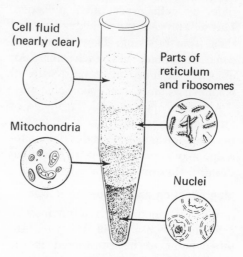

Fig. 1–4. Centrifuge tube showing various layers.

Microdissection Apparatus

Our knowledge has been increased by the manipulation and dissection of living cells. For example, biologists can remove the nucleus from a living *Amoeba* (a one-celled animal) to find

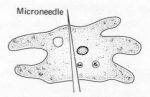

Fig. 1–5. An *Amoeba* being dissected by a microneedle. When separation is complete, the part without the nucleus will die after a while; the part with the nucleus will live.

out whether the cell can live and reproduce without it. The *micromanipulator*, the instrument used for this purpose, lets the user control the movement of very delicate instruments under a compound microscope.

Units of Measurement

The usual unit of measurement in cell studies is the micron (μ). For example, the human red blood cell is about 8.5μ in diameter and an average sized bacterium is about 2μ in length. To appreciate the size of this unit, let us start with the meter, which is a little more than a yard in length (actually 39.37 inches).

$^1/_{100}$ meter	= 1 centimeter (cm)
	2.54 cm = 1 inch
$^1/_{1000}$ meter	= 1 millimeter (mm)
$^1/_{1000}$ millimeter	= 1 micron (μ)
$^1/_{10,000}$ micron	= 1 Ångstrom unit (Å)

The Ångstrom unit is often employed in indicating the wavelength of light. For example, red light has a wavelength of about 6500 Å. The thickness of the cell membrane of a red blood cell is about 200 Å. This minute measurement gives an indication of the sensitivity of modern instruments.

How to Estimate Size
With Your Microscope

1. Low Power

    a. Place a transparent plastic metric ruler under the field of your low power objective as shown in the diagram.

    b. Estimate the diameter of the field in millimeters; for example, 1.3 mm.

    c. Convert the millimeters to microns; 1.3 mm to 1300μ. The

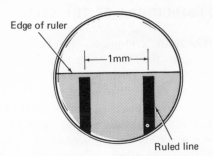

Edge of ruler

|← 1mm →|

Ruled line

diameter of the low power field is 1300μ.

Thus, an object that is about one-half the length of the diameter of the field, as seen under low power, has an approximate length of 650μ.

2. High Power

    Under high power, the thickness of one of the ruled lines occupies practically the whole field of view. Thus, a different method, the use of ratios, must be used.

    a. If the low power objective is $10\times$ and the high power objective is $40\times$, the ratio of magnification of the low power objective to the high power objective is 10/40, or one-fourth. The high power field is one-fourth of the diameter of the low power field; thus: $\frac{1}{4} \times 1300\mu$ is 325μ.

    b. If an object's length is about $\frac{1}{8}$ of the diameter of the field as seen under the high power, its length is approximately $\frac{1}{8} \times 325\mu = 41\mu$. Using this technique, you can make measurements that are accurate to 15μ. You are, therefore, able to obtain meaningful estimates of the sizes of various objects as, for example, the area of a cheek cell.

C STRUCTURE OF THE CELL

The term "protoplasm" was first used to refer to the "living substance" in the cell. It was then believed that protoplasm was the same in all living things. Today, we know that the cell is made up of many substances and structures, not just a single substance. All of these substances together are responsible for the characteristics of life. Just as an automobile is not made up of a basic "automobile substance," a cell is not made up of a basic "cell substance." Instead, it is a complex organization of many parts. Nevertheless, today the term "protoplasm" is still widely used to designate the living contents of the cell and we use it this way for convenience. *Cytoplasm* refers to the contents of the cell between the nucleus and the cell membrane. Cytoplasm often flows in a circular motion through the cell. This streaming motion is called *cyclosis.*

Using modern methods of cell study, biologists have developed new understandings of cell structure. We now know that the cell contains many different types of well-defined structures called *organelles.* Various specialized functions of the cell are performed by the organelles. The modern concept of cell structure is illustrated in Fig. 1–6.

Cell Organelles

Cell membrane. Every cell is enclosed by a living *cell membrane* (or *plasma membrane*) which separates the cell from its environment.

The cell membrane consists of one layer of lipid (fat) between two layers of protein. Since it permits some materials to pass through readily but not others, it is said to be a *selectively permeable membrane.* The cell membrane controls the transport of material into and out of the cell.

Wrinkles in the cell membrane, called *microvilli,* probably help cells cling together and increase the surface for passage of materials between cells. *Pinocytic vesicles* are sac-like inpushings of the cell membrane. The process by which these vesicles form and discharge their liquid contents into the cell is called *pinocytosis.* (See Chapter 6.)

The endoplasmic reticulum. Before the introduction of the electron microscope, it was thought that the cytoplasm lacked structural organization. Now we know that the cytoplasm has an extensive network of tubelike structures called the *endoplasmic reticulum* (*reticulum* means "network"). The tubes of the endoplasmic reticulum have a membrane which appears double under the electron microscope. Some of the tubes join with the nuclear membrane. The endoplasmic reticulum thus forms a passageway that probably functions in the transport of materials throughout the cell.

Ribosomes. Ribosomes are tiny dense particles which are attached to the walls of the endoplasmic reticulum or which move freely in the cytoplasm. Composed of ribonucleic acid (RNA) and protein, they play an important role in the synthesis of proteins by the cell.

The Golgi apparatus. The *Golgi apparatus* (also called the Golgi complex or Golgi bodies) was discovered by Camillo Golgi in 1898. As seen under the electron microscope, the Golgi apparatus appears in many shapes but most commonly as a cuplike stack of tiny flattened saccules. It has long been suspected that the Golgi apparatus functions in the secretory activities of

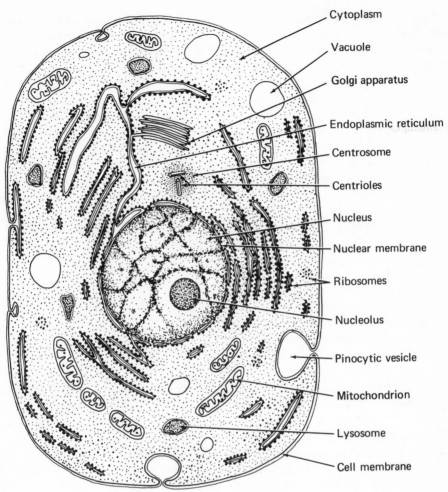

Cytoplasm

Vacuole

Golgi apparatus

Endoplasmic reticulum

Centrosome

Centrioles

Nucleus

Nuclear membrane

Ribosomes

Nucleolus

Pinocytic vesicle

Mitochondrion

Lysosome

Cell membrane

Fig. 1–6. Fine structure of the cell.

cells. Recent evidence indicates that this organelle *synthesizes carbohydrates* and adds them to the protein secretions funneled through the Golgi apparatus and *packaged for discharge from the cell.*

Mitochondria. *Mitochondria* (sing. — *mitochondrion*) appear as short rods or tiny spheres and are found in all plant and animal cells. The number of mitochondria varies with the amount of cellular activity being performed. Very active cells may contain more than a thousand mitochondria, while less ac-

tive cells contain fewer mitochondria.

Each mitochondrion is enclosed by a double membrane with an inner membrane folded into a system of shelflike ridges which increase the interior surface area of the membrane. The mitochondria have been called the "powerhouses of the cell" because it is here that much of cell respiration occurs, making the energy of food substances available to the rest of the cell.

Lysosomes. These oval bodies in the cytoplasm are centers of cellular digestion. *Digestion* is the process of breaking down large molecules such as proteins, nucleic acids and complex sugars into smaller, soluble materials. The digestive ability of a lysosome is an important mechanism for getting rid of old, wornout cells. In fact, a broken lysosome may digest the rest of the cell.

The nucleus. This spherical body is usually located in or near the center of the cell. The nucleus regulates the life activities of the cell and takes part in cell division. The nucleus is separated from the rest of the cytoplasm by a nuclear membrane. Usually there is one nucleus in a cell, but *Paramecium* (a one-celled animal) has two nuclei.

The nucleus contains the *chromosomes*, long, thin, coiled fibers that stain very deeply with certain stains. During cell division they form short, fat rods, whose number is characteristic of the species (there are usually 46 in human cells).

The nucleus contains two nucleic acids called *deoxyribonucleic acid* (DNA) and *ribonucleic acid* (RNA). (RNA is also present in the cytoplasm.) Chromosomes are composed of DNA and protein. By means of the activity of DNA, the nucleus controls cell growth and development, directs the activity of the cell, and passes on the cell's characteristics to succeeding generations of cells. (For a further discussion of RNA and DNA, see Chapter 21.)

Within the nuclei are one or more *nucleoli* (singular: *nucleolus*). The nucleoli, composed mainly of RNA, are involved in the passage of RNA to the cytoplasm.

In bacteria and in blue-green algae (simple forms of life), there is no nuclear membrane or organized nucleus. Newer methods of preparing cells for examination under the microscope indicate that the bacteria and blue-green algae have a single chromosome present in the cytoplasm rather than in an organized nucleus.

The centrosome. Almost all animal cells have a tiny spherical body in the cytoplasm, called the *centrosome*. This is generally absent in advanced types of plant cells. At the center of a centrosome are two deeply staining cylindrical particles called *centrioles*, at right angles to each other. As shown in Fig. 1–7, each centriole consists of 10 pairs of thin parallel strands (a circle of 9 pairs around a single central pair).

Fig. 1–7. Centriole.

Cilia and *flagella*, hairlike projections from the surface of certain cells, have the same basic structure as the centriole. The centrosome plays an important role in the division of animal cells, and may have a part in the formation and activity of cilia and flagella.

Plastids. These are small structures found in the cytoplasm of cells of higher plants and a few one-celled animals. Some plastids are colorless, and some contain pigments that give them color. *Chloroplasts* are plastids containing the green pigment *chlorophyll* which helps to manufacture food in plants.

Vacuoles. In certain cells, spherical sacs, called *vacuoles*, are present in the cytoplasm. Vacuoles of the cells of water plants are bubbles or globules filled with a fluid called *cell sap*. This is composed mainly of water and dissolved substances.

Plant cells may have many large

vacuoles, but most animal cells have only a few small vacuoles. Single-celled organisms may have specialized vacuoles; for example, *Amoeba* has food vacuoles and a contractile vacuole.

The cell wall. This is a fairly rigid envelope of nonliving material which surrounds the cell membrane of plant cells. It is composed mainly of cellulose. Unlike the cell membrane which controls the passage of materials into and out of the cell, the cell wall is penetrated by many tiny canals that permit most molecules to pass through.

The presence of a cell wall is an important characteristic used by biologists in deciding whether to classify an organism as a plant or an animal. The cell wall is the "plant skeleton" which protects and supports the individual cell and which helps to support an entire tree made up of many cells. The wood of a tree consists mainly of its cell walls.

CELL ORGANELLES AND THEIR FUNCTION

Organelle	Function	Organelle	Function
Cell wall	Protects and supports plant cell and maintains shape.	Vacuoles	Act as reservoirs for water, dissolved materials, and wastes; maintain internal pressure of cell.
Cell membrane	Controls transport of material into and out of cell.	Chloroplasts	Contain energy-accumulating pigments in plant cells; act as sites for food manufacture.
Nuclear membrane	Controls transport of material into and out of the nucleus.		
Cytoplasm	Provides an organized watery environment in which life functions take place by means of organelles contained in it.	Centrosome	Contains centrioles that are functional during reproduction of animal cells.
		Nucleus	Acts as information center for cell reproduction and other cell functions.
Endoplasmic reticulum	Provides channels through which transport of material occurs in cytoplasm.		
		Chromosomes	Contain the hereditary material (DNA); are the agents for distribution of hereditary information.
Golgi apparatus	Synthesizes carbohydrates and packages secretions for discharge from the cell.		
Ribosomes	Act as sites of protein synthesis.	Nucleolus	Acts as, a reservoir for RNA.
Mitochondria	Act as sites of cellular respiration and energy production.	Lysosomes	Act as centers of cellular digestion.

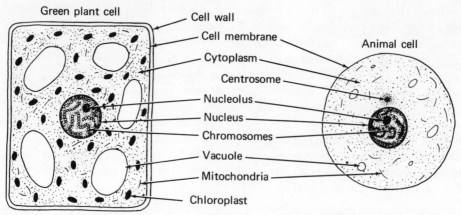

Fig. 1–8. Comparison of a generalized green plant cell and animal cell. Only structures which can be seen with the light microscope are shown.

Comparison of Plant and Animal Cells

Cells vary greatly in size, shape, and function. However, they have certain characteristic features. We have already noted some structures that can be studied under the electron microscope. Now we are ready to compare a "typical" animal cell and a "typical" plant cell as they would appear under the compound microscope. (See Fig. 1–8.)

Structures present in both plant and animal cells are the cell membrane, nucleus, mitochondria, and cytoplasm. Differences between the two kinds of cells are given in the table below.

DIFFERENCES BETWEEN GREEN PLANT CELLS AND ANIMAL CELLS

Structure	Green Plant Cell	Animal Cell
Cell wall	present	absent
Chloroplasts	present	absent
Centrosome	absent	present
Vacuoles	usually large	usually small

The Single-Celled Organism

A single-celled organism is very different from a cell in a many-celled organism. One of the important differences is that single-celled organisms can live without the immediate help of other cells.

A single-celled organism may be very complex and have many distinctive organelles which reflect its particular life style.

Euglena — Plant or Animal?

Most one-celled organisms have chiefly plant characteristics or chiefly animal characteristics. But this is not always the case. For example, *Euglena* is a microscopic organism which seems to be an animal because it has a whip-like flagellum for movement through the water and it lacks a cell wall. However, like a plant, it has chloroplasts which are used in the process of food manufacture. On the other hand, under experimental conditions the *Euglena* is known to lose its chlorophyll, give up the manufacture of food, and take in soluble nutrients from the water in the manner of an animal. From this, we can see that it is not always simple to classify a one-celled organism as either plant or animal. (See Chapter 4.)

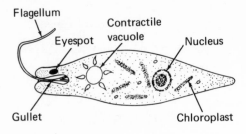

Fig. 1–9. *Euglena* — Plant or Animal?

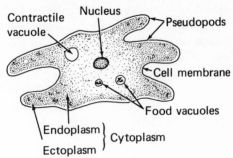

Fig. 1–10. *Amoeba*, a protozoan.

Amoeba

One-celled organisms which have animal-like characteristics are sometimes classified as protozoa.

Amoeba (Fig. 1–10) is an example of a protozoan. It is large enough to be seen by the naked eye and appears under the compound microscope as a transparent mass of protoplasm that slowly changes shape. It resembles a single cell of larger organisms because it consists of a single nucleus in a mass of cytoplasm surrounded by a cell membrane. Nevertheless, it can move freely through the water by the internal flow of its cytoplasm (*cyclosis*). As *Amoeba* moves by a streaming of protoplasm, *pseudopods* (outpushings of protoplasm) may extend in any direction. *Amoeba* takes in solid food by engulfing it. It surrounds the food and incorporates it into a food vacuole. Then it digests, assimilates, and oxidizes its food. *Amoeba* reproduces by splitting into approximately equal halves.

Paramecium

Another protozoan, *Paramecium* (Fig. 1–11), is much more complicated. It is smaller than *Amoeba*, and looks to the naked eye like a tiny thread. It has a flexible outer covering, the *pellicle*, which helps it hold its shape. Under the microscope it appears cigar-shaped,

with an indentation near the middle. (It is often called the "slipper-shaped animalcule.")

Paramecium has one or more small nuclei (*micronuclei*), depending on the species, and a much larger *macronucleus*. Its thick outer covering has thousands of tiny, hairlike *cilia* that beat in the water. Rows of cilia are connected with nervelike fibers that coordinate the beating movement to propel the animal through the water like a spirally thrown football. Food is taken in through a special *oral groove*, and en-

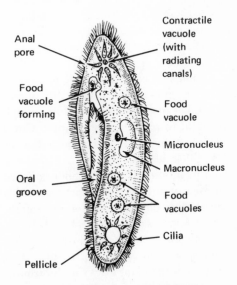

Fig. 1–11. Diagram of *Paramecium*.

closed in a food vacuole at the end of the groove. Other special structures, which we will study later, expel wastes, aid in digestion, and regulate water balance in the organism. Many biologists feel that such an enormously complicated, free-living creature should not be considered a cell, but rather a noncellular organism.

Investigations of the Cell

Cheek cells. If the inside of the cheek is gently scraped with the flat end of a toothpick and the dislodged cells are mounted on a microscope slide and stained with iodine, they appear as shown in Fig. 1–12.

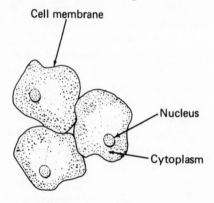

Fig. 1–12. Cheek cells.

Onion skin cells. It is easy to study a slice of onion skin under the microscope because the tissue is thin enough for light to pass through readily. Cut a wedge from an onion, remove one of the thick leaves, and then remove a thin membrane from the inner concave surface of that leaf. A portion of this membrane, observed under the microscope, appears as shown in Fig. 1–13.

In the onion skin cells in the diagram, chloroplasts are absent because this portion of the plant grows underground and is not green.

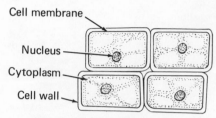

Fig. 1–13. Onion skin cells.

Anacharis, a green plant which has roots, stems and leaves, grows submerged in ponds and streams. A thin, delicate leaf removed from the tip of a stem has two layers of cells. A section of the leaf, stained with iodine, appears as shown in Fig. 1–14. *Anacharis* is also called *Elodea*.

If the plant has been growing in warm water exposed to sunlight, you can see the chloroplasts being moved in a circular pattern by the streaming motion of the cytoplasm (*cyclosis*).

The Nature of Science

In this chapter, you have investigated cells and have studied some of the work of famous biologists. From your studies, you should be ready to consider how this chapter illustrates the nature of science.

1. *Scientists observe.* Painstaking observation characterizes the work of the scientist. This is true of the early microscopists as well as of modern researchers who use the electron microscope.

2. *Scientists make hypotheses* and test them with experiments and observations. For example, after their early observations of cells in plants and animals, Schleiden and Schwann hypothesized that all plants and animals are composed of cells. However, they were not satisfied with their hypothesis until they had made many more observations.

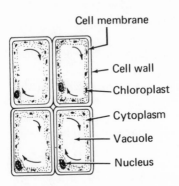

Cell membrane

Cell wall

Chloroplast

Cytoplasm

Vacuole

Nucleus

Fig. 1–14. Cells of *Anacharis* leaf.

3. *Scientists hold theories as tentative.* They are willing to change these theories if new evidence is presented. For example, even the respected cell doctrine is now being questioned.

4. *Science is international.* Men of many nationalities contribute to the solution of common scientific problems. The early history of the cell theory shows contributions by men from Holland, England, France, and Germany. Today in the United States, biologists are active in developing the modern concept of the cell.

5. Scientific progress is helped by *scientific societies* which encourage experimentation and the spread of information. The Royal Society of London was one of the early scientific societies which encouraged Leeuwenhoek and Hooke. In America, the American Association for the Advancement of Science (AAAS) is the major federation of scientific societies. However, each major

and many minor branches of science have their own organizations which usually publish research papers in their scientific journals.

6. Improvement in instruments is important in scientific development. The electron microscope opened up new microscopic worlds, and additional developments in *instrumentation* are constantly enhancing the powers of the research worker.

7. All of the sciences and mathematics are becoming increasingly *interrelated* by their individual contributions to each other. Physicists helped in the development of the microscope through their contributions to optics; chemists developed stains for cell observation; mathematicians created statistics for the analysis of records obtained by scientists; engineers developed better instruments for scientists; and biologists created new mathematical methods for the study of heredity.

8. *Facts and ideas are interrelated.* Robert Brown showed that plant cells contain nuclei. This told Schwann that where nuclei were present he could find cells. He looked for nuclei in animal tissue and discovered that animals are also made up of cells.

Today it is increasingly recognized that there is no universal "scientific method" which can serve as a sure guide for discovery in science. There is no prescription for creativity and imagination.

REASONING EXERCISES

1. Lysosomes are often called "suicide sacs." What reason can you give for this?
2. Could a cell live without a nucleus? Explain.
3. Of what materials are chromosomes composed? What are two functions of chromosomes?
4. Distinguish between a typical plant cell and a typical animal cell.

5. Distinguish between the cell membrane and the cell wall.

6. *Euglena* moves by means of a flagellum and *Paramecium* moves by means of cilia. If you were to examine a single cilium or flagellum in cross-section, what details would you expect to find?

7. Why do the cheek cells in Fig. 1–12 show so much less detail than the generalized cell of Fig. 1–6?

8. Observe cells from the inside of your cheek under low power. How does the width of one of these cells compare with the diameter of a red blood cell?

COMPLETION QUESTIONS

[A] 1. The energy present in food molecules is released by the life process of . . . .

2. The sum total of all life processes is called . . . .

3. The processes by which living things take in food and process it for use is . . . .

4. The removal of undigested food is known as . . . .

5. The first simple microscope was built by . . . .

6. The first person to see bacteria was . . . .

7. The concept that all plants and animals are composed of cells was stated by . . . . . .

8. The concept that cells arise from previously existing cells originated with . . . .

9. An example of a cell which contains many nuclei is . . . .

10. A "microorganism" that can be crystallized and that can function again at a later time is the . . . .

[B] 11. The amount of light reaching the objective lens of a compound microscope is regulated by the . . . .

12. If the total magnification of a microscope is $430\times$ and that of the eyepiece is $10\times$, that of the high power objective is . . . .

13. For a specimen to be in focus under high power, the objective lens must be (closer to, farther from) . . . . the specimen than under low power.

14. At magnifications of $970\times$, an oil immersion lens may be used to prevent the loss of . . . .

15. The ability of a microscope to distinguish between objects lying very close to each other is known as its . . . .

16. The microscope which enables the viewer to distinguish easily between structures in a living cell is the . . . .

17. In the electron microscope, the beam of electrons is focused by . . . .

18. Cell components can be separated into different layers according to their density by a (an) . . . .

19. 340 microns is equivalent to . . . . millimeters.

20. Human red blood cells average . . . . in diameter.

[C] 21. A membrane which permits certain substances, but not others, to pass through is said to be . . . .

22. The network of tubes in the region of the cytoplasm is called the . . . .

23. The synthesis of proteins in the cell occurs at the . . . .

24. The Golgi apparatus seems to have the functions of .... and ....
25. The organelles which are known as the "powerhouses of the cells" are the ....
26. The destruction of old wornout cells is accomplished by the ....
27. A tiny spherical body lying near the nucleus and functioning in cell division is the ....
28. Green pigmented plastids which are important in photosynthesis are the ....
29. The cell wall is composed mainly of ....
30. A plantlike characteristic of *Euglena* is the presence of ....
31. The movement of cell contents is called ....
32. The small nucleus of *Paramecium* is called the ....
33. *Paramecium* moves by means of its ....
34. The indentation through which *Paramecium* takes in food is called the ....
35. Digestion in the *Amoeba* takes place in ....

MULTIPLE-CHOICE QUESTIONS

A 1. One of the first men to observe living microorganisms was (1) Hooke, (2) Leeuwenhoek, (3) Virchow, (4) Von Mohl.
2. A contribution of Hooke to the development of the cell theory was that he (1) named cells, (2) defined cells on the basis of the presence of living protoplasm, (3) stated that all cells come from cells, (4) defined cells on the basis of the presence of nuclei.
3. The botanist who stated that all plants are composed of cells was (1) Schleiden, (2) Schwann, (3) Brown, (4) Dujardin.
4. The contribution of Virchow to the development of the cell theory was that he (1) named cells, (2) stated that all cells come from cells, (3) said cells contain nuclei, (4) defined cells on the basis of the presence of living protoplasm.
5. The cell theory states all of the following *except* (1) cells are units of structure, (2) cells are units of function, (3) cells have definite boundaries, (4) cells arise from living cells.

B 6. We can control the quantity of light entering the objective of the microscope by means of (1) mirror and ocular, (2) ocular and coarse adjustment knob, (3) mirror and coarse adjustment knob, (4) mirror and diaphragm.
7. The circle to the right indicates the position of the letter X as seen in the field of your microscope. To get the letter X in the *center* of the field you would move the slide (1) to the left and up, (2) to the left and down, (3) to the right and up, (4) to the right and down.
8. Which instrument has played a major role in permitting scientists to discover the chemical makeup of a mitochondrion? (1) electron microscope, (2) ultracentrifuge, (3) compound microscope, (4) oil immersion objective.
9. An object measures 16 mm. in length. Its length can be expressed as (1) 16 microns, (2) 160 microns, (3) 1600 microns, (4) 16,000 microns.
10. If the length of a *Paramecium* measures about ¼ the distance across the microscope field, and if the diameter of the field measures 1600μ, the length of the *Paramecium* is about (1) 400μ, (2) 0.4μ, (3) 0.0016μ, (4) 4000μ.

C Base your answers to questions 11 through 15 on the diagrammatic sketch of a cell below.

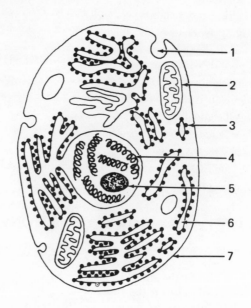

11. Which structure is composed primarily of lipid and protein? (1) 1, (2) 6, (3) 7, (4) 4.

12. Which structure is probably a major pathway in intracellular transport? (1) 5, (2) 6, (3) 3, (4) 7.

13. Which structure serves as the major site of protein synthesis? (1) 1, (2) 2, (3) 3, (4) 7.

14. Which structure is primarily concerned with the release of energy from nutrients? (1) 5, (2) 2, (3) 3, (4) 4.

15. Which cell structure is composed of DNA and protein? (1) 1, (2) 6, (3) 7, (4) 4.

CHAPTER TEST

1. A pupil observed a cell under the microscope. He identified it as a green plant cell and not a human cheek cell because he noted the presence of a (1) nucleus, (2) cell wall, (3) mitochondrion, (4) ribosome.

2. Viruses do not coincide with our concept of the cell because they (1) lack cytoplasm, (2) cannot multiply, (3) contain DNA, (4) are so small.

3. The organism *Euglena* shows that (1) the first cell was a plant cell, (2) the first cell was an animal cell, (3) intermediate forms exist between plants and animals, (4) some organisms exist on the borderline between the living and the nonliving.

4. The DNA of a cell is found mainly in its (1) membrane, (2) cytoplasm, (3) vacuoles, (4) chromosomes.

5. If the ocular is marked 10× and the high power objective is marked 43×, the magnification of the microscope is (1) 215, (2) 430, (3) 100, (4) 1000.

6. A transparent plastic ruler is placed on the stage of a microscope and observed under low power. Two divisions of the ruler can be seen across the width of the field. Each division of the ruler equals 1 millimeter. The diameter of the field is (1) 1 micron, (2) 2 microns, (3) 1000 microns, (4) 2000 microns.

7. The unit of structure and function in living things is the (1) mitochondrion, (2) organelle, (3) cell, (4) tissue.

8. Which structure regulates the entry and exit of dissolved materials in an animal cell? (1) cell wall, (2) cell membrane, (3) nucleus, (4) cytoplasm.

9. A structure present in most animal cells but absent from most plant cells is the (1) lysosome, (2) chromosome, (3) centrosome, (4) ribosome.

10. A microscope has a 10× eyepiece and a 10× and a 40× objective. The diameter of its low power field is 1300 microns. The diameter of its high power field is approximately (1) 30 microns, (2) 325 microns, (3) 1300 microns, (4) 5200 microns.

11. The phase-contrast microscope has aided in understanding the nature of the cell by permitting (1) magnification of cells up to 500,000 times their normal size, (2) observation of objects an inch or more in thickness, (3) detailed observation of unstained living cells, (4) chemical analysis of the parts of the cell.

12. Most cells lacking a cell wall would also lack (1) mitochondria, (2) chloroplasts, (3) cell membranes, (4) vacuoles.

13. A limitation of the electron microscope is that it (1) suffers loss of resolving power above 1000× magnification, (2) cannot magnify above 1000×, (3) uses short wave electrons, (4) cannot be used to study living specimens.

14. Which most closely represents the average diameter of a human red blood cell? (1) 7 Ångstrom units, (2) 7 centimeters, (3) 7 millimeters, (4) 7 microns.

15. Which statement concerning ribosomes is true? (1) They function in cell division. (2) They are the sites of protein synthesis. (3) They contain DNA. (4) They permit cells to contract.

For each question (16–20) write the number preceding the statement, chosen from the list below, which best applies to that question.

(1) *Both* statements are *true* and *B is the cause of A.*

(2) *Both* statements are *true* and *B is not the cause of A.*

(3) *Only one* statement is *true.*

(4) *Neither* statement is *true.*

Statement A	Statement B
16. The high power lens of a microscope should always be used before the low power lens.	A larger area of the object viewed can be seen with the high power lens.
17. The fine adjustment should be used for the final focusing of a microscope.	The diaphragm regulates the amount of light which enters the objective.
18. The total magnification of a microscope with a 10× eyepiece lens and a 4× objective lens is 40.	The total magnification is obtained by dividing the magnification of the objective lens by the magnification of the eyepiece lens.
19. When viewed under a microscope, a *Paramecium* may appear to move in the opposite direction of that in which it is actually moving.	A microscope reverses the image of an object seen under it.
20. If a smudge can be seen under both low and high power, the smudge cannot be on the eyepiece lens.	Marks on the eyepiece lens are not visible when using a microscope.

CHAPTER 2 | The Chemistry of Living Things

As research tools and techniques improved, biologists discovered finer details of structure within cells. The early microscope showed the cell as an undifferentiated mass of material. It was then discovered that the cell consists of organelles such as a nucleus, plastids, and mitochondria. In recent times, the electron microscope has revealed even smaller structures such as ribosomes, layers in membranes, and fibers in centrioles, which are not much bigger than large molecules. Even this was not the end, for within these, even finer substances have been discovered.

For the most part, these very fine structures are investigated with the tools of the biochemist. They are the molecules of which the cell is composed, and are generally too small to be seen even with the electron microscope. Therefore, an understanding of chemistry is necessary in order further to analyze the cell's structure down to smaller levels of organization.

A   BASIC CHEMISTRY

Matter is anything that has mass and occupies space. Even air, which is invisible, contains matter, for it consists of atoms and molecules which have mass. The basic form of matter is the *element*. Elements can combine in definite proportions to form *compounds*. Nearly all elements, under the proper conditions, will combine with other elements to form compounds.

Our study of chemistry starts with atoms — the smallest particles of an element that can take part in chemical reactions. The atom, once thought to be a solid unit, is now known to contain protons, neutrons, and electrons, along with other nuclear particles.

The Atom

Most particles in atoms have known properties. First, we will consider the property of charge. Protons have a unit positive charge $(+1)$ and electrons have a unit negative charge (-1). A neutron is an electrically neutral particle. It acts as if it were composed of a proton and an electron.

Now we will consider the property of mass. If the mass of a proton is considered to be 1, the mass of an electron $(1/1846$ of the mass of a proton) is virtually zero. Although the mass of a neutron is slightly greater than that of a proton, for convenience its mass is taken to be 1. The properties of mass and charge of subatomic particles are summarized in the table below.

PROPERTIES OF THE PARTICLES IN AN ATOM

Particle	Mass Number	Charge
Proton	1	$+1$
Electron	0	-1
Neutron	1	0

Because of the arrangement of the particles making up the atom, it is possible to write a symbol that indicates the structure of the atom. For example, the element carbon is written $_6C^{12}$. The meanings of the superscript $(^{12})$ and subscript ($_6$) in the formula are indicated below.

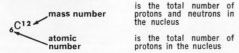

To show the structure of the carbon atom, we first draw a box representing the nucleus of the atom as shown in Fig. 2–1. Since the atomic number of carbon is 6, we write 6+ to represent 6 protons as shown in the diagram. However, the mass number of carbon is 12. The protons which we have already represented have a mass of 6. How many neutrons are needed to make up a total mass number of 12? We write in 6N to indicate the 6 neutrons which are needed to make up the total of 12 units of mass number. Fig. 2–2 shows the protons and neutrons in the nucleus of an atom of $_6C^{12}$.

Fig. 2–1   Fig. 2–2

Since the neutron has no charge, it is evident that the nucleus has a total of 6 positive charges. However, every atom is electrically neutral; that is, it has the same number of electrons as protons. Thus, six electrons with negative charges must also be present in the atom. They are located in electron shells ("rings", "orbits", or "energy levels" surrounding the nucleus). None of these shells ever has more than a specific number of electrons. The first electron shell can have a maximum of 2 electrons, the second shell a maximum of 8 electrons, and the third shell a maximum of 8 electrons.

The questions you are expected to deal with in this chapter are limited to these maximums.

In constructing a model of the electron shells for a particular atom, you first determine the number of electrons in the atom (carbon, as we have seen, has 6). Then, beginning with the first shell and moving outward from the nucleus, fill each shell to its maximum until you have used up all of the electrons present in the atom.

In our example of carbon, the first shell is filled with 2 electrons (its maximum) before we start to draw the next electron shell. The completed structure of the carbon atom is shown in Fig. 2–3.

Fig. 2–3. Carbon. Six electrons in two shells balance the charge of the protons in the nucleus. Elements with more protons in the nucleus may have as many as seven shells of electrons.

The atomic structure of oxygen, $_8O^{16}$, is shown in Fig. 2–4. The atomic structure of ordinary hydrogen, $_1H^1$, is shown in Fig. 2–5.

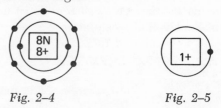

Fig. 2–4   Fig. 2–5

We are now ready to extend our study. Consider the atoms $_1H^1$, $_1H^2$, and $_1H^3$ shown in Fig. 2–6.

Fig. 2–6. Three isotopes of hydrogen.

Each of these atoms has one proton in the nucleus and one electron in the shell. Since the electrons in the outermost shell determine the chemical properties of an element, all of these atoms act alike chemically. Each is the element hydrogen. Each will combine with oxygen to form water. However, because of the varying number of neutrons in the nucleus, each has a different mass number.

Isotopes are atoms which have the same atomic number but different atomic masses ("weights"). This means that isotopes have the same number of protons but differ in the number of neutrons they contain. The three isotopes of hydrogen are called protium ($_1H^1$), deuterium ($_1H^2$), and tritium ($_1H^3$). "Heavy water" is water whose molecules contain the heavy hydrogen isotopes combined with oxygen.

The atomic number is often omitted in designating isotopes; for example, the two isotopes of carbon are referred to as carbon-12 and carbon-14. Since all atoms of carbon have an atomic number of 6, the atomic number 6 is understood although not written. Similarly, $_{92}U^{238}$ is referred to as uranium-238.

Radioactivity. Forces within some atoms make their nuclei unstable and they disintegrate or break apart. As they disintegrate, they emit radiations consisting of particles or electromagnetic waves. Atoms which emit radiations are called *radioactive isotopes.* For example, radium, discovered by Marie and Pierre Curie, is a radioactive element. As a radioactive element emits particles or waves, it changes to a different element. An isotope which does not emit radiations is called a *stable*

isotope. For example, two of the stable isotopes of oxygen are oxygen-16 and oxygen-18.

The presence of radioactivity can be determined by instruments such as the Geiger counter, Wilson cloud chamber, and photographic plate. In order to trace the pathway of an element through living organisms, scientists introduce a radioactive isotope to replace the stable atom. The atom is then called a *tracer* or *tagged atom.* It behaves like the normal atom in all chemical processes within the organism but its presence can be traced. For example, biologists have long known that plants use CO_2 (carbon dioxide) to produce sugar in the process of photosynthesis. However, they wished to determine some of the steps in this process of sugar production; therefore, they prepared CO_2 with a tagged carbon atom called radioactive carbon-14. The new compounds that plants produced from this radioactive CO_2 containing C-14 were also radioactive.

Stable isotopes can be traced by use of the *mass spectrometer,* an instrument which identifies an isotope by its mass rather than by its radioactivity.

Elements, Compounds, and Mixtures

Atoms may join together to form molecules. If a molecule contains two or more kinds of atoms, the molecule constitutes a *compound.* For example, a small amount of the compound glucose ($C_6H_{12}O_6$) contains billions of molecules, each composed of several kinds of atoms. Each molecule contains 6 carbon atoms, 12 hydrogen atoms, and 6 oxygen atoms. A small quantity of the element oxygen (O_2) may have billions of molecules, each of which is composed of the same kind of atom — 2 oxygen atoms. Thus, elements may exist as molecules; some of them may also exist as single atoms.

The atoms of molecules are in chemical combination. When several kinds of molecules are present together but are not chemically combined, this arrangement is called a *mixture*. For example, particles of sand (SiO_2) and particles of salt (NaCl) may be put together to form a mixture of sand and salt, but they are not chemically combined. Salt dissolved in water is another example of a mixture of two compounds, NaCl and H_2O. Air is a mixture of elements and compounds such as N_2, O_2, CO_2, H_2O, and others. Protoplasm, the material of the living cell, is a changing mixture of many compounds.

Bonding

When atoms are joined to form molecules, they are united by a chemical bond. Let us emphasize this idea. The chemical bond is the means by which two or more atoms are linked together to form molecules. Two methods of chemical bonding are by (1) transfer of electrons, and (2) sharing of electrons. First, we will investigate chemical bonding by the transfer of electrons.

Consider the atoms of sodium, $_{11}Na^{23}$, and chlorine, $_{17}Cl^{35}$, shown in Fig. 2–7.

In the diagram, the sodium electrons are drawn as dots, and the chlorine electrons are drawn as circles merely to distinguish between them. Actually, all electrons are the same.

You will recall that the third electron shell of an element can contain a maximum of 8 electrons. Note that in Fig. 2–7 neither the sodium atom nor the chlorine atom is shown with a complete outer electron shell. The concept of an incomplete outer shell is important to our understanding of chemical bonding. Let us see why by first considering the opposite case for purposes of contrast. Atoms which have a complete outer shell *rarely* enter into chemical combination. They are called "inert" atoms; some examples are the elements helium, neon, and argon, called the noble gases. On the other hand, some atoms with incomplete outer shells may transfer electrons in order to complete their outer shells and attain the electron structure of a noble gas. An example is the case of sodium atoms and chlorine atoms shown in Fig. 2–8. Sodium now has a complete outer or second shell since it transferred the one electron from the third shell to the outer third shell of the chlorine atom. Chlorine completes its third shell of 8 electrons by gaining that one sodium electron. After the transfer of one electron of the sodium atom to the chlorine atom, the two configurations are as shown in Fig. 2–8.

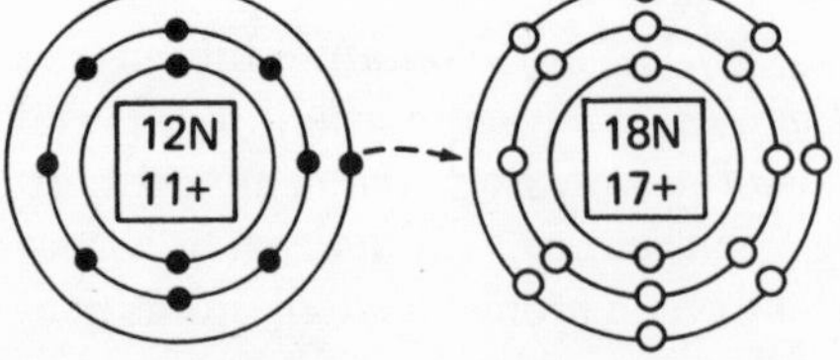

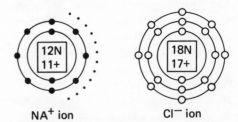

Fig. 2–7. Sodium atom and chlorine atom before the transfer of an electron from sodium to chlorine.

Fig. 2–8. Sodium ions and chlorine ions resulting from the transfer of an electron. Each now has a complete outer shell.

The sodium atom, which lost one electron, now has fewer electrons than protons. It now has a positive charge and is called a sodium ion. Similarly, the chlorine atom, with one electron more in its shells than protons in its nucleus, has a negative charge and is called a chlorine ion. An *ion* is an atom, or group of atoms, which has gained or lost one or more electrons. *Ionic bonds* result from the transfer of electrons. The resulting compound, called an *ionic compound,* is held together by the force of electrical attraction between the oppositely charged ions. Remember that positive ions (Na^+) were formed by a loss of electrons, while negative ions (Cl^-) were formed by a gain of electrons. The force of attraction between these oppositely charged ions is quite strong. This force of attraction results in the ionic bond. This force which holds the atoms together in a compound is called *chemical bond energy.* Since this energy is stored in a compound, it is a form of potential energy.

Some atoms achieve the effect of a complete outer shell by *sharing* rather than transferring electrons. The shared electrons move in a common pathway that includes the outer shells of both atoms. The force of attraction involved in the sharing of a pair of electrons constitutes a *covalent bond.* Many of the compounds produced in living things are formed by the sharing of electrons. An example is the molecule of the gas methane, CH_4, shown in Fig. 2–9.

A single molecule of the compound methane has one carbon atom (shown in the center of the diagram) and 4 hydrogen atoms. You will recall that carbon has 4 electrons in its *outer* shell (exactly the same number as the num-

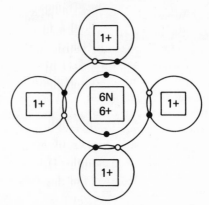

Fig. 2–9. Atomic structure of methane, CH_4.

ber of hydrogen atoms in the molecule). You will also recall that hydrogen has one electron in its outer shell.

Note that in methane each atom of hydrogen achieves a complete outer shell of 2 electrons by sharing an electron with the carbon atom. At the same time, the carbon atom achieves a complete outer shell of 8 electrons by sharing an electron with each of the four hydrogen atoms.

Since we are concerned now only with the electrons in the outer shell, we can simplify Fig. 2–9 by using an electron-dot diagram. In Fig. 2–10, dots are used to indicate only the electrons in the outer shells of both elements.

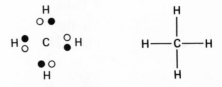

Fig. 2–10. Electron dot diagram of methane, CH_4.

Fig. 2–11. Structural formula of methane, CH_4.

Fig. 2–10 can be further simplified by substituting a short dash for each pair of shared electrons, as shown in Fig. 2–11.

A representation using dashes to show the bonds between atoms is called a *structural formula*. Structural formulas are valuable because they show the arrangement as well as the number and kinds of atoms in a molecule. In this sense, they perform a function for the scientist similar to that of our alphabet; for example, the arrangement of the three letters u, s, and e makes the difference between the words "use" and "sue." Re-examine Fig. 2–11 to note the number, kind, and arrangement of the atoms.

The *empirical formula* indicates the number of various types of atoms in the molecule; for example, H_2O, CH_4, $C_6H_{12}O_6$. Since it does not describe the arrangement of the atoms within the molecule, it does not indicate as much as the structural formula.

Organic Chemistry

Carbon atoms, with four electrons in their outermost shell, have a property that makes them outstanding among the atoms. They can combine to form long chains, which may be thousands of atoms long. It is this special property of carbon that makes possible the giant molecules in protoplasm. An example of a chain of carbon atoms is shown in the structural formula of the molecule of butane, C_4H_{10}, in Fig. 2–12. (Note how the hydrogen atoms are arranged around a "backbone" of carbon atoms.)

The same empirical formula of butane may be arranged in the structural formula shown in Fig. 2–13. Compounds with the same empirical formula but different structural formulas are called *isomers*.

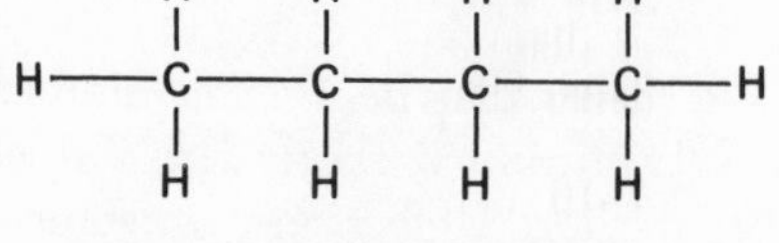

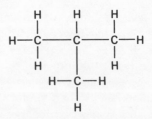

Fig. 2–13. Iso-butane.

Each of these arrangements of C_4H_{10} is a different compound with different properties. Accordingly, it is clear that the empirical formula does not provide a sure basis for identifying a compound. In dealing with carbon compounds, we must be able to use structural formulas. With larger molecules, a great many arrangements of the atoms are possible. This ability of carbon to form long chains with many arrangements results in a huge number of different carbon compounds.

The study of carbon compounds is known as *organic chemistry*. The word organic means "living." The term "organic chemistry" was originally applied to the chemistry of substances produced in living things. It used to be thought that such living substances were produced by a "vital force," and were entirely different from inorganic compounds — those not found in living things. In 1828, however, this distinction between living and nonliving compounds was proved false. Friedrich Wöhler produced the organic compound urea in the laboratory from inorganic substances. It had previously been found only in living organisms. The structural formula of urea is shown in Fig. 2–14.

Organic chemistry includes the study of all carbon compounds in living and

Fig. 2–12. Normal butane.

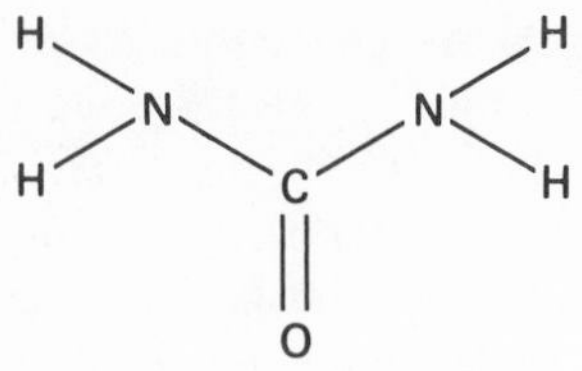

Fig. 2–14. Structural formula of urea. Its empirical formula is $CO(NH_2)_2$.

nonliving things. The study of the chemistry of living things is called *bio-chemistry*.

Ionization

While we may often write a formula for an ionic compound, such as NaCl for sodium chloride (table salt), these compounds do not exist as molecules of NaCl. A salt crystal consists of equal numbers of two kinds of ions, positive sodium ions (Na^+) and negative chlorine ions (Cl^-). If salt is dissolved in water, the ions separate from each other and move around among the water molecules as separate charged particles.

Many covalent compounds, water among them, separate into two ions to a minute degree. This process of separation of a compound into ions is called *ionization*. The ionization of water is represented by the following equation:

$$H_2O \rightleftarrows H^+ + OH^-$$

Neutralization

Acids are compounds which ionize to form hydrogen ions, H^+.

Hydrochloric acid   HCl $\longrightarrow H^+ + Cl^-$

Bases are compounds which ionize to form hydroxide ions, OH^-.

Sodium hydroxide   NaOH $\longrightarrow Na^+ + OH^-$

The typical properties of acids are due to their hydrogen ions, and the typical properties of bases are due to their hydroxide ions. When acids react with bases, the hydrogen ions combine with the hydroxide ions to form water.

$$H^+ + OH^- \longrightarrow H_2O$$

Neutralization is the removal of hydrogen ions and hydroxide ions from a solution to form water and a salt. Examples of neutralization reactions are shown in the tabulation below.

Base	+	Acid	yields	Water	+	A Salt
NaOH	+	HCl	$\rightleftarrows$	HOH	+	NaCl
KOH	+	HCl	$\rightleftarrows$	HOH	+	KCl
2KOH	+	H_2SO_4	$\rightleftarrows$	2HOH	+	K_2SO_4

A *salt* is the compound formed by the union of the positive ion of the base and the negative ion of the acid in a neutralization reaction. Ordinary table salt, NaCl, is just one example of a salt.

Neutralization reactions have many practical applications. For example, suppose someone accidentally spills lye (NaOH) on his skin. Since lye is a strong base, he should wash his skin with large amounts of water. He should then remove any remaining alkalinity by neutralizing the lye with a mildly acid solution of boric acid. As another example, if a soil is too acid, it is neutralized by adding limestone (calcium carbonate), which acts as a base when in water.

The pH Scale

Chemists have a scale to indicate the degree of acidity or alkalinity (basicity) of a solution. It is known as the *pH scale* as shown in Fig. 2–15.

The scale is expressed by numbers that lie between 0 and 14. The number

7, in the middle, represents the neutral point between an acid and a base. At this point, the concentration of hydrogen (H^+) ions equals the concentration of hydroxide (OH^-) ions. The smaller the number on the scale, the more acid is the solution. The larger the number on the scale, the more alkaline is the solution. Thus, the scale shows hydrochloric acid to be a stronger acid than boric acid; that is, it has a higher concentration of hydrogen ions.

The pH scale is used in agriculture to show the optimum (best) soil pH for various plants as shown in the table below.

Plant	pH
Eastern hemlock	4.4 - 6.0
Rhododendron	4.5 - 6.0
Gladiolus	6.0 - 8.0
Iris	6.0 - 8.0

Hemlock and rhododendron prefer a more acid soil than the gladiolus and iris. Human blood normally has a pH of 7.4, which is slightly alkaline (basic). Substances in plants and animals each have a specific pH value; some are acid and some are alkaline.

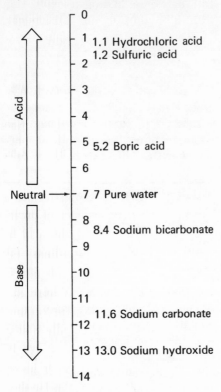

Fig. 2–15. pH scale.

EXERCISES

1. (*a*) Which particles in the atom have a mass number of 1?

   (*b*) Which particle can be considered to have no mass?

   (*c*) Which particle has a positive charge?

2. Draw the structure of the following atoms: $_2He^4$ (helium); $_3Li^7$ (lithium); $_4Be^{10}$ (beryllium); $_5B^{11}$ (boron); $_7N^{14}$ (nitrogen); $_{12}Mg^{24}$ (magnesium).

3. Name three isotopes of hydrogen. Define isotope. What instrument is used in identifying stable isotopes?

4. Identify each of the following as element, compound, or mixture: soup, salt, air, uranium, hydrogen, sodium chloride.

5. Name two kinds of chemical bonds. Define each.

6. How are carbon atoms unique among the atoms? What is the nature of an organic compound?

7. At a pH of 7, the concentration of H^+ ions as compared with the concentration of OH^- ions is (1) greater, (2) equal, (3) less.

B | CHEMICAL SUBSTANCES IN LIVING THINGS

The chemical elements present within the cells of living organisms are also present in the nonliving environment. Among the most important elements are:

carbon	(C)	nitrogen	(N)
hydrogen	(H)	sulfur	(S)
oxygen	(O)	phosphorus	(P)

In addition, there are a number of elements present in small concentrations, such as

iron	(Fe)	sodium	(Na)
magnesium	(Mg)	potassium	(K)
iodine	(I)	calcium	(Ca)
chlorine	(Cl)		

Besides these, there are at least fifty *trace elements* found in extremely small quantities (just a trace), such as

zinc	(Zn)	fluorine	(F)
copper	(Cu)	gold	(Au)

As previously indicated, compounds of carbon are called *organic compounds*. The relatively simple substances which do not contain carbon are known as *inorganic compounds*.

The elements present in living things are organized into the principal compounds listed below.

Inorganic compounds	Organic compounds
water salts, such as NaCl and KCl	proteins carbohydrates lipids nucleic acids

Water

When exploring the possibility of life on another planet, scientists try to find out if the planet has water, because life, as we know it, requires this important inorganic compound. At least 65 percent of the weight of the living cell is water. Since chemical reactions in living things depend upon the presence of water, it is evident that the life processes of "beings" on other planets, without it, would require entirely different chemical reactions. Some of the important properties of water are as follows:

Water is the best solvent known. Chemical reactions do not normally occur between large dry masses of matter but between molecules and ions. When a substance is dissolved in water, it may break up into molecules and ions which are continuously in motion. Chemical reactions can occur within the cell because the cell contains a large amount of the solvent water.

Water is a relatively stable compound. Water may be decomposed into hydrogen and oxygen only by the expenditure of a large amount of energy; it does not normally decompose. Accordingly, water provides a stable background for other reactions within the cell.

Water ionizes. The H^+ and the OH^- ions formed by the ionization of water are part of many chemical reactions in cells.

Proteins

Proteins contain nitrogen and are the most abundant organic compounds in living things. The main elements present in proteins are carbon, hydrogen, oxygen, nitrogen, and sulfur. Every cell contains large amounts of protein, which form the basic structure of membranes, chromosomes, and other cell parts. Muscle cells are largely protein and it is fibers of protein that produce their contraction. In addition, there are non-cellular proteins such as are found in blood plasma, cartilage, hair, and nails. We need proteins in our diet to provide the raw materials to make these structures. Among the foods that are rich in proteins are meat, fish, eggs, cheese, legumes (bean type plants), and grains.

The gigantic size and great complexity of proteins are indicated by their great molecular weight (weight of one molecule). Proteins range in molecular weight from about 5000 (insulin) to as high as 40 million (the tobacco mosaic virus). In comparison, the molecular weight of water is 18.

Amino acids. Proteins are composed of smaller building blocks called *amino acids*. There are about 20 different amino acids which occur in nature. The structure of the different proteins is determined by the variety and arrangement of their component amino acids.

The structure of amino acids is indicated in Fig. 2–16. This illustrates the structural formula of the simplest amino acid, glycine.

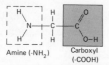

Fig. 2–16. Glycine.

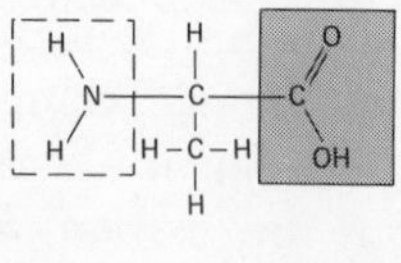

Shown in the diagram are the $\overset{H}{\underset{H}{\diagdown}}N-$

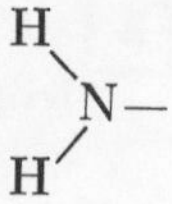

(the amino group)

and the $-C\overset{\displaystyle O}{\underset{\displaystyle O-H}{\diagup}}$

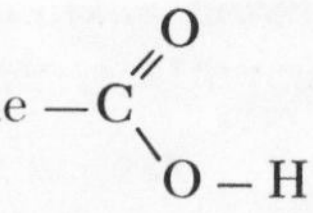

(or *carboxyl* group). These may be written in short form as $-NH_2$ and $-COOH$, respectively. The carboxyl group is also known as the *acid group*. Amino acids contain at least one amino group and one carboxyl group bonded to a central carbon atom. The number of bonds for each atom is: N (3), C (4), O (2), and H (1). Carbon and oxygen are linked by a double bond. This represents the sharing of two pairs of electrons.

One of the H atoms of the central carbon group of glycine may be replaced by various other groups (or radicals). For example, in alanine (Fig. 2–17), a $-CH_3$ (methyl) group replaces an H atom. For convenience, the letter R is used to represent any of a number of radicals or groups which may be present at the position indicated. The general formula for an amino acid is shown in Fig. 2–18.

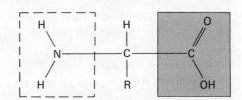

Fig. 2–18. General structural formula for an amino acid.

Dehydration synthesis. Synthesis means the putting together of substances. For example, "synthetic" fibers are artificial ones put together by man.

Fig. 2–17. Alanine.

When the synthesis of a large molecule from small ones is accomplished by the *loss* of a water molecule, the reaction is called *dehydration synthesis.*

Under proper conditions, two amino acid molecules will combine and will be synthesized into a larger molecule, as shown in Fig. 2–19.

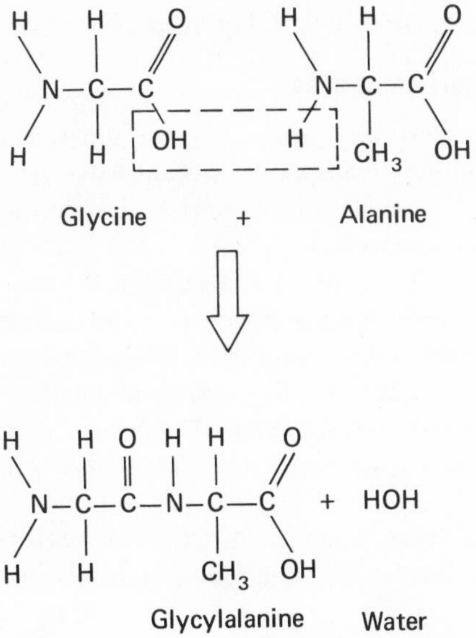

Glycine   +   Alanine

Glycylalanine   Water

Fig. 2–19. An example of dehydration synthesis. Glycine + alanine → glycylalanine + water (Glycylalanine is a **dipeptide**.)

The –OH portion of the carboxyl group of one amino acid molecule combines with the –H of the amino group of another amino acid molecule and forms HOH, or water. The remainder of the two amino acid molecules are bonded together to form a larger molecule composed of two amino acid units.

The region of linkage of the two amino acid units has the configuration

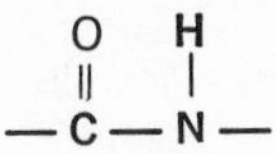

$$\underset{\text{O} \quad\text{H}}{-\underset{\|}{\text{C}}-\underset{|}{\text{N}}-}$$

This linkage is called a *peptide bond;* it is also known as a C–N bond. In Fig. 2–19, since *two* amino acids have been linked by a peptide bond, the synthesized molecule is a *dipeptide.* Its name is glycylalanine. Similarly, the dehydration synthesis of two molecules of glycine results in the dipeptide glycylglycine.

The ends of the dipeptide still have an $-NH_2$ and –COOH group. This can be seen in Fig. 2–19. Dehydration synthesis can thus continue further to produce a long chain of amino acid units, linked by many peptide bonds. This is a *polypeptide.*

Structure of proteins. Long-chain polypeptides, containing from 50 to 50,000 amino acid units, make up the huge protein molecules. The straight chains may be linked together, arranged as a helix, arranged as a globule, or folded. The thousands of different proteins found in cells result from an immense number of possible combinations of the amino acid building blocks. In cells, proteins are formed by a series of energy-consuming steps at the ribosomes. The proteins are assembled under "instructions" from the hereditary material in the nucleus — the DNA.

One of the great achievements of molecular biology was the discovery of the complete structure of a protein molecule. Frederick Sanger of Cambridge University was awarded the Nobel Prize in 1957 for his discovery of the structure of the insulin molecule. Insulin, one of the smaller molecular weight proteins, is a hormone secreted by the pancreas.

Sanger found that the insulin molecule consists of 15 different kinds of amino acid units arranged in two chains. He worked out the precise sequence of amino acids, as well as the kinds and positions of the connecting linkages between their chains. Another protein structure which has been determined is that of myoglobin, a muscle protein. The distance between the parts of the molecule, its shape, and the sequence of the building blocks are important factors in determining how proteins function in the cell. This illustrates on the molecular level the important biological principle that function is related to structure.

Early in 1969, scientists succeeded in synthesizing in the laboratory the protein known as *ribonuclease* (or RNA-ase), an enzyme that helps the cell to digest ribonucleic acid. These scientists built upon the work of other scientists who had isolated this enzyme, crystallized it, and discovered that it is composed of 124 amino acid units of 19 different kinds. The synthesis of this complex protein molecule was jointly announced by researchers from Rockefeller University and Merck, Sharpe & Dohme, Inc. Each team used a different method of synthesis.

Hydrolysis. We have noted that amino acids are synthesized to form polypeptides by the removal of water. The opposite reaction occurs when water combines with a polypeptide; the peptide bond is broken and amino acids are formed. *Hydrolysis* is the decomposition of large molecules into smaller units by combining them with water.

Fig. 2–20 is an illustration of the hydrolysis of the dipeptide glycylglycine to form two molecules of the amino acid glycine.

Certain conditions must be present for the dehydration synthesis and hydrolysis reactions to occur. These include proper temperature, pH, and the presence of enzymes, which will be discussed in detail on page 39.

Carbohydrates

Carbohydrates are organic compounds containing the elements carbon, hydrogen, and oxygen. Carbohydrate molecules have twice as many hydrogen atoms as oxygen atoms. A chemical abbreviation, but not an empirical formula, which expresses the relationship is $C(H_2O)$. Examples of carbohydrates are sugars and starches, which provide energy to the body when taken as food. Cellulose, another carbohydrate, is found in the cell walls of plants. Glycogen is a carbohydrate that is commonly called animal starch.

The sugar, glucose, is a building block of carbohydrates. Its empirical formula is $C_6H_{12}O_6$. Its structural formula, in a ring arrangement, is shown in Fig. 2–21.

The 24 atoms of $C_6H_{12}O_6$ can also be arranged to form the slightly different structures of the sugars galactose and fructose (Fig. 2–22). The different chemical properties of these three sugars is another example of the relation-

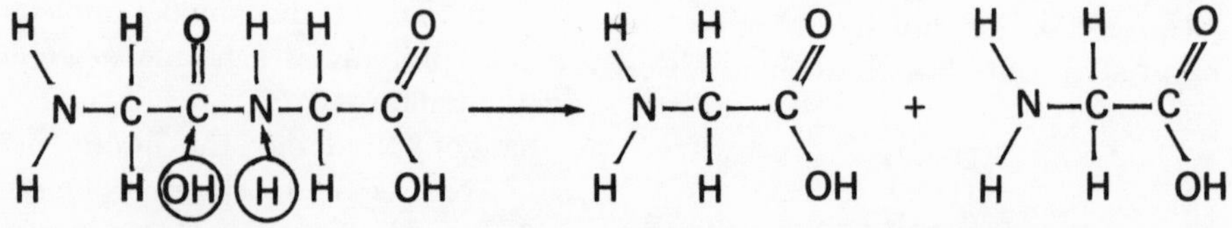

Fig. 2–20. Hydrolysis.   glycylglycine + $H_2O \longrightarrow$ glycine + glycine.

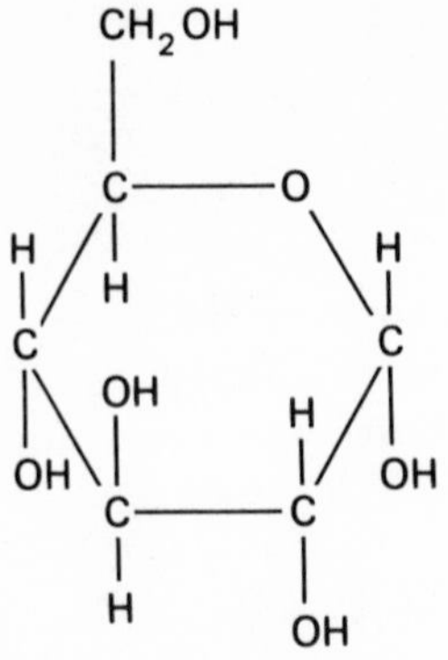

Fig. 2–21. Structural formula of glucose.

called *disaccharides*. Long chains of more than two glucose-like units are called *polysaccharides*. Examples of these sugars, with their formulas, are listed below.

$C_6H_{12}O_6$	monosaccharides	glucose, galactose fructose,
$C_{12}H_{22}O_{11}$	disaccharides	sucrose, maltose, lactose
$(C_6H_{10}O_5)_n$	polysaccharides	starches, glycogen, cellulose

ship between structure and function.

Considering glucose, $C_6H_{12}O_6$, as the single unit for a particular carbohydrate, you might expect the double structure to be $C_{12}H_{24}O_{12}$. However, this is not the case because the synthesis of the double sugar is a *dehydration* synthesis accompanied by the removal of a molecule of water. The double sugar is thus $C_{12}H_{22}O_{11}$.

$$2C_6H_{12}O_6 \longrightarrow C_{12}H_{22}O_{11} + H_2O$$
glucose + glucose   maltose   water

The structural formulas for this reaction are shown in Fig. 2–23.

The single sugars such as glucose, galactose, and fructose are called *monosaccharides* and the double sugars are

Glucose is a monosaccharide that is present in blood. It is also called dextrose. Common table sugar is the disaccharide, sucrose. Since polysaccharides have variable numbers of sugar units, the common expression of their empirical formula is $(C_6H_{10}O_5)_n$.

Hydrolysis of carbohydrates. A common test for "sugar" is to boil an unknown mixture with blue Benedict's solution. If this results in a greenish to brick-red precipitate, a "sugar" is present in the unknown mixture. However, this test works only with the monosaccharides, called "simple sugars" or "reducing sugars." For example, if this test procedure is applied to table sugar (sucrose), which is a disaccharide, the Benedict's solution remains clear blue

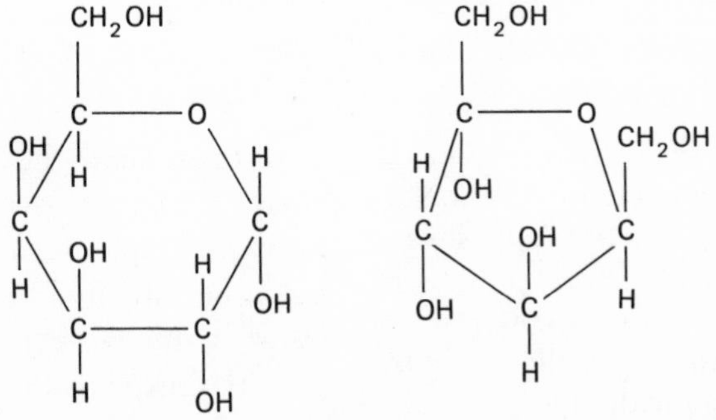

Fig. 2–22. Galactose and fructose, like glucose, have the empirical formula $C_6H_{12}O_6$. However, they differ in their arrangement of atoms and in their properties.

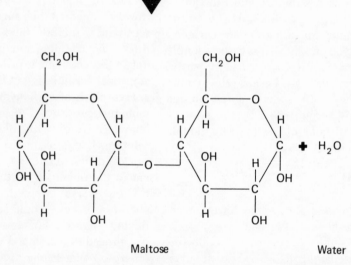

Fig. 2–23. Dehydration synthesis of two simple sugars (glucose molecules) to form a double sugar (maltose). This process occurs in several steps, each of which is controlled by a specific enzyme.

instead of turning green or red. If a drop of dilute hydrochloric acid is added to the sucrose before it is boiled, the sucrose hydrolyzes to simple sugars and yields a positive test. As previously noted, hydrolysis is the decomposition of large molecules into smaller units by combining them with molecules of water.

$$C_{12}H_{22}O_{11} + H_2O \longrightarrow 2C_6H_{12}O_6$$
Disaccharide   water   Monosaccharides

Another example of chemical hydrolysis is digestion in the alimentary canal, where starch molecules are broken down to simple sugars. In the living organism, the hydrolysis of carbohydrates to simple sugars happens at low temperatures with the aid of enzymes, rather than by boiling in an acid.

Hydrolysis and dehydration synthesis are compared in Fig. 2–24.

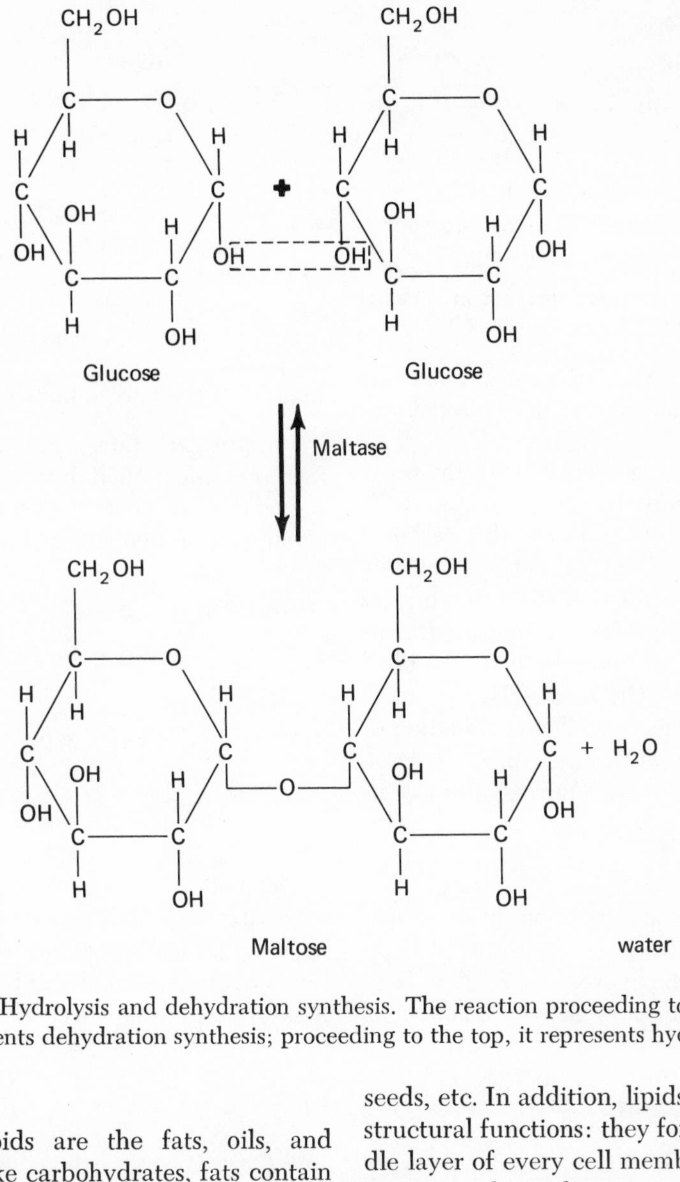

Fig. 2–24. Hydrolysis and dehydration synthesis. The reaction proceeding to the bottom represents dehydration synthesis; proceeding to the top, it represents hydrolysis.

Lipids

The lipids are the fats, oils, and waxes. Like carbohydrates, fats contain carbon, hydrogen, and oxygen, but the proportion of hydrogen to oxygen is not the same as in carbohydrates. Because lipids contain very little oxygen, they can yield large amounts of energy when combined with oxygen. Long-time storage of extra food in organisms is frequently in the form of lipids — the fats under the skin of animals, the oils in seeds, etc. In addition, lipids have some structural functions: they form the middle layer of every cell membrane. Food sources rich in fats are milk, butter, fatty meats, olives, and peanuts.

To understand the structure of fats, you must understand the structures of alcohols and fatty acids. First, we will consider alcohols. *Alcohols* are organic compounds that have the hydroxide (—OH) radical. Two examples of common alcohols are shown in Fig. 2–25.

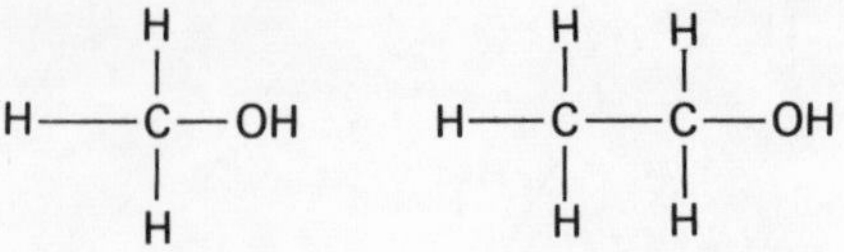

methyl alcohol
wood alcohol
$CH_3 \cdot OH$
poisonous—used as
shellac solvent

ethyl alcohol
grain alcohol
$CH_3 \cdot CH_2 \cdot OH$
present in wines,
beer, whiskey

Fig. 2–25. Two common alcohols are methyl alcohol and ethyl alcohol.

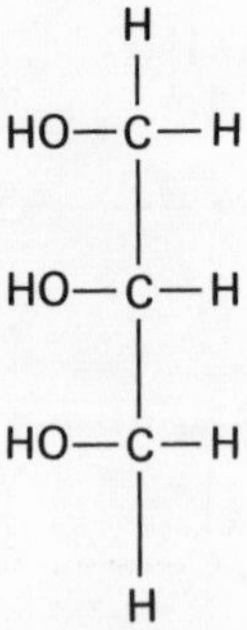

Fig. 2–26. Glycerol.

Glycerol is an alcohol with three hydroxide groups, as shown in Fig. 2–26.

Organic acids contain the carboxyl group (−COOH). A general formula for organic acids is R−COOH. In fatty acids, the R group is a long hydrocarbon chain. An example of a fatty acid is stearic acid, $C_{17}H_{35}COOH$.

A fat results from the combination of three fatty acids and one glycerol, united by dehydration synthesis, as shown in Fig. 2–27.

3 fatty acids + glycerol $\longrightarrow$ fat + $3H_2O$

The opposite reaction also occurs. The digestion of fats within the body consists of the hydrolysis of fats to fatty acids and glycerol. Enzymes, known as lipases, play a part in hydrolysis of fats.

Unsaturated fats. "Polyunsaturated fats" are mentioned by physicians as preventives of heart disease. A portion of an unsaturated fat is shown in Fig. 2–28.

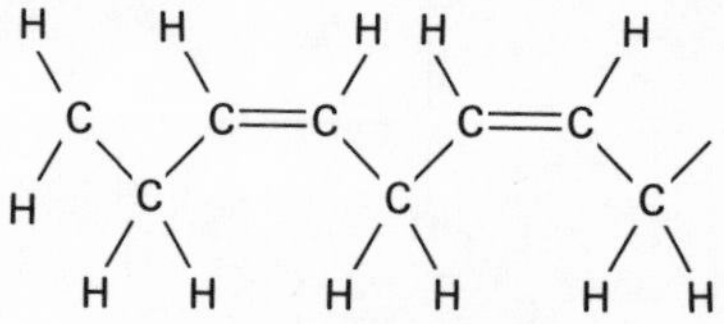

Fig. 2–28. A portion of the R group of an unsaturated fat.

The molecule of a fat is "unsaturated" when its R portion has double bonds between the carbon atoms where additional hydrogen atoms can be taken on. A saturated fat, however, has all the H

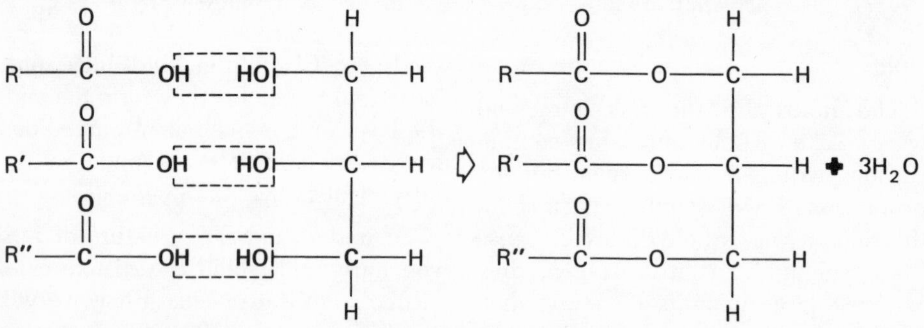

3 fatty acids +   glycerol $\longrightarrow$ fat   + $3H_2O$

Fig. 2–27. Formation of a fat by dehydration synthesis.

atoms that it can take on. The formula for a saturated fat shows no double bonds between the carbon atoms.

Saturated fats are believed to be converted to cholesterol (a complex organic substance) within the body and be deposited on the linings of blood vessels. When too much cholesterol is deposited within the coronary artery, the blood supply to heart muscle is stopped and portions of this muscle are injured. Many physicians advise people to use unsaturated fats instead of saturated fats to avoid possible heart attacks.

Milk and milk products, as well as meat, contain saturated fats. The oils of fish and vegetables are unsaturated.

Nucleic Acids

Nucleic acids are giant molecules of great molecular weight. They have a precise arrangement of five-carbon sugars, phosphate groups, and nitrogen groups. They are found in every living cell and are important in the mechanism of heredity. Two types of nucleic acids are DNA (deoxyribonucleic acid) and RNA (ribonucleic acid). Their structures will be considered in Chapter 21.

ENZYMES

Enzymes are organic catalysts. A common method for preparing oxygen in the school laboratory is to decompose potassium chlorate ($KClO_3$) by heating it in a test tube. The equation for this reaction is $2KClO_3 \rightarrow 2KCl + 3O_2$. Students normally add a small amount of manganese dioxide, MnO_2, which speeds up the reaction and permits it to proceed with less applied heat. The manganese dioxide acts as a *catalyst,* a substance that affects the speed of a chemical reaction

without itself being used up. Many catalysts speed the rate of reaction, but some may slow it down. *Enzymes* are organic catalysts which promote numerous reactions in every living cell.

Characteristics of Enzymes

1. *Enzymes are proteins.* Enzymes are giant protein molecules which regulate the rate of chemical reactions in living things. These molecules are folded and held together in a definite shape by chemical linkages and electrostatic forces. This shape is important in determining the activity of enzymes. The material acted upon by the enzyme is called the *substrate.* The substrate molecules are smaller than the enzyme molecules and fit into definite positions (reactive sites) on the *surface* of the enzyme molecules.

Like all proteins, enzymes are altered by heat so that most of them no longer function at temperatures above 60° C. One reason why high temperatures kill organisms is that heat destroys the enzymes needed for life. Lead and some mercury compounds are poisonous because these metals affect the reactive sites of vital enzymes. Manufacturers avoid painting children's toys with lead-based paint to protect children from lead poisoning. Mercury from a broken thermometer should not be handled because it is easily absorbed through the skin.

2. *Enzymes permit reactions to proceed under "milder" conditions of temperature and pH.* When a solution of the compound urea is heated in a test tube, the urea breaks down to liberate the gas ammonia (NH_3). The ammonia is detected by its odor and by its ability to turn moist pink litmus paper (a testing paper) to a blue color. If the enzyme urease is added to the solution,

the reaction proceeds at body temperature. This experiment is illustrated in Fig. 2–29.

Sucrose can be hydrolyzed to simple sugars in a test tube by boiling it in an acid medium. In the intestine, this hydrolysis occurs through enzyme activity at body temperature and at a pH only slightly below 7. If these enzymes were not present in the body, extreme conditions of temperature and pH would be needed in order for these chemical reactions to occur. Such extreme conditions of temperature and pH would destroy the cell.

3. *Enzymes are needed in small quantities.* A single enzyme molecule may be used over and over again because it is not used up during reactions. For example, one molecule of the enzyme catalase can catalyze the breakdown of 5 million molecules of hydrogen peroxide within one minute.

4. *Enzymes enter into the reaction with the substrate.* They form a temporary enzyme-substrate complex which then breaks apart. For example, consider a reaction in which enzyme E catalyzes the combination of molecules of A with molecules of B to form the

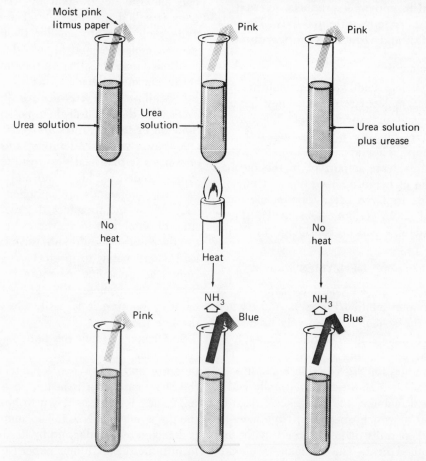

Fig. 2–29. Urease catalyzes the breakdown of urea to liberate ammonia at body temperatures.

new compound AB. This could be represented as:

E	A and B combine "in the
A+B→AB	presence of" enzyme E
	to form AB

However, the sequence of events is probably as follows:

(1) A+E→AE   The enzyme molecule forms a complex with a molecule of substrate A

(2) AE+B→AB+E   A and B unite and enzyme E is released. The enzyme is now free to help unite additional molecules of A and B.

5. *Enzymes are specific.* Here the word "specific" means that an enzyme usually enters into only one kind of reaction. As a result, a great number of different enzymes are needed in the cell for each of the many reactions to occur. The specificity of enzyme action is explained by the *lock and key* concept. Surface regions of the enzyme molecules have definite shapes which fit into the corresponding shapes of the substrate molecules. The enzyme molecule may thus bring together two different substrate molecules as shown in Fig. 2–30.

The action of an enzyme in bringing together two different substrate molecules is more rapid than the rate of combinations which result from the chance collision of molecules.

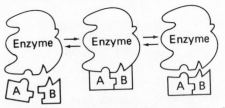

Fig. 2–30. The lock and key concept of enzyme and substrate explains the specificity of enzyme action.

6. *Enzyme actions are reversible.* The same enzyme which catalyzes the reaction A + B → AB also catalyzes the reverse reaction AB → A + B. Enzyme-controlled reactions are usually written with a double arrow, as A + B ⇌ AB.

7. *Enzymes have optimum pH and temperatures.* "Optimum" means the range over which the enzyme is most efficiently effective. For example, the optimum pH for salivary amylase, which acts in the mouth, is 6.2–7.0, and the optimum pH for pepsin, which acts in the stomach, is 1.5–2.2. The rate of activity or effectiveness of the enzyme drops off sharply on either side of the optimum range. The activity of salivary amylase is reduced when the food mass reaches the acid stomach.

The effect of temperature on the activity of one enzyme is indicated in Fig. 2–31.

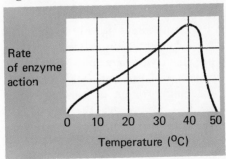

Rate of enzyme action

Temperature (°C)

Fig. 2–31. Effect of temperature on enzyme activity.

8. *Enzymes may be assisted by coenzymes.* A *coenzyme* is a molecule which assists an enzyme in its action. For example, when an enzyme removes hydrogen from a substrate, a coenzyme assists it by accepting and removing hydrogen atoms. For a long time scientists knew that vitamins were needed for good health and energy but did not understand their function. Now, the

vitamin called thiamine is known to be part of a coenzyme used in oxidation reactions which yield energy in the cell. If thiamine is not present in the diet, the needed coenzyme is not produced and the individual lacks the ability to produce the maximum cellular energy.

9. *The rate of enzyme action is regulated by the relative amounts of enzyme and substrate.* Fig. 2–32 illustrates what happens when the substrate concentration is held constant and increasing amounts of enzyme are added. At the left side of the graph, the enzyme concentration is very low, and there are numerous substrate molecules available. As each enzyme molecule completes a reaction, there are many additional substrate molecules available for it to react with. An addition of enzyme molecules increases the rate of the reaction and so the curve on the graph rises. However, when the enzyme concentration reaches a certain point, all enzyme molecules are actively engaged in "turnover" and an additional concentration of enzyme molecules does not increase the overall rate of the reaction. At this point, the curve levels off.

Note that Fig. 2–33 illustrates what happens when the enzyme concentration is held constant and increasing amounts of substrate are added. At the start, all the enzyme molecules are not actively engaged in turnover for lack of sufficient substrate molecules to work on. An increase in the concentration of substrate molecules causes an increase in the rate of the reaction, and so the curve rises. However, again a point is reached when all the enzyme molecules are actively engaged in turnover and additional sub-

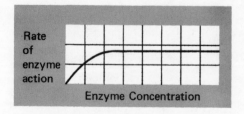

Fig. 2–32. The effect of varying the concentration of enzyme. The substrate concentration is constant.

strate molecules cannot involve more enzyme molecules. The curve levels off.

10. *Additional examples of enzyme action.* Most enzymes operate inside the cell. They are called *intracellular enzymes.* The activities of life within cells, such as respiration and photosynthesis, consist of a series of separate reactions, with each reaction catalyzed by an enzyme that is specific for it. Enzymes that function outside the cell, such as those in the alimentary canal, are called *extracellular enzymes.*

Enzymes are often named by rewriting the last suffix of the substrate on which they act as "ase." For example, reactions leading to the formation of the carbohydrate mal*tose* may be catalyzed by mal*tase.*

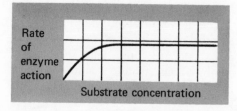

Fig. 2–33. The effect of varying the concentration of substrate. The enzyme concentrate is constant.

Lipase, proteases, and amylases are extracellular enzymes that act in human digestion. *Lipase,* found in pancreatic juice, catalyzes the digestion of lipids into fatty acids and glycerol. The juices of the stomach, small intestine, and pancreas contain at least five different *proteases* which assist in breaking proteins down into amino acids. *Amy-* *lases* in the saliva and pancreatic juice convert starch to the disaccharide maltose; other enzymes act on disaccharides to split them into monosaccharides.

REASONING EXERCISES

1. Why is water essential to life?

2. Why are amino acids important to life? Write the general formula for an amino acid. In the formula, what is the meaning of R? What is the meaning of the double bond? How many bonds does each carbon have?

3. Write the word equation for the dehydration synthesis of two amino acids. Write the word equation for the breaking down of a dipeptide. How do dehydration synthesis and hydrolysis show two different aspects of metabolism?

4. How does $C_6H_{12}O_6$ illustrate the important biological principle that structure is related to function?

5. Why does the synthesis of a double sugar from two glucose molecules result in the formula $C_{12}H_{22}O_{11}$ instead of $C_{12}H_{24}O_{12}$? What is the difference between hydrolysis of carbohydrates in the laboratory and in the organism?

6. Write the word equation for the synthesis of a fat. What is the difference between a saturated and an unsaturated fat?

7. Distinguish between enzymes and coenzymes. At what temperature does maximum enzyme activity occur in Fig. 2–31? What factors regulate the rate of enzyme action?

8. What is the difference between intracellular and extracellular enzymes?

COMPLETION QUESTIONS

A 1. The atomic particle with a mass number of one and no charge is the ....

2. Atoms with the same atomic number but different mass numbers are ....

3. A radioactive isotope used to study a series of chemical reactions is called a (an) ....

4. An isotope which is not radioactive is called ....

5. An atom which has gained or lost electrons is a (an) ....

6. A chemical bond formed by the sharing of electrons is a (an) .... bond.

7. A formula for a molecule which shows the arrangement of the atoms is called a (an) ....

8. Organic chemistry is the study of .... compounds.

9. The degree of acidity or alkalinity of a solution is indicated by the .... scale.

10. A pH of 4 indicates that the solution is ....

B 11. The building blocks of protein are ....

12. A molecule having two amino acid units is a (an) ....

13. The formation of a larger molecule from two smaller molecules, accompanied by the giving off of water is known as ....

14. The breaking down of a large molecule by the addition of water to form two smaller molecules is ....

15. An example of a common disaccharide is ....

16. The middle layer of every cell membrane consists of ....

17. Fats may be hydrolyzed to form fatty acids and ....

18. The material acted upon by an enzyme is the ....

19. Because an enzyme operates in only one kind of chemical reaction, enzymes are said to be ....

20. The specificity of enzyme action is explained by the .... concept.

MULTIPLE-CHOICE QUESTIONS

[A] 1. As the number of neutrons in the nucleus of an atom increases, the atomic number of the atom (1) decreases, (2) increases, (3) remains the same.

2. A helium nucleus contains two protons and two neutrons. The number of charges in the nucleus is (1) 0, (2) 2, (3) 8, (4) 4.

3. A substance that cannot be decomposed by a chemical change is a (an) (1) mixture, (2) compound, (3) element, (4) solution.

4. In a structural formula, a dash represents a (an) (1) transferred electron, (2) pair of transferred electrons, (3) shared electron, (4) pair of shared electrons.

5. The particles in NaCl exist as (1) related isotopes, (2) positive and negative ions, (3) separate molecules, (4) covalent bonds.

[B] 6. Which element is present in both starch and protein? (1) iron, (2) nitrogen, (3) calcium, (4) carbon.

7. An amino acid may be recognized if it has an amino group at one end and at the other end a (an) (1) sulfur group, (2) hydroxyl group, (3) carboxyl group, (4) saturated group.

8. Which best indicates the elements that are present in an amino acid? (1) CaPO, (2) CHO, (3) CHOFe, (4) CHON.

9. Which is a function of enzymes? (1) They provide energy for carrying on a chemical reaction. (2) They can speed up the rate of chemical reactions. (3) They become hydrolyzed during chemical reactions. (4) They serve as inorganic catalysts.

10. The "lock and key hypothesis" attempts to explain the mechanism of (1) vacuole formation, (2) enzyme specificity, (3) sharing of electrons, (4) pinocytosis.

CHAPTER TEST

1. Which is an example of an organic compound? (1) NaCl, (2) H_2O, (3) $(NH_4)_3PO_4$, (4) $C_{17}H_{35}COOH$.

2. Which of the following compounds has a bond formed by transfer of electrons? (1) NaCl, (2) CH_4, (3) H_2, (4) $C_6H_{12}O_6$.

3. Atoms with the same atomic number but different atomic masses are known as (1) ions, (2) radicals, (3) isomers, (4) isotopes.

4. If distilled water were tested with a pH meter, its pH would be (1) 2, (2) 7, (3) 13, (4) 4.

5. In living substance, sulfur and nitrogen are elements chemically combined in (1) proteins, (2) glycogen, (3) oils, (4) carbohydrates.

6. Which substance plays a major role in most of the chemical reactions occurring in a living cell? (1) amino acid, (2) water, (3) glucose, (4) fatty acid.

7. To tag glucose in an oxidation, it would be reasonable to use radioactive (1) carbon, (2) calcium, (3) cobalt, (4) iron.

8. The process by which amino acid molecules are joined together is (1) hydrolysis, (2) photosynthesis, (3) oxidation, (4) dehydration synthesis.

9. The number of bonds for each carbon atom in a structural formula is (1) 1, (2) 2, (3) 4, (4) 3.

10. Because urea is a nitrogen compound, it can *not* be derived from metabolism of (1) peptides, (2) proteins, (3) amino acids, (4) glucose.

For each of questions 11 through 15, write the letter preceding the formula, chosen from the list below, to which that question refers.

FORMULAS

11. This substance is hydrolyzed by the enzyme maltase.
12. This substance is an example of a fatty acid.
13. This substance is a building block used in the synthesis of proteins.
14. The metabolism of this substance is responsible for the creation of nitrogenous wastes.
15. This substance combines with fatty acids in fat synthesis.

16. Of which is the equation below a representation?

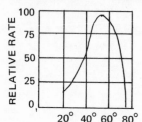

(1) dehydration synthesis   (3) protein digestion
(2) hydrolysis   (4) cellular respiration

17. Lack of certain vitamins in the diet results in weakness and lack of energy, as in the disease beri-beri. These vitamins are needed because (1) they are oxidized to yield energy, (2) they contain high energy phosphate bonds, (3) they have a high calorie content, (4) they are components of respiratory coenzymes.

18. Which of the following chemical formulas represents a carbohydrate? (1) $(C_6H_{10}O_5)_{100}$, (2) $CO(NH_2)_2$, (3) $C_{10}H_6Cl_2$, (4) $C_5H_{10}O_3N_2$.

19. The bonding of two amino acid molecules, glycine and alanine, to form one large molecule requires (1) addition of a nitrogen atom, (2) addition of a water molecule, (3) removal of a nitrogen atom, (4) removal of a water molecule.

20. Of the following chemicals, the one which is classified as an enzyme is (1) galactose, (2) lipids, (3) protease, (4) manganese dioxide.

Questions 21–25 are based on the following graphs which show the influence of temperature and pH on the rate of action of a certain enzyme.

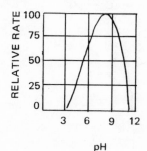

21. The temperature at which maximum enzyme action occurs is (1) 38° C, (2) 50° C, (3) 60° C, (4) 65° C.

22. Enzyme action ceases at a temperature of (1) 20° C, (2) 30° C, (3) 60° C, (4) 75° C.

23. The enzyme functions most effectively at a pH of (1) 6, (2) 7, (3) 8, (4) 9.

24. The rate of enzyme action will be *decreased* by increasing (1) concentration of the enzyme, (2) temperature from 50° C to 70° C, (3) concentration of the substrate, (4) pH from 3 to 6.

25. As the line of the pH graph approaches the value of 12, the medium has a greater excess of (1) OH− ions over H+ ions, (2) H+ ions over OH− ions, (3) Na+ ions over Cl− ions, (4) HCO_3^- ions over NO_3^- ions.

CHAPTER 3 | Taxonomy and Evolution

There are almost two million kinds of living things on earth today! The huge variety of presently living and extinct organisms poses the problems of naming and classifying them. The science of classification is called *taxonomy*. In this scheme, a worldwide name is provided for each kind of organism discovered, and the different kinds are organized into meaningful categories which show similarities and differences.

In the past, many ways of classifying living things have been used. In 350 B.C., the Greek philosopher Aristotle grouped the few hundred animals *then known* on the basis of their habitat as land-living, water-living, and air-dwelling. He classified plants mainly on the basis of their size. Later, scientists classified living things by their function, color, shape, and structure. Today, taxonomists have access to vast collections of preserved specimens of representative organisms. These preserved specimens are one of the basic working tools of the taxonomist.

A LIFE OF THE PAST

A good scheme of classification must include not only existing organisms but also extinct organisms (those that formerly existed). While we do not know as much about extinct organisms as we would like to, we do get considerable information from fossils.

Kinds of Fossils

A *fossil* is any record of an organism that lived in the geologic past. Remains, impressions, and traces of organisms of past geologic ages have been preserved in the earth's crust as fossils.

It is impossible to discover all of the previous forms of life that existed because not all of them became fossils. Another reason is that organisms are usually consumed or decomposed after death. While the number of fossils discovered somewhat limits the biologist's investigation of the past, the ones we do have throw great light upon it.

The *remains of hard parts of* organisms, such as bones, teeth, and shells can resist the action of weathering for long periods of time. Fossils are preserved much longer when their environment is dry. For example, the fossil bones of dinosaurs were preserved for 100 million years in the hot desert sands of Mongolia.

In some instances, the soft tissues of extinct organisms have been preserved by *refrigeration*. The frozen remains of ancient hairy mammoths (relatives of the elephant) have been found. The tissues were so well preserved that explorers ate the meat and became ill.

Imprints are the impressions left by plants and animals on the soft mud where they rested. Tracks and tunnels made in soft mud by prehistoric animals were preserved when the mud hardened into rock. Biologists can study the footprints left by dinosaurs 120 million years ago as they walked in regions that are now Texas and Connecticut.

When rock forms around a dead organism, which later disintegrates, a *mold* of the organism remains in the *rock*. Molds formed in this manner may preserve details of the dead organism with great accuracy. If the mold later becomes filled with mineral deposits, a *cast* is produced. Such a cast is a replica of the external details of the organism.

Another type of fossil is provided by *insects preserved in amber*. Amber is the hardened gum or resin given off by trees. Small insects were often caught in the sticky gum and preserved as the gum hardened. Some of the insects are so perfectly preserved that the lacy structure of their wings is intact, permitting biologists to compare them with insects that are living today.

Petrification resulted when the skeletons of animals and the cellulose of wood and plants were covered with water containing large amounts of dissolved minerals such as silica or lime. The cell cavities and intercellular spaces of the hard tissue were filled with the mineral matter, thereby preserving the structure of the original specimen. The Petrified Forest in Arizona has many trees that have turned to stone.

Coal is the fossilized remains of ancient plants. During the Carboniferous period, about 300 million years ago, deep layers of plant remains accumulated in the hot swampy regions where they grew. Movements of the earth later covered and compressed these plant remains, and coal gradually formed. Today, imprints of ferns and mosses are often found in coal.

Many ancient animals were trapped and preserved in large pools of gummy asphalt tar. The "saber-toothed tiger" is an example of a fossilized organism recovered from *tar pits* at La Brea in Los Angeles, California. Other fossil bone specimens taken from the tar pits include mammoths, horses, birds, and camels.

Where Fossils Are Found

Aside from the few special cases like frozen mammoths and creatures in tar pits and amber, fossils are generally found in sedimentary rock. All of the rocks of the earth can be classified into three general categories: igneous rock, sedimentary rock, and metamorphic rock.

Igneous rock is formed by the cooling and solidification of molten lava. This may occur either deep in the crust of the earth or on its surface when lava comes out of a volcano. It was presumably the original rock of the earth, for it is probable that the earth was entirely molten at some early stage in its history. Any animal or plant that fell into molten lava would be destroyed instantly and would not form a fossil. Therefore, fossils are not found in igneous rock. Examples of igneous rock are granite and basalt.

Sedimentary rock is formed from loose material such as sand or mud, obtained by the breaking up of previously existing rock. It is formed when this loose material is subjected to moderate pressure and chemical action that turns it to solid rock. Sedimentary rock is formed in layers on the bottom of bodies of water.

There are different types of sedimentary rock. Sand, washed out of continents and into oceans by rivers, turns to *sandstone* after many years. Clay may be changed to a smooth-textured rock called *shale*. Coarse gravel changes to *conglomerate*, in which the original pebbles are still easily seen. Many rivers carry calcium carbonate ($CaCO_3$) in solution, and this precipitates out when it meets salt water, forming a whitish material called limy mud. This may turn to *limestone*.

If an animal or plant dies and falls into mud, sand, or limy mud, it may leave an imprint which is found as a mold or cast when the loose sediment turns to sedimentary rock.

Sedimentary rocks occur in most parts of the world. They may even be found on some mountain tops, bearing fossils of prehistoric ocean-living organisms. The earth's crust is in constant movement. If a bed of sedimentary rock is

I. DEPOSITION OF SEDIMENTARY ROCK IN LAYERS

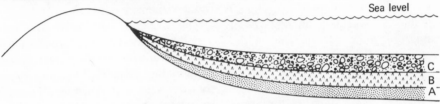

A, B, & C represent layers of sedimentary rock. Which is the oldest?

II. UPLIFT

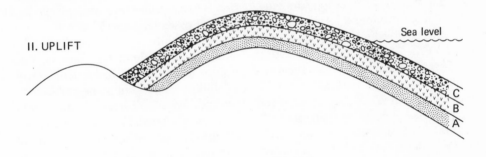

III. EROSION

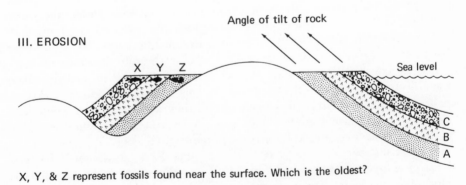

X, Y, & Z represent fossils found near the surface. Which is the oldest?

Fig. 3–1. Changes in the earth's surface. Sedimentary rocks, formed under water, are uplifted to form mountains, which are then worn away.

lifted only an inch a century, it will be very close to a mile above the level from which it started after 6 million years — not very long, by geological standards.

Igneous rock or sedimentary rock may be changed to *metamorphic rock*. How does this occur? Any kind of rock may be subjected to enormous pressure and high temperature due to movements of the earth's crust. If this occurs, the rock may undergo drastic changes in its crystal structure, producing metamorphic rock. Marble is an example of metamorphic rock. Metamorphism destroys almost all fossils.

The Ages of Fossils

In general, fossils found in the lower layers of sedimentary rock are older than those found in the upper layers. (See Fig. 3–1.) In the top drawing, the sediments are found forming layers. Since the new sediments are deposited on top of the older ones, A is the oldest and C is the youngest. When the sediments are turned to rock and uplifted, as in the second sketch, any fossil found in layer A must be older than one in B, and one in B is older than one in C. Often the layers are folded and eroded, so that the determination of age is more difficult. In the third sketch, for example, fossils X, Y, and Z are all found on the surface, but Z is the oldest.

The *relative age* of a fossil or layer of rock is its age stated on the basis of a comparison made between it and its immediate surroundings. From this, you should be able to see that relative age is not the same as age measured linearly for a definite number of years or centuries. Even if fossils are found in different parts of the world, their relative ages may be determined. In some cases, fossils of the same animal are found all

over the world. If the animal lived for only a short period of time, it can be used for dating any layer of rock in which it is found. By the use of such *index* fossils, the relative ages of layers of rock all over the world have been coordinated.

Radioactive Dating

Scientists use radioactive isotopes for dating rocks and fossils. We will consider the uranium-lead ratio and the radio-carbon method.

The *uranium-lead ratio* is a quite accurate method for finding the age of ancient igneous rocks which formed from molten lava. The disintegration of the atoms of a radioactive element to form new atoms occurs at an *unchanging* rate and may continue for a long period of time. The rate of disintegration of a radioactive isotope is expressed as its *half-life*. This is the time for the disintegration of half of the atoms in a sample of the element. The isotope uranium-238 has a half-life of 4.5 billion years (but some radioactive isotopes have half-lives of only thousandths of a second). Uranium-238 disintegrates into other unstable atoms and finally becomes the stable isotope lead-206. When scientists wish to judge the age of igneous rocks, they first determine the proportion of uranium and lead in the rocks. Then, because they know the rate at which uranium changes to lead, they can calculate how long change has been in progress. When they know this, they know how old the rock is. This method is a kind of "radioactive clock."

Uranium-lead ratios cannot be used to date sedimentary rocks, for the ratio will tell when the material first formed as an igneous rock rather than when sedimentation occurred. However, a lava flow on top of a sedimentary layer can be used. The lava is dated by the uranium-lead ratio, and the sediments just under it cannot be much older. The uranium-lead ratio places the age of the earth at about 6 billion years.

The uranium-lead ratio is a reasonably accurate method for dating fossils that are 30 million years old or older. Other methods are used for dating more recent specimens. The *radio-carbon method* is used to date very recent organic remains; it can be used effectively on organic remains as recent as a few thousand to about 40,000 years. Carbon in living organisms occurs in two varieties — ordinary stable carbon-12 and radioactive carbon-14. A living organism takes in a certain quantity of radioactive carbon-14 from the atmosphere and, when the organism dies, it cannot absorb any more carbon-14.

The half-life of carbon-14 is 5560 years. Thus, half of the atoms in a sample of carbon-14 disintegrate in this time period. In order to be sure that we understand what this means, let us consider the total amount of carbon-14 in a fossil to be a given sample. If half of this amount breaks down in 5560 years, half of the amount still remains in the fossil. In the next 5560 years, the remaining half is reduced by half. This process continues until the amount remaining is infinitesimally small. Accordingly, the radio-carbon method cannot be used effectively on fossils much older than 40,000 years.

There is a constant ratio between carbon-12 and carbon-14 in all living things. In fact, we assume that the concentration of carbon-14 in the atmosphere remains constant through the years. But how is this possible, since carbon-14 is constantly disintegrating? Let us try to answer this question.

As carbon-14 decomposes, it produces the stable isotope nitrogen-14. Carbon-14 is created continuously by the impact of cosmic rays on the nitrogen of the atmosphere, and its concentration in the air is assumed to remain constant through the years. Since plants take in CO_2 from the atmosphere, there is thus a constant ratio between carbon-12 and carbon-14 in all living things. But after they die, the carbon-14 disintegrates, at a known rate, while the carbon-12 does not. Therefore, the ratio of carbon-14 to carbon-12 in a fossil is an indication of how long ago the organism died.

Timetable of the Earth

Geologists have developed a timetable of the earth's history by dividing time into *eras*. Eras vary in length from many millions to a billion years. Eras are subdivided into units called *periods*.

The Azoic era. During this portion of time, the rocks of the earth were forming and conditions were developing for the later appearance of life. Geologists think no life existed in the Azoic era.

The Precambrian era. During this era the oldest exposed rocks on the continents were formed. Although there are no fossil remains, it is assumed that simple one-celled plant and animal life existed in the waters of the earth during the early part of this era. Subsequently, fossil remains of one-celled organisms, such as algae and protozoa were formed. In addition, there were simple, many-celled marine invertebrates (animals without backbones); examples are worms and sponges. Because the remains of these animals are very rare, it is assumed that organisms living at this time had not yet developed the hard parts that could be preserved as fossils.

Timetable of the earth showing development from single-celled animals to the highest kind of mammal-man.

	ERA	Dominant form of life
NOW	PSYCHOZOIC	Age of man; man dominates the earth; machines are developed to do the work formerly done by man and animals.
1 million B.C.	CENOZOIC	Age of Mammals. Primitive mammals appear: platypus, kangaroo. Followed by modern mammals, birds, insects. Finally, man appears.
70 million B.C.	MESOZOIC	Age of Reptiles. Water-inhabiting reptiles, flying reptiles (pterodactyl), dinosaurs (brontosaurus, stegosaurus). Appearance of bird-like form (archeopteryx).
190 million B.C.	PALEOZOIC	Appearance of shelled invertebrates. Trilobites dominant. Fish appear. First air breathers. Land plants, amphibians and insects.
360 million B.C.	PRECAMBRIAN	Simple marine invertebrates and algae.
2 Billion B.C.(?)		

The **Paleozoic era.** During this era great changes occurred in plants and animals. Many types of higher water-living invertebrates were present during this time, such as jellyfish, sponges, corals, and snails. In an early period of this era, the dominant form of life in the sea was a primitive relative of the lobster, called the *trilobite,* which is now extinct. The first jawless fish appeared. Then, sharklike and armored fish vertebrates (animals with backbones) came on the scene and became the predominant marine organisms by the middle of the Paleozoic era. Plant life appeared on land.

During the Paleozoic era, air-breathing organisms such as scorpions and insects appeared. Late in the Paleozoic era, other air-breathing animals, particularly amphibians, appeared in great numbers. The period is also noted for its vast amount of vegetation in the form of ferns, mosses, and eventually, seed-forming plants.

The end of the Paleozoic era is noted for its abundance of ferns, which turned into the coal used as fuel today. In addition, primitive reptiles appeared at the end of this era.

The **Mesozoic era.** During this era the earth was dominated by dinosaurs and other large reptiles. Some fossils from this era are remains of animals about 100 feet long and weighing almost 40 tons. Many of the reptiles of this era developed heavy protective armor. The first mammals appeared early in the Mesozoic era and the first birds a little later.

The trees and other plants of this era were quite similar to the plants of today. The ginkgo tree, for example, was the same then as it is today and for that reason is called a "living fossil." The insects of today are very much like the insects that developed during the Mesozoic era.

The **Cenozoic era.** The era of recent life, the *Cenozoic era,* began about 70 millions years ago. It is noted for the many changes produced by the great ice sheets that moved across the land. The primitive mammals present at the very beginning were eventually re-

STEGOSAURUS

TRICERATOPS

TYRANNOSAURUS

Fig. 3–2. Some dinosaurs of the Mesozoic era — stegosaurus, triceratops, and tyrannosaurus.

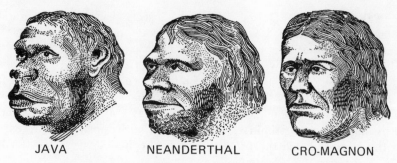

JAVA   NEANDERTHAL   CRO-MAGNON

Fig. 3–3. Three types of prehistoric man.

placed by modern mammals. Birds became common; flowering plants appeared and became more common than any other types. Manlike creatures appeared about a million years ago.

The Psychozoic era. This era begins with the dominance of man and is, of course, the era in which we now live.

EVOLUTION

Fossils show that the living things of the past were usually different from those of today. Since, so far as we know, the only way a living thing can originate is by the reproduction of some substance already living, it is clear that the form and structure of organisms change over many genera-

tions. Offspring are often a little different from their parents; if there is some way these small differences can accumulate from generation to generation, large changes can eventually result. Today's organisms are enormously complex, and the fossil record shows a general trend toward greater complexity as time goes by. This change of kind with time is called *organic evolution.*

Although fossils are scarce, there are a few cases in which a more or less continuous change can be traced for a long period of time. One of the best known is the series that produced the modern horse, *Equus* (a horseback rider is an "equestrian"). In recent layers of rock are fossils of animals which differ only slightly from *Equus.*

6 FT.
5 FT.
4 FT.
3 FT.
2 FT.
1 FT.

EOHIPPUS

EQUUS

Fig. 3–4. Comparison of *Eohippus* and *Equus.*

Progressively older layers of rock reveal fossils of horselike animals, each of which differs slightly from the later fossil. This series of horselike animals has been traced, step by step to *Eohippus* (The Dawn Horse), which lived about 50 million years ago. A comparison of *Eohippus* and *Equus* is given in the table below and in Fig. 3–4.

	Eohippus	**Equus**
Size	Size of fox	Present day horse (5'-6' high)
Toes	Four toes on front feet, three on hind feet.	One toe; the hoof is homologous with the toenail
Teeth	Short-crowned	Long-crowned

The changes in the legs of fossil horses show that the middle toe became progressively longer and the side toes progressively shorter. All that remains of these side toes in the modern horse is two "splint bones," as seen in Fig. 3–5.

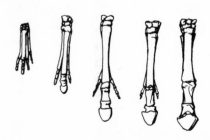

Fig. 3–5. Evolution of the foreleg of the horse.

The situation is actually far more complicated than is shown here, for the fossil history of the horse includes many side branches that died out. A few other nearly complete continuous series are known, such as the camel and the elephant, but in most cases there are not enough fossils to illustrate the whole story. Thus, the relationships between the older forms and the modern forms they generated must be deduced from indirect evidence.

The classification of living organisms from Aristotle's time until fairly recently consisted of methods that were more convenient than scientific. If one man chose to classify organisms by size, while another preferred to group them according to color, the main question was, Which is more convenient?

Today, the taxonomist uses a more meaningful approach. He attempts to create natural groups, based on *evolutionary relationships*. When he says that the lion, cat, seal, coyote, and wolf all belong to the same group, he does so because he believes that all of them have evolved from the same ancestor. All fossil forms that evolved from that same ancestor are included in the group as well. Wherever possible, the taxonomist identifies the common ancestor; in the group given above, the ancestor is a well-known fossil creature of the early Cenozoic era.

The Evidence for Relationship

It is hard to tell which organisms are most closely related through evolution. The fossil record of life in the past is incomplete. The classification of living things changes with new knowledge. Some of the bases which are used to determine evolutionary relationships are as follows:

Homologous structures. If two types of living things have many similar structures, they are probably closely related. Biologists distinguish between two kinds of similarity: *analogous* structures and *homologous* structures. Homologous structures are the only ones used as a basis of classification by today's scientists.

Two structures are *analogous* if they

have the same function but have different evolutionary origins and have different basic compositions. For example, the wing of a fly and the wing of a bird have the same function — both are used in flying. But the wing of a fly is a membrane, and the wing of a bird is bony and covered with feathers. The wing of a fly and the wing of a bird are analogous organs because their fundamental structures are so different in spite of their superficial similarity. Another example can be provided by comparing the horn of a rhinoceros and the antler of a deer. They have the same function, but the horn is hairlike and the antler is bone. These two structures have entirely different origins, and are analogous. The presence of analogous organs in two kinds of living things is only a superficial resemblance which cannot be used as a basis for classification.

Two structures are *homologous* if they have the same basic structure and evolutionary origin. They show the same pattern of early growth. Homologous organs evolved from the same an-

cestral structure. Homologous organs, however, do not necessarily share the same function. For example, the leg of man, the wing of a bat, and the flipper of a whale all have the same basic internal bone structure. They are homologous organs even though one is used for walking, another for flying, and the last for swimming. (See Fig. 3–6.)

Studies of the early development of living things show that the eustachian tube of man, which connects the throat with the middle ear, is homologous to one of the gill slits of fish. The three bones in the middle ear of man have been traced to certain jaw bones of fishes. The presence of these types of homologous structures is used as a basis for classification.

Homologies can be traced in every part of the body, and indicate close as well as distant relationships. All vertebrates have similar brains, as can be seen in Fig. 3–7. The fact that most organisms are composed of cells containing nuclei that divide in the same way is excellent evidence that all are

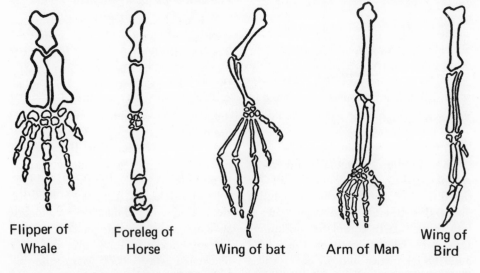

Flipper of Whale   Foreleg of Horse   Wing of bat   Arm of Man   Wing of Bird

Fig. 3–6. Homologous organs. Although these structures have different functions, their basic similarity in bone structure indicates common ancestry.

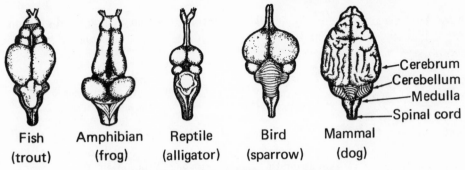

Fish (trout)   Amphibian (frog)   Reptile (alligator)   Bird (sparrow)   Mammal (dog)

—Cerebrum
Cerebellum
—Medulla
—Spinal cord

Fig. 3–7. Homologies of vertebrate brains.

related; that is, that all cellular living things have evolved from some far distant ancestor, the first cell.

Physiological homologies. The study of the structure of organisms is called *anatomy* or *morphology*. The study of the functions of organisms is called *physiology*. Similarity in the physiology of living things is another basis for classification. For example, the close relationship between birds and reptiles is indicated, in part, by the fact that their kidneys excrete the same kinds of wastes. Hormones extracted from sheep or pigs are enough like human hormones that they can be substituted for them when humans do not produce enough of their own. In our discussions of biochemistry in later pages, it is often unnecessary to state what kind of organism is involved, for many processes are almost universal in living things.

Immunological Homology. Similarities and differences in the blood proteins of various animals can be determined by the Nuttall precipitation technique. In this technique, a series of injections of human blood serum is put into the vein of a rabbit. (See Fig. 3–8.) The rabbit's blood then builds up a chemical called an antibody which reacts with the proteins of the human blood that were injected into the body

of the rabbit. These rabbit antibodies are very specific and will react only with human blood proteins, or with proteins that are very similar to human blood proteins. Some of the rabbit's blood is then drawn off; a liquid portion, called the *serum,* is separated from it. Next, the rabbit serum is poured into a series of test tubes, and serum from other animals is added to the test tubes. When chimpanzee blood serum is added to the prepared rabbit serum, a cloudy precipitate forms, but the precipitate is less dense than the precipitate created by the addition of human serum. When pig serum is added, there is no precipitate. These results show that there is a greater homology between the blood proteins of man and the chimpanzee than between those of man and the pig.

The Nuttall precipitation technique has been used to solve many problems in classification. For example, the horseshoe crab has many structural homologies with crabs and with spiders. Which is it most closely related to? The horseshoe crab was finally classified with the spiders after biologists used precipitation techniques. The biologists found that the chemicals in the horseshoe crab's blood were closest to those in the blood of the spider.

Homologies in the Embryo. It is not

1. Inject human blood serum into blood vessel of rabbit.

2. Rabbit's blood builds antibodies which are highly specific against the proteins in human blood.

3. Place serum of the rabbit into test tubes and add serums of man and of other animals.

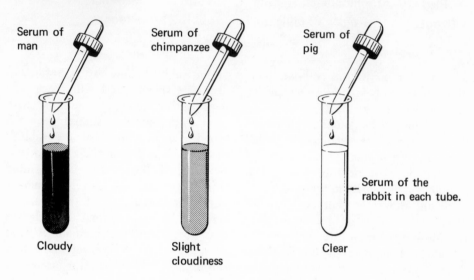

4. The fact that the serum of the chimpanzee produces a slight cloudiness whereas the serum of the pig produces no cloudiness shows that the protein molecules of man resemble those of the chimpanzee more than those of the pig.

Fig. 3–8. The Nuttall precipitation technique is used to indicate the degree of similarity in blood proteins of various animals.

always easy to tell whether a structure in one animal is analogous or homologous to a similar structure in another. Frequently, the question can be answered by studying the embryos rather than the adults, for embryos of related organisms resemble each other much more closely than the adults do. Moreover, the early embryos resemble each other more closely than the later ones, as shown in Fig. 3–9.

These homologies are very helpful. For example, adult mammals have three bones in the middle ear while adult reptiles have only one bone there. But if human and lizard embryos are compared, the two extra bones are found to form in both of them. In the lizard, they become the movable jaw joint; in mammals, they move inward and become part of the ear. Since we have found a series of fossils that shows

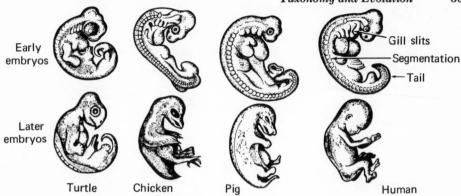

Fig. 3–9. Homologies of vertebrate embryos. All of the adult embryos show likenesses, but the early embryos have even greater resemblances.

a similar transition, it is clear that the human middle ear bones evolved from the reptilian jaw joint.

All the vertebrate embryos develop a series of gill slits in the early embryo. These become functional in fishes, but in land-living vertebrates, they disappear — except for one, which becomes the eustachian tube, running from the throat to the middle ear of man. The same structure has taken on two altogether different functions. The original function was respiration, as in the fishes. There were fishes in the water before there were vertebrates on land.

An early theory, advanced by E. H. Haeckel in the late nineteenth century, was that every organism repeats its evolutionary development in its own embryology. This idea is called the Theory of Recapitulation and is sometimes expressed as "Ontogeny recapitulates phylogeny."

We now know that this is true only if we confine our attention to the broadest generalizations; we do not repeat the details of our ancestry. What *is* true is that many of the structures in the embryo are much as they were in the embryos of our ancestors, but they may change beyond recognition as the embryo changes into an adult.

Vestiges. Structures present in modern organisms which do not appear to have any function are called *vestiges.* Many of the structures that appear in the human embryo, although presumably useful in our ancestry, are vestiges. A pouch that develops at the bottom of the large intestine grows into a large storage bag for half-digested food in some plant eaters. In us, this pouch, called the *appendix,* remains small and apparently useless. All vertebrate embryos have definite tails; in man, they never grow, and by the time a baby is born, the tail is reduced to a few tiny apparently useless bones at the bottom of the spine — the coccyx.

Other examples of vestiges are worth noting. The splint bones of a horse are the remains of two side toes. The python has legs in the embryo, but in the adult they are reduced to a couple of seemingly useless bones embedded in the muscles of its sides.

Vestiges may be used to indicate evolutionary relationships, as we have already noted in the case of the horse. Through evolution, existing structures are remodeled, sometimes for new and useful functions; if there are no functions, the remains of the structures may still persist as vestiges.

REASONING EXERCISES

1. What is the value of having a worldwide name for each kind of organism?
2. Why is the number of fossils limited?
3. Why are no fossils present in igneous rock?
4. How can the relative age of a fossil be determined? How can fossils be used to determine the relative age of a layer of rock?
5. How do scientists use the uranium-lead ratio to determine the age of the earth?
6. Carbon-14 disintegrates, while carbon-12 does not. Why does the ratio between the two remain in the atmosphere?
7. What is organic evolution? What is taxonomy? How is taxonomy based on evolution?
8. Distinguish between analogy and homology.
9. What is the nature of the homology between the flipper of the whale and the foreleg of the horse?

 ## THE CLASSIFICATION SCHEME

No two living things are exactly alike. Nevertheless, taxonomists try to group all known organisms into meaningful categories. The most primitive category is the *population*. All the robins in a park, for example, constitute a population. They can interbreed with each other, and might indeed do so, since they occupy the same region. But they do not interbreed with sparrows or earthworms or maple trees. Any given region has many populations of different kinds of living things.

The Species

The boundaries of a population are not sharp — a robin that was hatched in the park might move to another park or to someone's lawn. Adjacent populations breed with each other at the boundaries, if there are any. These populations, interbreeding wherever they meet, make up a *species,* the primary unit of classification. The species has a natural limit, since members of a species breed with each other, and only rarely with members of any other species. There are crosses (hybrids) between species, and these are often easily produced in captivity. But *in nature,* a species defines itself by breeding exclusively (or nearly so) within the group.

This definition of a species has limitations. Suppose a population is isolated on an island, so that it cannot breed with other populations. Does it belong to the same species as similar creatures on the mainland? Many island populations are quite different from anything else, and are classified as distinct species. In fact, in some instances, only one small population on an island is so classified. However, if an island population greatly resembles another population on the mainland or a nearby island, the similar populations are often considered to belong to the same species. The decision is quite arbitrary.

Another problem is that not all organisms reproduce sexually. For example, *Amoeba* reproduces asexually by cell division. How do we define a species in this case? Certainly the question of interbreeding cannot enter into it. Again, the decision is arbitrary. If a number of *Amoebas* live in the same environment and resemble one another closely, they are classified as a single species.

Subspecies. In a large species that occupies a large territory, there are often observable differences among populations from different parts of the territory. A group of robins from the West, for example, will average smaller and darker, and have less white on the tail, than a group from the East. When such average geographic differences in characteristics are found, the species may be divided into *subspecies*. There are about six subspecies of robin, for example. Those that meet each other will interbreed freely along the boundaries.

Such subspecies are geographically separated from the rest of the species. There is a population of robins isolated in the mountains of Lower California, which never interbreeds with any others. Under these conditions, the isolated subspecies is usually very different from the other subspecies, and it is a matter of choice whether to call it a separate species.

Speciation and the genus. A subspecies like the robin in the previous paragraph cannot interbreed with the rest of the species. The result is that it may evolve quite differently from the others. If the population is small, it can evolve very rapidly. Eventually, it may become so different from the rest of the species that it can no longer interbreed with it. It is now a different species. This process is called *speciation*.

Once a new species has formed, it may move back into the territory of its parent species. The two cannot reunite, however, because they can no longer interbreed. By repetition of this process, a species may divide into a number of more or less similar species, some of which live in the same territory (although probably in different habitats). A group of species of this kind, all recently descended from a single species, is called a *genus*.

Binomial nomenclature. In 1785, the Swedish naturalist Karl von Linné (also known as Carolus Linnaeus) published a book in which he classified all the then known animals and plants. He based his scheme on similarities of structure, but much of his scheme is still valid because similarity often means homology. He instituted the system, used by today's taxonomists, of naming organisms by their scientific classification. In this system, the first part of the organism's name is its genus and the second part of the name indicates its species. Man, then, belongs to the genus *Homo* and the species *Homo sapiens*. A genus of catlike animals known as *Felis* includes species whose scientific names are *Felis leo* (the lion), *Felis tigris* (the tiger), and *Felis domestica* (the house cat). The genus name always has a capitalized first letter, and the second part of the species name, a small first letter. This is a system of *binomial nomenclature. Binomial* means "two names," and *nomenclature* means "system of naming." (Genus and species names are usually italicized in formally printed material.)

Frequently, the Latin and Greek names which are given to organisms are descriptive. For example, the scientific name of the fruit fly, *Drosophila melanogaster,* means "black-bellied dew-lover" (*Dros* — dew, *phil* — like, *melano* — black, *gaster* — stomach). The description man took for himself, *Homo sapiens,* means "wise man."

A microscopic organism commonly studied in biology is *Paramecium.* However, this is only the genus name. Two species, under the genus, are *Paramecium caudatum* and *Paramecium aurelia.*

The scientific name may also include a third name, indicating the subspecies. The name or initial of the person who first described and named the organism is often added. Linnaeus took credit for naming the organisms in his system of classification and, therefore, his initial is included in many scientific names. For example, the red maple is *Acer rubrum, L.*

The Higher Categories

The members of a genus are closely related species having a common ancestor in the recent past. But in the more remote past, several genera may have come from the same ancestor. All such genera constitute a *family.* A still more remote ancestor may have given rise to several families of the same *order.* As the scheme is carried further, orders are grouped into *classes,* classes into *phyla* (sing. — *phylum*), and phyla into one of the three *kingdoms* that encompass all living things: Animalia, Plantae, and Protista. Thus, the main categories of classification are: kingdom, phylum, class, order, family, genus, and species.

Evolution: A Branching Tree

An easy way to understand the relationship between evolution and classifi-

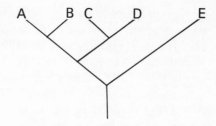

Fig. 3–10. Evolution, a branching tree.

cation is to think of evolution as a branching tree, as shown in Fig. 3–10.

A and B had the same ancestor and are, therefore, closely related. They would be placed in the same small grouping in the modern scheme of classification. Similarly, C and D would be placed in a small grouping. Going further back in time, we find that A, B, C, and D have the same remote ancestor; therefore, the two small groups previously designated would be placed together in a larger grouping. Going even further back in time, we find that A and B, C and D, and E all have the same ancestor; therefore, they are all included in a still larger grouping. Organisms that are closely related by evolution are placed in the same small group. Those organisms that are more distantly related are placed in a larger group.

However, classifying organisms is not quite this neat in practice because of the many gaps in our knowledge. For example, study the table (page 63) which shows the classification and relationships of the cat. Note that we know the common ancestors of some of the larger groups, whereas we do not know the common ancestors of the smaller groups. By the study of homologies and existing fossils, taxonomists are putting together as many of the details as possible.

CLASSIFICATION AND RELATIONSHIPS OF THE CAT

Category	Name	Members of Group	Common Ancestor	Date of Common Ancestor
Species	Felis domestica	House cat	House cat	10,000 B.C.
Genus	Felis	Cat, lion, panther, tiger, etc.	(unknown)	2 million B.C. (?)
Family	Felidae	Cats, lynx, saber-toothed tiger, etc.	(unknown)	40 million B.C.
Order	Carnivora	Cats, dogs, seals, foxes, hyenas, skunks, bears, walruses, raccoons, wolves, etc.	A Creodont	60 million B.C.
Class	Mammalia	All mammals (animals with hair and milk glands)	A Triconodont	120 million B.C.
Phylum	Chordata	All chordates (mammals, birds, reptiles, amphibia, fish)	A wormlike protochordate	500 million B.C.
Kingdom	Animalia	Chordates, worms, mollusks, insects, starfish, sponges, jellyfish, etc.	(unknown)	perhaps 2 billion B.C.

MULTICELLULARITY

The earliest fossil evidence clearly indicates that life was at first confined to the ocean. In addition, the record shows that more complex forms came later in time than simpler ones. The first living things were probably single-celled and marine.

While evolution can produce great specialization in single-celled creatures such as *Paramecium,* one-celled animals cannot grow beyond a few hundredths of an inch in length. The reason is that all parts of a cell need oxygen and must dispose of wastes. If a cell gets too large, its central region is too far from the surface, so oxygen coming in from the surface cannot reach the center before it is all used up. Furthermore, wastes produced at the center would poison the cell before they reached the surface.

Thus, living things were limited to small size until they evolved into many-celled creatures. This apparently happened by the formation of colonies. After a cell divided, the daughter cells remained together until, by repetition, a colony of several hundred cells may have formed. Such colonies are known in present-day primitive plants and primitive animals.

But real increase in size could occur only when the cells of the colony began to *specialize* or work together on a specific task. Thus, certain cells changed in structure as they developed the ability to do a particular job. As a cell becomes specialized to work together with other cells on a specific task, it loses its independence, for it can no longer perform all of the activities required for an independent existence. However, the colony as a whole benefits as different groups of specialized cells cooperate in the performance of different jobs. Once the cells of the colony are specialized, the colony can be considered a multicellular organism.

The fresh water colonial alga, *Volvox*, is an example of an early stage in specialization. Each cell of *Volvox* (Fig. 3–11) resembles the single-celled alga, *Chlamydomonas*. Each *Chlamydomonas* cell possesses two flagella, a chloroplast, an eyespot sensitive to light, cytoplasm, and a nucleus. *Volvox* is a sphere consisting of a single layer of from several hundred to four thousand Chlamydomonas-like cells. Each cell has its two flagella protruding outward, and is connected to its neighboring cells by a thin strand of cytoplasm. The cells are coordinated. Beating of the flagella propels the colony forward in a spinning manner. The cells toward the rear are larger than those at the front. Some of the posterior cells create daughter colonies. These are often seen in the interior space of the colony. Cells at the front end have larger eyespots. Thus, *Volvox* shows the development of different kinds of cells to carry on different functions.

As evolution proceeded, specialization reached high levels. A land animal of today has many different kinds of cells, each doing a special job. The cells do not operate alone, but as parts of *tissues,* or groups of cells specialized for a particular function. In addition, different kinds of tissues may function together in structures called *organs,* and groups of organs may cooperate in *organ systems.* The table on the facing page summarizes the main kinds of tissues in the human body.

LIFE ON LAND

Multicellular plants and animals evolved into a large variety of forms in the ocean before they were able to move into new habitats. Over a period of 200 million years they gradually moved up the rivers into fresh water, then into swamps, and finally to dry land. There were three main problems in adapting to land: (1) animals needed a waterproof skin so that they would

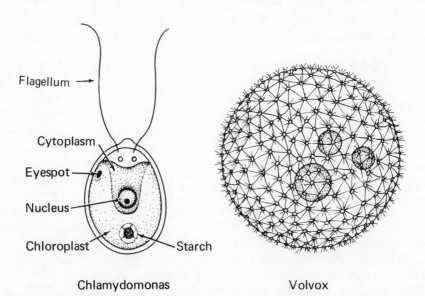

Flagellum →

Cytoplasm

Eyespot

Nucleus

Chloroplast   Starch

Chlamydomonas   Volvox

Fig. 3–11. Chlamydomonas and Volvox.

TISSUES OF THE HUMAN BODY

Tissue	Characteristics	Function	Adaptation of Structure for Function
Epithelial	Forms inner and outer linings for organs; lacks material between cells; groups of cells may form glands; special goblet cells may produce secretions.	1. Protection.	Closeness of cells prevents entry of germs.
		2. Secretion.	Goblet cells secrete mucus for lubrication of organs; glands secrete digestive juices.
Muscle	No intercellular material. Has contractile fibers. Found in the walls of the heart, digestive system, uterus, and arteries.	Muscle cells perform their function by contracting.	No intercellular material to interfere with contraction; contractile fibers aid contraction.
Cartilage	Has material between cells (matrix). Present on ends of bones, in nose, ear, and walls of windpipe (trachea).	1. Support.	Matrix is flexible.
		2. Flexibility.	Matrix is firm.
Bone	Interconnected cells in stiff matrix containing calcium and phosphorous salts.	Support.	Deposits of calcium and phosphorous salts add rigidity.
Fat (Adipose)	Cells have large vacuoles.	Storage of fats and oils.	Vacuoles for storage.
Connective	Found as ligaments connecting bones to bones, as tendons connecting muscles to bones, and also connecting skin to muscles. There are fibers present in matrix.	Connect parts of the body; bind together other cells.	Fibers in the matrix provide added strength.
Nerve	There are several types of neurons or nerve cells. May be very long and have greatly branched end regions. Have special cytoplasm and protective sheath.	Transmission of impulses.	The special cytoplasm can receive stimuli and transmit impulses. Transmission aided by great length of neurons, branchings to make connections between them, and the protective sheath.
Blood	The plasma is a liquid matrix. Three types of cells in plasma: red blood cells, white bood cells, platelets.	1. Transport.	Matrix is a liquid and functions in carrying carbon dioxide, food, and wastes. Red blood corpuscles contain hemoglobin which helps in transporting oxygen.
		2. Preventing disease.	White blood cells capable of ameboid movement aid in destroying disease organisms.
		3. Clotting.	Fragile platelets help cause clotting of blood.

not dry out; (2) they had to develop organs to extract oxygen from the air instead of taking dissolved oxygen from water; (3) they needed methods of reproduction that would not require them to lay eggs in water. The solutions to these problems came gradually. Most amphibians (when adult) have lungs and can breathe air, but their skin is moist and they lay eggs in water. Reptiles have dry skin, and most reptiles lay a large egg with a shell around it, on land.

At first, the land-living organisms were confined to the tropics — which was most of the world at that time. Before they could move into colder climates, they had to develop an internal mechanism to keep their bodies warm. This involved: (1) a more efficient digestive system to provide for the larger amount of food needed by a warm-blooded animal; (2) improved lungs to supply the larger amount of oxygen needed to oxidize this food; (3) improved circulation, to carry the oxygen to the cells; (4) insulation, to keep the heat in the body; (5) special adaptations to control the factors that influence body temperature. These problems were solved separately by two groups — the birds and the mammals — and these are the only land-living vertebrates that have colonized the Arctic.

The plants had other problems in colonization of the land. They needed waterproof coatings on the leaves, to prevent them from drying out. They had to have some way of taking in air without losing too much water. They needed a way to get water from their roots to their leaves. They needed stronger stems to support the plant body.

These are only a few of the evolutionary problems involved in colonizing new areas.

COMPLETION QUESTIONS

[A] 1. The science of classification is called . . . .
2. A system of taxonomy should embrace living and . . . . organisms.
3. An imprint of an extinct organism found in a rock is called a (an) . . . .
4. The hardened resin of trees forms . . . .
5. La Brea, Los Angeles, is famous for the fossils which are found in . . . .
6. Fossils are most often found in . . . . rock.
7. Hot molten material in the interior of the earth cools to form . . . . rock.
8. Scientists estimate the age of the earth by the ratio of . . . . in rocks.
9. Present scientific estimates of the age of the earth place it at about . . . . years.
10. Dinosaurs were dominant during the . . . . era.
11. In its evolutionary history, the horse showed a decrease in the number of toes but an increase in . . . .
12. Modern classification is based upon the concept of . . . .
13. Similar structures in different organisms which may have the same function but which had different origins are said to be . . . . structures.
14. A vestigial structure attached to man's large intestine is the . . . .
15. The splints in a horse's leg are vestiges of . . . .

B 16. Populations that interbreed in nature are considered to be a (an) . . . .

17. The process by which an isolated subspecies evolves into a new species is . . . .

18. A group of species recently descended from a single species is a (an) . . . .

19. The genus plus a second name is used to establish the . . . . name.

20. In our scheme of taxonomy, each phylum is divided into a number of . . . .

21. The adaptation of a structure for a special function is known as . . . .

22. A group of similar cells performing the same function is called a (an) . . . .

23. A group of tissues working together to perform a definite function is a (an) . . . . . .

24. The functions of epithelial tissue are protection and . . . .

25. The type of tissue specialized to contract and relax is . . . .

26. A tissue which provides support with flexibility is . . . .

27. The secretion of calcium salts is an important function of . . . . tissue cells.

28. An animal tissue whose cells have large vacuoles is . . . .

29. A tissue with a liquid matrix is . . . .

30. The only land-living vertebrates to become adapted to the Arctic are mammals and . . . .

MULTIPLE-CHOICE QUESTIONS

A 1. Petrification is best defined as the process by which (1) stone is weathered into soil, (2) the molecules of organisms are gradually replaced by minerals, (3) organic materials are converted into humus, (4) minerals are dissolved and carried down through the topsoil.

2. The age of organic remains less than 40,000 years old can be measured by the proportion of carbon-14 found in them because (1) there was more carbon-14 in the air in former times, (2) there was less carbon-14 in the air in former times, (3) carbon-14 undergoes radioactive decay at a known rate, (4) the half-life of carbon-14 changes at a known rate.

3. Fossil remains indicate that the evolution of organisms on the earth has advanced (1) in many different directions, (2) in large jumps, (3) in a continuous straight line, (4) at a known rate.

4. Although the arm of a man and the flipper of a dolphin serve different functions, there is great similarity in their structures and developments. This similarity is known as (1) analogy, (2) homology, (3) embryonic similarity, (4) biochemical similarity.

5. Many homologous structures found in pigs, frogs, and snakes indicate that these organisms originated from a (an) (1) protist, (2) invertebrate, (3) common ancestor, (4) fixed species.

B 6. Two animals belong to the same species if they (1) can live together in a similar environment, (2) can mate and produce fertile descendants, (3) have similar nutritional requirements, (4) show a very close resemblance.

7. The organism *Acer rubrum* is most closely related to (1) *Rubrum acer*, (2) *Acer cordis*, (3) *Rubrum cordis*, (4) *Cordis rubrum*.

8. Which is the correct classification arrangement? (1) phylum, genus, class, species, (2) genus, species, phylum, class, (3) species, genus, phylum, class, (4) phylum, class, genus, species.

9. Problems created by the change from a unicellular to a multicellular type of organism were overcome by (1) having all cells come in direct contact with the external environment, (2) having cells form into colonies of various shapes, (3) providing for specialization among body cells, (4) providing greater opportunities for diffusion of materials from cell to cell.

10. Discovery of marine fossils in sedimentary rocks of inland areas shows that these areas (1) were once covered by freshwater lakes, (2) were once covered by ocean waters, (3) have undergone glaciation, (4) have been recently depressed.

CHAPTER TEST

1. The main goal of taxonomy is to group organisms according to their (1) habitat, (2) evolutionary relationship, (3) shape, (4) appearance.

2. The scientific name of any organism consists of (1) its phylum and class, (2) its family and order, (3) its genus and species, (4) its species and variety.

3. In general, a major advantage of a multicellular organism over a unicellular organism is that the multicellular organism (1) reproduces faster, (2) reproduces sexually, (3) is always larger, (4) has specialized cells.

4. Which is the correct way to indicate the scientific name of man? (1) Homo Sapiens, (2) *Homo Sapiens,* (3) *Homo sapiens,* (4) Homo sapiens.

5. In higher animals, which tissue holds other tissues together in an organ? (1) epithelial tissue, (2) connective tissue, (3) muscle tissue, (4) nerve tissue.

6. In its fossil history, the horse showed (1) an increase in size and an increase in number of toes, (2) a decrease in size and an increase in number of toes, (3) a decrease in size and a decrease in number of toes, (4) an increase in size and a decrease in number of toes.

7. Which of the following terms includes all the others? (1) tissue, (2) organ system, (3) cell, (4) organism.

8. If fossils of two different organisms are found in two layers of rock that lie one above the other, it is most probable that the (1) two forms are closely related, (2) upper form lived before the lower one, (3) lower form is older than the upper one, (4) lower form descended from the upper one.

9. *Eohippus* is an early ancestor of the modern (1) pigeon, (2) hippopotamus, (3) man, (4) horse.

10. The flipper of a whale is homologous to the (1) wing of an insect, (2) tail of a fish, (3) arm of a man, (4) trunk of an elephant.

11. A vestigial structure is one that (1) is more useful than another similar structure, (2) is in the process of becoming useless, (3) is found only in organisms of no value to man, (4) has apparently lost its usefulness to the species.

12. The earliest known mammalian ancestor of the horse was about the size of a (1) cat, (2) fox, (3) cow, (4) dinosaur.

13. Of the following tissues, the one which contains the largest amount of mineral matter is (1) bone, (2) fat, (3) muscle, (4) nerve.

14. Glands are composed chiefly of (1) connective tissue, (2) nerve tissue, (3) muscle tissue, (4) epithelial tissue.

15. The cells that can easily be scraped from the lining of the mouth make up (1) connective tissue, (2) epithelial tissue, (3) supporting tissue, (4) voluntary tissue.

16. The binomial system for naming living things originated with the work of (1) Charles Darwin, (2) E. Haeckel, (3) Carolus Linnaeus, (4) Aristotle.

Below are listed four of the criteria which are used for indicating the possible common ancestry of organisms. For each of questions 17 through 20, indicate the letter of the criterion which best matches.

    A. Homologous structures

    B. Similar physiological characteristics

    C. Similar biochemical characteristics

    D. Similar embryonic development

17. At one time in their lives, both man and the bird have gill slits.

18. The wing of a bat is used for flying and the flipper of a whale is used for swimming.

19. The kidneys of birds and reptiles excrete the same kind of wastes.

20. If a rabbit is sensitized to human blood, the blood of the rabbit will react to chimpanzee's blood very much the way it does to human blood.

CHAPTER 4 | The Roll Call of Living Things

We have just studied (in Chapter 3) the scientific basis for the classification of organisms and fossils. This classification is based on evolution, with the species as the basic unit. The larger units are the genus, family, order, class, phylum, and kingdom. It is fair to conclude, after the study of Chapter 3, that an organism has a greater number of homologies with its species than with any of the larger units of classification. Now it is important to study some of the large taxonomic units in order to find out what their main characteristics are and to see how they fit into the broad scheme of evolution.

A | THE PROTISTA KINGDOM

Many biologists divide living things into two kingdoms, the plant kingdom and the animal kingdom. Recently, an idea for a third kingdom, first proposed in 1866, has been revived. This kingdom is called the *protista*. The protista kingdom includes organisms that are not like typical plants or typical animals. However, they may have characteristics of both plants and animals.

Although biologists disagree over which phyla should be included among the protista, these phyla generally include the protozoa, most algae, bacteria, fungi, lichens, and slime molds. Most protists are microscopic and single-celled. Since very few protists leave any fossils, little is known about their evolution. However, their rela-

tionships can often be deduced from homologies, which is how taxonomists arrived at their classification.

The Protozoa

The protozoa live in fresh water, salt water, moist soil, and as parasites in other organisms. Although a few kinds exist as colonies, the protozoa are mainly single-celled, microscopic organisms that have no chlorophyll. In fact, in the two-kingdom system, they are considered to be one-celled animals. The protozoa are classified by their method of movement.

Among the fresh-water protozoa are *Amoeba* and *Paramecium*, single-celled organisms large enough to be seen with the naked eye. These two protozoans were discussed on page 15. As noted, *Amoeba* forms pseudopods and *Paramecium* has cilia for movement.

The *Foraminifera* are salt-water protozoa that have an external skeleton of calcium carbonate (a rocklike substance). Undersea deposits of shells of these protozoa resulted in rock (chalk) that later rose to form the White Cliffs of Dover, England. *Radiolaria* also live in salt water; they have pseudopods that protrude from their external skeleton, which is made of silica.

Plasmodium is a spore-forming protozoan that causes malaria. The *Plasmodium* spends part of its life in the female *Anopheles* mosquito. If an *Anopheles* mosquito infected with a *Plasmodium* bites a human, the mos-

70

quito may transfer the protozoan into the human's body. When this happens, the protozoan causes malaria by producing poisons and by destroying red blood cells. People who have malaria get chills and fever.

Trypanosoma gambiense causes the disease called African sleeping sickness (often fatal). *Trypanosoma gambiense* is an example of a protozoan that has a *flagellum* — a whiplike tail that propels it. There are other protozoa which have flagella.

The Algae

Photosynthesis is a method of manufacturing food from inorganic chemicals by using energy from sunlight. Most photosynthetic plants have a green pigment, chlorophyll, which functions in photosynthesis. Plants possessing chlorophyll are called "green plants."

Algae are often called "one-celled green plants," but several phyla of algae include kinds that are many-celled. Some algae have other pigments as well as chlorophyll. Algae live in the ocean, fresh water, and moist places on land. *Protococcus* (Fig. 4–1) is a green alga which lives on the moist bark of trees and moist surface of rocks.

Among the common algae shown in Fig. 4–2 is *Spirogyra*, a green alga found in fresh-water ponds. Its single cells are attached to each other to form long filaments; they are held together by the filament sheath. Although the spiral-shaped chloroplast is not typical of algae, its shape makes this species interesting to study under the microscope. The *pyrenoids* store starch and stain dark with iodine solution.

Chlorella, a common single-celled green alga, has been suggested as a pos-

SPIROGYRA

CHLORELLA

DIATOM

RED ALGA

BLUE-GREEN ALGA

Fig. 4–2. Common algae.

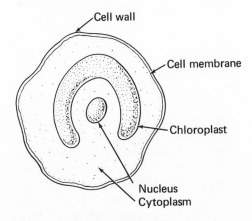

Cell wall

Cell membrane

Chloroplast

Nucleus
Cytoplasm

Fig. 4–1. Protococcus.

sible food and oxygen source for astronauts on space journeys.

As previously noted, *Chlamydomonas* is a one-celled, fresh-water alga which has two flagella. The colonial alga, *Volvox*, is a hollow sphere of cells, each resembling the single cell of *Chlamydomonas*. The many-celled colony is regarded as a single individual. *Chlamydomonas* and *Volvox* are compared in Fig. 3–11, page 64.

Chlamydomonas is very similar to many protozoa, and it has been suggested that the protozoa arose from flagellated algae that lost their chlorophyll. They also seem to be ancestors of the many-celled green algae from which all land plants eventually came.

Diatoms (Fig. 4–2) have a hard silica cell wall which consists of two overlapping halves. When they die, their silica shells deposit on the bottom of the sea as diatomaceous earth which is used as a fine abrasive. Diatoms constitute a large portion of the *plankton*, the mass of minute organisms floating near the surface of the sea. They are a major food source for nonphotosynthetic organisms in the sea.

Blue-green algae, brown algae, and red algae have pigments which mask the green of chlorophyll and give the plants their distinctive colors. The blue-green algae are primitive forms whose chlorophyll and nuclear material are scattered throughout the cell. They do not have chloroplasts or a formed nucleus. One of the red algae is the source of agar that is used to grow bacteria in biological laboratories. Brown algae include kelps and rockweeds. Rockweeds grow in the form of dense mats close to the shore. Note the form of the multicellular alga shown in the lower portion of Fig. 4–2.

Euglena has already been discussed on page 14. As noted, this unusual organism possesses characteristics of both animal and plant cells. *Euglena* resembles an animal in that it lacks a cell wall, possesses a flagellum for rapid locomotion, and has a light-sensitive, red-pigmented eyespot. Under certain conditions, it can live on food materials taken in from its watery environment. However, like a plant, *Euglena* possesses chloroplasts which are used for photosynthesis.

The Bacteria

Bacteria, barely visible under the ordinary microscope, are among the smallest one-celled plantlike protists. They are similar to blue-green algae in that they lack a formed nucleus. Bacteria have a single chromosome which is a ring-shaped structure in the cytoplasm. Most bacteria take in food materials from living or dead organisms, but a few types can make their own food by photosynthesis, using pigments other than chlorophyll. True bacteria have a cell wall.

Bacteria are found practically everywhere on earth. Although many kinds are disease-producing, some bacteria are essential to man's welfare; in the process of decay, they return into circulation the chemicals locked in the bodies of dead organisms. Nothing is known about the evolutionary origin of the bacteria.

The Viruses

Viruses are so tiny that they can pass through filters having the finest pores. They can be seen only with the electron microscope. Viruses are included with the protists although they lie on the borderline of life. Actually they live only within the cells of living hosts where they upset the metabolism of the cells, which then produce virus material instead of carrying on their normal activities. Viruses are known to cause some forms of cancer, although it has not been proved that they are involved in human cancers. They cause polio, yellow fever, influenza, smallpox, mumps, the common cold, and tobacco mosaic disease. Viruses which attack and destroy bacteria are called *bacteriophages*.

The Fungi

Fungi, a mixed group of non-green plants, do not have a single common ancestor. They are correctly classified in at least three phyla, each of which originated from a different group of algae. Fungi, which include the molds, mushrooms, and yeasts, do not carry on photosynthesis. Instead they live on other organisms as parasites or saprophytes. A *parasite* lives on living organisms (hosts); a *saprophyte* lives on dead organic matter. Fungi can exist as isolated single cells or as filaments called *hyphae* (sing. – *hypha*) and reproduce by reproductive cells called *spores*.

The *molds* are common fungi which may live on bread, fruits, vegetables, leather, fish, and living plants. The hyphae of the bread mold (*Rhizopus nigricans*) possess many nuclei instead of being composed of separate cells each with its own nucleus.

The *mushroom's* main portion grows below ground. Under suitable conditions, the above-ground portion develops as a mass of twisted hyphae which supports an umbrella-shaped structure. When ripe, the mushroom reproduces by spores. The shelf-fungi, puffballs, truffles, rusts, and smuts are grouped with the mushrooms. Wheat rust is a fungus disease causing great loss to the agriculture industry.

The *yeasts* obtain energy by the fermentation of sugars. The carbon dioxide and alcohol produced are the basis for the baking and brewing industries. These one-celled fungi reproduce by budding. (See Chapter 16.)

In the same class with the yeasts is *Penicillium notatum*, a fungus which produces the life-saving antibiotic, penicillin. Chestnut blight and Dutch elm disease, two serious tree diseases, are caused by other fungi in this class. Species useful to man add flavor to Camembert and Roquefort cheeses.

The fungi that cause ringworm and "athlete's foot" are also related to the yeasts. The ringworm fungus causes a raised ring to form under the skin, giving the appearance of a "worm under the skin" — yet it is caused by a fungus. "Athlete's foot" usually occurs between the toes and causes the foot to itch.

The Lichens

A lichen looks like a thin, gray, crust-like pancake growing on rocks or trees. It is a combination of a fungus and an alga which grow together and help each other in a relationship called *mutualism*. The cells of the alga produce food for both organisms, and the tough

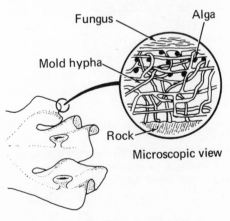

Fig. 4–3. Cross section of a lichen.

mycelium (mass of hyphae) of the fungus holds water and protects the alga. This combination of alga and fungus can grow in hostile places such as rocks and very cold regions. Acids produced by the fungus cause lichens to decompose rocks, leading to the formation of soil. The death and decay of lichens enriches soil by the addition of their organic matter to it.

Slime Molds

Slime molds are so named because at one stage of their existence they are a slimy mass of protoplasm. During this period in their life cycle they spread over damp, decaying vegetation. Some slime molds have a plant-animal life history. During one stage, they are animal-like flagellated cells; at another stage, they have amoeba-like cells; at other times, they have a moldlike structure that forms spore cases, a plant characteristic. Slime molds are a good example of protists because they cannot clearly be defined as plants or animals. A common type of slime mold that is often studied in the laboratory is *Physarum,* which is shown in Fig. 4–4.

Fig. 4–4. Plasmodium stage of *Physarum.* (A plasmodium is a spreading mass which is not separated into individual cells.) At another stage in its life cycle, *Physarum* consists of amoeba-like cells. Is it a plant or an animal?

There is little agreement among biologists on the origin of the slime molds.

THE PLANT KINGDOM

Placing plantlike organisms such as the algae and fungi among the protists leaves the plant kingdom with only two phyla, the bryophytes and the tracheophytes. Most of these are land plants.

Phylum — Bryophyta

The byophytes are small green plants that grow in moist places. They have simple leaves, but lack a stem or have a very simple stem. In their reproductive cycle, bryophytes alternate from one generation to the next between two types of plants. One generation reproduces sexually (by gametes) and is called the *gametophyte.* The next generation reproduces asexually (by spores) and is called the *sporophyte.* This alternation between an asexual reproductive stage and a sexual reproductive stage is called *alternation of generations.* In the bryophytes, the gametophyte generation is the larger of the two and the more easily seen. The bryophytes are divided into two classes, the liverworts and the mosses. Their name is derived from *bryon,* the Greek word for moss.

Class—liverworts. The plant body of a liverwort consists of simple branching leaves that grow flat on moist soil or water. (See Fig. 4–5.)

Class—mosses. The mosses have simple stems which bear tiny leaflets arranged in spirals. Rootlike structures, called *rhizoids,* anchor the plant to the soil and absorb water and minerals. Sphagnum, or peat-forming moss, grows in swamps and bogs. When dried, it is used in lawns and gardens to retain moisture and to keep the soil from becoming too tightly packed.

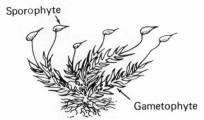

Sporophyte

Gametophyte

Polytrichum—a moss

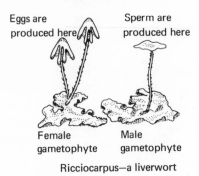

Eggs are produced here

Sperm are produced here

Female gametophyte

Male gametophyte

Ricciocarpus—a liverwort

Fig. 4–5. Bryophytes.

Fossils indicate that the bryophytes arose from green algae as an adaptation to life on land. But the bryophytes are only half free of the water. They cannot reproduce unless there is some water, for their sperm must swim to their eggs. They cannot grow very tall, for they have no efficient roots to absorb water from the soil and no tubes to carry water from rootlike structures to the leaves.

Phylum — Tracheophyta

The tracheophytes (tube plants) have a vascular (conducting) system of tubes for conducting water. This system of vessels allows these plants to grow taller than the bryophytes. Furthermore, the higher tracheophytes have true stems, leaves, and good root systems.

Subphylum—Lycopsida (club mosses). Club mosses grow close to the ground, and their leaves are tiny. They

have conelike reproductive structures at the top of their branches, as shown at the upper left of Fig. 4–6.

Subphylum—Sphenopsida (horsetails and scouring rushes). The branching stem of a horsetail resembles the tail of a horse. Horsetails have been used as an abrasive because of the hard silica in their stems. *Equisetum* is a common genus which grows in sandy places and along railroad embankments (Fig. 4–6).

Subphylum—Pteropsida. These include most of the organisms commonly recognized by the layman as "plants": the ferns, evergreens, and flowering plants. The scope of this subphylum can be outlined as follows:

Subphylum: Pteropsida

Class: Filicineae. The ferns.

Class: Gymnospermae. Evergreen conifers (pines, spruces, and others).

Class: Angiospermae. Flowering Plants.

Subclass: dicotyledons. Bean, peanut, maple, rose.

Subclass: monocotyledons. Grasses such as blue-grass, corn, oats, wheat.

Important Classes of Tracheophytes

Filicineae (ferns). When you examine a fern closely, you will see that both its roots and stem (rhizome) grow underground. Each spring, the familiar many-branched leaves called *fronds* appear. The roots, stems, and fronds of a fern have well-developed vascular tissue for conducting water. The undersurface of some fern leaves (Fig. 4–6) has brown spots called *sori*, which are clusters of numerous sporangia. The sporangia are the spore cases which contain the spores for the reproduction of the plant. Fern spores develop into an inconspicuous gametophyte generation which produces gametes (sex cells) that unite to produce the large plant, commonly considered by the layman to be the fern. Actually, this is only one of the generations of the fern; that is, it is the conspicuous sporophyte generation. As you can see, ferns undergo an alternation of generations. Some tropical ferns are tree-sized. Coal came mainly from the remains of prehistoric tree ferns.

All of the tracheophytes named until this point are limited to growth in moist places. Like the bryophytes, they require water for the fertilization of their eggs. Those that follow are true land plants, and they have special arrangements for fertilization that do not require water. Furthermore, their embryos are enclosed in a *seed,* which can survive long periods of dryness without losing the ability to produce a new plant.

Gymnosperms (naked seeds). The gymnosperm seeds are produced in cones such as the pine cone (Fig. 4–6). When the cone dries, its thick, hard scales spread out, and the seeds are exposed and carried away by the wind.

The gymnosperms include "evergreen" conifers, such as pines, spruces, hemlocks, and cedars. They also include the plants called the cycads and the ginkgo. The ginkgo is popular in cities because of its exotic shape and its ability to survive under the adverse conditions of city life. The ginkgo is called a "living fossil" because it is the last of a once prominent group of plants.

The gymnosperms have leaves which are often needle-like in shape, and they have well-developed roots and stems. Their vascular tissue enables them to grow to great heights. An example is

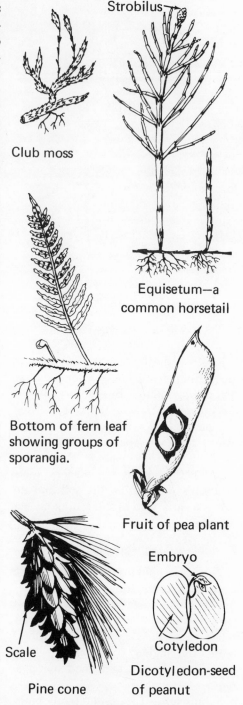

Club moss

Strobilus

Equisetum—a common horsetail

Bottom of fern leaf showing groups of sporangia.

Fruit of pea plant

Scale

Pine cone

Embryo

Cotyledon

Dicotyledon-seed of peanut

Fig. 4–6. Tracheophytes and their structures.

the giant redwood (*Sequoia gigantea*) of California which grows over 300 feet high. The coniferous forests are sources of lumber and paper.

Angiosperms (hidden seeds). These are the seed-producing "flowering plants," whose seeds are not exposed but held within a fruit. The flowers of some of the angiosperms are large and showy (tulip, gladiolus), but the flowers of others are relatively inconspicuous (maple tree, privet hedge, lawn grasses). Some angiosperms have lost the ability to produce chlorophyll and are saprophytes; for example, the colorless Indian Pipe which lives on humus on the forest floor. Although the angiosperms generally are adapted to terrestrial life, many (such as the water lily, *Anacharis,* and the duckweed) have returned to a watery environment.

The angiosperms are divided into two subclasses, the dicotyledons (or dicots) and the monocotyledons (or monocots), based upon the number of cotyledons present in the seed. A *cotyledon* is an organ which supplies food to the developing embryo plant. It may serve as the first photosynthetic organ of the seedling as it begins to grow. A familiar example of a dicotyledonous seed is the peanut. Each half of the nut is a cotyledon (Fig. 4–6). A corn kernel is a monocotyledonous seed.

DICOTS AND MONOCOTS COMPARED

Dicots	Monocots
Two cotyledons	One cotyledon
Veins of leaves in a network	Veins of leaves are parallel
Vascular bundles of stem in a radial pattern	Vascular bundles of stem scattered
Flower parts (such as petals) in fours or fives, or their multiples	Flower parts in threes, or their multiples

Oak and maple trees are dicots used in the construction of houses and manufacture of furniture. Many of our edible fruits such as the apple, peach, and cherry are dicots. Monocot grasses such as corn, wheat, rye, and oats are major sources of food for man.

REASONING EXERCISES

1. Distinguish between the protozoa and the algae in a basic way that would enable you to place them in separate taxonomic groups.
2. Protozoa and algae are classified in different groups. Yet, both groups include organisms with flagella. Give examples of flagellates from both groups.
3. What types of protists cause diseases?
4. Distinguish between a parasite and a saprophyte.
5. How did bryophytes arise?
6. What is meant by "alternation of generations?" Compare the gametophyte generation of the moss and the fern.
7. What is the importance of the following in the adaptation of plants for land living? (a) roots, (b) vascular system, (c) seeds.
8. Why is the environment in which an organism lives an inadequate basis for classification? Explain your answer.
9. What are some of the difficulties in classifying the following: (a) *Euglena,* (b) the slime molds, (c) lichens?
10. What do you consider the positive value of having some way of classifying organisms?

B THE ANIMAL KINGDOM

The placement of the protozoa in the protist kingdom leaves the animal kingdom only with multicellular organisms. Animals have neither chlorophyll nor cellulose cell walls. Most of them can move their bodies by means of contracting fibers. The nine main phyla of the animal kingdom are described.

Phylum — Porifera

The porifera, or sponges, are simple animals which spend their lives attached to solid objects under water. Most sponges are marine (live in salt water) but a few are fresh-water types. Their bodies have many pores, hence their phylum name, porifera (pore-bearers). The sponges feed by drawing water through their pores and filtering out food particles. The movement of water through the sponge is accomplished by cells whose flagella create a current. The filtered water passes out through the *osculum* (Fig. 4–7). Sponges have two layers of cells with a jellylike layer between. Many sponges have radial symmetry — parts of the body radiate from a central axis similar to the spokes of a wheel.

The sponge's skeleton is made of tiny, pointed granules of lime or silica or of a soft material called spongin. The adult sponge is sessile (nonmoving), but the larval form is free swimming.

This animal is in the simplest phylum of multicellular animals, but shows the beginning of true specialization. Special groups of cells are adapted to perform particular functions. The flagellated cells of the inside layer of the sponge move water through the pore cells. The tissues of the sponges are so different from those of any other animal that biologists believe this phylum arose separately from some kind of protozoan. Thus, the sponges have no direct relationship to other multicellular animals.

The common bath sponge is harvested in the shallow waters off Florida and Greece. The sponge is taken from the ocean floor and placed on land to dry and decay. The soft body skeleton which is left is the part that is used.

Phylum — Coelenterata

The coelenterates include *Hydra*, jellyfish, sea anemones, and corals.

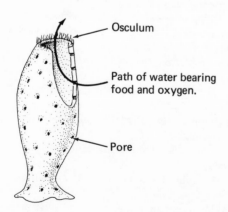

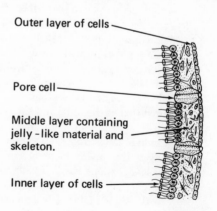

Osculum

Path of water bearing food and oxygen.

Pore

Outer layer of cells

Pore cell

Middle layer containing jelly-like material and skeleton.

Inner layer of cells

Fig. 4–7. Structure of a sponge.

Their bodies consist of a cuplike digestive sac with a wall composed of two layers of cells. The outer layer is the *ectoderm* and the inner layer is the *endoderm*. Between the two cell layers is a jellylike layer which is thin in some representatives of the phylum but greatly enlarged in others (such as the jellyfish). The opening to the cylindrical body of *Hydra* is the mouth which also serves as an anus; this opening is surrounded by a crown of tentacles. All coelenterates have radial symmetry. They also have stinging cells which they use to paralyze their prey. Most species are marine, but *Hydra* lives in fresh-water ponds and lakes. As shown in Fig. 4–8, the basic pattern of *Hydra* and the jellyfish are the same.

The Portuguese man-of-war looks like a gas-filled bag floating in the sea. Hanging from the animal's body are streamers which have stinging cells used to paralyze its prey. The sting can be very painful to man.

Coral is very common in tropical waters. The coral rock is formed in layers by the exterior skeletons of millions of tiny hydra-like animals. Some corals are used in jewelry. The most primitive coelenterates suggest affinities to colonial protozoa.

Phylum — Platyhelminthes

The platyhelminthes are the simplest of the three phyla of wormlike animals. The animals are flat and ribbonlike in shape as indicated by their name (*platy,* flat — *helminth,* worm). Some of the flatworms are shown in Figs. 4–9 and 4–10.

This is the first phylum whose members show *bilateral symmetry,* which means that they have a right side and a left side which are alike. They have

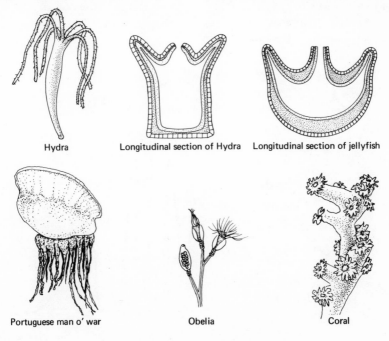

Hydra　　Longitudinal section of Hydra　　Longitudinal section of jellyfish

Portuguese man o' war　　Obelia　　Coral

Fig. 4–8. Coelenterates.

a dorsal (top) surface and a ventral (bottom) surface, as well as anterior and posterior ends. The flatworms are more advanced than animals discussed before because the body is composed of three layers of cells instead of two. The layers are the *ectoderm* (outer layer) the *endoderm* (inner layer), and a new layer called the *mesoderm* (middle layer). Flatworms have definite tissues, organs, and organ systems. However, the single body opening (mouth) which also serves as an anus suggests that flatworms evolved from the coelenterates.

Planaria, a common fresh-water flatworm, attaches to underwater objects and secretes a slimy mucus layer. It moves over objects by the beating of cilia on its ventral surface, and swims by undulating body movements. As representatives of a lowly form of life with a simple nervous system, planarians are used for experiments to test their ability to learn. According to some scientists, untrained, cannibalistic planarians

which have eaten trained planarians can learn a task more rapidly than those which have eaten untrained planarians.

Although *Planaria* is free-living, many species of platyhelminthes are parasites which cause great damage to man and other higher animals. In the life cycle of flukes and tapeworms, at different stages there are different life forms that may alternate between several hosts (the organisms in which parasites live). For example, the liver fluke, which is a dangerous parasite of sheep and cattle, also lives in snails during one stage of its life cycle.

A tapeworm grows as a long flat ribbon consisting of nearly square sections called proglottids. The tapeworm lacks a digestive system and a sensory system, which were presumably lost during evolution from free-living (nonparasitic) ancestors. The minute head of a tapeworm attaches to the lining of the small intestine of the host (man, cow, pig, or other animal), by means of hooks and suckers. Here the tapeworm

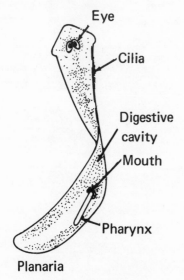

Planaria

Fig. 4–9. Planaria.

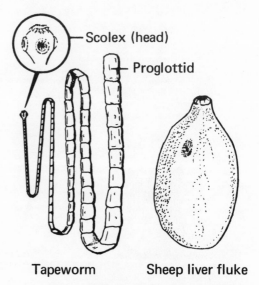

Tapeworm   Sheep liver fluke

Fig. 4–10. Tapeworm and liver fluke.

absorbs predigested food and may grow to a length of over 20 feet. Man may acquire tapeworm larvae by eating undercooked meat from infected animals such as the pig.

Phylum — Nematoda

The nematodes or roundworms have rounded bodies instead of flat bodies like the platyhelminthes. They show a great advance over their flatworm ancestors because they have a one-way digestive system that runs from an anterior mouth to a posterior anus.

Nematodes have bilateral symmetry and three layers of cells. Most of them are no larger than bits of thread and can be identified by their whiplike, thrashing movements. They are found in rich soil, fresh water, and salt water. Most roundworms are free living, but there are also many parasitic kinds such as those shown in Fig. 4–11.

The nematode *Trichinella* causes the disease trichinosis. These roundworms enter human bodies when people eat undercooked, infected meat (mainly pork). Unless there is strict local inspection, people should assume that all pork contains these parasites because the federal government does not inspect all meat for the larvae of this worm. The *Trichinella* larvae live in a cyst (or protective covering) in human muscle and cause pain and sometimes death. Postmortem (after death) examinations of people who died of other causes showed a substantial proportion of the American people to be infected with *Trichinella*.

The hookworm (*Necator americanus*) is a parasite which usually enters man's body through his feet. If a person walks barefoot on soil contaminated with human excrement (solid wastes), hookworm larvae, which are very small, may penetrate the soles of his feet and ultimately infect him. The hookworm travels through the body to the walls of the small intestine where it lives off blood. A large number of these worms in the human body makes a person weak and lethargic. Although hookworm disease was formerly common in the southeastern part of the United States, it has been controlled by increasing use of shoes and by improvements in methods of sanitation and waste disposal.

Elephantiasis is a disease, prevalent in tropical countries, caused by the minute nematode, *Filaria*. These worms cause enormous swelling of the feet, legs, arms, hands, or other parts of the body. Other parasitic nematodes in the soil cause severe losses to the orange, tobacco, and strawberry crops.

Trichinella encysted in muscle.

Necator americanus (hookworm)

Ascaris in human intestine, children are frequently infected with Ascaris.

Fig. 4–11. Nematodes.

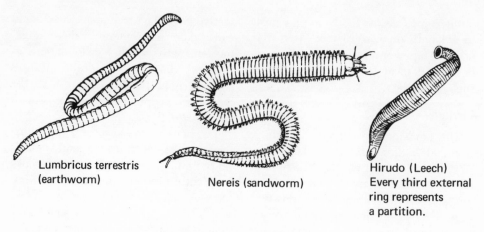

Lumbricus terrestris
(earthworm)

Nereis (sandworm)

Hirudo (Leech)
Every third external
ring represents
a partition.

Fig. 4–12. Annelid worms.

Phylum — Annelida

The annelid worms are the segmented worms. (See Fig. 4–12.) The visible sign of segmentation is the series of rings encircling the body, as seen in the common earthworm. These rings on the outside of the earthworm indicate partitions separating the body into segments. Most of the body segments have the same internal body organs repeated again and again. The annelid worms have bilateral symmetry, three cell layers, a body cavity, and a well-developed circulatory system of closed vessels. Annelids have excretory organs and a well-defined nervous system. In some ways, the body of an annelid suggests many roundworms placed end-to-end, but there is no real information about the relationship of the two phyla.

Earthworms are hermaphrodites; that is, each worm possesses both male and female sex organs and produces sperms and ova. It cannot, however, fertilize its own sex cells. The sperms of two earthworms are exchanged to start the reproductive cycle. As in the nematodes, the digestive tract is a long tube with a mouth and anus. Although the earthworm lives on land, most of the annelid

worms live in tiny tubes in the sea, or under rocks close to shore.

Leeches, which live in fresh-water ponds and streams, are in a separate group of the annelid phylum. They have suckers at each end of their bodies so that the posterior sucker can anchor the leech temporarily while the anterior sucker is used to suck the blood of the host animal. In some parts of the world, people still apply leeches to the body because they believe bloodletting is good treatment for many illnesses. Turtles and fishes, as well as man, are commonly attacked by these external parasites.

Phylum — Mollusca

The mollusks are soft-bodied, unsegmented animals which have three layers of tissue and well-developed organ systems. The bodies of many mollusks are protected by shells of calcium carbonate. The mantle, a body fold of the animal, forms these shells.

Clams, oysters, mussels, and scallops are called *bivalves* because they have two shells. Most of these mollusks are marine, although some of them live in fresh water. They feed by filtering food

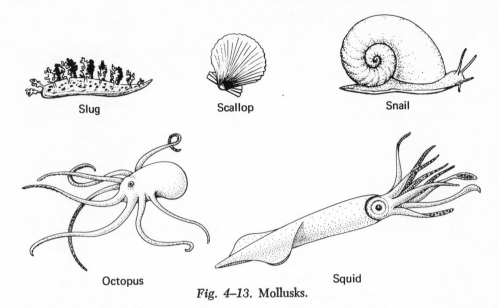

Slug   Scallop   Snail

Octopus   Squid

Fig. 4–13. Mollusks.

particles from the mouth into the interior of the organism. Clams move by extending a thick hatchet-shaped foot outside of their shells.

Snails are called *univalves* because they have only one shell. They feed by scraping their food to bits with a rasping tongue. Their thick foot, used for locomotion, bears elevated stalklike eyes.

Octopuses and squids are active shell-less mollusks with complex nervous systems. Their eyes function like man's but have a different evolutionary development. The octopus has eight arms, or tentacles, covered with suction disks; they are used for locomotion and grasping. The squid has ten arms that are used in the same manner. The squid moves by "jet propulsion" as it ejects a stream of water. It protects itself by giving off a screen of inky fluid when escaping from an enemy. Squids and octopuses hunt live prey.

Many mollusks (clams, oysters, scallops, and mussels) are used as food by man, but some are a source of trouble

for him. For example, the shipworm bores holes into the sides of wooden boats. The oyster drill, a univalve snail, bores through the oyster's shell and eats its flesh. A shell-less relative of the snail, the garden slug, causes great economic loss by eating vegetable crops.

The anatomy of mollusks is so different from that of any other animals that their relationship to other phyla is uncertain. We do know, however, that they most closely resemble the annelids because these two phyla have nearly identical larvae.

Phylum — Arthropoda

The arthropods are "the joint-legged animals" (*arthro*, jointed + *poda*, leg). This phylum contains over 90 percent of the species of the animal kingdom. More than 900,000 species have been identified, and the number of species is estimated to be in the millions. They are the most complex and successful of the animals without backbones. One class, the insects, are serious competitors of man for the world's food. Arthropods have segmented bodies, and there is little doubt that they have derived from annelids.

Several advanced features distinguish arthropods from annelids. The body is covered with a thick, flexible exoskeleton composed of *chitin.* Instead of the segments being nearly alike, as in the annelids, they are specialized in various ways, and form specific body divisions. Their chief distinction is their jointed appendages, which are highly modified for locomotion, food getting, sensation, reproduction, and protection. They have a ventral nerve cord as in annelid worms. The circulatory system is "open" so that the blood does not always travel through a system of tubes. They breathe through gills or by a system of tubes known as tracheae (*sing.* — trachea).

A group of fossil arthropods called trilobites appear to be the ancestors of the whole group. As shown in Fig. 4–14, they had jointed appendages, but every segment had a pair of appendages, and all were much alike, except in size. Much of the later evolution of the arthropods consisted in specialization of appendages in different parts of the body for different functions.

Arthropods live practically all over the earth — in fresh water, salt water, on the surface of the land, in the soil, and on or inside animals and plants.

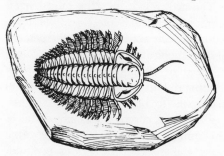

Fig. 4–14. A fossil trilobite.

The living arthropods are divided into five main classes.

Class — Chilopoda. The centipedes ("hundred leggers") are elongated, flattened, "wormlike" arthropods with little external difference in the segments. (See Fig. 4–15.) Each segment of the abdomen has one pair of walking legs, except for the first segment and the last two segments. The first body segment bears a pair of poison claws. Since most of the appendages do not show specialization for different functions, this class is primitive.

Class — Diplopoda. The millipedes ("thousand leggers") are rounder than the centipedes, and they have two pairs of walking legs per segment (except for the first one and the last two). Millipedes are vegetarians which are often found under logs, stones, and in damp basements.

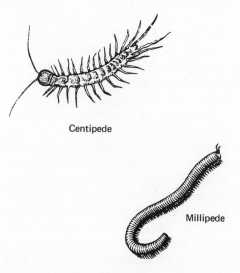

Centipede

Millipede

Fig. 4–15. The centipede and millipede. For the centipede (at the top) the general pattern is one pair of appendages per segment. For the millipede (at bottom) the general pattern is two pair of appendages per segment. There is little difference in each segment of either of these arthropods.

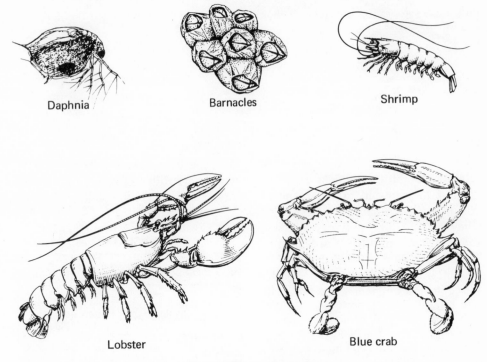

Daphnia   Barnacles   Shrimp

Lobster   Blue crab

Fig. 4–16. Crustaceans.

Class — Crustacea. Most crustaceans are water-dwelling animals that breathe by means of gills and have two pairs of antennae. The segments are commonly fused into two distinct body regions, the cephalothorax (fused head and thorax) and the abdomen. Specialized appendages are used for chewing, for walking, for swimming, and for reproduction. Some crustaceans are shown in Fig. 4–16. This class includes the pill bug, *Daphnia*, barnacle, shrimp, brine-shrimp, crayfish, lobster, and crab.

The barnacle, which might be mistaken for a mollusk because of its rocklike shell, is sedentary in its adult stage. It develops from a marine, shrimplike larval stage which attaches itself to ship hulls or wooden pilings. If a barnacle is placed in a pail of water, its shell opens and its appendages are seen to move rapidly. Ships must periodically be scraped of barnacles because their collective weight and water resistance greatly slows the vessels.

Lobsters, crabs, and shrimp are valuable foods for man. Minute crustaceans in the oceans serve as food for fish and for the giant blue whale.

Class — Arachnida. The arachnids include ticks, mites, spiders, daddy-long-legs, horseshoe crabs, and scorpions. One way to distinguish spiders from insects is that arachnids have eight legs but insects have only six. Count them in Fig. 4–17. Most arachnids are land-living and breathe through gills or "book lungs." Their bodies are divided into a cephalothorax and an abdomen.

Most spiders are helpful to man because they consume destructive insects, but the bite of the black widow spider can be fatal to humans. The sting of a

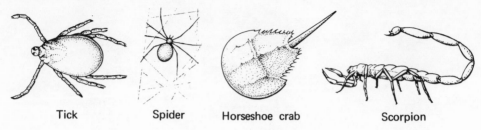

| Tick | Spider | Horseshoe crab | Scorpion |

Fig. 4–17. Arachnids.

scorpion is painful, but it is seldom fatal to man. Mites and ticks cause discomfort to man and to other animals when they live as parasites on the skin. The bite of an infected Rocky Mountain tick may transmit the disease called Rocky Mountain spotted fever. It is caused by a small organism injected into the host's bloodstream when the tick bites.

Class — Insecta. Insects make up the largest class of the phylum arthropoda. More than 625,000 living species have been classified and named. This is almost one-half of all the kinds of living organisms on earth. Insects include flies, mosquitoes, butterflies, moths, grasshoppers, wasps, ants, beetles, bees, and many others. Because all insects have six legs, this class is also known as *hexapoda.* The name "insecta" is derived from the fact that the body is "insected" (or cut) into three distinct divisions. These are the *head, thorax,* and *abdomen.* All insects have one pair of antennae on the head, and most adult insects have two pairs of wings on the thorax. They breathe by a system of tubes, called the tracheae, which have external openings, the spiracles. This method of breathing is not suitable for a large animal. Do you think this provides a reason why all insects are small?

When developing from an egg to an adult, insects undergo a remarkable series of transformations in body form known as *metamorphosis.* Some of these stages have different names in different insects as shown in the table.

METAMORPHOSIS IN DIFFERENT INSECTS

Stage	Fly	Mosquito	Moth	Butterfly
Egg	egg	egg	egg	egg
Larva	maggot	wriggler	caterpillar	caterpillar
Pupa	pupa	pupa	cocoon	chrysalis
Adult	adult	adult	adult	adult

Complete metamorphosis occurs if an insect passes through all four stages listed above. Grasshoppers pass through an *incomplete metamorphosis* in which the young, called nymphs, greatly resemble the adults but are just smaller versions of them. As they grow larger, the nymphs split and discard their exoskeleton in a process called *molting.* They undergo several molts before reaching the adult stage.

The class insecta has been divided into about 20 orders based upon differences in metamorphosis, mouth parts, and wing structure. Some of the orders of insects are illustrated in Fig. 4–18.

Insects are man's chief nuisance and competitor for food. Every crop that man plants for his own use feeds an army of insects. Grasshoppers eat plant parts, while other insects (such as aphids and scale insects) suck the

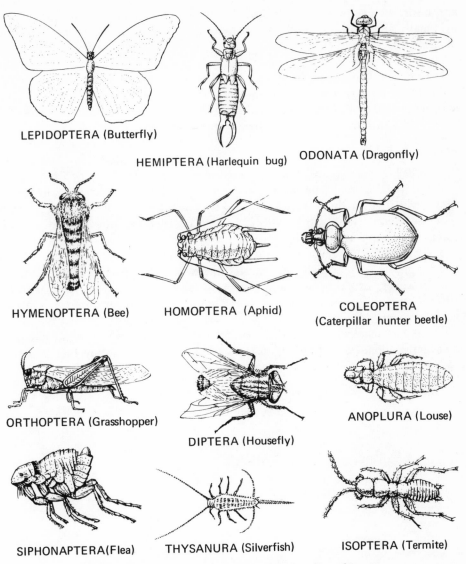

LEPIDOPTERA (Butterfly)

HEMIPTERA (Harlequin bug)   ODONATA (Dragonfly)

HYMENOPTERA (Bee)   HOMOPTERA (Aphid)   COLEOPTERA
(Caterpillar hunter beetle)

ORTHOPTERA (Grasshopper)   DIPTERA (Housefly)   ANOPLURA (Louse)

SIPHONAPTERA(Flea)   THYSANURA (Silverfish)   ISOPTERA (Termite)

Fig. 4–18. Representatives from the major orders of insects.

juices from plants. In moths and butterflies, it is the voracious caterpillar stage which is the most destructive. During this stage, the organism stores food which is to be used while it develops through the next stage. The "worm" in an apple is the caterpillar of the coddling moth. The European corn borer and the chinch bug are notorious destroyers of crops. The larva of the clothes moth, a common household pest, eats holes in woolens.

Insect bites spread the microorganisms which cause malaria, yellow fever, African sleeping sickness, elephantiasis, and typhus fever. Although man uses chemicals such as DDT to control harmful insects, his best allies are birds, other insects, fishes, small mammals, and other natural enemies of insects.

Many insects are useful to man. The silkworm moth produces silk in its cocoon. The honeybee produces beeswax and honey and pollinates plants which supply us with fruits and seeds for food.

Phylum — Echinodermata

The echinoderms are called the "spiny-skinned animals" because of the spines which cover their skins (*echino,* spine + *derm,* skin). The starfish, a common example of this phylum, has five arms radiating from a central disk, and it appears to be radially symmetrical. However, experiments show that all of its arms do not function in the same manner. Moreover, its larva is bilaterally symmetrical. A starfish has no segmentation. Its body is supported by an internal skeleton of hard plates; this is quite different from the external skeleton of an arthropod.

The water vascular system of a starfish (Fig. 4–19) draws sea water into a system of canals which ends in many *tube feet* under the radiating arms. The tube feet have cups; water is drawn from the cups to create a suction, thus enabling the animal to attach itself to solid objects. The starfish feeds on a bivalve (such as a clam or an oyster) by encircling it with its arms, attaching its suction cups to each half of the shell, and slowly pulling it open. When the shell is open, the starfish turns its stomach inside out, through its mouth. The mouth, which is in the center of the lower surface, is placed into the partly open shell of the oyster. Digestive juices dissolve the soft parts of the oyster which are then absorbed

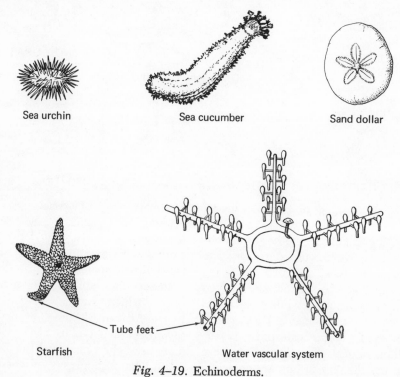

Sea urchin   Sea cucumber   Sand dollar

Tube feet

Starfish   Water vascular system

Fig. 4–19. Echinoderms.

as food by the starfish. The starfish retracts the stomach through the mouth when the meal is over. As you can imagine, starfish are a serious problem to oyster fishermen.

A starfish has great powers of regeneration. If it is cut up into parts, each arm that retains a portion of the central disk grows into a new, fully developed starfish.

The sea urchin, the sea cucumber, the brittle star, and the sand dollar are some of the other echinoderms. There is nothing in the structure, embryology, or chemistry of the echinoderms to suggest relationship to the annelids, mollusks, or roundworms. The only clue to their relationships is that their larvae, their internal skeleton, and some aspects of their body chemistry seem to suggest relationship to the chordates. The evolutionary line of descent that leads to the echinoderms and chordates appears to have separated from another line that leads to the roundworms, annelids, mollusks, and arthropods very early, perhaps at the flatworm stage.

Phylum — Chordata

The chordates take their name from the *notochord.* This is a stiffened supporting rod in the dorsal (upper) part of the body which is present at some stage of their lives. In higher chordates, the notochord is replaced at the adult stage by the vertebral column or backbone. Because most of the chordates have vertebrae, the term *vertebrates* is loosely applied to the chordates. All other phyla, lacking vertebrae, are called the *invertebrates.*

A few primitive marine chordates have no backbones and are, therefore, not true vertebrates. Chordates of this type are the acorn worm, the sea squirt,

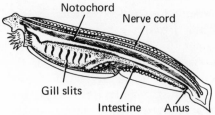

Fig. 4–20. The lancelet, *Amphioxus,* is a primitive chordate whose notochord remains throughout its life. Since it lacks vertebrae, jaws, or paired appendages, it is not truly a vertebrate. Such animals are sometimes called pre-vertebrate chordates or protochordates.

and the lancelet (*Amphioxus*). (See Fig. 4–20.) It is believed that the first chordates were very likely similar to these primitive chordates. However, there are no fossils to prove this.

All chordates, including the primitive ones, have the following characteristics at some stage of their life:

1. A dorsal notochord.
2. Paired gill slits connecting the pharynx and the outside.
3. A dorsal nerve cord.
4. A tail extending beyond the anus.

Vertebrates, the more advanced chordates, also have:

5. Vertebrae. These surround or replace the notochord. Vertebrae are composed of bone or of cartilage (a tissue which has no calcium deposits and is softer and more flexible than bone). The column of vertebrae is the spinal column, or backbone.
6. The cranium or skull, a boxlike structure which protects the brain.

The seven classes of the chordates, not including the primitive chordates, are the jawless fishes, cartilage fishes, bony fishes, amphibia, reptiles, birds, and mammals.

Fish are cold-blooded animals which inhabit fresh or salt water. The body temperature of cold-blooded animals is the same as the temperature of their environment. Warm-blooded animals, by contrast, maintain the same body temperature despite variations in their environment. Fish have two-chambered hearts and scales on their bodies. In addition, they have gills with many filaments that present a large surface area to the water. The gills have many microscopic blood vessels that absorb dissolved oxygen from the water. Many fish have air bladders which help them to move higher or lower in water.

Fish swim by paired appendages called fins and by undulations of the body and tail. The fins of certain prehistoric fish evolved into the legs of land-dwelling vertebrates — the amphibia, reptiles, birds, and mammals. The paired fins of fish are, therefore, homologous to the arms and legs of man.

Class — Agnatha (jawless fishes). The earliest vertebrates (early Paleozoic era) were free-swimming fishes that had no movable jaw. They fed by sucking in mud (nourishing themselves on the microorganisms in the mud). The only survivors of the class are two species that live as parasites.

The lamprey eel is a jawless fish which caused great economic loss by its attacks upon other fish in the Great Lakes. The lamprey's head has a sucking funnel which surrounds the mouth. It has hooklike teeth on the sucker and a rasping tongue. (See Fig. 4–21.) It attaches itself to the body of its prey by its sucking mouth. Next, it extracts tissue fluids from its prey with the aid of its rasping tongue. Lampreys were so numerous that long-distance swimming contests off Toronto, in Lake Ontario,

have been canceled because of their attacks on swimmers.

The jawless fishes gave rise to a large group of hunting fishes with bony, movable jaws. These also died out, but first generated two other classes — the cartilage fishes and the bony fishes.

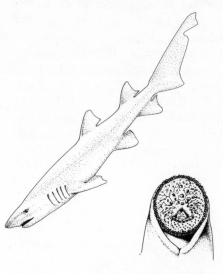

Fig. 4–21. Lower forms of fish. Shark, a cartilage fish (top). Head of lamprey (bottom).

Class — Chondrichthyes (cartilage fishes). The sharks, dogfish, and sting rays have no bones in their bodies. Their skeletons are of cartilage, and their scales are of the same materials that compose teeth. Their five to seven pairs of openings (Fig. 4–21) and the enormous amount of urea in their blood further distinguish them from the other fishes.

Class — Osteichthyes (bony fishes). This is the largest group of marine and fresh-water vertebrates, and includes the trout, seahorse, catfish, perch, bass, pickerel, flounder, and swordfish. Both their skeletons and their scales are made of bone, and their gills are covered by a single bony plate.

Class — Amphibia. The amphibia (*amphi*, on both sides + *bios*, life) get their name from the fact that nearly all of them go through a metamorphosis involving a water-dwelling stage and a land-dwelling stage. A typical amphibian passes its early life in water and usually its adult life on land.

The amphibians are derived from the bony fishes through a group of lung-fishes, a transition that is well-documented by fossils. The fossil series shows a gradual change from a bony fin to legs with toes, and rearrangement of the nose to permit air breathing. The lung is an ancient structure, found in some of the earliest fishes, but the amphibians have developed a special circulation of blood to the lung to use it more effectively as an oxygen supply. A part of this was the development of a third chamber in the heart; fishes have only two. Modern amphibians are degenerate forms; they have lost a great deal of their skeleton, and they have no scales in the skin, as the earliest amphibians had. They have a naked, moist skin which can be used to absorb oxygen from air or water.

Amphibians such as the frog and the toad have both a water-dwelling and a land-dwelling stage. They begin life in the water when they have gills and a tail, but in their adult stage they have lungs for breathing on land and no tail. The mudpuppy, which remains in the water all of its life, retains both its gills and its tail.

Frogs deposit their eggs in the water. When fertilized, these develop into larvae, the tadpoles. The tadpole breathes by gills, swims with its tail, and feeds on plants. When a tadpole metamorphoses to an adult frog, it absorbs its gills and tail, develops lungs and legs, and changes to a diet mainly of insects.

Fig. 4–22. The frog — an amphibian.

Class — Reptilia. The modern reptiles are the turtles, lizards, snakes, crocodiles, and alligators — not an especially prominent group today. However, in the Mesozoic era, this class dominated the earth. There were giant dinosaurs, flying reptiles, creatures of the sea and of the swamps as well as of the deserts. The earliest reptiles are scarcely distinguishable from an early amphibian, since the reptiles arose from the amphibia soon after the latter came out of the water.

In the transition from amphibian to reptile, the main changes that occurred are as follows: a dryer skin; an improved skeleton and musculature for walking on land; and a heart in which the main pumping chamber is partially divided so that air with oxygen is separated from air without oxygen. (The crocodiles and alligators have a complete division instead of a partial one, so their heart is four chambered.) Reptiles skip the amphibian's larval stage by laying a large egg on land; the egg, which is covered with a leathery, waterproof shell, hatches into a land-living creature.

As you know, the turtle has a hard shell on its body. This shell on the turtle is not exoskeleton like that in the arthropods; instead, it has many greatly enlarged ribs which have fused into bony plates. Although snakes have no limbs, vestigial remnants of hind limbs are present in the boa constrictor and python. A snake's jaw can be temporarily unhinged to swallow large prey. Most snakes in America are nonpoisonous, but some species deadly to man are the copperheads, the water moccasins, the rattlesnakes, and the coral snakes. The Gila monster and the beaded lizard of Mexico are the only poisonous lizards. One difference between the American alligator and the American crocodile is the more pointed snout of the crocodile.

Sphenodon, a lizard of great biological interest, is almost extinct, but specimens can be found on the islands off New Zealand. This pineal gland in the brain of this lizard is a light-sensitive organ, like a third eye. Another lizard is the chameleon which is noted for its ability to make rapid changes in color. It lives in the dry regions of our western states.

Reptiles are not popular animals though many are valuable to man. Some species of snakes and turtles are eaten as a delicacy. The skins of lizards, snakes, alligators, and crocodiles are used to make shoes, belts, and handbags. Land snakes aid the farmer by consuming large numbers of destructive insects and rodents.

Class — Aves (birds). A fossil called *Archaeopteryx* (middle Mesozoic era) has a skeleton almost identical with that of a small dinosaur, but it also has feathers. If this is not the first bird evolved from a reptile, it is surely close to it. The transition from reptile to bird is an adaptation to flight and to warm-bloodedness.

The most characteristic feature of birds is their feathers, which provide the insulation that a warm-blooded creature needs, and airflow surfaces like the wings of a plane. Birds have a highly efficient lung system, and a four-chambered heart that enables them to keep oxygen flowing from lungs to tissues without mixing with low-oxygen blood. A special adaptation of the bird for flight is its hollow bones.

Birds are important to man in many ways. For example, chickens and their eggs are such an important part of man's diet that, in this country, chickens are now raised in highly mechanized "factories." Wild birds are also important to man because they eat harmful insects and the seeds of weeds. Owls and hawks are valuable to man because they eat rats and mice.

Class — Mammalia. Mammals have three main characteristics which set them apart from other vertebrates: (1) the females have mammary glands to supply their young with milk; (2) mammals have hair on their bodies; and (3) the females give birth to living young. Mammalian birth consists of the separation of the young from the uterus of the mother, where it had been attached to a placenta. Some fish and snakes "give birth" to living young, but they differ from mammals because their young were never attached to the mother.

Mammals arose from a fairly early reptile in the early Mesozoic era. They are warm-blooded, like birds, but they have entirely different adaptations to provide the necessary conditions for their style of life. Their hearts are four-chambered, but the arrangement of vessels is entirely different from that of birds. Insulation is provided by hair, not feathers. Air is pumped in and out of the lungs by a muscular *diaphragm* stretched across the bottom of the chest.

Mammals are widely distributed on earth. Most of the mammals live on the land, but some (whales, dolphins) have returned to water. Bats are mammals which fly with wings, but these wings do not have feathers like the wings of birds.

Some vertebrates which are classified as mammals are so primitive that they do not have all the characteristics of mammals. For example, the duck-billed *Platypus* of Australia (Fig. 4–23) is a mammal that lays eggs much like those of reptiles. The marsupials, such as the kangaroo and the opossum, bear their young in such an immature state that the tiny newborn young climb into the mother's pouch (the marsupium) on the abdomen and fasten to the mammary glands. The opossum is the only marsupial which is native to this country; the others, such as the Koala bear and the wombat, are native to Australia.

The more advanced members of the class mammalia are divided into 14 orders.

SOME REPRESENTATIVES OF MAMMALIA

Rodents	These are the gnawing mammals, such as the rat, mouse, squirrel, and beaver.
Bats	The wing of the bat is a modified arm.
Carnivora	These meat eaters include the lion, cat, dog, walrus, and sea lion.
Hoofed Animals	Some have an odd number of toes, and others have an even number. Most are vegetarians. Examples of these animals are the horse, zebra, rhinoceros, camel, giraffe, sheep, and hippopotamus.
Primates	These animals are "first" in importance because of their brain development and because they stand on two legs. The other two appendages are free for grasping and manipulating objects. The opposable thumb of man's hand permitted his ascendancy as a tool-using animal. His brain is the most highly developed of all animals. Other examples of primates are apes, monkeys, and chimpanzees.

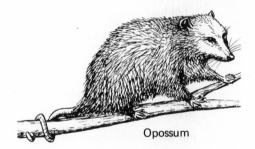

Opossum

Platypus

Fig. 4–23. Primitive mammals.

COMPLETION QUESTIONS

[A] 1. Some taxonomists divide living things into three kingdoms: plants,, animals, and . . . .

2. *Amoeba* moves by means of outpushings of protoplasm called . . . .

3. *Paramecium* moves by beating its . . . .

4. Malaria is caused by a protozoan transmitted by a female mosquito of the genus . . . .

5. Algae with a spiral-shaped chloroplast have been classified in the genus . . . .

6. Chloroplast-containing cells having two flagella are found grouped together in a colony called . . . .

7. Photosynthesis is carried on by a group of single-celled and colonial protists called . . . .

8. Two types of organisms lacking a formed nucleus are bacteria and . . . .

9. An organism that feeds on a living host is known as a (an) . . . .

10. A lichen is a combination of a fungus and a (an) . . . .

11. A mold whose life cycle seems to be partly plantlike and partly animal-like is the . . . .

12. Plants with a well-developed vascular system for transport of water are placed in the phylum . . . .

13. Dark brown spots on the underside of a fern frond are places where . . . . are produced.

14. Plants whose seeds are *not* hidden in a fruit are classified as . . . .

15. Angiosperms with parallel-veined leaves are the . . . .

[B] 16. The skeleton of sponges may be composed of lime, silica, or . . . .

17. *Hydra* has . . . . symmetry, and a single opening called the . . . .

18. The middle layer in flatworms is called . . . .

19. The mouth of *Planaria* also serves as a (an) . . . .

20. The tapeworm attaches itself to the . . . . of man.

21. Flatworms and roundworms have . . . . symmetry.

22. A parasitic worm whose larvae enter man's body through the soles of the feet is the . . . .

23. In a single earthworm, the body organs in one . . . . may be repeated again in the next.

24. The long tube running the length of the body in roundworms and annelids is their . . . .

25. A relative of the snail which has no shell is the . . . .

26. The ancestor of all modern insects, crustacea, and arachnids appears to be an extinct arthropod called the . . . .

27. During metamorphosis, the larva of a fly develops into a (an) . . . .

28. Starfish can reproduce by the process of . . . . as well as by sperm and eggs.

29. At some time of their lives, all chordates have a dorsal . . . . for support.

30. Birds and mammals seem to have evolved from . . . .

MULTIPLE-CHOICE QUESTIONS

[A] 1. *Euglena* is a single-celled organism which possesses chlorophyll but has no cell wall. It carries on photosynthesis, but it has a mouth and swims by means of a flagellum. Therefore, it is most reasonable to classify it among the (1) plants, (2) animals, (3) protists, (4) protozoa.

2. Mushrooms and the Indian pipe can be distinguished from a fern by their lack of (1) flowers, (2) roots, (3) seeds, (4) chlorophyll.
3. The vertical growth of mosses is comparatively limited because mosses lack (1) rootlike structures, (2) chlorophyll, (3) vascular tissue, (4) seed-forming organs.
4. Which green plant does *not* reproduce by means of seeds? (1) conifer, (2) fern, (3) grass, (4) ginkgo.
5. An angiosperm with its vascular bundles arranged in rings within the stem would have (1) veins of leaves parallel, flower parts in threes or their multiples, two cotyledons, (2) veins of leaves in a network, flower parts in fours or fives or their multiples, one cotyledon, (3) veins of leaves in a network, flower parts in threes or their multiples, two cotyledons, (4) veins of leaves in a network, flower parts in fours or fives or their multiples, two cotyledons.

B 6. Animals with a mouth and two layers of cells are classified as (1) coelenterates, (2) sponges, (3) protists, (4) annelids.
7. A chitinous exoskeleton is characteristic of a (1) clam, (2) beetle, (3) fish, (4) sponge.
8. One advantage of the internal skeleton, as compared with the external skeleton, is that the internal skeleton (1) permits greater variation in the form of vital internal organs, (2) furnishes the organism with more protection against changes in climate, (3) gives the organism bilateral symmetry, (4) need not be discarded as the rest of the body grows larger.
9. Which class of vertebrates contains animals with a two-chambered heart? (1) birds, (2) fish, (3) reptiles, (4) amphibia.
10. Which term includes the others? (1) marsupial, (2) carnivore, (3) mammal, (4) primate.

CHAPTER TEST

1. An example of an invertebrate is the (1) guinea pig, (2) salamander, (3) sponge, (4) eel.
2. Which term includes the others? (1) insect, (2) crab, (3) crustacea, (4) arthropod.
3. The kingdom containing organisms which cannot easily be classified as either plants or animals is the (1) slime molds, (2) protista, (3) lichens, (4) fungi.
4. Grain rusts are classified as (1) algae, (2) bacteria, (3) viruses, (4) fungi.
5. The presence of well-developed vascular tissue for conducting liquids is characteristic of (1) mosses, (2) algae, (3) ferns, (4) slime molds.
6. The earthworm is a (an) (1) annelid worm, (2) flatworm, (3) ringworm, (4) nematode roundworm.
7. Which protists do *not* have true roots, stems, or leaves but do contain chlorophyll? (1) molds, (2) fungi, (3) algae, (4) ferns.
8. The members of which pair of organisms are *least* related? (1) jellyfish and tuna fish, (2) whale and elephant, (3) frog and snake, (4) *Trichinella* and the hookworm.
9. The pouched mammal which is native to the United States is the (1) porcupine, (2) skunk, (3) beaver, (4) opossum.
10. Which pair of characteristics is found only in mammals? (1) three-chambered heart and lungs, (2) four-chambered heart and milk glands, (3) four-chambered heart and lungs, (4) two-chambered heart and milk glands.

11. The two classes of chordates which are warm-blooded and have a four-chambered heart are the (1) amphibians and birds, (2) amphibians and reptiles, (3) birds and mammals, (4) reptiles and mammals.

12. An animal that is covered with scales, is cold-blooded, and usually reproduces by laying eggs is a (1) reptile, (2) bird, (3) mammal, (4) coelenterate.

13. Which combination of traits would identify an animal as probably belonging to the phylum Arthropoda? (1) dorsal nerve cord, endoskeleton, closed circulatory system, (2) nonsegmented body, ventral nerve cord, (3) exoskeleton, segmented body, jointed appendages, (4) ventrally located heart, dorsal nerve cord.

14. Two animals belong to the same species if they (1) can live together in a similar environment, (2) have similar nutritional requirements, (3) have many physical analogies, (4) can mate and produce fertile descendants.

15. Any animal which has a notochord and gill arches, at least in its early embryonic stages, belongs to the phylum of (1) invertebrates, (2) chordates, (3) carnivores, (4) reptiles.

16. Which pair of organisms belongs to the same genus? (1) jellyfish and tuna, (2) whale and shark, (3) horse and cow, (4) lion and tiger.

17. A newspaper article stated, "The fallout of radioactive materials from bomb tests threatens *the whole human race*." To be scientifically precise, the italicized phrase should read (1) the whole race of man, (2) the whole human species, (3) the whole human class, (4) the whole human family.

18. To which organism is the whale most closely related? (1) fish, (2) turtle, (3) dinosaur, (4) horse.

19. Which group contains organisms that are most closely related? (1) swordfish, shellfish, silverfish, (2) flying squirrel, bat, whale, (3) tree toad, lizard, frog, (4) garter snake, eel, alligator.

20. *Hydra* and the sea anemone are classified as coelenterates because both organisms (1) live in water, (2) can serve as food for other organisms, (3) have a tube-within-a-tube digestive tract, (4) have a body wall composed of two layers of cells.

Column I lists various organisms, Column II lists various classifications. For each of the items in Column I write the letter of the item in Column II which is most closely related to it. (A letter may be used only once.)

	Column I		Column II
. . . . . .	21. Grasshopper	A.	Echinodermata
. . . . . .	22. *Hydra*	B.	Protista
. . . . . .	23. Man	C.	Annelida
. . . . . .	24. Earthworm	D.	Arthropoda
. . . . . .	25. *Paramecium*	E.	Coelenterata
		F.	Chordata
		G.	Tracheophyta

In questions 26–30, write the number of the word or expression which includes all the others in the group.

26. (1) invertebrate, (2) clam, (3) sponge, (4) worm.

27. (1) elephant, (2) rat, (3) camel, (4) mammal.

28. (1) fish, (2) chordate, (3) mammal, (4) acorn worm.

29. (1) primate, (2) monkey, (3) man, (4) ape.

30. (1) green alga, (2) protist, (3) protozoa, (4) slime molds.

UNIT TWO

ANIMAL MAINTENANCE

We know that an animal, like an automobile, must have what it needs to maintain it (keep it in good working order). The maintenance of animal life is dependent upon universal requirements:

- obtaining essential organic and inorganic substances
- processing them, as needed, into usable form
- distributing usable substances
- synthesis of usable substances into the composition of the organism
- a steady flow of energy to support the activities of the body
- removal of waste products
- regulation of processes

In this unit we will see how animals satisfy these requirements.

CHAPTER 5 | Nutrition

A source of energy for animals is the food which they take in. *Nutrition* consists of those activities by which organisms obtain and use food in carrying out their life functions.

The usable portions of foods are known as *nutrients,* which are used mainly as (1) energy sources, (2) the basis for the building or repair of cell structures, (3) regulators of metabolic processes.

The portion of a food that does not serve as a nutrient is the *waste.* Thus, food consists of nutrients plus wastes. Although wastes aid digestion as roughage, they are not utilized by the cells of the body but eliminated as feces.

A THE CONCEPT OF NUTRITION

It is possible to think of nutrition as embracing all activities and destinies of consumed materials — a very broad concept that could even include excretion. Although this is correct, nutrition is commonly limited to ingestion and digestion — the processes discussed in this chapter.

The Nutrients

A summary of the nutrients and their uses is given in the table on page 98. Among them are fats, carbohydrates, and proteins, which are sources of energy. The energy value of foods is measured in calories. A *Calorie* is the amount of heat required to raise the temperature of one kilogram of water one degree Celsius. The calorie content of foods is measured by burning them in a calorimeter. The number of calories in one gram of nutrients is: fats (9 calories); carbohydrates (5 calories); proteins (4 calories). The chemical structure of the nutrients was discussed in Chapter 2.

THE NUTRIENTS

Nutrient	Composition	Uses	Sources
Carbohydrates (sugars and starches)	Carbon, hydrogen, and oxygen.	Source of energy.	Cereals, potatoes, bread, fruit, candy, ice cream.
Proteins	Carbon, hydrogen, oxygen, nitrogen. Many proteins also contain sulfur, phosphorus, and iron.	Synthesis of compounds needed for growth and repair. Proteins are also oxidized for energy.	Beef, fish, liver, peas, beans, nuts, milk, cheese, eggs.
Fats and oils	Carbon, hydrogen, and oxygen.	Source of energy.	Butter, cream, lard, bacon, fats in meats, vegetable oils, nuts.
Water	Hydrogen and oxygen.	Transport of materials. The solvent in which chemical reactions take place.	Drinking water, milk, and other beverages. Fruits and vegetables. Most foods contain water.
Mineral salts	1. Calcium and phosphorus.	For strong bones and teeth.	Milk, cheese, eggs.
	2. Iodine.	To manufacture thyroxin, the secretion of the thyroid gland.	Sea foods, iodized salt.
	3. Iron.	In the manufacture of hemoglobin, the red pigment in the blood that carries oxygen.	Liver, meats, eggs, vegetables.
	4. Fluorine.	Fluorides prevent tooth decay by making the enamel hard.	Present naturally in some drinking water. "Fluoridated water" has controlled amounts of fluorides added.
	5. Sodium.	Needed for proper functioning of cells.	Table salt (sodium chloride).
	6. Chlorine.	Part of the hydrochloric acid needed for digestion of food in the stomach.	Table salt (sodium chloride).
Vitamins	Carbon, hydrogen, oxygen, nitrogen, and other elements.	Lack of vitamins results in "deficiency diseases." Some vitamins are parts of enzymes (coenzymes) that participate in cellular respiration.	About ten important vitamins known. A varied diet is needed to include all of them.

Tests for Nutrients

Test for sugar. You will recall that glucose and other simple sugars are monosaccharides. To test a food for the presence of simple sugar, place the food in a test tube, adding enough clear Benedict's solution to cover it. Heat to boiling. An orange-red cloudiness indicates the presence of a reasonable amount of simple sugar, and a green color indicates a small amount of simple sugar. If no simple sugar is present, Benedict's solution retains its clear blue color. This test does not work with a disaccharide such as sucrose.

Test for starch. Add amber-colored iodine solution to a food as a test for starch. The iodine solution will turn blue-black if starch is present. In the laboratory, Lugol's solution is used for the starch test. Lugol's solution is a water solution of iodine and potassium iodide.

Test for protein. Place a small amount of food in a test tube and cover the food with dilute nitric acid. Heat gently. (*Caution! Hot nitric acid is dangerous.*) After pouring off the nitric acid, rinse the contents of the tube with water and add ammonium hydroxide. If protein is present, the food turns yellow after the nitric acid treatment and turns orange after the ammonium hydroxide treatment.

Another test for protein can be demonstrated by placing some raw egg white into a test tube. Add one drop of a 1% solution of copper sulfate and five drops of a 10% solution of potassium hydroxide. A purple color indicates the presence of protein.

Test for fat. There is a simple test for fat. Rub food on a piece of unglazed paper and hold it to the light. If the paper becomes translucent, fat is present.

Test for water. To test a food for water, heat a sample of food in the bottom of a dry test tube. If droplets of water form at the top of the test tube, the water was present in the food sample.

Test for vitamins. Although the tests for most vitamins are complex, the test for ascorbic acid (vitamin C) is relatively simple. This test utilizes indophenol which, when dissolved in water, forms a clear blue solution. Ascorbic acid causes this solution to become colorless.

To perform the test, first add ten drops of blue indophenol to a test tube and then add, one drop at a time, a solution of the substance being tested. To interpret the test correctly, keep an accurate count of the drops. The greater the number of drops needed to decolorize the indophenol, the less the concentration of ascorbic acid present.

THE NEED FOR DIGESTION

All cells require certain usable materials. They must have:

1. Raw materials to build themselves (particularly amino acids, fatty acids, and glycerol).

2. Energy sources (simple sugars, as well as fatty acids and amino acids)

3. Water as a solvent and as a medium for transporting materials.

4. Dozens of different minerals in tiny amounts, chiefly for the control of chemical processes.

5. Certain organic compounds to keep their chemistry going.

These basic organic compounds and minerals consist of relatively small molecules that can dissolve (or emulsify) in water, which also has relatively small molecules. All of these molecules can be distributed throughout the body and throughout the cell, and are small enough to pass through cell membranes by processes that will be discussed in the next chapter.

Every cell has a complex system of enzymes that enables it to transform many of these chemicals into each other. For example, excess glucose may be converted to fats, a process familiar to everyone who watches his weight. The one basic limitation of this process is that it is *impossible to make high-energy compounds from low-energy compounds without adding energy.* Since cells are built of high-energy compounds that constantly tend to break down, every cell needs a steady supply of energy-rich materials (fats, glucose, etc.), to serve as an energy source.

Autotrophic and heterotrophic nutrition. Some organisms make their own high-energy foods from simple chemicals. This is known as *autotrophic nutrition.* Of course, they need a source of energy to do this. If the energy source is light, the synthesis of foods is known as *photosynthesis,* the food-production process of green plants. Certain bacteria obtain the energy for autotrophic nutrition from the oxidation of sulfur, iron, or of any of several other simple materials.

Since animals cannot make food, they must eat plants (or other animals that have eaten plants) to obtain their high-energy organic chemicals. This is called *heterotrophic nutrition.* The chemicals obtained in this way are not the simple sugars, amino acids, etc.,

that the cell can use, but extremely large molecules (proteins, starches, lipids) which must be broken down into usable form (glucose, fatty acids, etc.). How is this accomplished?

Definition of digestion. By the process of *digestion,* large food molecules are hydrolyzed into small usable molecules. This breaking down by hydrolysis is accomplished by special digestive enzymes. In protozoa, digestion is mainly *intracellular;* that is, the food particle is taken into the cell and there is acted on chemically by digestive enzymes. In most multicellular animals, digestion is *extracellular.* The food is retained within a digestive cavity of some kind (stomach, intestines) and juices are secreted into it. In the broad category of plants and plantlike protists known as *saprophytes* (molds, mushrooms, many bacteria), the food is dead organic matter digested completely outside the body. The plant secretes juices into the food and then absorbs the digested material.

Chemical digestion, in any case, is limited to the outer surface of the food mass, for the digestive juices cannot penetrate very far. Active animals, therefore, must have some way of breaking up the large food particles to allow the juices to penetrate. Accordingly, they break the food down by the physical actions of grinding, tearing, and cutting. This physical breakdown of food particles is called *mechanical digestion.* A snake does not use mechanical digestion; it swallows its victim whole and digests it slowly, over many days or weeks.

There are many digestive enzymes, each with a special function, but they may be classified into three general groups as indicated in the table that follows:

GROUPS OF DIGESTIVE ENZYMES

Substrate	Enzyme	End Products of Digestion
proteins	proteases	amino acids
carbohydrates	amylases	glucose
fats	lipases	fatty acids and glycerol

NUTRITION IN LOWER ANIMALS

Although protozoa are classified by many biologists as protists, in this section they are compared with animals. They exhibit animal-like characteristics and are thought to be related to multicellular animals.

The Protozoa

Ingestion is the process by which organisms take in food. *Amoeba* ingests smaller protists by surrounding them with protoplasmic projections called *pseudopods.* The food particle is then enclosed in a small globule (food vacuole) which is freed into the cytoplasm. Enzymes enter and digest the food as the vacuole circulates through the streaming cytoplasm.

Paramecium feeds on microorganisms drawn into the oral groove by the beating of cilia lining the groove. (See page 15.) The food is forced to the end of the gullet where food vacuoles form. Then they are detached and circulated in a definite pattern as digestion occurs. Eventually, the food vacuole breaks open at the anal spot where undigested solid materials are discharged.

In most animals, food-getting requires motion either in search, pursuit, or capture of food. *Amoeba* and *Paramecium* seem to be attracted to food by chemical stimuli. When cultured (grown) in dishes in the laboratory, they congregate in large numbers around decaying food particles.

Hydra

Hydra, usually found on submerged vegetation in ponds or lakes, is a cup-shaped animal about one-fourth inch in length, consisting of two layers of cells arranged like a hollow cylinder. The outer layer of cells is the *ectoderm,* and the inner layer of cells, the *endoderm.* The foot at one end of the body is used for attachment or locomotion. (*Hydra* moves by a slow gliding motion of the foot, by a slow somersaulting movement or by an "inching" process.) The mouth, at the other end of the body, is surrounded by a circular arrangement of about six *tentacles* bearing stinging cells called *nematocysts.*

Hydra takes in (ingests) only living food such as minute crustacea and worms which it accidentally contacts with its tentacles. Threads expelled from the nematocysts (cells on the tentacles) penetrate the prey and paralyze it. A substance released from the prey initiates the feeding reflex (automatic reaction) in *Hydra,* and the tentacles contract, drawing the food to the mouth which surrounds and engulfs it. The food enters the digestive cavity and is digested. Ten minutes to one-half hour later the undigested food is expelled (egested) through the mouth.

Digestion in *Hydra* is both extracellular and intracellular. Let us see why we say this. First, visualize the undigested food in the digestive cavity surrounded by the endoderm cells lining the digestive sac. These endoderm cells secrete enzymes into the digestive cavity where proteins and fats are hydrolyzed. Since this digestion occurs *outside* cells, it is extracellular.

Another type of digestion also occurs. Cells of the endoderm can form outpushings of protoplasm (similar to the pseudopods of *Amoeba*) and thereby engulf fairly large particles of food. The food particles are hydrolyzed *inside* the cells of the endoderm (intracellular digestion). Biologists think that *Hydra* does not digest carbohydrates.

The end products of digestion are absorbed; they then pass by diffusion from the endodermal cells to the other cells of the body. Within the cells, energy-rich materials are oxidized and energy is released for the activities of life.

Hydra can be maintained in the laboratory. They are fed live *Daphnia* (water fleas) or the freshly hatched, washed larvae of brine shrimp but they thrive only if the dish contains pond water or demineralized tap water which is changed daily to remove waste products.

The Earthworm

The earthworm is an annelid with

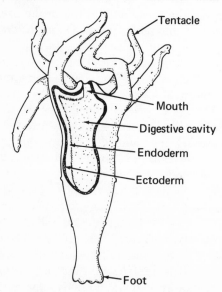

Fig. 5–1. Body plan of *Hydra*.

Labels: Tentacle, Mouth, Digestive cavity, Endoderm, Ectoderm, Foot

more than 100 segments, indicated by grooves extending around the cylindrical body. Behind the first segment is the *mouth,* and the last has the *anus.* Every segment but the first and last has four pairs of bristles (setae) used in locomotion. The earthworm's food consists of fragments of decaying vegetation and animal matter present in the soil. As the earthworm burrows through the ground, it takes in soil and food particles. The soil particles and undigested food pass from the mouth, through the food tube, and out of the anus.

Earthworms are valuable to farmers. By turning over the soil, they make it porous and, by mixing decaying leaves with it, they enrich its content of humus.

The earthworm is dissected by making a lengthwise slit along the dorsal (upper) body wall. *The earthworm has a tube-within-a-tube body plan.* The digestive tract is the inner tube and the muscular body wall is the outer tube (Fig. 5–2). The cavity between these tubes is divided at each segment by thin partitions which are connected to the grooves seen on the outside surface. Many of the internal organs, for example, those for excretion, are repeated in almost every segment in the body of the earthworm.

The digestive system of animals is composed of two main parts: (1) the *alimentary canal,* the tube through which the foods pass, and (2) the *digestive glands,* such as the liver and pancreas of man. The digestive juices pass from the digestive glands of the earthworm into its alimentary canal, which has specialized regions. Fig. 5–3 shows the specialized regions of the earthworm's alimentary canal.

Food particles are sucked into the mouth cavity (ingested) when the

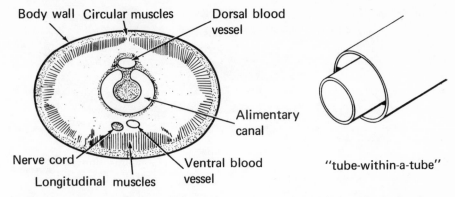

Fig. 5–2. Cross-section of an earthworm showing tube-within-a-tube body plan.

pharynx is enlarged by muscular activity. The food is forced along the alimentary canal by waves of muscular contractions called *peristalsis*. The muscular walls of the esophagus push the food along to the thin-walled *crop* where it is stored before being moved to the *gizzard*. The gizzard has thick muscular walls that grind the food mass into smaller particles before passing it on to the intestine where enzymes (secreted by the walls of the intestine) digest the food. Undigested food is eliminated at the *anus* as feces.

Comparison between earthworm and *Hydra*. Digestion in *Hydra* and the earthworm may be compared as follows:

1. In *Hydra*, food and wastes pass through the mouth — a two-way digestive system. The earthworm has a *one-way digestive tube* (with a mouth and anus), which permits stages of digestion to be done in succession by different parts of the tube. Thus, the earthworm has more division of labor (a higher degree of specialization) than *Hydra*.

2. In *Hydra*, digestion is extracellular and intracellular, whereas, in the earthworm, it is extracellular (carried on in the digestive tube).

3. *Hydra* has no blood transport system, whereas the earthworm does. In the earthworm, end products of digestion are absorbed through the cells lining the digestive tube and pass into the blood.

The Grasshopper

We study the grasshopper as a representative of the insects. Like all ar-

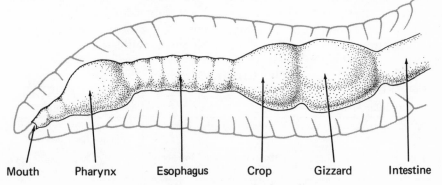

Fig. 5–3. Alimentary canal of earthworm.

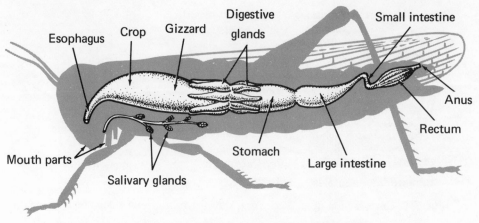

Fig. 5–4. Digestive system of the grasshopper.

thropods, insects have an exoskeleton (external skeleton) of chitin. The body of an insect is divided into three major regions: the *head, thorax,* and *abdomen.*

The head bears one pair of antennae, three pairs of mouth parts, a pair of simple eyes, and a pair of compound eyes composed of hundreds of lenses. These compound eyes permit an insect to see in many directions and to detect moving objects readily.

On the thorax are six legs (three pairs) and, as in most insects, two pairs of wings. The abdomen of the grasshopper is divided into numerous segments, none of which has legs.

The grasshopper moves from plant to plant by walking, hopping, or flying, and is very destructive to crops. It has complicated mouth parts including appendages for sensing and grasping food, and strong mandibles for chewing it.

The digestive system of the grasshopper is shown in Fig. 5–4. Food, mechanically broken down in the mouth, is sucked through the esophagus into the *crop,* a large thin-walled sac where it is temporarily stored before entering the muscular gizzard where it is further broken down by the action of tooth-like chitinous plates. From the gizzard the food passes to the stomach. Large branching digestive glands on the outside of the stomach pour their juices into this organ where digestion is completed. Digested food is absorbed from the stomach into the blood, and undigested food passes through the intestine and rectum to be eliminated through the anus.

Like the earthworm, the grasshopper has a tube-within-a-tube body plan, a one-way digestive tract, and extracellular chemical digestion.

EVOLUTIONARY TRENDS

Unicellular Animals	Hydra	Earthworm, Grasshopper, Man
No digestive tract (food vacuoles)	Two-way digestive tract	One-way digestive tract (specialized food tube)
Mostly intracellular digestion	Extracellular and intracellular digestion	Mostly extracellular digestion

REASONING EXERCISES

1. What are nutrients? How are they used? What use do animals make of fats, carbohydrates and proteins?
2. What is nutrition? Distinguish between autotrophic and heterotrophic nutrition.
3. What is digestion? Why is digestion necessary?
4. Distinguish between intracellular and extracellular digestion. Give a chemical reason why water is essential to hydrolysis. What is the value of mechanical digestion?
5. Compare digestion in *Amoeba* and *Hydra*.
6. Demonstrate that digestion in the earthworm is more specialized than in *Hydra*.
7. Place one drop of a *Paramecium* culture into some methyl cellulose on a slide, and then add a drop of Yeast-Congo Red preparation. Observe under high power, looking specifically at one *Paramecium*, to see how it takes in food. What is the significance of the color changes in the food vacuoles? Make a sketch, showing structures involved in nutrition, and movements of the food vacuoles.

(B) DIGESTION IN MAN

Although the digestive system of man appears complicated, it is essentially a tube-within-a-tube. It is similar in this respect to that of the earthworm and grasshopper. The tube is greatly lengthened so that portions of it are much coiled. Specialization here has increased beyond that of the lower animals.

The upper portion of the human di-

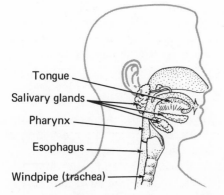

Fig. 5–5. Upper portion of the digestive system of man.

gestive system is shown in Fig. 5–5. Three pairs of salivary glands lead into the *mouth* (oral cavity). The *pharynx* (throat) is a region where the respiratory system and the digestive system cross. The two tubes leading from the pharynx are the *esophagus* (gullet) and the *trachea* (windpipe). Food passes down the esophagus and air passes down the trachea.

Fig. 5–6 is a diagrammatic scheme of the lower digestive system. The diagram shows that the alimentary canal in man is one continuous tube. The *small intestine* in man is a long and narrow tube about 22 feet long. The *large intestine* is divided into an ascending colon, transverse colon, and descending colon. There is a pouch where the small intestine joins the large intestine, and the *appendix* is a slight extension of this pouch. The *rectum* and *anus* are at the end of the large intestine.

Glands passing digestive juices and enzymes into the alimentary canal are:

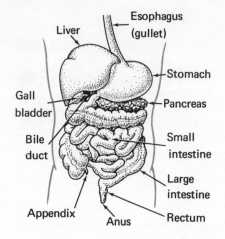

Fig. 5–6. Lower portion of the digestive system of man.

1. *Salivary glands* — secrete saliva into the mouth.

2. *Gastric glands* — secrete gastric juice into the stomach. These minute glands are present in the wall of the stomach.

3. *Pancreas* — secretes pancreatic juice into the small intestine.

4. *Liver* — secretes bile into the upper portion of the small intestine. The gall bladder stores bile.

5. *Intestinal glands* — secrete intestinal juice into the small intestine. These microscopic glands are present in the walls of the small intestine.

Thus, the human digestive system consists of a continuous tube, the alimentary canal, and the accessory glands and organs which function in conjunction with the alimentary canal.

Digestion in the Mouth

The mouth and alimentary canal contain many mucous glands in the epithelial linings. *Mucus,* the secretion of these glands, acts as a lubricant for food.

The digestive juices of the mouth are in the saliva secreted by the salivary glands. When you chew food, you mix it with saliva and break it up into smaller particles, thus providing more surface for the action of digestive juices secreted by the salivary glands. Saliva contains an amylase called *ptyalin* which hydrolyzes starch to a double sugar (disaccharide) called *maltose.* However, it does not break down starch completely to simple sugars (monosaccharides). When the food and saliva reach the stomach, the acid pH in that organ gradually stops the action of ptyalin.

Swallowing is a coordinated reflex action which forces food from the mouth to the esophagus. The esophagus and trachea lead from the pharynx. During swallowing, the top of the trachea rises against a flap called the *epiglottis.* This diverts the food mass toward the esophagus so that it does not block the trachea and the passage of air.

The Esophagus

The esophagus leads the food to the stomach by a series of wavelike muscular contractions and relaxations known as *peristalsis.* These peristaltic waves start in the esophagus and continue along the entire alimentary canal. Muscular contractions of the stomach and small intestine churn the food and mix enzymes with it.

The esophagus, stomach, and small intestine contain an outer covering of fibrous connective tissue, middle layers of smooth muscle, and inner linings of epithelial tissue. Some of the smooth muscles are circular while others are longitudinal. They are thus well adapted to producing the ringlike contractions and relaxations of peristalsis. Some of the lining epithelial tissue is specialized for the production of mucus.

Digestion in the Stomach

Proteins are the only important nutrients acted upon by the stomach's *gastric juice*. In the stomach, proteins are only partially hydrolyzed, forming the intermediate products of digestion known as *peptones* and *proteoses*. The protease (enzyme) in gastric juice is called *pepsin*. A weak (0.5%) solution of hydrochloric acid provides the proper acid pH for the action of pepsin. This acid also helps to kill germs and to dissolve some minerals. Most other digestive enzymes function best in a nearly neutral or slightly alkaline environment. The temporary storage of food is an important function of the stomach.

Rennin is an enzyme present in the gastric juice of many infant animals (for example, calves), but not in the gastric juice of human infants. Rennin curdles milk proteins; that is, it changes them into a solid state, thus separating them from the watery portion of the milk. The water passes along the alimentary canal, but the proteins remain longer in the stomach for more complete digestion. In man, milk proteins are curdled by pepsin. When housewives add "rennet" to milk in the preparation of custards, they use a commercially prepared form of rennin. Curdled milk can also be seen in the form of the "cheese" regurgitated by infants through reverse peristalsis.

Minute amounts of gastric juice are always present in the stomach, and the flow of this juice increases when protein is in the stomach. How do the gastric glands "know" when to secrete gastric juice? Three types of stimuli are involved:

1. *Nerve action.* Food in the mouth stimulates nerve endings which pass "messages" to the gastric glands, causing them to secrete.

2. *Contact with food in the stomach.* Moderate amounts of gastric juice are secreted in response to the contact of food with the lining of the stomach.

3. *Hormone action.* The copious flow of gastric juice needed for the digestion of meat is begun in a special way.

Let us see how this occurs. The gastric juice in the stomach hydrolyzes proteins to smaller molecules which stimulate certain cells lining the stomach to secrete the hormone *gastrin*. *Hormones* are "chemical messengers" secreted by glands and carried by the blood throughout the body. Although gastrin reaches all parts of the body in the blood, it stimulates only the gastric glands, the "target organ" for this hormone. Thus, the copious flow of gastric juice is initiated.

In 1883, Dr. William Beaumont, an American Army surgeon, performed studies on the human stomach which are the basis for modern understanding of stomach function. His patient, Alexis St. Martin, had received a shotgun wound which resulted in a permanent opening leading from the outside body wall into his stomach. Into this organ Beaumont inserted pieces of meat tied to a string, and later removed them for observation. Beaumont's reports of digestive action of gastric juice are classics of scientific experimentation.

Saliva and gastric juice add water to the stomach which churns and changes the food to a liquid mass called *chyme*, which is of the proper consistency to enter the small intestine. The lower end of the stomach is separated from the small intestine by a ring-shaped muscle known as the *pyloric sphincter,* which surrounds the point where the stomach connects to the small intestine. Although the sphincter completely closes the passageway when contracted, it opens for short periods, allowing the chyme to enter the small intestine a little at a time.

Digestion in the Small Intestine

Proteases, amylases, and lipases are secreted into the small intestine where they continue the action of hydrolysis. Finally, the end products of digestion are formed. The juices of the small intestine are in an alkaline medium. Secretions from three kinds of glands function in the small intestine:

1. *Liver.* Bile, secreted by the liver, and stored in the gall bladder, has no digestive enzymes. However, it aids in the digestion of fats by emulsifying them. *Emulsification* is the breaking up of large globules of fats or oils into tiny globules, which present more surface for the action of lipase.

Bile is strongly alkaline. It neutralizes the acid of the stomach, and provides the slightly alkaline medium required by the enzymes secreted into the small intestine.

2. *Pancreas.* Pancreatic juice contains *proteases, amylases,* and *lipases.* There are several kinds of proteases, each of which is very specific in its action. For example, one enzyme acts on the bonds between the amino acids aspartic acid and arginine, and another enzyme on the bonds between the amino acids valine and tyrosine. In addition to proteases, pancreatic juice contains amylase, which continues the hydrolysis of starch to maltose, and lipase, which hydrolyzes fats to fatty acids and glycerine.

3. *Intestinal glands.* These tiny glands in the wall of the small intestine secrete the following enzymes: protease, which completes the breakdown of proteins; amylases, which complete the breakdown of carbohydrates; lipase, which completes the breakdown of fats.

The hormone *secretin* is produced by cells lining the small intestine. It stimulates the flow of pancreatic juice, and affects to a lesser degree the flow of bile and intestinal juice. (The copious flow of pancreatic juice originates with the introduction of acid food from the stomach into the small intestine.)

Absorption in the Small Intestine

In the small intestine, the end products of digestion are absorbed into the blood mainly by diffusion. An enormous number of tiny projections called *villi* (sing. − *villus*) absorb these end products of digestion. Villi cover the lining of the small intestine and are so numerous that they look like a fuzz. As shown in Fig. 5–7, each villus has an outside lining of epithelial cells through which absorption occurs. Inside are a network of blood capillaries and a projection of the lymph system called a *lacteal.* Amino acids and simple sugars pass into the capillaries. Fatty acids and glycerol enter the lacteals where fats are again formed. The suspension of fat droplets in the lacteals gives them a milky appearance. The lacteals are part of the lymph system. The largest lymph vessel, the thoracic duct, empties into a large vein and adds the absorbed fats to the circulating blood.

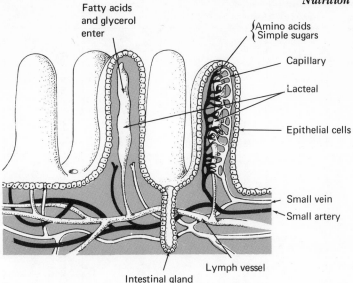

Fatty acids
and glycerol
enter

Amino acids
Simple sugars

Capillary

Lacteal

Epithelial cells

Small vein

Small artery

Lymph vessel

Intestinal gland

Fig. 5–7. Diagram of villus and portion of lymph system. (At left, capillaries cut away from lacteal for ease of study.)

The small intestine is an excellent example of adaptation of structure for function:

1. Presence of muscles permits contractions which mix the liquids and bring them in intimate contact with the villi.
2. Great length (about 22 feet) provides greater surface for absorption.
3. Folds in the wall provide greater surface for absorption.
4. Projections in the form of villi provide greater surface for absorption. The small intestine of man has a surface area about ten times greater than his skin surface.
5. Thin membranes of the epithelial cells, the capillaries, and the lacteal permit absorption of digested food.

The Large Intestine

No digestion of food occurs in the large intestine. The functions of the large intestine are:

1. *Absorption of water.* During the process of digestion, water has been added to the alimentary canal at each step. The absorption of water from the alimentary canal back into the bloodstream is an important mechanism for the conservation of water by land animals.

2. *Bacterial action.* Bacteria which normally inhabit the large intestine produce important vitamins which are used by the body. When a patient is given strong doses of antibiotics, the bacteria of the large intestine are destroyed. The patient may then have symptoms of vitamin deficiencies. Bacteria in the large intestine also break down the small quantities of incompletely digested proteins which reach the large intestine, liberating sulfur-containing gases.

If the wastes pass too rapidly through the large intestine, insufficient water is absorbed and diarrhea results. Excessively slow passage of wastes causes too much absorption of water, resulting in constipation.

SUMMARY OF HUMAN DIGESTION

Organ	Glands	Juice	Enzymes	Action	Miscellaneous
Mouth	Salivary	Saliva	Amylase (ptyalin)	Starch → maltose	Mechanical as well as chemical digestion Mucus flow, started here, continues throughout alimentary canal
Esophagus	Mucous	Mucus	None	Lubrication	Peristalsis begins
Stomach	Gastric	Gastric juice	Protease (pepsin)	Proteins → peptones and proteoses	Gastrin stimulates gastric glands HCl provides acidity and kills germs Storage of food
Small Intestine	1. Liver	Bile	None	Emulsifies fats	Neutralizes stomach acid
	2. Pancreas	Pancreatic juice	Proteases Amylase Lipase	Proteins → peptones and proteoses, and amino acids Starch → maltose Fats → fatty acids and glycerol	Secretin stimulates flow of pancreatic juice
	3. Intestinal glands	Intestinal juice	Protease Amylases Lipase	Peptones and proteoses → amino acids Disaccharides → simple sugars Fats → fatty acids and glycerol	Absorption of end products occurs in small intestine Villi facilitate absorption
Large Intestine	Mucous	Mucus	None	Lubrication	Absorption of water Bacterial action Defecation

The feces, or wastes, are stored in the lower part of the large intestine (the rectum) and eliminated (defecated) through the anus. This opening is controlled by both involuntary and voluntary muscles. Bacteria constitute from 10% to 50% of the bulk of the feces. The remainder is undigested food material, mainly the cellulose of plant cell walls. The remains of bile pigments, minerals, mucus, and epithelial cells are also present.

COMPLETION QUESTIONS

[A] 1. The usable portions of foods are known as . . . .

2. The energy value of foods is measured in units called . . . .

3. If starch is present, Lugol's solution will turn . . . .

4. A heterotrophic protist that digests dead organic matter outside its body is known as a (an) . . . .

5. Chemically, the process of digestion consists of . . . .

6. Enzymes which hydrolyze proteins are called . . . .

7. The structure in *Amoeba* that corresponds in function to the stomach in higher animals is the . . . .

8. *Paramecium* ingests food by the action of . . . .

9. *Hydra* egests food through the . . . .

10. The grasshopper stores food in the thin-walled . . . .

(B) 11. The fluid which lubricates the alimentary canal of man is . . . .

12. The digestion of starch begins in the . . . .

13. The series of wavelike contractions of the alimentary canal is called . . . .

14. The first organ in the alimentary canal of man in which proteases react chemically with food is the . . . .

15. The main functions of the stomach of man are digestion and . . . .

16. In suckling calves, milk is curdled by . . . .

17. The hormone which starts the flow of gastric juice is . . . .

18. The opening from the stomach is controlled by a type of muscle known as a (an) . . . .

19. Bile aids in the digestion of fats by the process of . . . .

20. Two kinds of glands which secrete proteases, amylases, and lipases are the intestinal glands and the . . . .

21. The part of the alimentary canal in which most of the digested food is absorbed is the . . . .

22. Special structures for absorption in the small intestine are the . . . .

23. Amino acids pass into the . . . . of the villi.

24. Large amounts of water are absorbed from the . . . .

25. The undigested food wastes of man are called the . . . .

MULTIPLE-CHOICE QUESTIONS

A 1. When a sucrose solution was boiled with Benedict's solution, a negative result was obtained. However, when a sucrose solution was first boiled in the presence of a little hydrochloric acid and then tested by boiling with Benedict's solution, a positive result was obtained. The explanation is: (1) The Benedict's solution test works only at a low pH. (2) Repeated boiling is needed to carry out the test with Benedict's solution. (3) Hydrolysis occurred when the sucrose was boiled with dilute acid. (4) Dehydration synthesis occurred when the sucrose was boiled with dilute acid.

2. Which organism is a heterotroph? (1) *Chlorella,* (2) bean plant, (3) moss, (4) grasshopper.

3. A carbohydrate that does not need to be digested is (1) starch, (2) cellulose, (3) glucose, (4) amino acid.

4. In *Hydra,* the digestion of food is (1) intracellular, (2) extracellular, (3) not necessary, (4) intracellular and extracellular.

5. The tube-within-a-tube body plan is characteristic of (1) flatworm and frog, (2) grasshopper and earthworm, (3) tapeworm and sponge, (4) *Hydra* and sea anemone.

B 6. The digestion of proteins in the stomach by pepsin requires (1) a low pH, (2) a high pH, (3) the presence of bile, (4) the presence of rennin.

7. A human whose gallbladder is removed may at first have difficulty in (1) carbohydrate digestion, (2) fat emulsification, (3) protein oxidation, (4) starch assimilation.

8. Enzymes acting on starch are found in (1) gastric juice, (2) secretin, (3) pancreatic juice, (4) bile.

9. Which is secreted by glands in the walls of the digestive tract? (1) intestinal juice, (2) bile, (3) ptyalin, (4) pancreatic juice.

10. Digested fats are absorbed through structure called (1) lacteals, (2) capillaries, (3) root hairs, (4) cilia.

CHAPTER TEST

1. In order to be utilized by cells, carbohydrates can be in the form of (1) gelatin, (2) glucose, (3) glycerin, (4) glycogen.

2. The chemical process which occurs during the digestion of food is (1) hydrolysis, (2) dehydration synthesis, (3) hydration, (4) dehydration.

3. The end products of protein digestion consist of (1) glucose, (2) fatty acids, (3) glycerol, (4) amino acids.

4. Which process prepares nutrients for transportation through the human body? (1) digestion, (2) excretion, (3) circulation, (4) assimilation.

5. An end product of fat digestion is (1) glucose, (2) starch, (3) amino acids, (4) fatty acids.

6. Food is moved along the alimentary canal by the process of (1) digestion, (2) osmosis, (3) peristalsis, (4) diffusion.

7. The result of testing an unknown solution with Benedict's solution was a blue color. Which is the most reasonable conclusion from this evidence? (1) There was no glucose present. (2) There was protein present. (3) The unknown solution was starch. (4) The unknown solution contained fatty acids.

8. The digestive system consists of (1) the alimentary canal, (2) the glands which send fluids into the canal, (3) the organs through which food passes, (4) all of the above.

9. Which secretion does *not* contain enzymes? (1) bile, (2) saliva, (3) gastric juice, (4) pancreatic juice.

10. The formation of a brick-red color when a substance is boiled with Benedict's solution indicates the presence of (1) an acid, (2) monosaccharide, (3) sucrose, (4) disaccharide.

11. Heavy muscular walls for grinding food are present in the earthworm's (1) crop, (2) esophagus, (3) mouth, (4) gizzard.

12. Bile is used to (1) digest fat, (2) emulsify fat, (3) digest glycerol, (4) emulsify glycerol.

13. A nutrient that undergoes partial hydrolysis during its stay in the stomach of man is (1) starch, (2) glucose, (3) protein, (4) fat.

14. The alkaline medium required for the action of enzymes in the small intestine is provided by (1) pancreatic juice, (2) intestinal juice, (3) ptyalin, (4) bile.

15. Which structures in the small intestine of humans serve to increase the area for absorption? (1) intestinal glands, (2) villi, (3) pseudopods, (4) cilia.

In questions 16 to 25, select the letter of the item, chosen from the list below, which best matches the statement.

A. Protozoa   C. Earthworm, grasshopper, man
B. *Hydra*   D. All of the above

16. Intracellular and extracellular digestion.
17. Cilia are used for the capture of food.
18. Digestion is mainly intracellular.
19. Tentacles are used in the capture of food.
20. Two-way digestive tract.
21. Digestion occurs in food vacuoles.
22. One-way digestive tract.
23. Food is hydrolyzed.
24. Saclike digestive system.
25. Eliminates undigested food.

CHAPTER 6 / Transport

Transport is the intake and distribution of materials throughout an organism. Transport may occur across membranes of cells, within the cell, and between parts of multicellular animals. Among the materials carried are water, oxygen, digested foods, the wastes of metabolism, and hormones.

Solubility is important for transport because insoluble materials do not circulate readily. The usable materials found in the blood are soluble materials (for example, glucose, amino acids, minerals, and vitamins). Since glucose is a soluble nutrient, a patient can receive it directly into the bloodstream intravenously, and no digestion will be required. In fact, digestion is really a way of changing insoluble nutrients into soluble form for absorption and circulation.

A | MEANS OF TRANSPORT

If a soluble material such as sugar is placed in water, it will dissolve. This occurs by a process known as diffusion. Let us see what this term means.

Diffusion

Diffusion can be observed in all liquids and gases. This is the process by which any gas will gradually spread out into another gas with which it is in contact. Many liquids will interpenetrate each other also, and any material dissolved in a liquid tends to spread throughout the liquid as well. This phenomenon is explained on the basis of the theory that liquids and gases are composed of separate molecules moving at random in all directions. If there is a high concentration of one kind of molecule in a particular place, the random motions cause the molecules to spread apart from one another until they are uniformly distributed in whatever space is available to them.

A small amount of ammonia gas released in the front of a classroom will soon be distributed throughout the room. This indicates that the molecules of gas are gradually spreading among the air molecules. The diffusion of molecules in a liquid is seen when a crystal of blue copper sulfate is placed at the bottom of a tall cylinder of water. The crystal dissolves to form a deep blue layer at the bottom of the cylinder. After several weeks, the blue color has spread to the top of the cylinder of water. The molecules of copper sulfate have diffused throughout the cylinder of water. Diffusion is an important means of transport in living things.

Diffusion through a membrane. Molecules of ammonia gas may diffuse through a membrane as shown in Fig. 6–1. At the beginning of the demonstration, the test tube contains a solution of colorless phenolphthalein. Phenolphthalein is an indicator which is colorless when neutral or acid, and pink

114

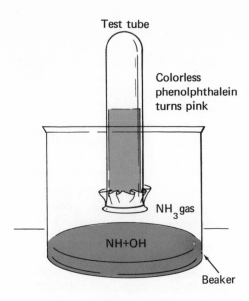

Test tube

Colorless
phenolphthalein
turns pink

NH₃ gas

NH+OH

Beaker

Fig. 6–1. Diffusion of a gas through a
membrane.

when basic. The ammonium hydroxide
in the beaker liberates ammonia gas:

$$NH_4OH \longrightarrow NH_3 + H_2O$$
ammonium hydroxide   ammonia gas   water

Molecules of NH_3 (ammonia) diffuse
through the membrane and combine
with water in the test tube to form

the base NH_4OH, turning the phenolphthalein pink.

The diffusion of liquids through a
membrane is illustrated in Fig. 6–2.
Note that in setups A and B the starch
suspension and iodine solution are in
different containers. When iodine and
starch unite, a blue-black color is produced. In apparatus A, the bottom liquid (in the beaker) turns blue-black.
In apparatus B, however, the top portion (in the test tube) turns blue-black.
This indicates that the molecules of
the iodine solution pass through the
membrane, regardless of whether they
must travel downward or upward. The
starch, composed of undissolved particles and large molecules, does not pass
through the membrane.

A theoretical explanation of diffusion is diagrammatically shown in Fig.
6–3. The circles represent molecules of
water and the x's represent molecules of
glucose. The membrane that separates
the two compartments of the container
permits passage of both the water
molecules and the glucose molecules.

Side A is a dilute solution of glucose;
on this side there is a relatively *high*

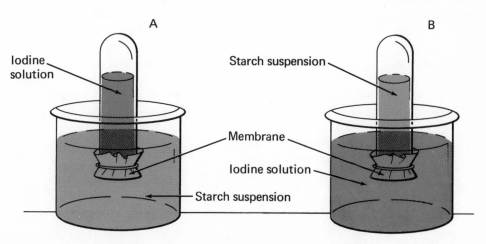

A   B

Iodine
solution

Starch suspension

Membrane

Iodine solution

Starch suspension

Fig. 6–2. Diffusion of a liquid through a membrane.

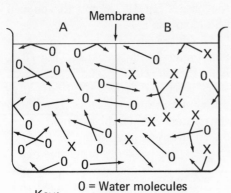

Key: 0 = Water molecules
      X = Glucose molecules

Fig. 6–3. Theoretical explanation
of diffusion.

concentration of water molecules. Side B has a more concentrated solution of glucose; here there is a *low concentration of water molecules.* Consider the movement of the water molecules. As they bounce about at random, water molecules will pass through the membrane in both directions. However, there are more water molecules in A than in B; therefore, more water molecules will pass from A to B than the other way. This continues until the concentration is the same on both sides, and there will then be equal numbers of molecules moving in both directions.

Now consider the movement of glucose molecules. At the start, more glucose molecules pass through the membrane to side A than to side B. Finally, the concentration of glucose molecules on each side of the membrane is equal. In any process of diffusion, *there is a net movement of a substance from a region of higher concentration of its molecules to a region of lower concentration of its molecules.*

Osmosis. Diffusion is a general term for migration of materials as a result of random molecular motion. *Osmosis* is a special kind of diffusion which re-

fers to the passage of *water molecules* through a membrane.

Often substances are present which do not diffuse through the membrane. An example is provided by the large protein molecules in protoplasm. These molecules may be too large to pass through the "openings," or they may be stopped by other factors. When this happens, the direction of diffusion is determined only by the concentration of the molecules diffusing through the membrane. For example, consider a cell in the leaf of *Anacharis,* a freshwater plant.

If *Anacharis* is placed in distilled water (which lacks any salts), water passes from a high concentration of water molecules (outside the cell) to a low concentration of water molecules (in the protoplasm). This causes the cell to swell and then burst. On the other hand, if a leaf is placed in a strong salt solution, water molecules pass from a region of higher concentration of water (inside the cell) to a region of lesser concentration of water molecules (outside the cell). The cell loses water and shrinks.

Plasmolysis is the shrinkage of cell contents due to outward osmosis. Fig. 6–4 illustrates plasmolysis of a cell in an *Anacharis* leaf which has been placed in salt water.

The bursting of cells because of the *inward osmosis* of water is seen when a drop of blood is placed in distilled water. The water molecules are more concentrated outside the blood cells than inside these cells. By osmosis, water molecules pass into the red blood cells causing them to swell and then burst.

Membranes which permit some materials to pass and not others are *selectively permeable* or *semi-permeable.*

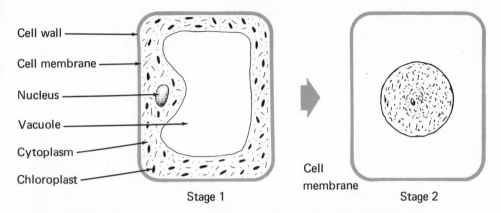

Cell wall

Cell membrane

Nucleus

Vacuole

Cytoplasm

Chloroplast

Stage 1

Cell membrane

Stage 2

Fig. 6–4. Plasmolysis of *Anacharis* cell when placed in strong salt solution. The cell contents shrink as water passes out by osmosis but the rigid cell wall maintains its shape. The cell membrane, which is normally adjacent to the cell wall, can now be identified.

Living cell membranes are selectively permeable. *Osmosis* is the diffusion of water through selectively permeable membranes. This property of selective permeability is lost when the cell dies. For example, when Congo Red dye is added to a culture of yeast cells, the dye does not penetrate the cell membranes and the yeast cells appear colorless. However, if the mixture is boiled, the living cell membranes lose their selective ability. Then the dye penetrates the cell membrane, and the yeast cells now appear red.

Active Transport

The diffusion of molecules from a region of high concentration to a region of low concentration does not require additional energy and is called *passive transport.* Diffusion may therefore be compared to a ball rolling down a hill as shown in Fig. 6–5.

The difference in concentration of molecules on each side of the membrane is the *concentration gradient* and may be compared to the slope of the hill in the figure. The ball rolls down the hill without the input of extra en-

ergy. However, rolling the ball up the hill requires energy in going against the gradient.

There are numerous situations in living cells in which molecules pass through a membrane in the direction opposite from what we would expect

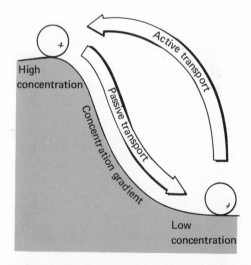

Fig. 6–5. Active and passive transport. It requires no added energy to roll a ball down a hill. Raising the ball up the hill, against the gradient, requires the input of energy.

on the basis of the principle of diffusion: the molecules pass from a region of *low* concentration of their molecules to a region of *high* concentration. This process is called active transport. *Active transport is the process by which molecules move against the concentration gradient.* Active transport requires the contribution of energy by the cell. Consequently, it is not surprising to find large accumulations of mitochondria ("the powerhouses of the cell") in cells which carry on much active transport. For example, many mitochondria are present in certain cells of the human kidney where materials seem to pass "the wrong way."

Certain seaweeds accumulate iodine in their cells in a concentration which is a million times greater than that of the surrounding ocean. The continued passing of iodine into these cells cannot be explained on the basis of simple diffusion. These cells force iodine molecules in a direction opposite to (against) the concentration gradient and use energy in the process.

Pinocytosis

The electron microscope reveals the ability of the cell to form tiny inpocketings (pinocytic vesicles). The "pockets" contain water and water-borne substances. These "pockets" break off inside the cell to form vacuoles. Large molecules, such as proteins, which cannot diffuse through the cell membrane, may be admitted to the cell in this way. *Pinocytosis* is the process whereby large molecules are admitted to the cell by the formation of pockets which break off to form vacuoles. (See Fig. 6–6.) Like active transport, pinocytosis requires cellular energy.

Dissolved solids (such as glucose), gases, ions, and water molecules diffuse through the cell membrane. Large molecules, such as proteins, fats, and starch, cannot diffuse into or out of cells. However, size does not appear to be the determining factor for the passage of molecules in all cases.

Phagocytosis

When portions of the cell engulf prey, large particles, or chunks of matter by flowing around them and enclosing them in a vacuole, the process is called *phagocytosis*. This is the method by which *Amoeba* ingests its food and by which certain cells in the inner cell layer of *Hydra* take in food fragments. White blood cells which engulf bacteria are known as *phagocytes*.

From the viewpoint of the size of the particles taken into the cell, these processes may be compared as follows:

diffusion — usually small molecules, such as oxygen, carbon dioxide, mineral salts, glucose, and amino acids.

pinocytosis — large molecules such as proteins, or very small particles.

phagocytosis — undissolved large particles, chunks of matter, or living microorganisms.

Particles taken into the vacuoles formed by pinocytosis or phagocytosis are broken down into small molecules which can diffuse through the membrane which surrounds the vacuole. In this process of intracellular digestion, lysosomes fuse with the vacuole and digestive enzymes from the lysosomes carry on hydrolysis in the combined structure.

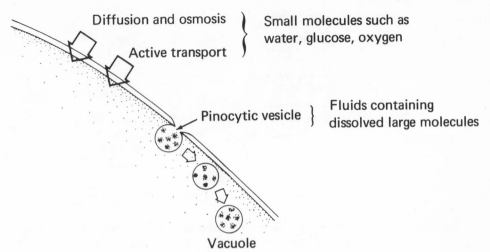

Fig. 6–6. Methods of transport across the cell membrane. Pinocytic vesicles occur as outpocketings or inpocketings.

PATTERNS OF CIRCULATION

Absorption is a general term for the passage of materials through membranes. Once materials have entered a cell, they are transported to all regions of the cell. *Circulation* is the transport of materials within cells or between parts of a many-celled organism.

Diffusion (molecular motion) accounts for some of the spread of materials within the cell. However, this is a relatively slow process and does not account completely for intracellular circulation. Another type of circulation can be observed when the leaf from the tip of an *Anacharis* plant which is actively growing in light is placed under the microscope. The chloroplasts are seen to follow a circular motion around the periphery of the cell. As previously noted, this streaming movement of the cytoplasm is called *cyclosis.*

Other examples of cyclosis are the streaming in *Amoeba*, the route taken by the food vacuoles of *Paramecium*,

and the movement of cytoplasm in the bread mold and slime mold. Cyclosis helps to distribute materials throughout the cell. In addition, the branched network of tubes called the endoplasmic reticulum probably serves as a means for transporting materials within the cell. Thus, circulation within the cell is by diffusion, cyclosis, and the use of the endoplasmic reticulum.

Circulation in Protozoa

In single-celled organisms, oxygen enters the cell by diffusion and wastes leave in the same way. No part of the cell is so far from the surface that it cannot receive sufficient oxygen and give up its wastes to the surface by means of ordinary diffusion and cyclosis. Fresh-water protists like *Amoeba* and *Paramecium* are continually taking in water by osmosis, and this must be pumped out by the contraction of contractile vacuoles.

Circulation in *Hydra*

Hydra is built like a hollow sac

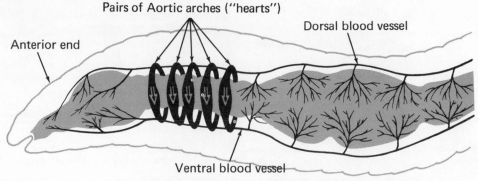

Pairs of Aortic arches ("hearts")

Dorsal blood vessel

Anterior end

Ventral blood vessel

Fig. 6–7. Circulatory system of the earthworm.

whose wall is composed of two layers of cells, the ectoderm and the endoderm. Most cells are in contact with water — either with the outside water or with the water in the digestive sac. This facilitates exchange of materials.

Food which is taken in by the endoderm cells that line the digestive sac is circulated within these cells by diffusion and cyclosis. The endoderm cells pass soluble food substances to the ectoderm cells by diffusion through the cell membranes. In *Hydra*, there is no special transport system for the distribution of materials.

Circulation in the Earthworm

The single factor that most sharply limits the size of organisms is their transport system. An *Amoeba* cannot grow to be a quarter of an inch wide, for its center would then be so far from the surface that diffusion and cyclosis could not carry sufficient oxygen to the center and remove the wastes. Animals larger than *Hydra* must all have some sort of circulatory system to carry oxygen to the interior and to carry wastes to the external environment.

The earthworm has a special circulatory system for the transport of materials. It is a "closed" circulatory system, which means that the blood is trans-

ported within a closed system of tubes. In the "open" circulatory system, by contrast, the blood leaves the tubes and passes into large open spaces. Grasshoppers have an open circulatory system.

Fig. 6–7 shows the main blood vessels in the closed circulatory system of the earthworm. These are the *dorsal* (upper) blood vessel, the *ventral* (lower) blood vessel, and five pairs of *aortic arches*. The dorsal blood vessel carries blood forward (anteriorly). The ventral blood vessel carries blood backward (posteriorly). The aortic arches carry blood from the dorsal to the ventral vessel.

The earthworm's blood is pumped by contractions of the larger blood vessels and the five pairs of aortic arches. These arches are the "hearts" of the earthworm. The smallest branches of the blood vessels are the thin-walled *capillaries*, where the exchange of materials between the blood and the cells takes place.

Digested foods are picked up by the blood capillaries in the intestine, and are then delivered to all the cells of the body. Oxygen is taken into the blood through the moist outer skin, and carbon dioxide is given off at the skin. Nitrogenous wastes are transported by the blood to special excretory structures.

The earthworm's blood is red because of a hemoglobin-like pigment, which is dissolved in the blood. The earthworm's blood does not have blood cells as does the blood of man.

Few cells of the annelid worms are in direct contact with the external environment. A closed circulatory system has evolved in these organisms which indirectly brings the materials of the external environment to the cells.

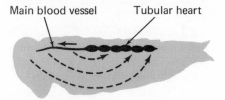

Main blood vessel   Tubular heart

Fig. 6–8. Circulatory system of the grasshopper.

Circulation in the Grasshopper

The grasshopper has an open type of circulatory system. In an open circulatory system, the blood is not always enclosed in tubes. During its circuit, blood passes out of the tubes into spaces (*sinuses*) between the tissues. The colorless blood of the grasshopper is pumped by a heart which is little more than a beating tube (Fig. 6–8). One blood vessel carries blood toward the head. It then passes through the spaces between the tissues and returns to the heart. As the blood bathes the tissues, it gives up nutrients and takes up the products of metabolism.

The circulatory system of the grasshopper does not transport oxygen and carbon dioxide. A separate system of tracheal tubes carries these gases.

Circulation in Man

In the earthworm and grasshopper, the blood in the dorsal vessel flows to the anterior. In vertebrates, however, blood flows through the dorsal vessel toward the posterior. These two plans of circulation are compared in Fig. 6–9.

Man has a closed circulatory system and a blood flow conforming to the overall pattern of circulation in vertebrates. In man, blood is pumped by a four-chambered heart through a system of arteries, capillaries, and veins. The transport systems of earthworm, grasshopper, and man are compared in the table on the next page.

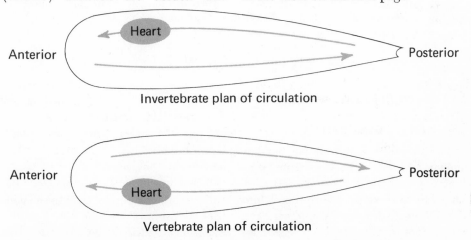

Invertebrate plan of circulation

Vertebrate plan of circulation

Fig. 6–9. Comparison of overall circulatory pattern in invertebrates and vertebrates.

CIRCULATORY SYSTEMS OF EARTHWORM, GRASSHOPPER, AND MAN

	Earthworm	Grasshopper	Man
Open or closed system	Closed	Open	Closed
Pumping organ	5 aortic arches and main vessels	Tubular heart	4-chambered heart
Direction of flow of dorsal vessel	Anterior	Anterior	Posterior
Does circulatory system carry O_2 and CO_2?	Yes	No	Yes

Blood circulates more rapidly in a closed than in an open system.

REASONING EXERCISES

1. What is transport? Why is solubility important to transport? How is a large food particle changed into soluble form?
2. What is diffusion? Show that osmosis is a form of diffusion. Why must a salt solution containing an *Anacharis* cell be *concentrated* to produce outward osmosis in the cell?
3. Distinguish between passive transport and active transport. What is the role of the living cell membrane in transport?
4. Distinguish between pinocytosis and phagocytosis.
5. What is absorption? What is circulation? How does circulation occur within the cell?
6. By what type of transport does *Paramecium* expel excess water? How does it accumulate this excess water? If it did not constantly expel it, what would happen to the organism?
7. Distinguish between the open and closed circulatory systems. Which system circulates blood more rapidly?
8. How do nutrients in the circulatory system of an earthworm get into one of its cells? How do nutrients in the circulatory system of a grasshopper get into one of its cells?
9. How do the circulatory systems of the earthworm and grasshopper differ in respect to the transport of oxygen and carbon dioxide?
10. Compare the overall circulatory pattern in invertebrates and vertebrates. What are some differences in heart structure found among the vertebrates? (Consider the fish, the frog, the reptile, and man.)

(B) CIRCULATION IN MAN

The Heart and Blood Vessels

The structure of the human heart is shown in Fig. 6–10. A partition separates the right side from the left side. On each side, veins empty blood into the thin-walled upper chambers (the *atria*). The atrium on each side contracts, forcing blood through valves leading into the thick-walled lower chambers (the *ventricles*). Contraction of the ventricles forces the blood through other valves leading into arteries (on each side).

When the ventricles contract, the valves leading from the atria to the ventricles close and prevent return of the blood to the atria. Then, after the contraction of the ventricles, the valves

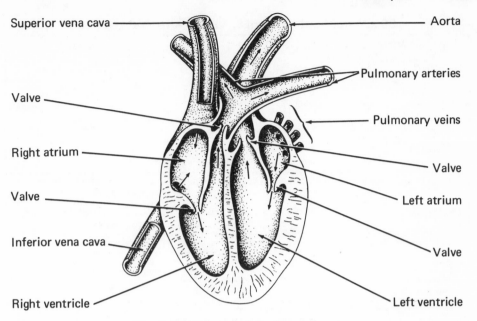

Superior vena cava

Aorta

Valve

Pulmonary arteries

Pulmonary veins

Right atrium

Valve

Valve

Left atrium

Inferior vena cava

Valve

Right ventricle

Left ventricle

Fig. 6–10. Diagram of the human heart.

leading into the two arteries close and prevent return of the blood from the arteries to the heart. The closing of these two sets of valves is heard through a stethoscope as a "lub-dub," the first and softer sound being produced when the valves leading out of the atria close. Unusual heart sounds can indicate defects in the heart, such as the constriction of the passageway or the improper closing of a valve.

In laboratory glassware, heart muscle tissue contracts slowly and rhythmically by itself, without outside stimulation. In the heart, the speed and extent of the contraction is regulated by the *pacemaker,* a region which acts as both muscle and nerve tissue. The pacemaker is located near the top of the right atrium. Branches from the pacemaker start contractions of muscle fibers in the two atria and then in the two ventricles. The pacemaker is regulated by two nerves originating in the medulla (lower part) of the brain. The

accelerator nerve stimulates the pacemaker and the *vagus nerve* slows down the pacemaker.

Arteries carry blood from the heart, and *veins* carry blood to the heart. After an artery leaves the heart, it branches repeatedly into ever smaller arteries, until the smallest branches are microscopic. From these smallest arteries branch the still smaller capillaries, which have walls only one cell thick and are wide enough to allow red cells to pass through. It is while flowing through the capillaries that blood takes on materials from body tissues and gives up materials to body tissues. It is at the capillaries that materials enter and leave the bloodstream. The capillaries join together to form tiny veins, which connect with larger veins until the very largest veins empty the blood into the atria of the heart. A comparison of the three types of blood vessels is given in the table that follows:

COMPARISON OF ARTERIES, VEINS, AND CAPILLARIES

Artery	Vein	Capillary
		(not drawn to scale)
Carries blood from the heart	Carries blood to the heart	Connects small arteries and small veins
Thick wall	Thin wall	Wall is a single layer of epithelial cells
Much muscle present	Little muscle	
Elastic	Not elastic	
Muscles in wall controlled by nerve endings	No nerve endings	
No valves (except at heart).	Valves present	

Flow of Blood in the Arteries

The muscles in the walls of the smallest arteries (*arterioles*) contract or relax in response to nerve and chemical stimulation. The constriction of an arteriole reduces the supply of blood to the capillaries it feeds, and in this way the supply of blood to a tissue is controlled.

When the body is too warm, the supply of blood to the capillaries under the skin is increased. This permits loss of body heat to the outside. The increased supply of blood to the skin causes the individual to appear flushed. When the outside temperature is too low, the supply of blood to the capillaries near the skin is reduced. This reduces the heat loss. The individual may appear "blue with cold." On the other hand, red ears in cold weather indicate that blood has been diverted to the surface of the ear to prevent freezing of the cells at the surface.

The elasticity of artery walls permits them to accept the increased flow of blood resulting from the contraction of a ventricle. If the arteries were not elastic, there would be greater strain on the heart as it pumped blood into a rigid container. Hardening of the arteries (arteriosclerosis) leads to heart strain. The *pulse* is a wave of alternate stretchings and contractions which proceed from the aorta along all arteries and arterioles. The frequency of the pulse rate is the same as the heart beat. Most arteries are deep within the body, but the pulse rate may be detected at a few places, such as the wrist, where an artery is close to the surface.

Flow of Blood in the Veins

Contractions of the left ventricle are not sufficient to force the blood through all the arteries, through the whole bed of capillaries, and through the veins. The flow of blood in the veins is helped in two ways:

1. During body movements, skeletal muscles press upon veins and tend to force blood in both directions. How-

ever, the valves in the veins open to permit blood to flow only toward the heart and blood is blocked from flowing in the opposite direction. For example, when a person sits motionless for some time, the blood may collect in the veins of his feet. Moving or stamping the feet starts the blood moving again, accompanied by a tingling feeling.

2. During breathing, air pressure is reduced in the chest cavity at each inhalation. This lowered air pressure helps venous blood to flow toward the heart.

The Path of Circulation

A warm-blooded animal must supply large amounts of oxygen to its tissues. For this reason, all birds and mammals have a special arrangement for pumping blood to the lungs, where it can receive oxygen. In mammals, the right side of the heart pumps blood to the lungs, and the left side supplies the rest of the body.

Circulation in the human being is shown in Fig. 6–11. Note that the wall of the right ventricle is thinner than that of the left ventricle, for the right ventricle does not develop as much pressure as the left. The human circulatory system has three main pathways:

Pulmonary circulation. Pulmonary means "of the lungs." The large *pulmonary artery* leaves the right ventricle, and then divides into two arteries that carry the blood to the lungs. Here the arteries divide to form capillaries. Blood from the right side of the heart has little oxygen, and is dark red in color. When it acquires oxygen in the capillaries of the lungs, it turns bright red. It enters the *pulmonary veins,* which return it to the left atrium. Thus, the blood is dark red (deoxygenated)

in the right side of the heart and bright red (oxygenated) in the left side. Division of the heart of man into right and left sides helps to separate oxygenated and deoxygenated blood.

Systemic circulation. The systemic circulation is the circulation of blood through the major systems of the body. The left atrium contracts, forcing blood through a valve into the left ventricle. The left ventricle contracts, forcing blood into the *aorta,* the largest artery in the body. The blood thus passes from the left side of the heart through the aorta with its branches to all parts of the body except the lungs.

The arteries subdivide to form capillaries in the various organs and tissues. Exchange of materials occurs between the blood capillaries and the tissues, including loss of oxygen from the blood. The capillaries unite to form veins, which become progressively larger. The *superior vena cava* from the head region and the *inferior vena cava* from the trunk region empty into the right atrium.

The muscle cells of the heart itself are supplied by the *coronary arteries,* which branch from the aorta.

Most cold-blooded animals have only one ventricle in the heart, so that the oxygenated and deoxygenated blood mix to some degree in the ventricle. In most reptiles, there is an incomplete wall between the two sides. This provides sufficient separation for the needs of an animal that does not require as much oxygen as a warm-blooded animal. Occasionally, a human baby is born with the wall between the ventricles incomplete, and the child dies at an early age for lack of oxygen in his tissues, unless the wall is closed surgically. A baby with this condition is called a "blue baby."

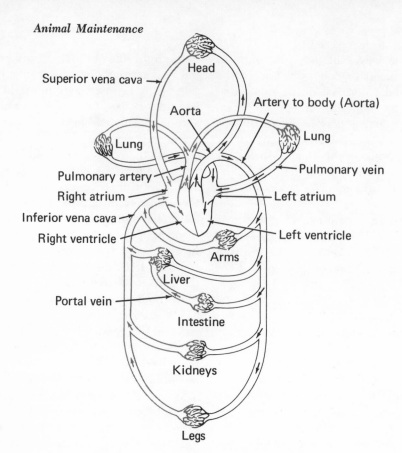

Fig. 6–11. Circulation in the human being.

Portal circulation. The portal circulation consists of veins which carry blood from the stomach, the small intestine, and the large intestine to the *liver* and from the liver to the inferior vena cava. The *hepatic portal vein* begins as veins usually do by the joining together of capillaries – in this case, the capillaries from the major organs of the digestive system. However, it ends in a unique fashion. Instead of emptying into a larger vein, it subdivides again to form a network of capillaries in the liver. All the capillaries of the liver feed into the *hepatic vein*, which empties into the inferior vena cava. Thus, the liver interrupts the course of the blood from the digestive organs to the inferior vena cava. Let us examine this situation further.

The pathway of the portal circulation is shown in Fig. 6–12. The blood leaving the small intestine has picked up large quantities of glucose in the capillaries of the villi. (The small intestine is where absorption of digested food occurs.) If this glucose passed into the general circulation, there would be a great increase of blood glucose shortly after meals. This would violate the principle of *homeostasis* which is that "the internal environment of the body must remain stable." Excess glucose is changed to glycogen as the blood passes through the liver. Glycogen is a form of insoluble polysaccharide which is stored in the liver (and also in muscle). When the chemical

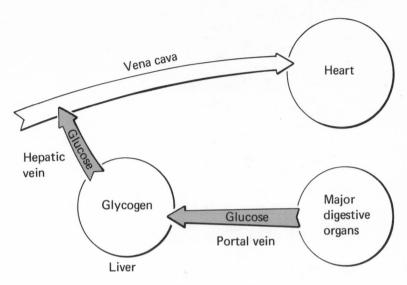

Fig. 6–12. Portal circulation.

energy of glycogen is needed by the body, the glycogen of the liver is changed back to soluble glucose and returned to the blood. Hormones called *insulin* and *adrenaline* control this homeostatic mechanism. (See Chapter 11.)

Harvey and the circulation of the blood. For 15 centuries before Harvey's time, biology was dominated by the views of Galen. The 2nd-century Greek physician said that blood ebbed through the veins to and from the heart. Blood supposedly passed between the two sides of the heart by invisible pores in the wall. Throughout the Middle Ages no one dared to challenge the views of "Authority."

William Harvey, an English physician of the 17th century, showed that veins carry blood to the heart and that the heart pumps the blood by arteries to the various organs. He therefore is credited with discovering the pathway of blood circulation. One major portion of this pathway, the role of capillaries, was beyond Harvey's powers of observation because the microscope had not

yet come into general use. Marcello Malpighi, an Italian microscopist, discovered the capillaries in 1661 (33 years after Harvey's report). An understanding of the basic plan of circulation was complete with this new discovery.

Harvey's reasoning was based on observation acquired through dissection and experiment. He discovered that the arrangement of valves in the veins and in the heart would permit blood to flow in one direction only. When he cut an artery, he noticed that blood spurted only from the end which is nearer the heart. Therefore, he concluded that arteries carry blood *from* the heart. When he cut a vein, blood flowed only from the end which was farther from the heart. Therefore, he reasoned veins carry blood *to* the heart.

Harvey's work illustrates (1) that reasoning based upon observation and experiment can challenge the views of "Authority"; (2) that some discoveries must await the development of needed instruments.

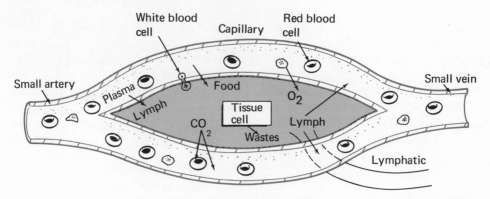

Fig. 6–13. Diagram of tissue cell surrounded by capillaries. Plasma and white blood cells pass out of the capillaries, as lymph forms. Lymph passes back into the capillaries or into microscopic lymph vessels.

Lymph and Its Circulation

The lymph network is a system of tubes carrying a colorless fluid called *lymph* through the body. As indicated in Fig. 6–13, lymph is mainly blood plasma which has filtered through the blood capillaries into the spaces surrounding cells. Lymph bathes all the cells. It is also called *tissue fluid* or *intercellular fluid*. Lymph is a watery fluid that differs from blood plasma in having about half the protein concentration. The composition of lymph varies depending upon materials added to it or taken away from it by the surrounding cells. Lymph acts as the middleman in the transport of materials between the blood and the cells. White blood cells which pass between the cells of the capillary walls are also present in lymph. The liquid in a blister is lymph.

Lymph vessels (lymphatics) are present in all parts of the body. The lacteals in the villi of the small intestine are part of the lymphatic system. The microscopic lymphatic capillaries are open at the end; they drain lymph from the tissue spaces and empty it into larger lymphatics. The flow of lymph is caused primarily by the contraction and relaxation of surrounding skeletal muscles.

Lymphatic vessels have valves which prevent backward flow. The flow of lymph from the region of the intestines is aided by the motion of villi. The small lymphatics unite many times to form larger vessels, all emptying into the *thoracic duct*, which is the largest lymph vessel in the body. The thoracic duct empties the lymph into a large vein on the left side of the body near the neck, returning it to the general circulation.

The functions of lymph are as follows:

1. Fluid which has diffused out from the capillaries is collected by the lymph vessels (lymphatics) and eventually returned to the blood system.

2. Lymph in the lacteals of the intestinal villi takes up fats and transports them to the blood system.

3. The lymphatic system aids in the destroying of microorganisms. Masses of cells called *lymph nodes* (or sometimes lymph glands) are present where smaller lymph vessels join to form larger ones. White blood cells are pro-

duced in these areas and chemical substances called *antibodies* are formed. Lymph nodes also filter out bacteria and cancer cells which may be circulating in the lymphatic system. Physicians detect some forms of infection by noting swollen lymph nodes. The tonsils and adenoids are lymph glands.

4. Lymph transports materials between the capillaries and the cells.

Blood

Blood is a tissue composed of cells in a liquid matrix. Plasma is the liquid part of the blood. Three types of solid components are suspended in the plasma: the *red blood cells* (corpuscles), the *white blood cells*, and the *platelets*. The general functions of plasma are as follows:

1. Serves as the liquid matrix of blood tissue.
2. Transports products of digestion: amino acids, simple sugars, fats, vitamins, salts, water.
3. Transports nitrogenous wastes (urea) and carbon dioxide (CO_2).
4. Transports hormones.
5. Helps to regulate body temperature.

Plasma is composed of 90 percent water, plus inorganic and organic substances. The blood proteins in plasma are of special importance because they maintain osmotic pressure of the blood. *Fibrinogen* is a protein which helps the blood to clot. *Gamma globulin* is a fraction (portion) of the blood proteins which includes the antibodies that control disease.

The *red blood cells* of man are disc-shaped structures which are concave on each side (biconcave discs). During their formation in bone marrow, they lose their nuclei. Red blood corpuscles contain the red pigment, *hemoglobin,* which gives blood its red color. Hemoglobin contains iron, and helps to transport oxygen and carbon dioxide. The red blood cells buffet about in the circulatory system, survive for about 120 days, and then break apart (usually in the liver). New red blood cells are constantly produced in large numbers in the bone marrow of the shaft regions of the long bones in our body.

The *white blood cells* are produced in the bone marrow of the shafts of long bones, in lymph nodes, and in the spleen. They increase in large numbers when an infection is in the body. As previously noted, white blood corpuscles escape from the capillaries by passing between the cells which line the capillary walls. White blood cells called *phagocytes* travel by ameboid motion toward bacteria and die after engulfing many bacteria. The process of engulfing large particles is called *phagocytosis. Pus* is a suspension in tissue fluid of the

RED BLOOD CELLS AND WHITE BLOOD CELLS

	Red Cells	White Cells
Function	Carry oxygen and CO_2	Engulf bacteria, produce antibodies
Number	5 million/mm³	7000-8000/mm³
Where produced	Bone marrow	Bone marrow, lymph nodes, spleen
Nucleus	No nucleus	Nucleus present

living or dead bacteria and white blood cells.

The *platelets* are living, small, oval cell fragments that usually lack a formed nucleus. When they break at the rough surface of a wound, they liberate an enzyme. This enzyme starts the complicated mechanism of blood clotting. When a blood clot shrinks and hardens, the straw-colored liquid which remains is *serum*, which is similar to plasma but lacks the protein fibrinogen.

REASONING EXERCISES

1. In the "lub-dub" sound of the heartbeat, the first sound is softer than the second. Why do you think this is so?
2. How are the walls of the human heart adapted for their functions?
3. How is the structure of arteries adapted for their functions? How is the structure of veins adapted for their functions?
4. What use does glycogen serve in the body?
5. What is lymph? Distinguish between lymph and plasma. Distinguish between serum and plasma.
6. What are the functions of blood proteins?
7. Distinguish between red blood cells, white blood cells, and platelets. What are the functions of each?

(c)   THE PROTECTIVE FUNCTIONS OF BLOOD

The circulatory system provides protection to the body from loss of blood and from disease. These protective functions include (1) blood clotting, (2) phagocytosis, and (3) immunological reactions.

The clotting of blood is an example of an enzyme-controlled reaction that protects the body and aids in maintaining homeostasis. The clotting of blood is outlined in Fig. 6–14.

When a blood vessel is cut, the ruptured platelets and damaged tissue cells in the area release substances that react together to produce the enzyme *thromboplastin*. Prothrombin, another enzyme present in the plasma, is in an inactive form. The thromboplastin from the ruptured platelets, with the aid of calcium ions, converts the prothrombin to active *thrombin*. This coagulates the *fibrinogen* (a protein present in the plasma) to form threads of *fibrin* and a network of fibrin threads is formed. This network traps red blood cells, shrinks, and forms a clot which stops the flow of blood from the cut vessel.

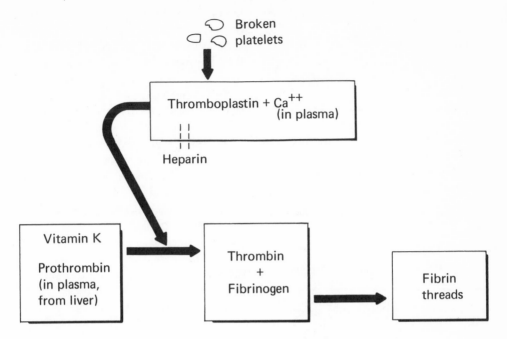

Fig. 6–14. Outline of blood clotting.

The liver has many functions relating to the activity of the blood: (1) It removes broken red cells, excreting the waste products in the bile; (2) it makes *heparin,* a material that interferes with blood clotting so that undesirable clots are not formed; and (3) it makes prothrombin. Vitamin K is needed for the manufacture of prothrombin, and if this vitamin is deficient in the diet, the blood loses its ability to clot readily.

Another protective function of the blood is phagocytosis: it is one of the body's important defenses against disease.

The body has another defense against microorganisms causing disease; blood forms *antibodies,* which are proteins produced in the host as a reaction to proteins from outside the organism. The foreign protein is an *antigen.* The antibody links to the antigen in a "lock and key" arrangement. This destroys or neutralizes the effectiveness of the antigen.

An antibody is very specific — it combines only with a particular antigen. Generally, the body produces antibodies in response to a foreign material in the blood, and the antibody reacts only to the particular antigen that caused the antibody to be produced. Substances which act as antigens include bacteria, bacterial toxins, pollen dust, and foods. Serum and cells from another organism can also act as antigens. The building of antibodies is known as the *immunological reaction.*

Hypersensitivity occurs when the body responds excessively in the antibody-antigen reaction. *Allergy* is a mild hypersensitivity to foreign materials. Allergic responses usually affect only particular regions of the body. They are the skin (rashes and hives), the respiratory tract (asthma), the upper respiratory membranes and eyes (hay fever).

Immunological reactions help to explain (1) immunity to disease, (2) blood types, (3) the Rh factor, and (4) rejection of grafts.

Immunity to Disease

An infectious disease in the body resembles a condition of warfare between attacking and defending forces as shown below:

Attack	Defense
1. Bacteria	Phagocytes
2. Bacteria and their toxins	Antibodies

The attacking forces are bacteria and other microorganisms. Damage to the body is caused not only by bacteria but also by their *toxins,* or poisons given off by bacteria during their metabolism. The body uses *phagocytes* (white blood cells) to defend itself against bacteria, and another bodily defense, *antibodies,* to fight the bacteria and their toxins. Some antibodies are listed below:

antitoxins — combine with toxins

agglutinins — cause clumping of large particles such as bacteria or blood cells

precipitins — cause soluble antigen molecules to become insoluble and form a precipitate

lysins   — break up (or dissolve) cells

opsonins   — ("appetizers") cause bacteria to be more readily engulfed by phagocytes

Immunity is the resistance of an organism to disease, and *susceptibility* is the organism's lack of immunity. Everyone is exposed repeatedly to mild infections during his lifetime. Every time this happens, the body responds by making antibodies to destroy the germs and immunity may be developed. That is why "children's diseases" do not affect adults. Populations completely free of a given disease develop no immunity to it. When measles was first introduced into Polynesia, it killed adults by the thousands.

Immunity may be classified as active immunity and passive immunity. In *active immunity,* the individual builds his own antibodies from having had the disease in a strong or mild form. In *passive immunity,* antibodies are given to the individual. Active and passive immunity are compared in the table on page 133.

Antibodies which the body builds against some killed microorganisms are effective against live microorganisms. For example, active immunization against typhoid fever consists of injecting killed typhoid fever bacteria. The body then is immune to the live typhoid fever bacteria. Similarly, the *Salk vaccine* against polio consists of viruses which have been killed by the addition of formaldehyde. The body then produces antibodies which give immunity against live polio virus. The *Sabin vaccine* for polio, consisting of weakened live virus, is administered orally.

Preparation of toxin and antitoxin. Toxin is prepared from bacteria grown in the laboratory in tissue culture or in eggs containing chick embryos. When the toxin is weakened by adding formaldehyde or by heating, the product is a *toxoid.* If dilute diphtheria toxin is injected into a horse, the horse pro-

Active Immunity	Passive Immunity
Acquired by having the disease or a mild form of it. Inject: 1. Weakened or killed bacteria or viruses. 2. Weakened toxin. (**Toxoid** is toxin weakened by a chemical such as formaldehyde; **toxin-antitoxin is** toxin weakened by the addition of antitoxin.)	Given by an inoculation. Inject: 1. Antibodies (for example, a serum containing antitoxin). 2. Gamma globulin from pooled blood.
Takes a relatively long time to acquire but lasts a long time.	Immunity is conferred rapidly but does not last long.
Administered usually before exposure to the disease, as a preventive.	Administered when the person has the disease, to help the body's defense.

duces antitoxins. As increasingly strong injections of toxin are given to the horse at set intervals, it then develops a high concentration of diphtheria antitoxin in its blood. A sample of blood taken from the horse is allowed to clot and the liquid which remains after the clot has formed is horse serum which contains diphtheria antitoxin. This serum may be administered to build passive immunity in a child who is sick with diphtheria. The antitoxin may also be combined with toxin to form toxin-antitoxin for developing active immunity.

Gamma globulin. Pooled blood is blood donated by a large number of healthy persons and mixed together. During their lifetimes, these persons may have recovered from a variety of diseases. Pooled blood, therefore, usually has a wide variety of antibodies. When the proteins are extracted from this blood, the fraction (portion) of these proteins which contains the antibodies is the *gamma globulin*. When a physician administers gamma globulin, the patient may get passive immunization against a disease.

Active immunization of infants is given by the injection of DPT "shots." This inoculation leads to active immunity against diphtheria, pertussis (whooping cough), and typhoid (and paratyphoid) fever.

The word *vaccine* originated with Edward Jenner, an English physician of the early 19th century. He used the cow (*vacca,* cow) to prepare the first safe immunization against smallpox. Jenner saw that milkmaids who were exposed to the blisters of cows which had cowpox became slightly ill with the mild disease of cowpox; however, they seemed to be immune to the dangerous smallpox. When Jenner introduced cowpox virus into people, they developed active immunity to smallpox.

Since this discovery was made before viruses were known to exist, it was not possible for Jenner to understand why he obtained these results. Today, we know that the viruses causing cowpox and smallpox are sufficiently alike that the antibodies produced against cowpox act effectively against both diseases. Accordingly, the smallpox virus was weakened in the body of the cow before it was transferred to the milkmaids.

Blood Groups

When the practice of blood transfusion was begun, it was soon found that the mixing of some bloods would result in death due to the formation of clumps. The mystery was solved in 1900 by Karl Landsteiner, a Viennese physician, who discovered that each human has one of the four major blood groups called *O, A, B,* or *AB.* The blood groups are based upon the presence or absence of antigens on the surface of red blood cells.

There are two antigens, called *A* and *B.* A person who has type *A* antigen is said to have blood which is type *A;* a person who has type *B* antigen has blood type *B;* a person whose red blood cells contain both type *A* antigen and type *B* antigen has blood type *AB.* A person who has neither antigen on his cells has blood type *O* (to indicate zero).

Present in the plasma may be antibodies which agglutinate (or clump) red cells containing these antigens. The antibody which agglutinates cells containing antigen *A* is called anti-*A;* the antibody for antigen *B* is called anti-*B.* An individual with type *A* antigen could not have antibody anti-*A* or he would clump his own red cells; however, for unknown reasons, he does have antibody anti-*B.* Antigens and antibodies of the four types of blood groups are summarized below.

Blood Type	Antigen on Cells	Antibody in Plasma
O	—	anti-A anti-B
A	A	anti-B
B	B	anti-A
AB	A,B	—

Blood typing. This is the laboratory procedure which determines an individual's blood group. Though blood typing is a simple procedure which can be performed as an exercise in school, only the results from certified laboratories should be used. The materials required for blood typing are two bottles of antibodies, one marked *Anti-A serum* which contains anti-*A* antibodies. The other bottle is marked *Anti-B serum* and contains anti-*B* antibodies. A drop of each serum is placed at opposite ends of a microscope slide as shown in Fig. 6–15 and a drop of the diluted blood of the "patient" is then added to each of these serums.

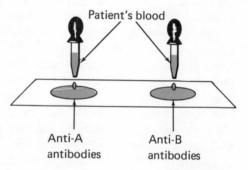

Fig. 6–15. Blood typing. A drop of the patient's blood is added to each kind of antibody.

The contents of each drop are stirred either by use of separate toothpicks or by tilting and rotating the slide, and the two drops are examined after a few minutes. The examiner looks for signs of clumping which appears as heavy dots of agglutinated cells. Clumping occurs when cells carrying an antigen are mixed with the corresponding antibody. The four possible kinds of observations are shown in Fig. 6–16.

Blood transfusions. When clumping occurs in the blood vessels of a person

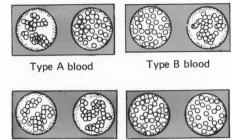

Type A blood   Type B blood

Type AB blood   Type O Blood

Fig. 6–16. Blood typing. On each slide, the anti-*A* antibodies were placed on the left side and anti-*B* antibodies were placed on the right side.

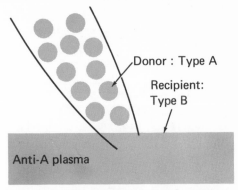

Fig. 6–17. Donor is type *A* and recipient is type *B*. The results could be fatal.

receiving a blood transfusion, it may block flow to a vital organ. In any blood transfusion, it is important that the *cells of the blood of the donor not be agglutinated by the plasma of the recipient.* (Remember the antigens are on the red blood cells and the antibodies are in the plasma.) The possibility that the antibodies in the plasma of the donor may agglutinate the cells of the recipient is not so dangerous because *the donor's plasma is greatly diluted* as it enters the recipient's blood stream. In practice, the laboratory cross-matches the two bloods to make sure that there is no possibility of agglutination even though the donor's plasma is diluted.

Consider the case where the donor is type *A* and the recipient is type *B* as in Fig. 6–17. In such a transfusion, the anti-*A* antibodies in the plasma of the recipient would agglutinate the type *A* red cells of the donor, which could be fatal to the person receiving the blood. Similarly, if the donor is type *B* and the recipient is type *A*, the donor's red cells agglutinate; here the anti-*B* antibodies in the plasma of

the recipient agglutinate the type *B* red cells of the donor.

Now consider the case of a type *O* donor and a type *B* recipient as shown in Fig. 6–18. Here, the donor has *no* antigen on his cells; therefore, no matter what antibodies are present in the recipient's plasma, the donor's red cells will not be agglutinated. Type *O* is called the *universal donor* because this type can give blood to any other type, usually with little or no ill effect. Type *O* does contain anti-*B* antibodies in the plasma which tend to agglutinate the red cells of the type *B* recipient, but the donor's plasma is too diluted to cause serious trouble. Remember, basically,

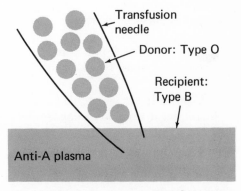

Fig. 6–18. Donor is type *O* and recipient is type *B*.

only the cells of the *donor* must not be agglutinated.

If the recipient is type *AB*, he can receive blood from any donor because the plasma of type *AB* has no antibodies. It will not clump the cells of any donor. Type *AB* is called the *universal recipient.* Persons of all blood types can receive blood from donors who have the same blood type as theirs. The antigens and antibodies involved in blood types are also known as *agglutinogens* and *agglutinins*, respectively.

Blood plasma. Plasma is used in place of whole blood for some types of emergency situations such as accidents and battle wounds. It restores the volume of circulating fluid, and keeps the blood pressure high enough to sustain circulation to the vital organs. Plasma is advantageous — it lacks blood cells; therefore, it has no agglutinogens and is not agglutinated. Plasma can be dried to a powder, sterilized, and stored indefinitely. It can be shipped wherever needed, and restored to its original state by adding sterile distilled water.

Rh Factor

The *Rh factor* is an antigen present in the red blood cells of about 85 percent of people. The *Rh* factor was discovered by experiments on Rhesus monkeys in 1940, and its name comes from the first two letters of Rhesus. *Rh positive* (*Rh+*) individuals have this antigen; people who lack this antigen are *Rh negative* (*Rh−*).

It is possible for a pregnant woman to be *Rh* negative and have an *Rh* positive fetus (embryo), provided that the father is *Rh* positive. The blood of the mother and the fetus have separate circulations, but there may be some mixing of the two bloods in the placenta. If the unborn child inherited the *Rh+* fac-

tor from the father, the *Rh* antigen in the baby's red blood cells acts as a foreign protein to the mother's blood. The mother's blood builds antibodies to destroy the red blood cells of the fetus. The mother does not build sufficient antibodies to cause much damage during the first pregnancy. By the second or third pregnancy, the mother may have built enough antibodies to destroy the fetus' red blood cells and kill the fetus. This results in a miscarriage. (The dead fetus is rejected from the mother.) The baby may be born alive if it has not had too much destruction of its red blood cells, and its life can be saved by replacing all or part of its blood with fresh, *Rh* positive blood.

When a person is to receive a blood transfusion, it is important to know the *Rh* factor as well as the *ABO* blood types of the recipient and of the prospective donor. If the recipient is *Rh* negative and the donor is *Rh* positive, the recipient will build antibodies that may destroy the red blood cells in future transfusions. If the recipient is an *Rh* negative woman, she may build antibodies that may harm her first *Rh* positive fetus.

The *ABO* types are distinct from the *Rh* factor. An individual who is blood type *A* and is *Rh* negative is described as A− (A negative). There are many other factors present in the blood, but they do not ordinarily have much significance for blood transfusions.

Rejection of Grafts

Successful transplants of tissues have been made between humans who are identical twins, and also between mice of the same strain. However, the transplanting of tissues and organs is usually not successful, unless the donor and recipient have very much the same heredity. When heredity is different, the transplanted tissue acts as an antigen. This arouses the immunity mechanism of the recipient which produces antibodies and causes the graft to slough off or be rejected.

Recent studies show that the thymus gland is involved in this immune response. The removal of the thymus gland in newly born mice greatly reduces the amount of gamma globulin in their plasma; as a result, the immune mechanism has been muted or lessened. Although these mice are susceptible to infection, they are able to accept grafts from *unrelated* strains of mice. Moreover, the grafts remain attached for long periods of time. Chemicals and irradiation are also used to mask or mute the immune mechanism.

The immune mechanism is helpful to man in overcoming infection but harmful to him when it causes rejection of organ and tissue grafts.

An outstanding recent achievement of medical science is the successful transplantation of the human heart which is removed from the donor immediately upon his death. The surgical procedures for this operation have long been known, but only recently have physicians learned to overcome partly the immune mechanism of the recipient. Further advances may overcome the antigen-antibody reaction to permit tissue and organ grafts from "banks" of spare parts. There are many methods used to preserve these parts; for example, refrigeration, quick freeze, dehydration, saline solution, irradiation, and chemical immersion.

COMPLETION QUESTIONS

[A] 1. If a lump of sugar is placed in a container of water, the sugar molecules and water molecules will become uniformly distributed by the process of . . . .
2. The diffusion of water through a membrane is called . . . .
3. The shrinkage of cell contents because of outward osmosis is known as . . . .
4. Membranes which permit only certain substances to pass are said to be . . . .
5. Active transport is the movement of materials against the . . . .
6. The formation of pockets in the cell membrane for the intake of materials is known as . . . .
7. Two classes of compounds present in the cell membrane are fat and . . . .
8. The streaming of cytoplasm observed in *Anacharis* is known as . . . .
9. The branching network of tubes in the cytoplasmic region is the . . . .
10. Food is passed from cell to cell of *Hydra* by the process of . . . .
11. In the dorsal blood vessel of the earthworm the blood flows in the . . . . direction.
12. The "hearts" of the earthworm are the . . . .
13. The earthworm's blood picks up oxygen at the . . . .
14. An example of an organism with colorless blood is the . . . .
15. Blood circulates slowly in a (an) . . . . circulatory system.

(B) 16. The structure in man's heart that controls the rate of its beating is the . . . .
17. Hardening of the arteries is called . . . .
18. Blood moves in veins because of pressure exerted by . . . .
19. In man, the heart chamber that pumps blood to the lungs is the . . . .

20. The largest artery in the body is the ....
21. The heart chamber that pumps blood to most parts of the body is the ....
22. The liver becomes more active in converting glycogen to glucose when the amount of glucose in the blood (increases, decreases) ....
23. The circulation of the blood was discovered by ....
24. The liquid which bathes the cells of the body is ....
25. The largest lymph vessel is the ....
26. Lymph vessels present in the villi are the ....
27. The liquid portion of the blood is called ....
28. The red pigment in the blood is ....
29. Red blood cells are produced in the ....
30. Phagocytes move by a method called ....

(C)31. Cells in the blood which aid in blood clotting are the ....
32. Proteins in plasma which assist blood clotting are prothrombin and ....
33. Failure of blood to clot properly may be caused by lack of vitamin ....
34. A foreign protein which stimulates the formation of antibodies is called a (an) ....
35. An allergic response of the respiratory tract is the condition called ....
36. Antibodies which cause the clumping of bacteria are called ....
37. Two types of acquired immunity are .... and ....
38. The injection of toxin-antitoxin results in .... immunity.
39. Two men who prepared vaccines for polio are .... and ....
40. Blood proteins containing antibodies are present in the fraction of the blood called ....
41. An individual with blood type *B* has .... antibodies.
42. In blood typing, if both drops of serum show clumping, the individual is blood type ....
43. The antibodies present in the plasma of the donor do not cause much clumping of the recipient's cells because the donor's blood is .... as it enters the recipient's blood stream.
44. Blood plasma can be used for transfusions without fear of causing clumping because it contains no ....
45. Destruction of a fetus' red cells can result when the mother has blood type .... and the fetus has blood type ....

MULTIPLE-CHOICE QUESTIONS

[A] 1. Which part of the cell is most directly concerned with the exchange of materials between the cell and its environment? (1) nucleus, (2) centrosome, (3) ribosomes, (4) cell membrane.
2. Which of the following can readily pass through the cell membrane? (1) proteins, (2) glucose, (3) fats, (4) starch.
3. Diffusion is the process by which (1) molecules or ions move from a region of low concentration to a region of higher concentration, (2) molecules or ions move from a region of high concentration to a region of lower concentration, (3) the same number of molecules or ions move in opposite directions, (4) the molecules or ions do not move.

4. Plasmolysis occurs by (1) movement of salt into the cell, (2) active transport, (3) osmosis, (4) cyclosis.

5. Which is a characteristic of the cell membrane? (1) It is present only in animal cells. (2) It is nonliving. (3) It is permeable only to water. (4) It selectively regulates the passage of materials.

6. Molecules pass through a membrane from a low concentration of molecules to a high concentration of those molecules by the process called (1) hydrolysis, (2) active transport, (3) passive transport, (4) diffusion.

7. Molecules which are too large to pass through the cell membrane may be taken in by (1) pinocytosis, (2) active transport, (3) homeostasis, (4) osmosis.

8. The major function of the contractile vacuole is (1) removal of nitrogenous wastes, (2) removal of excess amino acids, (3) removal of solid wastes, (4) regulation of water content of the cell.

9. An organism lacking any special transport system is (1) *Hydra*, (2) earthworm, (3) grasshopper, (4) man.

10. A closed circulatory system is characteristic of (1) *Hydra*, (2) grasshopper, (3) earthworm, (4) *Amoeba*.

Ⓑ 11. The exchange of materials between the cells of the body and the blood occurs principally at (1) the heart, (2) capillaries, (3) veins, (4) arteries.

12. The formula of a carbohydrate which is absorbed into the bloodstream is: (1) $(C_5H_{10}O_5)_n$, (2) $C_{12}H_{22}O_{11}$, (3) $C_6H_{12}O_6$, (4) CH_2O.

13. A blood vessel which contains thick walls and much muscle is a (an) (1) artery, (2) lacteal, (3) capillary, (4) vein.

14. The backward flow of blood in veins is prevented by (1) muscles, (2) valves, (3) the heartbeat, (4) lymphatics.

15. For the blood of a normal man to pass from the right side of the heart to the left side, it must (1) diffuse through the partition between the two sides, (2) pass through the valve which connects the two sides of the heart, (3) enter the lymph vessels, (4) pass through the lungs.

16. The circulation of blood between the heart and the lungs is called (1) portal circulation, (2) systemic circulation, (3) pulmonary circulation, (4) coronary circulation.

17. Which adaptation has made possible the separation of oxygenated and deoxygenated blood? (1) development of lungs in air-breathing vertebrates, (2) appearance of aortic arches in the annelids, (3) single ventricle in the frog's three-chambered heart, (4) four-chambered double heart in mammals.

18. Much of the glucose absorbed in the villi is removed from the blood in the (1) liver, (2) gall bladder, (3) pancreas, (4) large intestine.

19. Which function of the human blood includes the other three? (1) transporting nutrients within the body, (2) transporting oxygen, (3) assisting in the maintenance of homeostasis, (4) collecting wastes.

20. A radioisotope of iron could best be used to measure the life span of a (an) (1) epithelial cell, (2) red blood cell, (3) white blood cell, (4) platelet.

Ⓒ 21. A chemical substance in the blood which aids in the formation of a clot is (1) fibrinogen, (2) ammonium hydroxide, (3) hemoglobin, (4) heparin.

22. A blood clot consists of (1) serum and corpuscles, (2) plasma and corpuscles, (3) fibrin and corpuscles, (4) plasma and platelets.

23. The ability of white blood corpuscles to attack bacterial invasions of the body is known as (1) plasmolysis, (2) homeostasis, (3) pinocytosis, (4) phagocytosis.

24. Antibodies are chemicals that are (1) produced by the body in response to an antigen, (2) synthesized from carbohydrates, (3) nonspecific, (4) transported by red blood cells.

25. Which are most directly involved in the human body's defense against disease? (1) white blood cells and antibodies, (2) white blood cells and red blood cells, (3) red blood cells and antibodies, (4) red blood cells and platelets.

26. When a person has appendicitis, his blood shows an increase in the number of (1) enzymes, (2) platelets, (3) white corpuscles, (4) red corpuscles.

27. Assume that a drop of anti-*A* serum and a drop of anti-*B* serum are placed side by side on a slide. A drop of blood of unknown type is added to each drop of serum. If blood cells clump in anti-*A* and not in anti-*B*, the blood of unknown type is probably type (1) *A*, (2) *B*, (3) *AB*, (4) *O*.

28. A blood transfusion which would probably not cause difficulty is (1) *AB* donor with *B* recipient, (2) *A* donor with *AB* recipient, (3) *B* donor with *A* recipient, (4) *B* donor with *O* recipient.

29. An individual with type *A* blood may receive a blood transfusion from types (1) *A* and *O*, (2) *A* and *AB*, (3) *A*, *AB*, and *O*, (4) *O* and *B*.

30. Blood plasma does not have to be typed because (1) it does contain agglutinogens, (2) it does contain agglutinins, (3) it does not contain agglutinogens, (4) it does not contain agglutinins.

CHAPTER TEST

1. A dorsal tubular heart is present in the (1) earthworm, (2) grasshopper, (3) fish, (4) amphibian.

2. Materials enter the transport medium of a closed circulatory system through the (1) heart, (2) arteries, (3) veins by way of lymphatics, (4) capillaries.

3. The cell membrane is composed mainly of (1) protein and fatty materials, (2) proteins and carbohydrates, (3) cellulose and fats, (4) fats and carbohydrates.

4. When bacteria of decay digest a piece of meat, the end products of digestion generally enter the bacteria by diffusion because the (1) bacteria of decay have completely permeable cell membranes, (2) bacteria have no cell wall, (3) digestive enzymes cannot pass through the cell membranes of the bacteria, (4) end products are in higher concentration outside the bacteria than inside.

5. Several hours after the apparatus was set up as shown in the diagram, the outer water was tested with iodine.

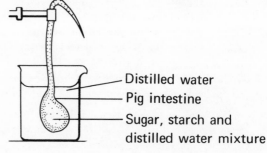

Distilled water

Pig intestine

Sugar, starch and distilled water mixture

A possible reason for a negative result is that the (1) starch molecules were too large to pass through the membrane, (2) sugar molecules were too large to pass through the membrane, (3) starch was diluted too much by the water, (4) sugar was diluted too much by the water.

6. The passage of water molecules through a membrane is known as (1) hydrolysis, (2) diffusion, (3) osmosis, (4) dehydration synthesis.

7. Oxygenated blood enters the heart of man at the (1) left atrium, (2) right atrium, (3) left ventricle, (4) right ventricle.

8. Which statement concerning movement of white corpuscles is true? (1) They are moved from place to place only as the blood circulates. (2) They are moved by means of the action of molecules moving within the same liquid. (3) They are able to move independently from place to place by means of pseudopodia. (4) They are carried from place to place by toxins within the organism.

9. Carbohydrates are stored in the liver in the form of (1) saturated fats, (2) bile, (3) glucose, (4) glycogen.

10. Foreign proteins which stimulate antibody production are called (1) agglutinins, (2) antigens, (3) enzymes, (4) antibiotics.

11. Which statement concerning the immunological reactions of blood is generally true? (1) The introduction of foreign protein *A* results in the production of antibodies effective against foreign proteins *A*, *B*, and *C*. (2) Antibody concentration of plasma cannot be increased artificially. (3) People with type *B* blood produce an antibody against antigen *A*. (4) Type *O* blood is the universal recipient.

12. A crystal of blue copper sulfate is placed at the bottom of a cylinder of water. In a few days, a blue color has reached partly up the cylinder. This movement is an example of (1) active transport, (2) diffusion, (3) osmosis, (4) pinocytosis.

13. The movement of materials against a concentration gradient is known as (1) active transport, (2) fermentation, (3) diffusion, (4) osmosis.

Base your answers to questions 14 and 15 on the following information:

A clean toothpick was dipped into a sample of human blood. The blood was then mixed with a drop of anti-*A* serum. No agglutination was produced.

14. The blood type is (1) *A* or *B*, (2) *A* or *O*, (3) *B* or *O*, (4) *AB* or *B*.

15. Using blood from the same sample, the procedure described above was repeated using anti-*B* serum instead of anti-*A*. Agglutination was produced. The blood type is (1) *O*, (2) *A*, (3) *B*, (4) *A* or *O*.

For each of questions 16 through 20, write the letter preceding the means of transport in living organisms, chosen from the list below, which best applies to that question.

Means of Transport

A. An open blood system.
B. A closed blood system.
C. Both an open and a closed blood system.
D. Neither an open nor a closed blood system.

16. Materials are transported very rapidly.

17. *Hydra* is able to transport materials to all its cells.

18. Blood moves out from the heart to the sinuses.

19. Digested nutrients are transported in a liquid medium.

20. A pumping mechanism enables the blood to circulate.

CHAPTER 7 | Respiration at the Cellular Level

Energy is a basic requirement for life, and the ultimate source of energy for living things is the sun. During the process of photosynthesis, the sun's energy is captured by green plants and is locked into foods as chemical energy. Animals eat these foods and then release this energy for their own use by the process of respiration.

This reverse relationship between photosynthesis and respiration is indicated by the double arrows in the equation below:

$$\text{Energy+carbon dioxide+water} \underset{\text{respiration}}{\overset{\text{photosynthesis}}{\rightleftharpoons}} \text{glucose+oxygen}$$

The equation shows that photosynthesis stores energy in glucose, whereas respiration makes energy available for the chemical reactions in living things.

An overall equation for respiration, as it occurs in most cells, is:

$$\underset{\text{glucose}}{C_6H_{12}O_6} + \underset{\text{oxygen}}{6O_2} \longrightarrow \underset{\text{carbon dioxide}}{6CO_2} + \underset{\text{water}}{6H_2O} + \text{energy}$$

This equation indicates the raw materials and the end products of the process, but gives the impression that respiration occurs as a simple one-step chemical reaction. Actually, respiration within the cell consists of a complex series of reactions in which numerous enzymes (mainly in the mitochondria) play their parts. The above equation shows the oxidation of glucose; however, the cell may also obtain energy by the oxidation of other energy-rich compounds.

A | THE CONCEPT OF CELLULAR RESPIRATION

Cellular respiration is the oxidation of food molecules in the cell with the release of energy. This oxidation is not exactly the same as the oxidation that occurs in burning. When wood is burned, oxygen combines with the wood as a chemical change occurs and energy is rapidly released. Most cellular respiration occurs with the use of oxygen (*aerobic*), but it may occur without the use of oxygen (*anaerobic*).

Even though cellular respiration may occur without the use of oxygen, it is still oxidation. We will be better able to understand this when we study the modern definition of oxidation on page 144.

Burning and respiration are processes which release the energy of organic molecules by oxidation. The differences between burning and cellular respiration are summarized in the table on the facing page.

Role of ATP

As you can see from studying the table, energy release in cellular respiration occurs in measured amounts. The controlled release of energy is part of the proper metabolism of your cells. However, it does not really explain the amazing ability of your body to draw upon energy quickly even if you have not eaten recently.

An understanding of ATP (adenosine triphosphate) is basic to the mod-

COMPARISON OF BURNING AND CELLULAR RESPIRATION

Burning	Cellular Respiration
Uncontrolled, rapid release of energy with accompanying high temperatures.	Slower release of energy in measured amounts at temperatures which do not injure the cell.
Not controlled by enzymes.	Controlled by enzymes.
Most of the energy released in the form of heat and light.	Most of the energy used to create new chemical bonds; only a small amount of heat energy liberated.
Single chemical reaction.	Series of interrelated chemical reactions.
Uses oxygen (aerobic).	One form of cellular respiration occurs **with** the use of oxygen (aerobic); one form of cellular respiration occurs **without** the use of oxygen (anaerobic).

ern concept of cellular respiration. ATP is a compound in which energy is stored temporarily as a result of biochemical reactions. The ATP then supplies the energy to other reactions whenever it is needed. ATP is present in relatively small concentrations in the cell and is therefore not really a means of energy storage. Rather, it is a means for making energy immediately available for chemical reactions. Since every organism requires a constant supply of energy and yet does not take in food constantly, it is evident that some *immediate* source of energy is needed. This source of energy is ATP, which acts as the "energy currency" of the cell by carrying energy from one biochemical reaction to another.

ATP is continually being made and used up. In the energy economy of cells, fats and polysaccharides are long-term investments; glucose is a checking account for current expenses; and ATP is the cash in your pocket.

ATP is defined in Fig. 7–1. If the adenosine portion of the molecule is represented by the letter A and each of the phosphate groups by the letter P, adenosine triphosphate may be represented as $A-P{\sim}P{\sim}P$. The wavy lines linking the phosphate groups are *high-energy bonds*. Most of the energy of the ATP molecule is in the high-energy bonds of the two phosphate groups at the end. When a molecule of ATP gives up one phosphate group from its end, the molecule which accepts this group gains energy and can react more readily with other molecules.

When adenosine is attached to only

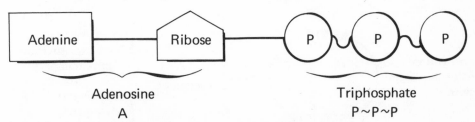

Fig. 7–1. Adenosine triphosphate. Adenine and the 5-carbon sugar ribose together form adenosine, which is abbreviated as A. Each P represents a *phosphate group* consisting of phosphorus and atoms of hydrogen and oxygen.

two phosphate groups, it is *adenosine diphosphate,* or *ADP.* The equation below shows ATP giving up one phosphate group to form ADP, with the release of the energy of the terminal bond:

$$ATP \underset{ATP\text{-}ase}{\rightleftharpoons} ADP + P + energy$$

This may also be represented as follows:

$$A - P{\sim}P{\sim}P \underset{ATP\text{-}ase}{\rightleftharpoons} A - P{\sim}P + P + energy$$

ATP-ase is an enzyme regulating this reaction. The double arrow shows that the reaction is reversible. As the reaction proceeds to the left, ADP takes on energy and a phosphate group to form ATP. Now the energy in ATP is available for use by the cell. When ATP later releases this energy, it reverts to ADP. ADP and ATP together carry energy from those reactions in the cell that release energy to those that use energy. This cycle is shown in Fig. 7–2.

Oxidation and Reduction

When a substance combines with oxygen, we say that *oxidation* has occurred. For example, when we burn a strip of magnesium ribbon, the compound magnesium oxide is formed and energy in the form of heat and light are liberated. The chemical equation is $2Mg + O_2 \longrightarrow 2MgO$. Similarly, the rusting of iron is a slow oxidation of iron in which oxygen combines with the iron to form iron oxide. In the process of combining with oxygen, atoms of the metal undergo only one change: they give up electrons to the oxygen.

Today, the word *oxidation* is used to mean *the loss of electrons* by an atom or molecule, whether the electrons are received by oxygen or any other substance. The substance that *receives the electrons* is said to undergo *reduction.* The entire process is called an *oxidation-reduction reaction.*

In biological systems, there are two very common kinds of oxidation-reduction reactions. In one, material becomes oxidized by reducing oxygen; that is, it gives up electrons to oxygen. This may or may not result in the substance combining chemically with oxygen. The second common kind of reaction is that in which an organic substance becomes reduced by receiving an electron from hydrogen, that is, by oxidizing hydrogen. Usually, this results in the combination of the material with the hydrogen. Other reactions also occur: a substance is oxidized by losing hydrogen atoms or reduced by gaining hydrogen atoms.

Thus, *oxidation* may consist of a *loss* of electrons or a loss of hydrogen atoms, and *reduction* may consist of a

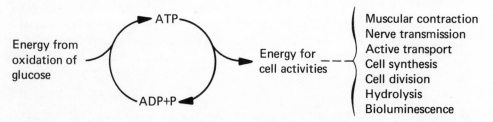

Fig. 7–2. The ADP-ATP cycle makes energy from glucose available for cell activities.

gain of electrons or a gain of hydrogen atoms.

Aerobic and Anaerobic Respiration

Cellular respiration is essentially an oxidation of high-energy organic compounds; however, this oxidation need not involve oxygen. So long as electrons or hydrogen atoms are lost by a compound, it is oxidized and energy is released. Those stages of cellular respiration which use oxygen are called *aerobic respiration,* and those stages which do not use oxygen are called *anaerobic respiration.*

Many organisms carry on respiration without the use of oxygen, and are called *anaerobes,* e.g., yeast (at times) and some bacteria. Other organisms use anaerobic respiration during the early stages of their respiratory process but use oxygen during the later stages. These are called *aerobes.* Examples of aerobes are protozoa, algae, higher plants, and animals. In both types of respiration, much of the potential chemical energy in C–C and C–H bonds of organic molecules, such as glucose, is transferred to the high-energy bonds of ATP.

REASONING EXERCISES

1. What does cellular respiration have in common with burning? How does cellular respiration differ from burning?
2. What is ATP? What does each P in ATP represent? In what way does the body use ATP?
3. What part of the ATP molecule has most of the energy? How does ATP help other molecules in the cell to gain energy?
4. What is ATP-ase? What is its function?
5. What is the ADP-ATP cycle? What is its function in the body?
6. How is oxidation defined today? What is oxidation-reduction?
7. Define cellular respiration. Distinguish between aerobic respiration and anaerobic respiration.

B   GENERAL PATTERN OF CELLULAR RESPIRATION

It is important for you now to acquire an understanding of the broad pattern of cellular respiration. In studying this section, you will need to concentrate closely on what you are reading and to inspect the diagrams intently. You will be considering both anaerobic respiration and aerobic respiration. Therefore, during the course of your study, bear in mind that some organisms carry on only anaerobic respiration. Also bear in mind that other organisms (aerobes) carry on anaerobic respiration during the early stages of cellular respiration.

The general pattern of cellular respiration is shown in Fig. 7–3. This figure shows that the reactions of aerobic and anaerobic respiration are interrelated. The reactions on one side of the figure (at your left) are anaerobic and those on the other side are aerobic.

Anaerobic Respiration

The first phase is the oxidation of glucose, through several steps, to *pyruvic acid.* Glucose is a 6-carbon mole-

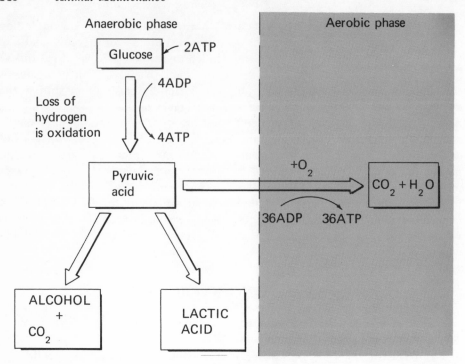

Fig. 7–3. Overall pattern of cellular respiration. One glucose molecule is oxidized anaerobically to two pyruvic acid molecules, liberating energy for the synthesis of ATP. Pyruvic acid is oxidized aerobically to carbon dioxide and water, liberating much more energy for the synthesis of a greater number of ATP molecules.

cule ($C_6H_{12}O_6$) and pyruvic acid is a 3-carbon molecule ($C_3H_4O_3$). When one glucose molecule is broken down to form two molecules of pyruvic acid, there is a *loss* of 4 atoms of *hydrogen*. You can see that this is so if you compare the number of hydrogens in each of the formulas above. Since four atoms of hydrogen are lost, this reaction represents an *oxidation* even though no oxygen is involved. (Recall the definition of oxidation.) Energy liberated by this oxidation is used to synthesize ADP into ATP.

The pyruvic acid which has been formed in this reaction is an important starting point for several other reactions. (See Fig. 7–3.) In anaerobic organisms, depending upon the en-

zymes present, pyruvic acid may be changed into (1) *alcohol and carbon dioxide* or (2) *lactic acid*. No additional oxidation occurs during these stages, and no additional ATP's are formed. There still remains much additional chemical energy in pyruvic acid (and in alcohol and lactic acid) which has not been released by the above stages of anaerobic respiration.

All of the steps of anaerobic respiration described above, starting from glucose, are also known as *fermentation*. Yeast cells possess the necessary enzymes for the fermentation of sugars to alcohol and carbon dioxide and are thus the foundation for the brewing and baking industries which depend upon these end products. Lactic acid bacteria possess enzymes for the steps in fermentation which lead to the production of lactic acid. They are important in the production of cheese, butter, milk, yogurt, sauerkraut, and pickles.

Aerobic Respiration

The potential energy remaining in pyruvic acid is utilized by aerobic organisms which break down this compound further to carbon dioxide and water. Since they use oxygen in this process, these phases in the process of respiration constitute aerobic respiration. This aerobic oxidation is far more efficient than the anaerobic respiration of glucose to pyruvic acid and yields far more ATP molecules.

Strictly speaking, the aerobic phase of respiration consists only of the oxidation of *pyruvic acid* to carbon dioxide and water. However, aerobes also carry on the first stages in which pyruvic acid is itself formed. Consequently, the term "aerobic respiration" is also generally applied to the entire pathway, starting from *glucose* and ending with carbon dioxide and water.

At times, cellular respiration occurs in animal cells under conditions where there is an insufficient supply of oxygen. For example, when an athlete is engaged in a sprint, there is an insufficient supply of oxygen to the muscle cells which are actively using up oxygen in the liberation of energy. Under these conditions, anaerobic respiration is producing pyruvic acid faster than it can be used up in aerobic respiration; therefore, it is converted to lactic acid. As a result, there accumulates in the tissues an "oxygen debt." The heavy breathing of a sprinter after a race supplies sufficient oxygen to oxidize the lactic acid. *Muscle fatigue* is associated with an accumulation of lactic acid.

Yield of ATP

In order to activate the oxidation of one molecule of glucose, the cell must first "invest" 2 molecules of ATP to get the reaction started. (See Fig. 7–3.) The anaerobic oxidation of glucose to pyruvic acid yields 4 ATP's. Thus, there is a net gain of only 2 ATP's during anaerobic respiration.

During the aerobic phase, the oxidation of pyruvic acid continues with the aid of oxygen. During this process, 36 ATP molecules are synthesized. Starting with one molecule of glucose, aerobic respiration thus yields 38 ATP's as compared with only 2 ATP's for the anaerobic pathway.

An overall summary equation for the aerobic respiration of glucose is:

$$\underset{\text{molecule})}{\overset{(1}{}}\text{glucose} + \text{oxygen} + 38\ \text{ADP} + 38\ \text{P} \xrightarrow[\text{enzymes}]{\text{numerous}}$$

$$\text{carbon dioxide} + \text{water} + 38\ \text{ATP}$$

Efficiency of Respiration: Evolution

We have seen that the yeast cell normally gains useful energy in the form of 2 molecules of ATP from one molecule of glucose. However, experiments show that in the presence of sufficient molecular oxygen, a yeast cell can follow the aerobic pathway and gain a total of 38 molecules of ATP from one molecule of glucose. The aerobic pathway is thus 38 : 2 or almost 20 times as efficient as the anaerobic method. Accordingly, aerobic organisms had a distinct survival advantage over anaerobic organisms in the evolutionary struggle for existence. Since the simpler organisms today use anaerobic respiration and the more complex ones use a pattern of aerobic respiration built upon anaerobic stages, it seems obvious that the anaerobic method was the first to evolve on earth.

Not all of the chemical energy of the glucose molecule is transferred to ATP, even in the aerobic method of respiration. Compared with the total energy that can be released from glucose by burning in a calorimeter, only about 55 percent of the energy of glucose is stored in the useful form of ATP during aerobic respiration. Nevertheless, a 55 percent efficiency seems high when compared with the efficiency of many man-made devices, particularly if we consider the temperature limitations placed upon cellular respiration because of its occurrence within the living cell.

COMPARISON OF ANAEROBIC AND AEROBIC RESPIRATION

	Anaerobic	Aerobic
Oxygen requirement	None	Oxygen required for aerobic phase
Yield of ATP per molecule of glucose	2 ATP	38 ATP
Efficiency		Almost 20 times as efficient as the anaerobic method

REASONING EXERCISES

1. Define anaerobic respiration. How does the oxidation of glucose occur without oxygen?
2. What are the end products of fermentation? What is the economic importance of fermentation?
3. Define the aerobic phase of aerobic respiration.
4. Why does lactic acid build up in the muscles during vigorous exercise?
5. Where does the cell get the energy to activate cellular respiration?
6. Why do we believe that anaerobic respiration evolved first?
7. Discuss the potential for survival of aerobes and anaerobes in a world where air pollution is steadily increasing.
8. Compare anaerobic respiration and aerobic respiration from the standpoint of oxygen requirement, yield of ATP, and efficiency.

ⓒ ADDITIONAL REACTIONS IN CELLULAR RESPIRATION

One of the characteristics of living things is the complex of biochemical reactions which proceed simultaneously within their cells. It is beyond the scope of this book to present an exhaustive treatise on biochemistry. However, it is instructive for the student who wishes to comprehend the living state to adventure somewhat deeper into the reactions which take place in cellular respiration. Such an exploration serves as career guidance — to give the potential biologist an inkling of how biology can help him to understand the mysteries of the living state. We thus consider NAD, the Krebs cycle, and the electron transport chain.

NAD

During oxidation-reduction reactions in the living cell, hydrogen atoms may be removed from one substance (oxidation) and picked up by another substance (reduction). Such reactions are carried forward with the aid of enzymes and coenzymes. Let us see how this occurs.

When an enzyme catalyzes an oxidation reaction that results in the liberation of hydrogen atoms, it is aided by a *coenzyme* which picks up these hydrogen atoms. A coenzyme is a nonprotein molecule which is essential to the reaction controlled by an enzyme and which, like the enzyme, is not used up in the reaction. The removal of hydrogen atoms in cells is accomplished primarily by the coenzymes *NAD* (nicotinamide adenine dinucleotide) and *NADP* (nicotinamide adenine dinucleotide phosphate).* NAD and NADP are *hydrogen acceptors.* When they take on two hydrogen atoms, they are reduced to $NADH_2$ and $NADPH_2$, which are high-energy compounds. However, they can also give up these hydrogen atoms to other compounds and revert to NAD and NADP, releasing energy. Because NAD and NADP are so similar in their actions, we will refer only to NAD and $NADH_2$, the oxidized and reduced forms of the coenzyme NAD.

Recall that a hydrogen atom is composed of one electron and one proton (or hydrogen ion). NAD combines directly with two electrons from the substance being oxidized and the partners of these electrons — the protons — are associated with NAD which now becomes $NADH_2$.

Anaerobic Breakdown

A summary of anaerobic respiration is shown in Fig. 7–4. In the breakdown

---

* Under an older terminology NAD is also known as DPN (diphosphopyridine nucleotide) and NADP is TPN (triphosphopyridine nucleotide).

of glucose to pyruvic acid, only a small part of the energy released goes into the two extra ATP molecules produced. Much of the rest goes into two molecules of $NADH_2$, which also accounts for the 4 hydrogen atoms lost in the conversion of $C_6H_{12}O_6$ (glucose) into $2C_3H_4O_3$ (pyruvic acid). While the $NADH_2$ has a lot of energy, it cannot be used unless there is free oxygen present. Under anaerobic conditions, the energy is not released as the hydrogen is transferred to the pyruvic acid which is then converted to lactic acid.

Some bacteria have enzymes which produce acetic acid CH_3-CH_2-COOH (vinegar). Louis Pasteur saved the French wine industry when he found that bacteria which had gotten into the vats were fermenting the grape juice to vinegar instead of to alcohol. His suggestion to destroy the bacteria by heat began the process of *pasteurization* which is now applied to milk and other substances.

Aerobic Breakdown

The potential energy of the glucose molecule has been only partially liberated during the conversion of glucose to pyruvic acid. Aerobic organisms continue this process further until only the waste products *carbon dioxide* and *water* remain. Two groups of reactions are involved. One group of reactions removes carbon atoms from the 3-carbon molecule of pyruvic acid until only the 1-carbon molecule of carbon dioxide remains. This is (1) *the carbon pathway.* The other group of reactions transfers the hydrogen atoms of pyruvic acid to oxygen, forming water. This is (2) *the hydrogen pathway.*

The carbon pathway: Krebs cycle. The carbon pathway is shown in Fig.

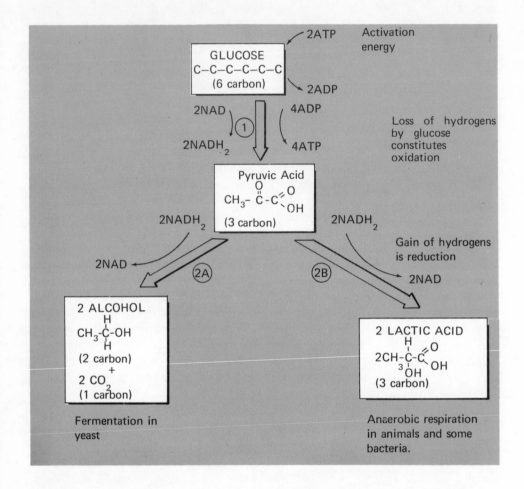

Fig. 7–4. Summary of anaerobic respiration showing number of ATP molecules formed and action of NAD as a hydrogen acceptor and NADH₂ as a hydrogen donor.

7–5. This pathway utilizes the Krebs cycle (also called the citric acid cycle), named in honor of Sir Hans Krebs of Oxford University, England. He was awarded the Nobel Prize for discovering this route.

A molecule of pyruvic acid is first broken down to one molecule of CO_2 and a 2-carbon molecule. With the ad-

dition of a coenzyme, this 2-carbon molecule becomes "active acetic acid" which now enters the Krebs cycle. The *Krebs cycle* is a cyclical chain of enzyme-controlled reactions which permits the initial substance to be formed again at the end of the sequence. It is thus a cycle, or wheel. The "first" molecule in this wheel is a 4-carbon com-

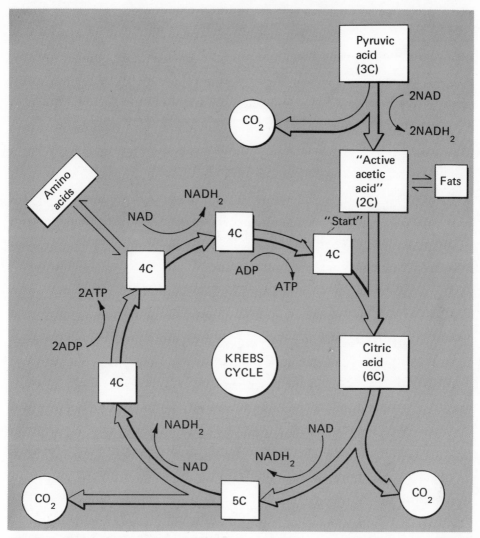

Fig. 7–5. The Krebs cycle showing the carbon pathway. In addition to the removal of CO_2, hydrogens are taken up by NAD, water is utilized, and ATP is formed. Since one glucose molecule yields two pyruvic acid molecules, two turns of the wheel are required for each molecule of glucose oxidized. (See text for explanation.)

pound which reacts with the 2-carbon "active acetic acid" to form the 6-carbon citric acid (which gives its name to the cycle).

As shown in Fig. 7–5, there are additional stages in which 5- and 4-carbon compounds are formed until the initial

4-carbon compound is formed again to complete the cycle. Each complete turn of the wheel liberates two molecules of CO_2. These two molecules of CO_2 formed during the Krebs cycle, together with the one molecule of CO_2 liberated from pyruvic acid at the

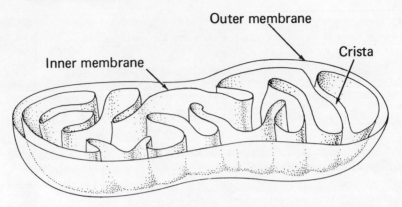

Fig. 7–6. The mitochondrion. This diagram of the internal structure shows the folded inner membrane on which are present enzymes of the Krebs cycle, of the electron transport chain, and for ATP production.

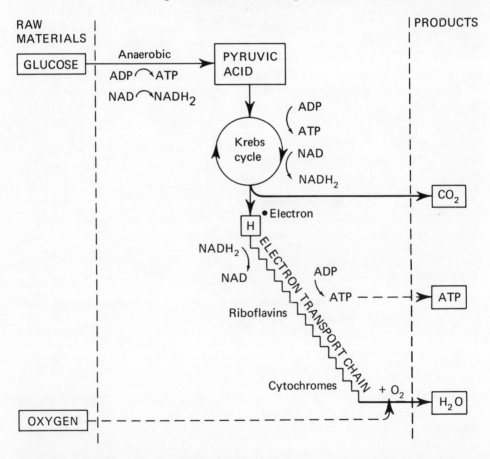

Fig. 7–7. Overview of aerobic respiration. At the left are shown the raw materials, glucose and oxygen. At the right are the products, carbon dioxide and water, and energy associated with ATP molecules.

start, account for the three atoms of carbon present in one molecule of pyruvic acid ($C_3H_4O_3$). During each cycle, 3 molecules of ADP are converted to ATP and 3 molecules of NAD are reduced to $NADH_2$. These compounds carry away the energy stored in the pyruvic acid.

Hydrogen transport: the electron transport chain. The removal of CO_2 from pyruvic acid is accompanied by the loss of hydrogen, mainly to NAD. This hydrogen is now oxidized *by oxygen*, thereby liberating the greatest amount of energy during cellular respiration. However, this is not a one-step oxidation, for the one-step combination of oxygen with high-energy hydrogen would result in a tremendous liberation of heat which would destroy the cell. Rather, there is a whole series of reactions and it is only at the end of this series that oxygen combines with hydrogen. This entire process takes place mainly in the mitochondria. (See Fig. 7–6.) As the pair of "hydrogen atoms" from a single $NADH_2$ molecule passes down the chain, it loses enough energy to meet the requirements of the process.

It is not actually the hydrogen that is transported, but a pair of electrons. $NADH_2$ ionizes to form NAD^{--} and $2H^+$. The hydrogen ions ($2H^+$) simply remain in the cell fluid, but the excess electrons of the NAD pass to a whole series of protein compounds, located mainly in the walls of the mitochondria. Each transfer of electrons is controlled by a special enzyme, and the process requires several coenzymes, including riboflavin (vitamin B_2). At every transfer, the electron loses some energy. At the end of the chain, the electrons transfer to oxygen molecules, which then unite with hydrogen ions in the fluid to form water. In the transfer process, each pair of electrons of an NAD^{--} ion loses enough energy on its way down to the final union with oxygen to convert 3 molecules of ADP into 3 molecules of ATP. This series of reactions, which provides a very large part of the energy of aerobic cells, is called the *electron transport chain.*

An overview of aerobic respiration is given in Fig. 7–7. We can count up the results of the whole process of oxidation of glucose as follows:

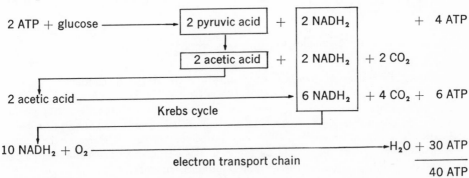

Fig. 7–8. Summary of respiration. Subtracting the 2 ATP used to activate the process from the 40 ATP produced gives 38 ATP, the net production resulting from the oxidation of one glucose molecule.

COMPLETION QUESTIONS

A 1. The gas released by animals during respiration is . . . .
2. Many of the enzymes involved in cellular respiration are located in organelles known as . . . .
3. When ATP gives up energy and a phosphate group, it forms . . . .
4. A compound which transports energy in biochemical reactions is . . . .
5. Oxidation is the loss of . . . . or . . . .
6. Reduction is the gain of . . . . or . . . .
7. Organisms which use molecular oxygen during respiration are called . . . .

B 8. Two waste products of aerobic respiration are . . . . and . . . .
9. When yeast ferments sugar, the end products are . . . . and . . . .
10. During strenuous muscular activity, man's cells may carry on anaerobic respiration and produce . . . .
11. In the anaerobic respiration of one molecule of glucose, the net gain of ATP is . . . .
12. In the aerobic respiration of one molecule of glucose, the net gain of ATP is . . . . . .
13. Aerobic respiration is almost . . . . times as efficient as anaerobic respiration.

C 14. A coenzyme which transports hydrogen during the stages of cellular respiration is . . . .
15. The energy released as glucose is broken down to pyruvic acid goes into molecules of ATP and . . . .
16. Reduction occurs as . . . . is transferred to pyruvic acid.
17. A cyclic series of enzyme reactions which occurs during aerobic respiration is known as the . . . .
18. Two turns of the wheel in the citric acid cycle result in the production of . . . . molecules of ATP.
19. A pair of . . . . from $NADH_2$ moves down the electron transport chain where it participates in the conversion of 3 molecules of ADP to 3 molecules of ATP.
20. Ten molecules of $NADH_2$ provide ten pairs of electrons which produce . . . . molecules of ATP.

CHAPTER TEST

1. During the process of respiration, energy from the oxidation of glucose is temporarily stored in molecules of (1) ATP, (2) DNA, (3) ADP, (4) RNA.
2. Aerobic respiration and fermentation are similar in that both processes (1) liberate energy, (2) utilize light energy, (3) produce carbohydrates, (4) require oxygen.
3. Mitochondria are organelles which are known to (1) be necessary for the process of diffusion to take place, (2) be found in the nucleus of some cells, (3) initiate cell division in living cells, (4) contain respiratory enzymes.
4. The greatest amount of energy is released by the (1) oxidation of glucose to lactic acid, (2) oxidation of glucose to carbon dioxide and water, (3) conversion of glucose to pyruvic acid, (4) conversion of carbon dioxide and water to glucose.
5. An organism which converts pyruvic acid to ethyl alcohol and carbon dioxide is a (an) (1) yeast, (2) alga, (3) protozoan, (4) virus.

6. Muscle cells engaged in vigorous activity build up a relatively high concentration of (1) lactic acid, (2) pyruvic acid, (3) oxygen, (4) alcohol.
7. The process by which green plants produce carbon dioxide and water is called (1) excretion, (2) fermentation, (3) respiration, (4) photosynthesis.
8. The brewing and baking industries depend on the activity of organisms which obtain energy by (1) aerobic respiration, (2) dehydration synthesis, (3) fermentation, (4) hydrolysis.
9. Oxidation may be considered as (1) loss of oxygen, (2) loss of hydrogen, (3) gain of hydrogen, (4) gain of electrons.
10. Cellular respiration *differs* from burning in that, in cellular respiration (1) CO_2 is a waste product, (2) energy is obtained from C–C bonds and C–H bonds, (3) activation energy is required, (4) energy is yielded mainly in the form of chemical bonds of ATP.
11. The souring of grape juice by bacterial action is an example of (1) aerobic respiration, (2) anaerobic respiration, (3) immunity, (4) susceptibility.
12. Which is represented by the following equation?

$$\text{Glucose} \xrightarrow{\text{enzymes}} \text{lactic acid} + \text{ATP}$$

(1) hydrolysis, (2) dehydration synthesis, (3) anaerobic respiration, (4) aerobic respiration.
13. A lack of certain vitamins in the diet results in muscular weakness because these vitamins (1) are oxidized to yield energy, (2) are a part of certain respiratory enzymes, (3) have a high calorie content, (4) contain high-energy phosphate bonds.
14. In aerobic respiration, the final hydrogen acceptor is (1) chlorophyll, (2) carbon dioxide, (3) molecular oxygen, (4) water.
15. Reduction may be considered the (1) gain of electrons, (2) gain of oxygen atoms, (3) loss of hydrogen atoms, (4) loss of electrons.
16. The process by which green plants produce glucose and oxygen is (1) respiration, (2) fermentation, (3) photosynthesis, (4) secretion.

For each process in questions 17 through 20, write the letter preceding the type of respiration, chosen from the list below, to which that process is most closely related.

      A. Anaerobic respiration
      B. Aerobic respiration
      C. Both anaerobic and aerobic respiration
      D. Neither anaerobic nor aerobic respiration

17. Process by which glucose is utilized by a cell.
18. Process which liberates carbon dioxide and alcohol.
19. Process known as fermentation.
20. Process which utilizes CO_2 as a raw material.

CHAPTER 8 / Respiration at the Organism Level

In multicellular animals, respiration is aerobic. This means that oxygen must be brought from the environment to every living cell. Accordingly, respiration includes not only the oxidation that occurs in the individual cells but also the mechanisms for the transport of oxygen to the cells and the removal of carbon dioxide. Transport systems are thus intimately involved in respiration. The multicellular animal faces the problem of using its limited body surface in such a manner as to meet the respiratory needs of all the internal body cells.

In this chapter we will compare the respiratory activity of different animals. We will become aware of the variety of adaptations which animals have for the exchange of oxygen and carbon dioxide with the environment.

A   ADAPTATIONS FOR RESPIRATION

Respiratory Surfaces

Multicellular animals generally have a thick outside covering (or skin) that presents a barrier to the passage of oxygen to the cells in the interior. Since oxygen is used in aerobic respiration, the presence of such a barrier presents a problem. Multicellular animals solve this problem in a variety of ways, but the solution usually involves the presence of a special *respiratory surface*. Consider, for example, the earthworm which has a thin, moist skin adjacent to the circulatory system, as shown in Fig. 8–1.

Although numerous variations of this pattern are present in animals, this diagram illustrates the characteristics of an ideal respiratory surface:
(1) thin
(2) moist
(3) close to a source of oxygen
(4) close to a gas transport system
In protozoa, these characteristics are met by the cell membrane.

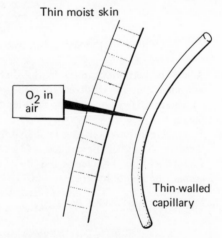

Fig. 8–1. Respiratory surface of the earthworm.

The gills of a fish follow the general pattern indicated above. Capillaries close to the surface of the gill absorb dissolved oxygen from the water. The blood then transports the oxygen to all the cells. When a fish is taken from the water to "open air," the fish dies for lack of oxygen even though the air contains more oxygen than does the water. This happens because the moist membranes of the gills dry and do not allow passage of oxygen to the blood.

Survey of Adaptations in Animals

Protozoa. The protozoa absorb dissolved oxygen from the surrounding water. They oxidize food within the cytoplasm and give off carbon dioxide and water. The exchange of gases takes place through the moist cell membrane. Surplus water is removed by active transport at the cell membrane and at the contractile vacuoles, when present.

Hydra. The thin, two-cell layered structure of *Hydra* permits almost all of the cells to be in contact with the oxygen-bearing water. As with unicellular organisms, oxygen diffuses into the cells, and carbon dioxide diffuses out through the cell membranes. No special system for the transport of respiratory gases is present in *Hydra*.

The earthworm. The skin of the earthworm has a rich supply of blood capillaries. Oxygen diffuses readily into the blood through the thin skin, which is kept moist by the secretion of mucus. The oxygen is carried by the blood to all the cells of the body. The blood transport system also carries carbon dioxide back to the capillaries of the skin where it diffuses out through the moist skin. Earthworms trapped on a sidewalk after a rain die when the moist skin dries.

The blood transport system of many animals serves to link the thin moist membrane, which is exposed to the external environment, with the cells. During the course of evolution, transport systems have evolved respiratory pigments which can carry more oxygen and carbon dioxide than can ordinary water. Hemoglobin is a general name for respiratory pigments that are red in color because of the iron they contain. In the earthworm, a kind of hemoglobin is dissolved in the liquid part of the blood. In man, a hemoglobin is present in red blood cells. A comparison of the oxygen-carrying capacity of these fluids is given in the table below.

Note that the earthworm's blood carries 13 times as much oxygen as does water, and man's blood carries 50 times as much oxygen as does water.

The grasshopper. The grasshopper's body possesses pairs of openings, called *spiracles,* for the taking in of air. The spiracles lead to a system of branched tubes, called *tracheae,* which subdivide into smaller and smaller tubes reaching to all parts of the body. Pulsating contractions of the body wall draw air into the tubes and force it out. The inner end of the tracheal tubes contains a fluid, thus providing a thin moist membrane for the exchange of gases with the tissues.

Air is brought directly to the cells by the system of tracheal tubes, and the blood of the grasshopper does not contain hemoglobin for the transport of oxygen. However, it is interesting that larval forms of some insects do contain hemoglobin. This indicates that the ancestors of the arthropods possessed hemoglobin in their blood and supports

CAPACITY OF FLUIDS TO CARRY OXYGEN

Fluid ⟶ Water	Blood of Earthworm (hemoglobin dissolved in blood)	Blood of Man (hemoglobin in red blood cells)
Oxygen carried ⟶ 5 ml by 100 ml	65 ml	250 ml

the hypothesis that the arthropods descended from annelid-like ancestors.

Man. From an external opening branch respiratory tubes which extend throughout the lungs. The lungs come in contact with the circulatory system by capillaries and other blood vessels. Air sacs in the lungs provide much surface area for the diffusion of oxygen into the blood, and for the removal of carbon dioxide. Unlike the earthworm (which has an *external* respiratory surface), man has an *internal* respiratory surface — thin moist membranes surrounding the air sacs within the lungs.

The adaptations for respiration among animals are summarized in Fig. 8–2. Only the movement of oxygen is indicated in these diagrams, but the reverse movement of carbon dioxide is implied. In each case, a thin moist membrane is involved.

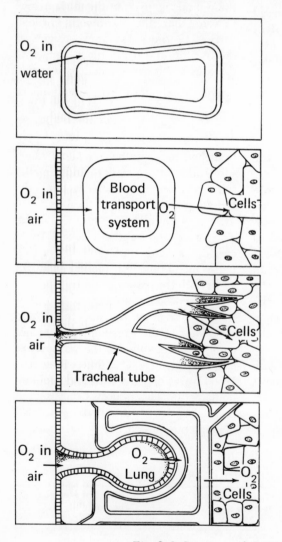

A. **Unicellular organisms and Hydra.** Oxygen in water diffuses through cell membrane.

B. **Earthworm.** Oxygen is carried by blood transport system to cells of the body. Hemoglobin is dissolved in blood. Moist external respiratory surface.

C. **Grasshopper.** Tracheal tubes carry oxygen directly to the cells where it diffuses through a moist membrane into the cells. A blood transport system, though present, is not used for carrying oxygen.

D. **Man.** Small openings in the exterior surface lead to the lungs which possess a great surface area for the diffusion of oxygen into the blood. Hemoglobin is present in red blood cells. Moist internal respiratory surface.

Fig. 8–2. Summary of respiratory patterns.

REASONING EXERCISES

1. What characteristics are common to the respiratory surfaces of an earthworm and a man?
2. Distinguish between the respiratory surfaces of an earthworm and a man.
3. Why does a fish die when it is taken from water, even though there is more oxygen in the atmosphere than in the water?
4. Why does the blood of an earthworm carry more oxygen than an equivalent volume of water does?
5. What adaptation of the grasshopper helps to compensate for the absence of hemoglobin in the grasshopper's blood?

RESPIRATION IN MAN

The structure of the respiratory system is shown in Fig. 8–3. Respiratory gases from the environment enter the blood through this system.

Structure of the Respiratory System

The nose. Air enters the two nostrils and passes into the nasal passages. The nasal passages are lined with mucous membranes, containing many cilia. The cilia and the mucus secreted by the membranes trap some bacteria and dust in the inhaled air, thus filtering it. As the air passes through the nasal passages, it is also warmed and moistened. People who breathe through the mouth lose these advantages.

The pharynx (throat). This cavity at the back of the mouth is a passageway for both air and food. The adenoids and tonsils are found in the pharynx.

The larynx. This is often called the voice box, or *Adam's apple*. It is made of cartilage and lies at the upper end of the trachea (windpipe). The vocal cords which it contains produce sound when vibrated by exhaled air as we speak or sing.

During the act of swallowing, food is prevented from passing from the pharynx to the larynx by the *epiglottis*. This is a cartilaginous flap which diverts food and liquid toward the esophagus, and prevents them from entering the larynx. During swallowing, the epiglottis lowers, and the larynx rises upward toward the epiglottis, thus preventing food from entering the trachea.

The trachea. This large tube is about 4½ inches long and extends from the larynx to the bronchi. The walls of the trachea contain rings of cartilage which keep the tube open for the passage of air.

The trachea is lined with a ciliated epithelial membrane which sweeps foreign particles back up to the pharynx. Deposits from cigarette smoking interfere with the protective action of this membrane.

The bronchi. The trachea divides at its lower end into two smaller cartilage-ringed tubes called the bronchi (sing. — bronchus). The bronchi extend into the lungs, where they divide into smaller and smaller tubes called *bronchial tubes* which also have cartilaginous support.

The bronchioles. When the subdivisions of the bronchial tubes reach a diameter of 1 mm. or less, the cartilage is lacking and the tubes are now called *bronchioles*. Their walls contain smooth muscle and nerve endings. Contractions of the bronchioles occur during attacks of asthma.

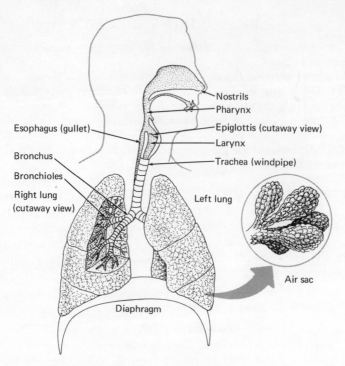

Fig. 8–3. Respiratory system.

The air sacs and alveoli. (See Fig. 8–4.) The bronchioles lead to the functional units of the lung, the *air sacs*. Each air sac resembles a cluster of grapes. An air sac contains several microscopic outpocketings called *alveoli* (sing. – alveolus). The walls of the thin, moist alveoli are composed of a single layer of cells, and the alveoli are surrounded by a network of capillaries. *The exchange of gases between the external air and the blood occurs at the alveoli.*

The total number of alveoli in both lungs is estimated to be about one billion. The total surface of the alveoli is about 100 square meters. Thus, two small openings in the thick, dry skin surface – the nostrils – lead the external air to a thin, moist membrane (within the lungs) whose surface area is 50 times that of the body surface.

Breathing

Breathing is the process of drawing air into the lungs (inspiration) and expelling it (expiration). The bell-jar model of the chest (Fig. 8–5) helps to explain this process.

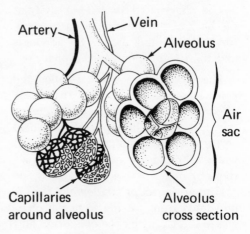

Fig. 8–4. Air sacs.

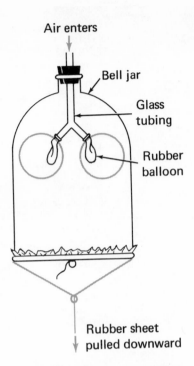

Air enters

Bell jar

Glass tubing

Rubber balloon

Rubber sheet pulled downward

Fig. 8–5. Bell-jar model of chest.

In this model, the glass bell jar represents the wall of the chest cavity, the glass tubing represents the trachea and bronchi, and the rubber balloons represent the lungs. The rubber sheet represents the diaphragm, which is a sheet of smooth muscle below the lungs.

Inspiration. When the "diaphragm" is pulled down, the volume of the chest cavity is increased. This decreases the pressure in the space surrounding the lungs. Since the pressure of the outside air is now greater than the pressure in the chest cavity, air moves into the lungs, inflating them.

Expiration. The diaphragm is allowed to return to the flat position, which causes the volume of the chest cavity to decrease. As a result, the pressure inside the chest cavity increases, forcing air from the lungs to the outside.

The bell-jar model is valuable in showing that the flow of air into and out of the lungs is caused by changes in the pressure in the chest cavity. It also shows that the lungs play a passive role in the flow of air; there are no muscles in the lungs which cause the air to move in and out. However, there are a number of inaccuracies in this model: (1) The chest wall is not a rigid container. It contains ribs which move during breathing to change the pressure in the chest cavity. (2) The diaphragm in the relaxed position is arched upward in a dome shape. At inspiration, this sheet of muscle contracts to a more flattened position, thus increasing the volume of the chest cavity. (3) The lungs are closely pressed to the chest wall; there is no big space in the chest cavity as shown by the model.

Exchange of Gases

The pathway of gases may be considered in two parts: (1) external respiration, and (2) internal respiration. These are diagrammed in Fig. 8–6.

External respiration. This consists of (1) breathing, and (2) the exchange of gases between the alveoli and the blood.

1. *Breathing.* This merely consists of moving air between the outside environment and the alveoli.

2. *The exchange between the alveoli and the blood.* Oxygen diffuses into the capillaries through the single layer of cells making up the walls of the alveoli. In the capillaries the hemoglobin in the red blood cells takes up the oxygen. Carbon dioxide and water diffuse in the opposite direction from the blood into the alveoli.

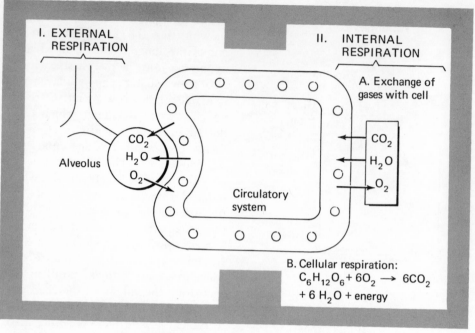

Fig. 8–6. Pathway of the gases in respiration.

Internal respiration. Internal respiration includes (1) the exchange between the blood and the cells, and (2) cellular respiration.

1. *The exchange between the blood and the cells.* Oxygen diffuses from the blood into the cells where it is used in cellular respiration. Carbon dioxide and water, the products of cellular respiration, diffuse into the blood.

2. *Cellular respiration.* Aerobic respiration in the cells results in the release of energy — largely in the form of chemical bond energy in ATP molecules — and the formation of carbon dioxide and water.

Transport of Oxygen by the Blood

Oxygen is carried mainly by the red blood cells, not by the plasma. *Hemoglobin* is a protein compound containing iron which is present in the red blood cell. It combines with oxygen to form the compound *oxyhemoglobin.*

Hemoglobin may be symbolized as Hb, and oxyhemoglobin as HbO_2 (these are not chemical formulas). The combination of hemoglobin and oxygen is shown by the equation:

$$Hb + O_2 \rightleftarrows HbO_2$$

This reaction is reversible.

In the capillaries which surround the alveoli, there is a high concentration of oxygen, and the reaction goes to the right, forming oxyhemoglobin. The oxygen travels through the blood to the tissues in the form of oxyhemoglobin. At the capillaries near the tissues, oxygen is being removed from the blood by the tissue cells. Consequently, the oxygen concentration in the capillaries is low. In the presence of a low concentration of oxygen the reaction goes to the left, and oxyhemoglobin releases oxygen for use by the tissue cells. The uptake and release of oxygen by hemoglobin is diagrammed in Fig. 8–7.

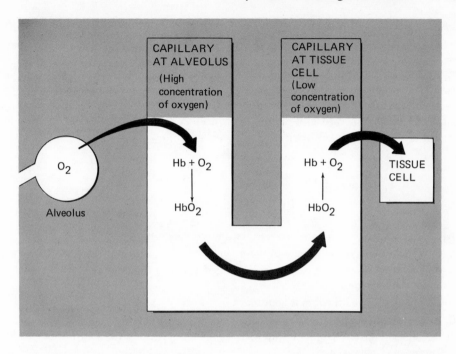

Fig. 8–7. Reversible nature of the reaction between hemoglobin and oxygen. In the presence of a high concentration of oxygen (at the alveoli), the hemoglobin becomes oxygenated. In the presence of a low concentration of oxygen (at the tissues), oxyhemoglobin liberates oxygen and is deoxygenated.

Blood containing oxyhemoglobin is bright red in color; blood containing hemoglobin is dark red in color. When a vein is cut and the dark red blood comes to the surface, the hemoglobin becomes oxygenated when the blood meets the air. Thus, the appearance of venous blood after it meets the air is bright red.

Transport of CO_2 by the Blood

Carbon dioxide is transported from the tissue cells to the lungs in three different states:

1. As the bicarbonate ion (HCO_3^-). — 85% of the carbon dioxide.

2. As carboxyhemoglobin — a combination of carbon dioxide with the hemoglobin of the red blood cells. — 10%.

3. As CO_2 dissolved in the plasma. — 5%.

The major portion of the carbon dioxide travels in the blood as the bicarbonate ion. This is formed with the aid of enzymes in two steps:

1. $CO_2 + H_2O \rightleftarrows H_2CO_3$

2. $H_2CO_3 \rightleftarrows H^+ + HCO_3^-$

These are reversible reactions. At the alveoli, the reactions proceed backward, releasing CO_2 and water.

Regulation of Rate of Breathing

When the body's cells increase their rate of activity, they add more carbon dioxide to the blood. An increase in the amount of dissolved carbon dioxide in the plasma indirectly stimulates the *respiratory center* in the medulla of the brain. The respiratory center sends impulses to the diaphragm and rib muscles to increase the rate and depth of breathing. This increased breathing removes the additional carbon dioxide from the blood and brings more oxygen to all the cells. These steps are an example of *homeostasis* or self-regulating feed-back. The respiratory center is also stimulated by lactic acid which is produced in muscles during strenuous exertion.

COMPLETION QUESTIONS

A 1. An ideal respiratory surface has a structure which is .... and ....

2. A fish takes in oxygen through capillaries in the ....

3. In protozoa, respiratory gases are exchanged between the organism and its environment through the ....

4. The skin of the earthworm is kept moist by ....

5. Hemoglobin contains the metal ....

6. Blood can carry a larger supply of .... than water can because of the presence of hemoglobin.

7. The openings of the grasshopper's respiratory system are called ....

8. The "thin moist respiratory membrane" of the grasshopper is present at the ends of the ....

9. The earthworm has a (an) .... respiratory surface but man has a (an) .... respiratory surface.

10. In man, hemoglobin is present in .... blood cells.

B 11. The cavity behind man's mouth is called the ....

12. During swallowing, food is prevented from entering the trachea by the ....

13. Dust and germs in the trachea are swept up to the mouth by the action of ....

14. The wall of man's trachea is supported by rings of ....

15. The sheet of muscle which aids man in breathing is the ....

16. The exchange of gases between the cells and the blood is known as .... respiration.

17. Oxygen is carried by red blood cells in the form of ....

18. Most of the carbon dioxide is carried by the blood in the form of ....

19. The respiratory center of the brain is indirectly stimulated by .... in the blood.

20. The part of the brain which controls respiration is the ....

CHAPTER TEST

1. Oxygen enters the cells of *Hydra* by (1) cyclosis, (2) hydrolysis, (3) osmosis, (4) diffusion.

2. Respiration in earthworms is directly dependent upon large surface areas of (1) lungs, (2) moist skin, (3) tracheae, (4) gills.

3. Which statement about the respiratory surface in large animals is *not* true? (1) The respiratory surface must be moist. (2) The respiratory surface must be thin. (3) The respiratory surface must be in touch with a proportionately large blood supply. (4) The respiratory surface must be external.

4. Spiracles are used in insects to (1) excrete solid wastes, (2) discharge sperm cells, (3) breathe, (4) aid in detecting vibrations.

5. The respiratory gases of the grasshopper are carried by the (1) blood vessels, (2) tracheae, (3) hemoglobin, (4) lungs.

6. *Paramecium* gets oxygen through the (1) anal spot, (2) cilia, (3) contractile vacuole, (4) cell membrane.

7. Which is a direct result of the increase of carbon dioxide in the blood? (1) ability to expend more energy, (2) increased rate of breathing, (3) increased rate of heartbeat, (4) decreased rate of respiration.

8. In man, a thin, moist respiratory membrane is present at the (1) trachea, (2) bronchi, (3) diaphragm, (4) air sacs.

9. Oxygen is carried by the blood of man mainly (1) in the form of bicarbonate ions, (2) dissolved in plasma, (3) as oxyhemoglobin, (4) by white blood cells.

10. Most of the carbon dioxide is transported by the blood of man as (1) hemoglobin ions, (2) bicarbonate ions, (3) oxyhemoglobin, (4) carboxyl ions.

11. The exchange of gases in human lungs takes place at the (1) trachea, (2) bronchi, (3) diaphragm, (4) air sacs.

12. The exchange of gases between the cells and the blood is called (1) exhalation, (2) internal respiration, (3) external respiration, (4) aerobic respiration.

13. As blood passes through the lungs, gases are exchanged through blood vessels called (1) capillaries, (2) arteries, (3) arterioles, (4) veins.

14. A structure which helps to keep food out of the trachea is the (1) larynx, (2) epiglottis, (3) eustachian tubes, (4) air sac.

15. During inhalation, the diaphragm is (1) lowered, (2) raised, (3) stationary, (4) lowered and then raised.

As an organism carries on its life processes, it produces metabolic wastes. These wastes tend to upset the homeostatic state and can act as poisons by altering the chemical environment needed for enzyme-controlled reactions. *Excretion* is the process by which an organism gets rid of metabolic wastes.

In this chapter, we shall consider the wide variety of adaptations which animals possess for the removal of harmful wastes.

A ADAPTATIONS FOR EXCRETION

An overview of the general pattern by which metabolic wastes are produced is given in Fig. 9–1. Note that *elimination* (also called defecation or egestion) refers to the removal of undigested food and is not included as a form of excretion.

The Protozoa

The wastes of metabolic activities include carbon dioxide, water, nitrogenous wastes, and inorganic salts.

Carbon dioxide. This gas passes by diffusion directly into the watery environment.

Water. Water is produced in cellular respiration. In addition, water from the environment enters the cell by osmosis. This is due to the fact that the water concentration in the fresh-water environment is greater than that within the cell. Cells require water, but they can swell and burst when they have an excess. Removal of water occurs against the concentration gradient by active transport and by the pumping action of the contractile vacuoles. Both of these methods use energy supplied by ATP. Salt water protozoa have no contractile vacuoles, as they do not absorb water.

Small amounts of carbon dioxide,

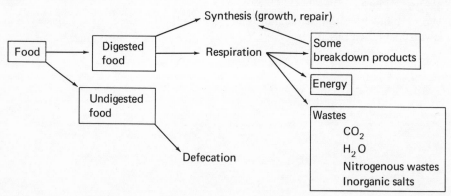

Fig. 9–1. Production of excretory wastes.

166

mineral salts, and nitrogenous waste may be present in the water expelled by the contractile vacuoles of *Paramecium* and *Amoeba*. However, the contractile vacuole is chiefly a means of regulating the cell's water balance.

Nitrogenous wastes. Excess amino acids from the digestion of proteins in foods are not stored in cells. They are broken down, as shown in Fig. 9–2.

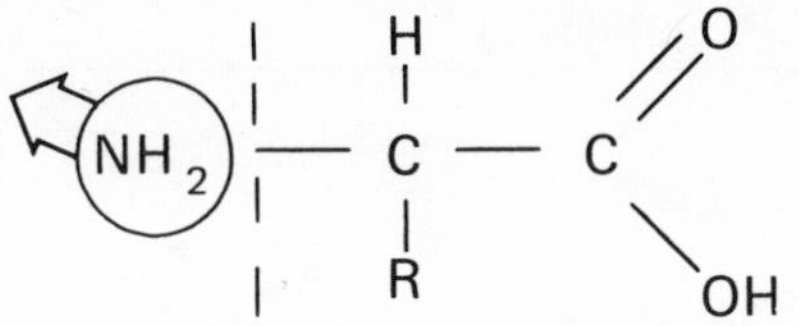

Fig. 9–2. Deamination of an amino acid to form ammonia.

The amino group ($-NH_2$) is converted to *ammonia* (NH_3) and the rest of the molecule is oxidized to release chemical energy. This removal of an amino group is *deamination.*

Ammonia is a highly toxic (poisonous) substance. It diffuses easily from the surface of cells which are in a watery environment. If a contractile vacuole is present, some ammonia (with water) is expelled from this organelle. For the most part, multicellular organisms, with less surface of their cells exposed to water, convert ammonia to the less toxic compounds *urea* or *uric*

acid. Uric acid is not poisonous because it is insoluble. Fresh-water fish absorb much water through their gills; they excrete ammonia directly.

A comparison of the forms in which nitrogenous waste is excreted by animals is given in the table below.

Inorganic salts. Inorganic salts of metabolism may include sodium chloride and potassium chloride. They also include the sulfates and phosphates of ammonia, calcium, and magnesium. They are excreted by diffusion through the semipermeable cell membrane. Active transport may play a part under certain environmental conditions.

Excretion in unicellular organisms is summarized in the table on page 168.

The *Hydra*

As a multicellular organism, *Hydra* has few problems of excretion because its cells are in contact either with the external water or with the water in the digestive sac.

Water. The cells of *Hydra* excrete water through the cell membranes. Since *Hydra* lives in fresh water, the concentration of water molecules is higher outside the cells than inside the cells, and water tends to diffuse into the cells. To remove excess water in the direction opposite to this concen-

FORMS OF NITROGENOUS WASTES

Waste	Toxicity	Water Relationship	Where Found
NH_3	Very toxic	Requires large amounts of water for removal	Microorganisms; many invertebrates living in water; fresh-water fish
Urea	Less toxic	Requires large amounts of water for removal	Man; adult stages of amphibia
Uric acid	Comparatively harmless — insoluble	Present in land animals with limited supply of water	Insects; adult stages of reptiles and birds

SUMMARY OF EXCRETION IN UNICELLULAR ORGANISMS

Waste	How Formed	How Excreted
CO_2	Cellular respiration	Diffusion through cell membrane
H_2O	Cellular respiration; dehydration synthesis	Osmosis; contractile vacuole (if present); active transport
Nitrogenous waste	Removal of —NH_2 groups (deamination) of excess amino acids leads to formation of NH_3	Unicellular organisms remove NH_3 dissolved in much water, by diffusion (Contractile vacuole may play a role)
Inorganic salts	General metabolism	Diffusion

tration gradient, energy is used in active transport. The cells of *Hydra* thus excrete water in a manner similar to that of protozoans. However, the cells of *Hydra* do not have contractile vacuoles such as are found in *Paramecium.*

Nitrogenous wastes. *Hydra* excretes nitrogenous waste in the form of ammonia. The ammonia diffuses through the cell membrane of ectodermal and endodermal cells.

The Earthworm

The body cavity of the earthworm contains a fluid in which metabolic wastes collect. Most of the segments have a pair of excretory tubules called *nephridia* (sing. — nephridium), one on each side. These tubules carry wastes from the body fluid to the outside of the body.

Each nephridium takes in the wastes through a funnel-shaped opening containing cilia. The tubule then passes to the succeeding segment of the body where it is much coiled and intimately associated with the blood supply. While the fluid is passing through the nephridium, useful materials are absorbed out of it back into the blood; thus, only the wastes are carried to the outside through a pore in the body wall. (See Fig. 9–3.)

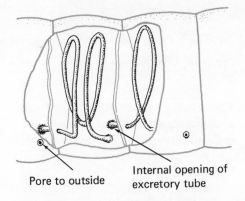

Pore to outside   Internal opening of excretory tube

Fig. 9–3. Diagram of excretory system of earthworm, showing nephridia in successive segments.

Water serves as a solvent for the nitrogenous wastes ammonia, mineral salts, and urea which are excreted through the nephridia. Carbon dioxide diffuses through the moist skin.

The Grasshopper

Excretion in the grasshopper is complicated by the fact that it lives in a dry environment. Thus this organism cannot "spend" water in dissolving and excreting ammonia or urea. Instead, it excretes nitrogenous waste in the form of insoluble crystals of uric acid. This is a water-conservation adaptation of the grasshopper for life in a dry environment.

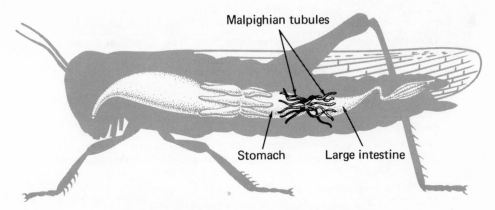

Fig. 9–4. Malpighian tubules of the grasshopper.

The *Malpighian tubules* of the grasshopper, shown in Fig. 9–4, are specialized structures for excretion. These are fine tubules which lead from the sinuses of the open circulatory system into the large intestine. The Malpighian tubules pick up nitrogenous wastes, mineral salts, and water, but most of the water is resorbed. The minerals and uric acid are expelled along with the fecal material through the anus.

Carbon dioxide passes by diffusion from the grasshopper's blood into the tracheae, which carry it to the atmosphere via the spiracles.

Man

Included among the excretory organs of man are the kidneys, lungs, skin, and liver. The major excretory wastes are water, urea, carbon dioxide, and mineral salts.

Water is excreted by man with the air exhaled by the lungs, with the sweat given off by sweat glands in the skin, and with the urine excreted by the kidneys. Nitrogenous waste is given off largely in the form of urea, which is present in the urine and in sweat. Carbon dioxide is removed as the blood passes through the lungs, and is expelled into the atmosphere.

The kidneys of man contain microscopic tubules, called *nephrons*. The nephrons remove urea, water, and mineral salts from the blood and pass these wastes, in the form of urine, through a system of tubes to the outside.

COMPARISON OF NITROGENOUS WASTES

	Protozoa	Hydra	Earthworm	Grass-hopper	Man
Waste	Ammonia	Ammonia	Urea; ammonia	Uric acid (solid)	Urea, and some uric acid
Structure for removal	Cell membrane; contractile vacuole	Cell membrane	Paired nephridia	Malpighian tubules; intestine	Nephrons of kidney

REASONING EXERCISES

1. Distinguish between excretion and egestion. Why is excretion necessary for life?
2. How are the wastes carbon dioxide and metabolic water produced in the body?
3. What is deamination? What are the forms of nitrogenous wastes?
4. What is the function of the contractile vacuole in protozoa? How do protozoa excrete ammonia?
5. Why would it be disadvantageous for man to excrete nitrogenous waste in the form of ammonia? How is the fish adapted for the excretion of ammonia?
6. How is the earthworm adapted for excretion?
7. How do the Malpighian tubules of the grasshopper function in excretion?
8. What are the excretory organs of man? In what form is urea excreted from man? What is the function of the nephron?

(B) THE HUMAN EXCRETORY FUNCTION

In considering the human excretory function, we shall discuss the (1) liver, (2) kidneys, (3) lungs, (4) skin.

The Liver

In addition to its function in digestion and food storage, the liver participates in excretion. It does this in three ways:

Discharge of bile salts. One of the functions of the liver is the discharge of bile salts. As previously noted, bile salts function in the digestion of fat. These salts, and portions of hemoglobin from worn-out red blood cells, pass from the liver to the small intestine. They are egested with the feces.

Inactivation of chemicals. Hormones and many other biologically active chemicals are converted to inactive form in the liver. These inactive chemicals are returned to the blood for excretion by the kidneys.

Deamination and urea formation. Some excess amino acids result from the intake of protein food. These excess amino acids are *not* stored by the body. If not used for growth or repair, they are broken down to form pyruvic acid and ammonia in the process of deamination. The pyruvic acid is oxidized for energy, and the ammonia is converted to urea (in the ornithine cycle).

The Ornithine Cycle

The ornithine cycle is a process by which carbon dioxide and toxic ammonia are converted to the less toxic urea. The overall chemical equation is as follows:

$$2NH_3 + CO_2 + \text{ornithine} \xrightarrow[\text{ATP} \longrightarrow \text{ADP}]{\text{enzymes}}$$

$$\begin{array}{c} NH_2 \\ | \\ C = O + H_2O + \text{ornithine} \\ | \\ NH_2 \\ \text{urea} \end{array}$$

Ornithine is at the start of this process and is produced again at the end of the process. Since ornithine is used over and over without being destroyed, it is evident that the biochemical reactions involved constitute a *cycle*. As you will recall, we previously studied another cycle, the Krebs cycle (page 149).

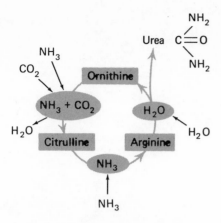

Fig. 9–5. The ornithine cycle

The ornithine cycle is shown in Fig. 9–5. It involves three amino acids: ornithine, citrulline, and arginine. *Ornithine* picks up a molecule of ammonia and a molecule of carbon dioxide, and is converted to *citrulline*. A molecule of water is given off in this reaction. Citrulline picks up another molecule of ammonia and is converted to *arginine*. Arginine takes up a molecule of water and splits into two portions: a molecule of urea (the product) and a molecule of ornithine. In this cycle, or wheel, the ornithine is formed again at each "turn of the wheel." The net result is that one molecule of urea is formed from two molecules of ammonia and a molecule of carbon dioxide. The urea formed in the liver passes into the blood capillaries and is carried by the plasma to the kidneys and sweat glands.

Deamination of Amino Acids

Ammonia, which is used in the production of urea, is formed in the liver from excess amino acids. Since amino groups are removed, this process is deamination. An example of the deamination of an amino acid is given in Fig. 9–6. Note that pyruvic acid, formerly encountered in our study of cellular respiration in Chapter 7, is also produced. Pyruvic acid may be oxidized to yield ATP.

Summary of Nitrogen Excretion

Fig. 9–7 summarizes the steps by which excess amino acids, derived from the breakdown of proteins, are removed from the body. Deamination in the liver forms pyruvic acid and ammonia. The pyruvic acid is oxidized to carbon dioxide and water, releasing energy. Toxic ammonia is changed to less toxic urea by means of the ornithine cycle. The urea is removed from the body by the kidneys and sweat glands. The lungs remove carbon dioxide and water, and the skin and kidneys remove water.

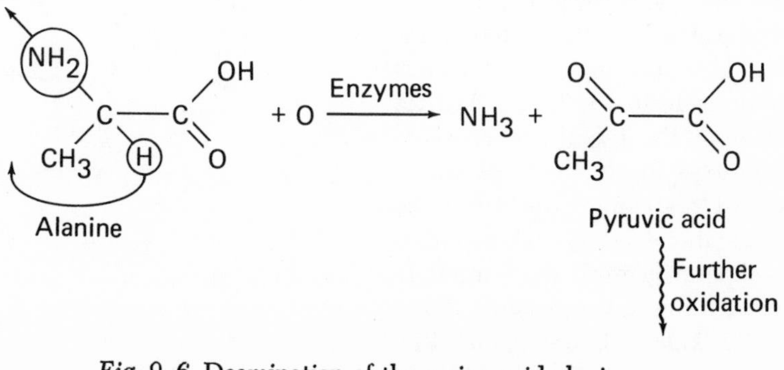

Fig. 9–6. Deamination of the amino acid alanine.

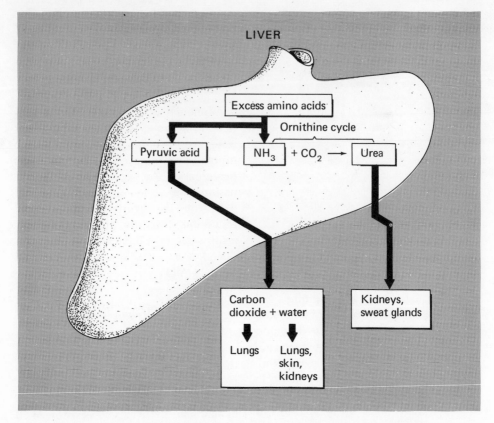

Fig. 9–7. Scheme of excretion of some products of protein metabolism.

The Kidneys

Wastes are brought to the two *kidneys* by the renal arteries which branch from the aorta. (See Fig. 9–8). Within the kidneys, the arteries branch into capillaries which are intimately associated with the microscopic nephrons.

From the kidneys the renal veins carry blood to the inferior vena cava which returns it to the heart. Wastes removed from the blood are excreted by the kidneys in the form of urine. The two *ureters* carry urine from the kidneys to the *bladder* where it is stored temporarily until discharged to the exterior through the *urethra*.

When the kidney is cut open (Fig. 9–9), we can see how the ureter ex-

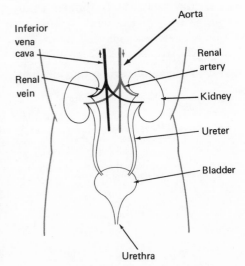

Fig. 9–8. Urinary system of man.

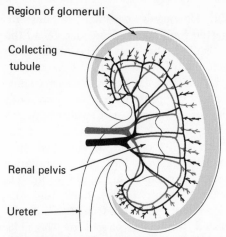

Region of glomeruli

Collecting
tubule

Renal pelvis

Ureter

Fig. 9–9. Longitudinal section of kidney.

the kidney are numerous *collecting
tubules,* which collect urine from
smaller tubules, the microscopic *nephrons.* The outer region of the kidney
is the location of knots of capillaries,
the *glomeruli* (sing. — glomerulus)
which are visible only with the microscope.

Microscopic Structure of Kidney: the Nephron

The nephron, shown in Fig. 9–10, is
the functional unit of the kidney. It is
also called the *renal tubule* or the *kidney tubule.* The nephron is a microscopic thin-walled elongated tubule
about $1\frac{1}{4}$ inches in length. Each human kidney contains approximately one
million nephrons. One end of the nephron is expanded into a double-walled

pands to form a hollow interior portion
of the kidney, the *renal pelvis* (*renal*
means "kidney"). Leading into the
renal pelvis from the solid portion of

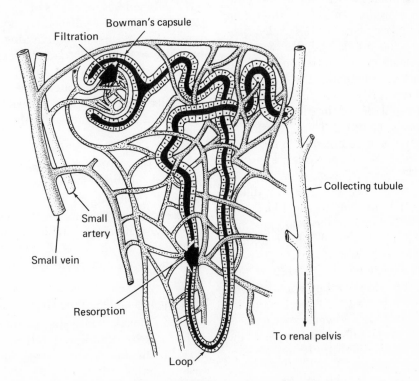

Bowman's capsule

Filtration

Small
artery

Small vein

Resorption

Loop

Collecting tubule

To renal pelvis

Fig. 9–10. Diagram of a nephron.

cup, the *Bowman's capsule*. From each Bowman's capsule a highly coiled portion descends to form a loop, followed by an ascending coiled portion of the tubule. The nephron then joins a larger collecting tubule. Blood is supplied to the nephron by a small branch of the renal artery. This branch forms a knot of capillaries in the cup of Bowman's capsule. The knot of capillaries is called a *glomerulus*. From the glomerulus a small artery leads to a highly branched system of capillaries which covers the coiled loop. These capillaries lead to small veins which eventually form the renal vein.

Formation of Urine

The removal of wastes by the nephron involves two distinct processes: (1) filtration, which occurs at Bowman's capsule, and (2) resorption, which takes place at the coiled portions of the tubule and the loop. The main features of filtration and resorption are summarized in Fig. 9–11.

Filtration. Materials diffuse from the blood capillaries of the glomerulus into Bowman's capsule. This diffusion is aided by the high pressure in the arteries leading to the capillaries. Diffusing into the capsule are water, glucose, amino acids, salts, and urea. Not diffusing, and remaining in the blood, are the cells and most of the blood proteins. The filtrate (material which diffuses through) is essentially the same as plasma except for the absence of proteins. This was shown by scientists who were able to insert a fine glass micropipette into the Bowman's capsule of amphibians and obtain a sample of the filtrate passing into the Bowman's capsule.

Resorption. The filtrate passing into the Bowman's capsule contains digested food materials needed by all the cells of the body. It also contains large amounts of water, critically needed by land animals. However, these materials do not all pass out of the body with the urine. Instead, as the filtrate passes along the coiled tubules and the loop, much of its content is *resorbed* (absorbed again) into the capillaries which surround the nephron in this region. Water passes back into the blood by passive osmosis, but digested foods are passed by the cells of the tubule into the blood by *active transport*. Energy for active transport is supplied by ATP. This resorption process is highly selective. Glucose, amino acids, and mineral ions are resorbed back into the bloodstream to a greater extent than urea.

If the blood has an excess of glucose (as in the disease diabetes), active transport is incapable of returning all the glucose to the blood and glucose appears in the urine.

The urine contains water, urea, uric acid, salts, some hormones, breakdown products of hemoglobin, and other organic materials.

Homeostasis. The kidneys remove from the blood the waste products of protein metabolism. They also remove salts, water, and other substances present in excessive amounts. In this manner, they maintain a stable internal environment for the cells of the body. The kidneys are important homeostatic organs.

The Lungs as an Excretory Organ

Carbon dioxide and water, produced in the cells by aerobic respiration, are carried by the blood to the lungs. These waste materials diffuse into the alveoli and pass out of the body with the ex-

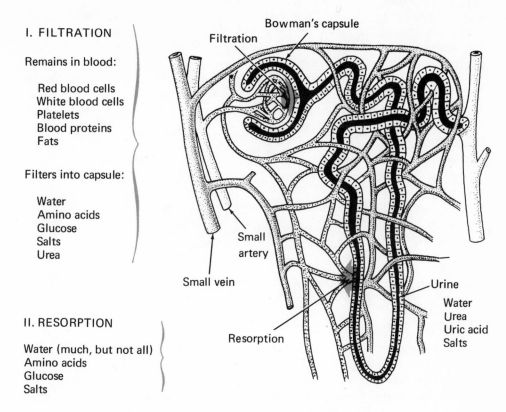

I. FILTRATION

Remains in blood:

Red blood cells
White blood cells
Platelets
Blood proteins
Fats

Filters into capsule:

Water
Amino acids
Glucose
Salts
Urea

Bowman's capsule

Filtration

Small artery

Small vein

II. RESORPTION

Water (much, but not all)
Amino acids
Glucose
Salts

Resorption

Urine
Water
Urea
Uric acid
Salts

Fig. 9–11. Summary of filtration and reabsorption.

haled air. The lungs have been described in Chapter 8.

The Skin

Structure of the skin. As shown in Fig. 9–12, the skin consists of two layers: (1) the epidermis, which is the outer layer, and (2) the dermis, which is the heavier inner layer, or true skin.

The outer layer, the *epidermis*, consists of many layers of closely packed epithelial cells. The inner layers are living and continuously dividing to form new cells. Cells gradually die and become filled with protein as they move toward the surface, until they are converted to thin flakes. These dead cells are constantly being rubbed off and replaced by the inner cells. The epidermis is watertight and covered with a fine oil film. The main function of the epidermis is to protect the active tissues beneath it.

The inner layer, the *dermis*, lies under the epidermis. It is a thick, heavy layer of connective tissue. Present are capillaries, lymph vessels, nerve endings, oil and sweat glands, and hair follicles.

The *sweat glands*, found in the dermis, have an opening to the surface of the skin. Sweat or perspiration, which is about 98 percent water, is excreted by these glands, and it usually evaporates soon after it reaches the surface of the skin. As it evaporates, the water produces a cooling effect on the skin. This process aids in maintaining a constant body temperature. The remaining

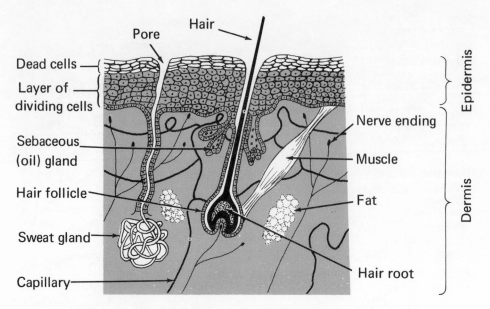

Fig. 9–12. Cross section of the skin.

2 percent of the sweat usually consists of nitrogenous wastes and dissolved salts. It is important to keep the skin clean to prevent dirt from clogging the pores and causing body odors.

Functions of the skin. The functions of the skin are as follows:

1. *The skin is a protective organ.* When unbroken, the skin protects the body from mechanical injury and bacterial infections. In addition, it protects the inner tissues from drying out.

2. *The skin is an organ of sensation.* The skin has nerve endings that detect changes in temperature, pressure, pain, or touch.

3. *The skin is an organ of excretion.*

It assists the kidneys in the excretion of urea and uric acid.

4. *The skin helps to regulate the temperature of the body.* It is richly supplied with tiny blood capillaries. As the sweat glands excrete perspiration, evaporation occurs. In this way, heat is withdrawn from the outer tissues, lowering the temperature of the blood. This in turn lowers the body temperature. On the other hand, as the body temperature begins to drop, the surface blood vessels become smaller (constrict). Accordingly, less blood flows through these blood vessels, which reduces the amount of heat lost from the skin.

COMPLETION QUESTIONS

[A] 1. Wastes produced by an organism as it carries on its life processes are called .... wastes.

2. The removal of the wastes from metabolism is ....

3. The removal of food which has passed through the alimentary canal is known as ....

4. Carbon dioxide is excreted through the cell membrane of protozoa by the process of ....

5. The energy for the active transport of wastes is supplied by ....

6. Excess amino acids are changed to .... wastes.

7. Ammonia results from deamination of .... acids.

8. Uric acid is not poisonous to cells because it is ....

9. Uric acid is excreted from insects, reptiles, and ....

10. *Hydra* excretes nitrogenous wastes in the form of ....

11. In the earthworm, nitrogenous wastes are excreted through pairs of tubes called .... and carbon dioxide diffuses through the ....

12. In the grasshopper, nitrogenous wastes are conveyed to the large intestine through .... and carbon dioxide is carried to the environment through ....

13. Two wastes excreted through the lungs of man are .... and ....

14. Man excretes urea in the urine and in the ....

15. The microscopic structures within the human kidney which remove urea from the blood are the ....

(B) 16. The liver acts as an organ of excretion by breaking down excess .... and by forming ....

17. Ammonia is converted to urea in the .... cycle.

18. Urea formed in the liver is removed from the blood in the .... and ....

19. Excess amino acids in cells are broken down to form ammonia and pyruvic acid by the process of ....

20. Urine is carried from the kidneys to the bladder by the ....

21. The tube in man which discharges urine from the bladder is the ....

22. The functional unit of the kidney of man is the ....

23. Filtration of materials from the blood into the nephrons occurs at the ....

24. A tiny mass of capillaries in Bowman's capsule is known as a (an) ....

25. The removal of wastes at the nephrons involves the processes of .... and ....

26. Glucose and amino acids pass from the loop portion of the nephron back into the blood by the process of ....

27. The presence of glucose in the urine is a possible indication of the disease ....

28. Oil glands are found in the layer of the skin called the ....

29. Wastes are excreted through the skin by the .... glands.

30. As sweat evaporates, the body's temperature tends to ....

MULTIPLE-CHOICE QUESTIONS

A 1. If the organism is a protozoan, the substances which might be excreted are (1) carbon dioxide and lactic acid, (2) end products of extracellular digestion, (3) carbon dioxide and ammonia, (4) digestive enzymes and hormones.

2. The major function of the contractile vacuole is (1) removal of nitrogenous wastes, (2) homeostatic regulation of water content, (3) digestion of food, (4) removal of excess amino acids.

3. Which of the following is most likely to excrete most of its nitrogenous waste as ammonia? (1) *Paramecium*, (2) grasshopper, (3) pig, (4) man.

4. Which is characteristic of organisms which excrete uric acid as their main nitrogenous waste? (1) They usually live in water. (2) They are usually land dwellers. (3) They can carry on only anaerobic respiration. (4) They cannot metabolize proteins.

5. A certain red dye turns yellow when treated with a mild acid. If a small amount of this red dye is added to 10 cubic centimeters of water and a small aquatic animal is placed in the solution, the reddish solution will turn yellow after about 20 minutes. The most reasonable explanation is that (1) the animal's body is composed of acids, (2) the animal is secreting gastric juice into the solution, (3) products given off by the animal have created acid conditions, (4) a reaction has taken place between the water and the oxygen in the air.

B 6. The bladder of man empties to the outside by means of the (1) urethra, (2) ureter, (3) capillary network, (4) kidney tubule.

7. All of the following diffuse from the blood into the kidney tubules of man *except* (1) red blood cells, (2) urea, (3) glucose, (4) amino acids.

8. An important function of the human kidney is to (1) store glycogen, (2) remove carbon dioxide from the blood, (3) remove nitrogenous wastes from the blood, (4) produce white blood cells.

9. The kidneys of man have a function similar to the function of the contractile vacuole of protozoa in that both (1) synthesize and excrete urea, (2) control the excretion of all metabolic wastes, (3) control the secretion of hormones, (4) help to maintain the water balance of the organism.

10. In man the passage of glucose from the kidney tubules into the blood is an example of (1) diffusion, (2) digestion, (3) active transport, (4) hydrolysis.

CHAPTER TEST

1. A structure used by *Paramecium* for the excretion of liquid wastes is the (1) anal spot, (2) food vacuole, (3) trichocyst, (4) contractile vacuole.

2. One of the kidney's functions is to (1) complete the digestion of urea, (2) destroy old red blood cells, (3) maintain a supply of glycogen, (4) maintain the normal composition of the blood.

3. The process by which metabolic wastes are removed from an organism is (1) excretion, (2) elimination, (3) egestion, (4) secretion.

Below is a diagram of the human circulatory system. For each of the questions 4 through 10, write the number of the structure which is most closely related to the function given.

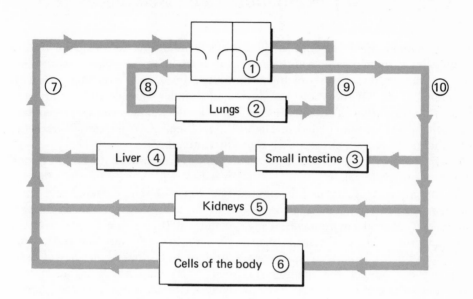

4. Urea leaves the circulatory system.
5. Carbon dioxide enters the circulatory system.
6. Amino acids enter the circulatory system.
7. Oxygen enters the circulatory system.
8. A thin, moist membrane is in contact with the external environment.
9. Urea is formed from excess amino acids.
10. A vein carries oxygenated blood.

In each of questions 11 through 15, write the letter of the organism, chosen from the list below, which is most closely related to the characteristic described.

A. *Hydra*   C. grasshopper
B. earthworm   D. man

11. Paired nephridia.
12. Major nitrogenous waste is uric acid.
13. Malpighian tubules.
14. Intestine is used to eliminate nitrogenous waste.
15. Has a kidney containing nephrons.

CHAPTER 10 / Regulation—The Nervous System

The various organ systems of the multicellular animal are coordinated so that the organism can function as a whole. This coordination is also known as *regulation*. The activities of the animal body are regulated through (1) the nervous system, and (2) the endocrine system. These two systems are themselves interrelated.

In the process of regulation, organisms receive information from the external and internal environment. They respond to it by a movement, a secretion, or growth. In this chapter, we will consider one of the methods of regulation—the nervous system.

A | THE NERVOUS SYSTEM

Basic Terminology

Stimulus and response. A *stimulus* is a change in the environment that affects the sensitive cytoplasm. Stimuli may be physical or chemical in nature. Examples of stimuli are changes in temperature, in pressure, in electromagnetic frequencies (light), in compression of nearby air molecules (sound), and in the chemical environment.

An *impulse* consists of electrical and chemical changes that travel along a nerve fiber. What passes over a nerve from the eye is not light but an impulse. Similarly, sound, taste, heat, and pain do not pass along the nerve fiber. The nature of the impulse is the same regardless of the stimulus.

The action or movement resulting from a stimulus is a *response*. One-celled organisms may respond directly to a stimulus, as when *Paramecium* moves away from a crystal of salt. In multicellular animals the impulse carried by nerve fibers causes (1) contraction of muscle cells, or (2) secretion by a gland.

Basic structural features. "Receptor" is a broad term used to cover the various types of sense organs. A *receptor* may be defined as specialized tissue in contact with nerve cells and sensitive to a specific stimulus. Some receptors are merely the naked endings of nerve cells in the skin which are sensitive to changes in temperature. Other receptors are complicated structures such as the eye or ear. Each receptor is specialized to be particularly sensitive to a specific stimulus. Receptors can be stimulated by chemicals (taste and smell), mechanical stimuli (touch and pressure), light (sight), etc. Some receptors provide information about the degree of contraction of muscles or about the position of parts of the body.

Effectors are the parts of the body that respond. These are the muscles that contract or the glands that secrete.

The *neuron* or nerve cell is the basic unit of which the nervous system is composed. Although visible only with the microscope, these cells may be very long; some may extend from the toes to the spinal cord. A "typical" (motor) neuron is diagrammed in Fig. 10–1.

The three main portions of the neuron are (1) the *cyton* or cell body, with its branched extensions, the *dendrites,*

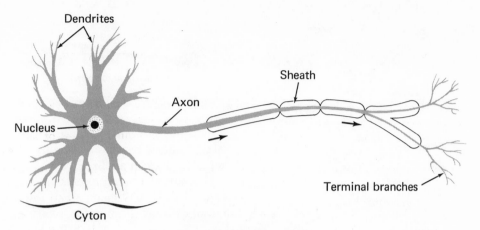

Fig. 10–1. Motor neuron.

(2) a long *axon* covered by a protective *sheath*, and (3) the *terminal branches*. The direction in which the impulses travel is from the dendrites toward the terminal branches.

A neuron is microscopic, but a nerve can be seen as a thin white string. A *nerve* is composed of a bundle of axons covered with connective tissue. As in the wires in a telephone cable, each axon is insulated from its neighbors by the protective sheath composed of myelin. Nerves are specialized for the transmission of impulses.

A *synapse* is the region between the terminal branches of one neuron and the dendrites of a second neuron to which the impulse is carried. As revealed by the electron microscope, a slight gap exists between the two neurons. The mechanism by which an impulse is conducted across the synapse will be discussed shortly.

Adaptations Among Animals

Protozoa. Since it is a one-celled organism, it is evident that *Paramecium* has no nervous system. Yet its behavior is coordinated, and it makes responses to stimuli. When *Paramecium* bumps

into an object, it backs up by reversing the direction in which its cilia beat. It moves away from intense heat or light and irritating chemicals such as strong acids. It is attracted by oxygen and weak acids. The cilia of *Paramecium* are coordinated by bundles of tiny fibers which connect the bases of the cilia. Cutting these fibers destroys the coordinated activity of rows of the cilia.

Hydra. The nervous system of *Hydra* (Fig. 10–2) consists of a network of cells throughout the body, called a *nerve net*. This nerve net lies between the two layers of the body wall. Receptor cells are present in both the ectoderm and endoderm. There is no *specific pathway* for the carrying of impulses; instead, impulses gradually spread over the body. Impulses can pass in either direction over the neurons.

When the nematocysts (stinging cells) of *Hydra* are struck by living prey, the nematocysts release tube-like threads that penetrate the prey. The threads inject a chemical that paralyzes it. At the same time, the tissues of the prey release the substance *glutathione*, which acts as a stimulus for

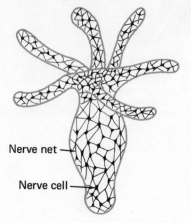

Fig. 10–2. Nerve net in *Hydra*.

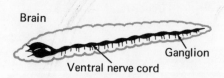

Fig. 10–3. Nervous system of earthworm.

initiating the feeding reflex in *Hydra*. Experiments have been performed in which glutathione has been added to water containing a living *Hydra*. The animal periodically shortened its tentacles, opened its mouth, and inserted the tentacles into the mouth. This activity shows coordinated behavior by a simple organism lacking a well-developed nerve center.

The earthworm. The nervous system of the earthworm is more complex than that of *Hydra*. Groups of neurons are organized into clusters called *ganglia* (sing. – ganglion). Ganlia are enlarged structures containing several neurons. Ganglia may serve as relay centers for directing incoming impulses in several directions.

The earthworm represents an early stage in the evolutionary trend toward a well-developed central nervous system. (See Fig. 10–3.) It has a simple central nervous system that permits impulses to take a definite pathway from receptors to effectors.

A large paired ganglion at the front end of the earthworm is the "brain." Well-defined pathways which link receptors and effectors are centralized in the *nerve cord*. In invertebrates the nerve cord is a solid paired structure along the ventral (under) side of the body.

The grasshopper. The grasshopper has a well-developed nervous system. This consists of two solid nerve cords, the ganglia, and a "brain." The nerve cords are located on the ventral side of the body (Fig. 10–4A).

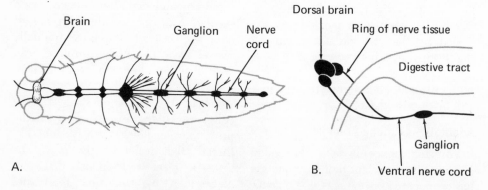

Fig. 10–4. Nervous system of grasshopper. A. Viewed from above. B. Side view showing dorsal brain and ring of nerve tissue around the digestive tract.

The grasshopper's nervous system is basically like that of the earthworm, but it has a greater variety of sense organs. In both the earthworm and the grasshopper, the ventral nerve cord is joined to the dorsal "brain" by a ring of nerve tissues surrounding the diges- tive tract as illustrated in Fig. 10–4B.

Vertebrates. In vertebrates, by contrast, the nerve cord is dorsal to (above) the digestive tract and is a *hollow* tube. The hollow nerve cord of vertebrates is expanded into a frontal enlargement, the brain.

REASONING EXERCISES

1. Distinguish between receptors and effectors.
2. Distinguish between a neuron and a nerve.
3. How does *Paramecium* respond to stimuli? How does it coordinate its behavior?
4. How is the feeding reflex initiated in *Hydra?*
5. Why do we say that the nervous system of the earthworm is more complex than that of *Hydra?*
6. How does the nervous system of the grasshopper resemble that of the earthworm? How does it differ from it?

Ⓑ NERVOUS SYSTEM OF MAN

The nervous system of man consists of two divisions: (1) the central nervous system, and (2) the peripheral nervous system.

Central Nervous System

The central nervous system is composed of the brain and spinal cord (Fig. 10–5).

The *spinal cord* of man is a hollow tube composed of nerve fibers that carry impulses to and from the brain. The brain is an expanded hollow portion of the spinal cord. Composed of millions of neurons and their synapses, the brain has been likened to a giant telephone switchboard. However, the brain does more than merely connect neurons. In some as yet unexplained

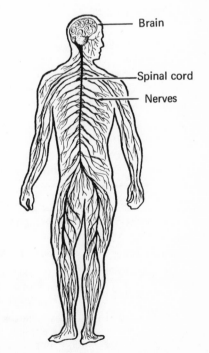

Fig. 10–5. Central nervous system of man.

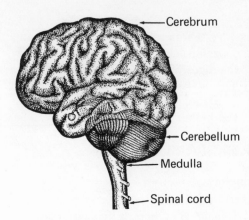

Fig. 10–6. Parts of the brain.

The superiority of man over other animals has been attributed to the extensive development of the gray matter of his cerebrum. Somehow, as impulses are passed from neuron to neuron, new pathways come into use that result in thought and voluntary action.

Although the eye receives the stimulus of light, the *sensation* of sight occurs in the cerebrum of the brain. For example, if the optic nerves leading from the eyes are stimulated in some manner (such as by pressure or electrical stimulus), the sensation of sight occurs in the cerebrum. Each of the sense organs, and each part of the skin, has a special part of the cerebral cortex where its impulses are received and interpreted. Furthermore, there are other regions of the cerebral cortex that control specifically the voluntary muscles of the body: a right arm region, a left toe region, etc. The higher functions of coordination — reasoning, memory, emotion — are controlled mainly by the front part of the cerebrum.

manner it enables man to have sensations, to make judgments, to initiate action, and to think creatively. The parts of the brain and their functions are indicated in Fig. 10–6 and the table below.

The *cerebrum* is more highly developed in man than in any other animal. The surface of the cerebrum has ridges and folds (convolutions) that greatly increase its surface area. The *cortex,* the outer portion of the cerebrum, is composed of cytons and numerous synapses, which comprise the *gray matter.* The inner portion consists chiefly of axons with fatty sheaths, which make up the *white matter.*

Peripheral Nervous System

The peripheral nervous system is composed of (1) *spinal nerves* to and from the spinal cord, and (2) *the autonomic nervous system.* The au-

THE BRAIN AND THE SPINAL CORD

Part	Function
Cerebrum	Creative thought. Memory. Judgment. Control of activities. Sensation.
Cerebellum	Coordination of voluntary activities. Control of balance.
Medulla	Controls reflexes in upper regions of the body involving heart action, blood pressure, breathing, coughing, swallowing, sneezing, movements of digestive system.
Spinal cord	Regulates reflexes from the neck and lower regions such as: pulling a finger from a hot stove or the knee jerk. Carries nerve impulses between various parts of the body and the brain.

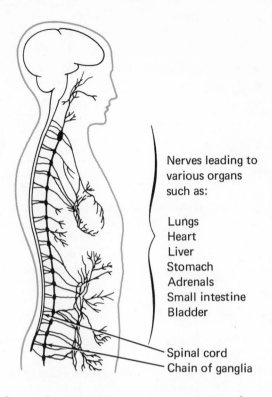

Nerves leading to
various organs
such as:

Lungs
Heart
Liver
Stomach
Adrenals
Small intestine
Bladder

Spinal cord
Chain of ganglia

Fig. 10–7. Relation of autonomic nervous system to central nervous system.

tonomic nervous system controls many of the automatic actions of the body. It is linked by nerves to the central nervous system (Fig. 10–7). Thus, the central nervous system and the autonomic nervous system are interdependent. The autonomic nervous system includes two vertical chains of ganglia, one on each side of the spinal cord. From these ganglia nerves lead to many involuntary muscles and glands of the body. A large group of ganglia constitutes a *plexus* (for example, the solar plexus which lies just above the stomach). A plexus resembles a ganglion in serving as a relay center containing numerous associative neurons. However, a plexus is larger and is a sheet-like mass of nervous tissue.

The autonomic nervous system contains two sets of nerves running to the same organs but having opposite effects. (1) *Sympathetic nerves* are connected with ganglia near the spinal cord where they receive impulses from the middle portion of the spinal cord. (2) *Parasympathetic nerves* are connected with ganglia scattered throughout the body. They receive impulses from nerves coming directly from the brain and the lower portion of the spinal cord. Sympathetic and parasympathetic nerves are both connected to the internal organs. They act in antagonistic pairs, one kind speeding up activity and the other slowing it down. This provides a homeostatic system of checks and balances for controlling the involuntary responses of many internal organs.

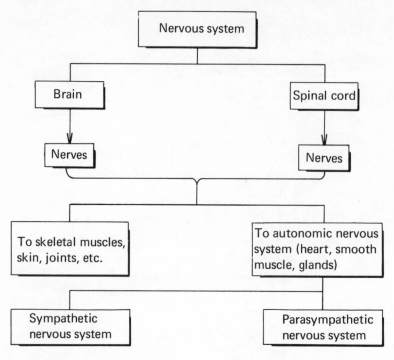

Fig. 10–8. Organization of the nervous system in man.

The autonomic nervous system regulates many of the automatic activities of the body such as heartbeat, rate of breathing, peristalsis, secretion of digestive juices, and the flow of blood to capillaries. Familiar examples of the operation of the autonomic nervous system are blushing, increased heart beat, and "goose pimples."

The general organization of the nervous system is given in Fig. 10–8.

Transmission Across the Synapse

The synapse is the region where the terminal branches of one neuron are separated by a slight gap from the dendrites of another neuron. As shown in Fig. 10–9, a nerve impulse is conducted across this gap from the terminal branches to the dendrites, in one direction only.

The mechanism by which an impulse "crosses the gap" was disclosed through the pioneering experiments of Otto Loewi who won a Nobel Prize for his discoveries. Loewi experimented on the hearts of frogs. The frog's heart beats at its own slow rhythm when removed from the body and placed in saline solution. Attached to the heart are two nerves, the *accelerator nerve* which speeds heart beat and the *vagus nerve,* which slows the heart beat. In one of his experiments, Loewi immersed two beating hearts in separate dishes of saline solution (Ringer's solution). With a mild electric current, he stimulated the vagus nerve which was attached to the first heart. This heart began to beat more slowly, as expected. Loewi then transferred saline solution from the dish containing the first heart to the dish containing the second heart. He

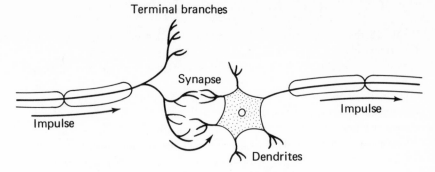

Fig. 10–9. The synapse.

found that the second heart also began to beat more slowly! Thus, the first heart, which received impulses from its vagus nerve, had released into the water *a chemical substance* which affected the second heart so that it also beat more slowly. This was the first indication that a chemical substance was involved in neural pathways. The chemical substance was later identified as *acetylcholine*.

The conduction of an impulse across a synapse is explained as follows: When an impulse reaches the ends of the terminal branches of one neuron, these nerve endings secrete acetylcholine. The acetylcholine diffuses across the tiny gap and then acts as a stimulus to the dendrites of the second neuron. Diffusion of the acetylcholine across the synapse is slower than the passage of an impulse along an axon. Thus, an impulse moves relatively slowly along a pathway which includes many synapses.

Another chemical secreted by nerve endings is *noradrenaline* (also commonly called sympathin). Acetylcholine and noradrenaline are called *neurohumors* (or neurohormones). A neurohumor is a substance released by the terminal branches of neurons. Such a substance permits impulses to be conducted from neurons to other neurons or to muscle cells and gland cells. Acetylcholine and noradrenaline are opposite in their effects on muscles and glands.

Acetylcholine and noradrenaline help to explain how two different nerves carrying impulses to the same organ can have different effects. The nerve in the sympathetic nervous system which carries impulses to the small intestine releases noradrenaline at the muscles, causing an increase in peristalsis. The nerve in the parasympathetic nervous system which leads to the small intestine releases acetylcholine, causing a decrease in peristalsis.

Cholinesterase. Acetylcholine released at a synapse continues to stimulate the second neuron until the neurohumor is destroyed. Its rapid destruction is achieved with the aid of *cholinesterase*, an enzyme that is widely distributed in the body. Cholinesterase hydrolyzes (destroys) the neurohumor. Otherwise there might be continuous activity of the muscles and glands such as to cause spasms or paralysis. Deadly nerve gases which inhibit cholinesterase have been prepared. These have the effect of generalized continuous stimulation of the parasympathetic nervous system. Fortunately, they have not been used.

REASONING EXERCISES

1. What are the two parts of the central nervous system of man? How are they related?
2. Distinguish structurally between the gray matter and white matter of the brain.
3. What is the importance of the gray matter of the brain?
4. What is the autonomic nervous system?
5. Distinguish between sympathetic and parasympathetic nerves.
6. Why can an impulse travel across a synapse in one direction only?
7. Distinguish between the roles of acetylcholine and cholinesterase in the transmission of impulses.

(c) NATURE OF THE NERVE IMPULSE

The *nerve impulse* may be defined as an electrochemical change which proceeds along the neuron. The speed of the nerve impulse is about 0.6 miles per second (100 meters per second, or 350 ft. per second). The passage of a nerve impulse along a nerve fiber may be compared to the travel of a burning spark along a fuse cord. In both the case of the nerve fiber and the fuse cord, a change at one point causes a change at the subsequent adjoining point. However, the fuse cord does not restore itself whereas the nerve fiber does return to its original state, ready to accept the next impulse.

Electrical Changes

Using the giant neurons of the squid, scientists have been able to measure the charge on the inside and outside of the membrane of the axon. This has enabled them to compare the electrical properties of the two sides of the membrane. When no impulse is being carried, the membrane is considered to be a resting membrane. The outside of the membrane has a positive electrical potential as compared to the inside, of about 0.07 volts. Since each side of the membrane is like the poles of a battery, the membrane is said to be *polarized*. (See Fig. 10–10.)

As the impulse arrives at any section of the axon, there is a movement of electrons through the membrane. A small section (0.1 cm to 10 cm in length) obtains an opposite charge: the outside becomes negative and the inside becomes positive. Thus, for a brief period of time, a section of the membrane is *depolarized*. The depolarized section causes changes in the adjacent section of the membrane so that it also becomes depolarized. Shortly after the impulse passes a section of the neuron, it returns to its original polarized state

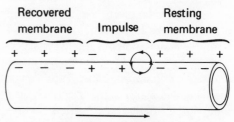

Direction of impulse

Fig. 10–10. Change in electrical charge on each side of the membrane during the passage of an impulse.

and is then ready to pass another impulse.

The nerve impulse has been called "a wave of depolarization," "a wave of negativity," and "a wave of chemical and electrical changes."

Chemical Changes

We have noted the difference in potential on each side of the membrane of an axon. It is explained by the difference in concentration of *sodium* and *potassium* ions on each side of the membrane. Depolarization is caused by the movement of ions through the membrane.

The resting cell is more permeable to positive potassium ions than to positive sodium ions. Therefore, sodium ions (Na^+) are in higher concentration outside the resting membrane, whereas potassium ions (K^+) are in higher concentration inside it (Fig. 10–11). Furthermore, inside the membrane, there is a high concentration of large negative organic ions which do not diffuse readily outward. (Negative chlorine ions are present outside the membrane.)

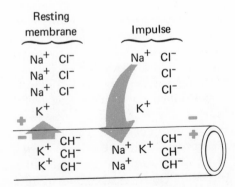

Fig. 10–11. Movement of ions during passage of an impulse. The membrane becomes permeable to the inward diffusion of sodium ions. (Large negative organic ions are represented as CH^-.)

There is a slight outward diffusion of positive potassium ions, which causes the inside to be negatively charged. This is characteristic of the resting state. In short, the slight outward diffusion of K^+ explains the difference in potential between the two sides of the membrane.

As the impulse passes a section of the axon, the membrane becomes much more permeable to Na^+. Thus, *the positive sodium ions pass into the cell.* For a brief period of time, the membrane loses its potential (becomes depolarized) and the inside becomes positive. After the impulse passes, the membrane becomes impermeable to the sodium. A small quantity of potassium diffuses out of the membrane to restore the original potential: positive on the outside and negative on the inside.

As succeeding impulses pass over a fiber, this process continues to add sodium to the inside of the fiber and to remove potassium to the outside. Metabolic processes which reverse this process and remove sodium from the inside of the fiber are collectively called the *sodium pump.* This is a form of active transport which requires energy from ATP. The nerve cell carries on active respiration to yield this energy currency of the cell.

"All or None" Response

The stimulus must be of sufficient strength to exceed the threshold, or initial barrier to its passage. Once this threshold is exceeded, each impulse is of the same strength and speed, regardless of the strength of the stimulus which initiated the impulse. Thus, the impulse has an "all or none" characteristic. The action in a fuse cord provides an analogy to this "all or none" response. Thus, initially, a fuse cord is

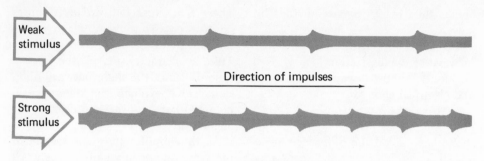

Fig. 10–12. Strength of stimulus. Each impulse is represented as a bump in the axon. A strong stimulus causes more impulses to be sent per second. However, each impulse is of the same strength and travels at the same speed in any one neuron.

either lighted or not, but once lighted, the spark travels at a constant speed which is not affected by the heat of the match which lighted the fuse.

Strength of Stimulus

A stimulus of greater strength initiates a train of more frequent impulses. When these impulses result in the production of sufficient acetylcholine to exceed the threshold of the next neuron, the impulse has been carried across the synapse. It is the number of impulses, not any variation in their speed or strength, which determines the intensity of a response. This is illustrated in Fig. 10–12.

(D) TYPES OF BEHAVIOR

Reflexes

A *reflex* is a relatively simple type of behavior. Some examples of reflexes are coughing, sneezing, blinking, and pulling one's hand quickly away from a hot object. Characteristics of reflexes are:

1. Reflexes are inborn (present from birth).
2. Reflexes are automatic (do not require thought; not voluntary).
3. Reflexes protect the body (adapt

the organism to the environment).

Reflexes in a pithed frog. Pithing is a procedure by which a trained individual inserts a needle through the base of a frog's brain. If the needle is inserted forward, the brain is destroyed and the frog is a *cranial frog*. After the brain is destroyed, the needle may be partially removed and then inserted down the spinal column to destroy the spinal cord. Such a frog, lacking a brain and a spinal cord, is a *spinal frog*.

If an object is brought near to the eyes of a cranial frog it does not show the blinking reflex. However, if a small piece of paper dipped in dilute acetic acid is placed in contact with the leg, the leg rises by reflex action. (The acid should quickly be removed by placing the frog into a previously prepared jar of water.) When the same procedure is performed on a spinal frog (whose brain *and* spinal cord are destroyed), the leg does not move. This shows that reflexes occurring in the lower portions of the body can be performed without the brain. However, they do require the spinal cord. Because of the importance of the spinal cord in lower reflexes, we will now consider its structure.

Structure of the spinal cord. The spinal cord is a tubular structure con-

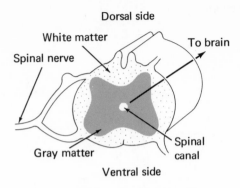

Dorsal side

White matter

Spinal nerve

To brain

Gray matter

Spinal canal

Ventral side

Fig. 10–13. Spinal cord. One spinal nerve is shown. Numerous spinal nerves are present in pairs. In man, with his erect position, the "ventral" side is anterior and the "dorsal" side is posterior.

taining a hollow space, the *spinal canal.* In man, the spinal cord is protected by 26 vertebrae, which make up the spinal column, and by three membranes. A cross section of the spinal cord is shown in Fig. 10–13.

The diagram reveals a central H-shaped darker region, the gray matter. This is the region where synapses occur between neurons. The white matter, in the outer areas, contains bundles of nerve tracts which carry impulses up and down the spinal cord.

The reflex arc. The reflex arc, diagrammed in Fig. 10–14, is in its simplest form.

The diagram shows only two neu-

rons. The six major steps of this simple reflex arc are:

1. *Stimulus.* In this case, the pressure of a tack on the skin is the stimulus (the change in the environment).

2. *Receptor.* The skin is the receptor (sense organ) which detects the stimulus and changes it to an impulse.

3. *Sensory neuron.* This neuron receives the impulse and conducts it to the spinal cord. Since sensation may later result, this neuron is called the *sensory neuron* (or *afferent neuron*). Note that the sensory neuron is different in shape from the neuron previously diagrammed in Fig. 10–1. Its cyton is outside the spinal cord.

4. *Synapse.* The synapse is present in the gray matter of the spinal cord. A neurohumor, secreted by nerve endings, carries the impulse across the gap between the terminal branches of the sensory neuron and the dendrites of the motor neuron.

5. *Motor neuron.* Another neuron carries an impulse from the synapse to the effector. Since the action of this neuron may result in motion, it is called the *motor* or *efferent neuron.*

6. *Effector.* The motor neuron releases a neurohumor which causes (*a*) a muscle to contract *or* (*b*) a gland to secrete. The muscle or gland is called the effector.

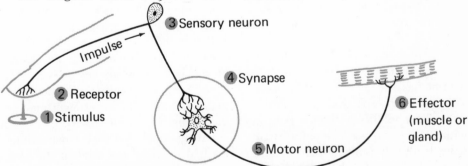

Impulse

③ Sensory neuron

④ Synapse

② Receptor

① Stimulus

⑥ Effector (muscle or gland)

⑤ Motor neuron

Fig. 10–14. Reflex arc, showing six major steps.

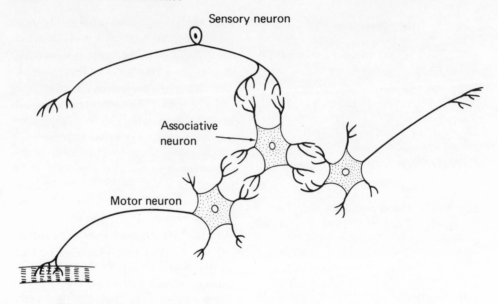

Fig. 10–15. A more complicated reflex arc involving an associative neuron.

Associative neurons may be present between the sensory and motor neurons, as shown in Fig. 10–15. Branches of the associative neurons may permit impulses to pass to various parts of the spinal cord.

A person who touches his finger to a hot stove does more than merely pull his finger away by reflex action. In addition to the reflex, he experiences the sensation of pain and engages in thought and voluntary activity.

Impulses are conducted to the brain and back to various muscles and glands. Nerve tracts in the white matter of the spinal cord carry impulses to and from the brain. Associative neurons play a part in these pathways; however, by the time a person feels the pain he has already removed his finger. This shows the speed of the reflex action which follows a shorter pathway with fewer synapses. The pathway of a reflex is already built into an organism at birth because of its heredity. *Reflexes are inborn, automatic patterns of be-* havior which are important in controlling everyday behavior.

The Conditioned Reflex

In 1900 the Russian physiologist Ivan Pavlov experimented with dogs to find out whether a simple reflex could be changed. Pavlov knew that dogs were born with a simple reflex causing the flow of saliva when food was given to them. Pavlov placed a dog in a soundproof room and rang a bell just before feeding it. Each time the bell was sounded the dog's meal followed. After numerous trials, Pavlov rang the bell but did not feed the dog. However, the dog responded to the sound of the bell by producing saliva. This type of experiment was performed with many animals. The experiment may be diagrammed as in Fig. 10–16.

Pavlov concluded that a new stimulus had replaced the original stimulus. As noted, a reflex is an *inborn* automatic act. A *conditioned reflex* is a re-

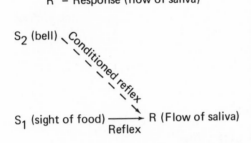

S_1 = Original stimulus (sight of food)

S_2 = Second stimulus (bell)

R = Response (flow of saliva)

Fig. 10–16. Conditioning.

flex in which a substitute stimulus results in the same response as the original stimulus.

Much of the process of training and acquiring of habits in animals and man is a matter of conditioning in which a whole series of new stimuli are associated with a given response.

Operant Conditioning

Operant conditioning is a branch of psychology which attempts to use the results of experiments with animals to explain human behavior. This type of conditioning can be illustrated by a pigeon that receives a pellet of food when it randomly makes a partial turn to the right. Each time it repeats this behavior, it receives the same reward so that the response is "reinforced." After a time, the pigeon receives a reward only when it goes still farther to the right.

By continuing this process, the experimenter can "shape" the pigeon's behavior so that it makes complete turns during a brief period of training.

Many psychologists today are wary about drawing any conclusions about human behavior from such studies. Our mental processes work on a vastly more complex level than those of even the most "intelligent" animals, and must be studied in humans, not lower animals.

Other psychologists, however, find definite evidence of the importance of operant conditioning in man. How much of our behavior is shaped by rewards and punishments administered by family, friends, media of mass communication, and forces in society?

Instincts

Many animals are able to perform highly complex acts in the same way as their parents did, without ever being trained to do so. The first web a spider spins is just like all others of its species, and birds migrate in the same paths as their parents, without previous experience.

Little is known about the development of these *instincts*, although it is certain that the mechanisms of heredity are somehow involved. There is much disagreement as to how far the concept of instinct can be applied to human behavior, complicated by the fact that many psychologists use the word with an entirely different meaning.

Habits

Habits are complicated acts which, when learned, become automatic through constant repetition. Habits may be good or bad. Good habits are useful, for they make us more efficient by saving time and energy. The following are involved in the formation of a good habit:

1. A desire to form the habit.

2. Performing the act in the correct manner from the start.

3. Repeating the act many times.

4. Satisfaction. Each performance of the act should result in satisfaction. Satisfaction can be achieved through reward or through withholding of punishment. An *intrinsic* reward is one which follows naturally from the performance of the act; for example, brushing one's teeth results in healthy beautiful teeth. An *extrinsic* reward is an artificial reward, such as giving a child a toy for brushing his teeth regularly.

5. Persistence. Never allow an exception to the performance of the act.

Voluntary Acts

A voluntary act is one which is done consciously and purposefully, under the control of the cerebrum. Voluntary behavior is neither automatic nor inborn.

TYPES OF BEHAVIOR

Type	Description	Examples
Reflex	Inborn response (unlearned). Automatic. Controlled by spinal cord or medulla	Blinking Knee-jerk Peristalsis
Instinct	Complex act performed without training.	Spider spins web Bird builds nest
Conditioned reflex	A reaction to a stimulus which has replaced the original stimulus.	Dog's saliva flows upon hearing a bell ringing
Habit	A learned response which has become completely automatic.	Biting finger nails Typing Changing classes
Voluntary action	Requires thought, reasoning, will, and memory to bring about the desired response.	Sewing Cooking Gardening

COMPLETION QUESTIONS

[A] 1. The activities of an animal are regulated through the nervous system and the . . . .

2. A change in the environment is a (an) . . . .

3. A series of chemical and electrical changes which pass along a nerve fiber is a (an) . . . .

4. Effectors may be muscles or . . . .

5. Three major portions of a neuron are the cyton, terminal branches and . . . .

6. The gap between two neurons is the . . . .

7. The type of nervous system present in *Hydra* is known as a (an) . . . .

8. Clusters of nerve cell bodies are . . . .

9. The earthworm's nerve cord is on the . . . . side of the body.

10. Man's nerve cord is on the . . . . side of the body.

(B) 11. The two main divisions of man's nervous system are the peripheral nervous system and the .... nervous system.

12. The portion of the brain which controls balance is the ....

13. Reflexes in the upper regions of the body are controlled by the .... and reflexes in lower regions are controlled by the ....

14. The sensation of hearing occurs in the ....

15. Peristalsis and blushing are controlled by the .... nervous system.

16. Two chemicals which can carry impulses across the synapse are ...... and ......

17. The enzyme that destroys acetylcholine is ....

(C) 18. When the inside of the membrane of an axon becomes negative, the membrane is said to be ....

19. "A wave of depolarization" is another name for a (an) ....

20. The stronger the stimulus the more .... are the impulses which pass along the neurons.

(D) 21. The hollow space inside the spinal cord is the ....

22. Synapses are present in the .... matter of the spinal cord.

23. Neurons which carry impulses away from the spinal cord are .... neurons.

24. Neurons present between sensory and motor neurons are .... neurons.

25. The simplest pathway, in such behavior as a frog brushing acid from its skin, is called a (an) ....

26. The conditioned reflex was discovered by ....

27. An inborn automatic act is known as a (an) ....

28. The type of behavior in which a substitute stimulus replaces the original stimulus is the ....

29. A complex unlearned act is a (an) ....

30. Complex learned acts which become automatic through repetition are ....

CHAPTER TEST

1. The axon of a neuron is protected by (1) dendrites, (2) acetylcholine, (3) a sheath, (4) terminal branches.

2. At the synapse (1) there is contact between two neurons, (2) an electric current jumps a gap, (3) static electricity jumps a gap, (4) chemicals are released.

3. A definite pathway from receptors to effectors is characteristic of (1) nerve net, (2) central nervous system, (3) hormones, (4) neurohumors.

4. The ability of living things to respond to changes in their environment is called (1) pathology, (2) evolution, (3) sensitivity, (4) assimilation.

5. In the cell diagram of a motor neuron below, an important part is missing. The missing part is the (1) contractile vacuole, (2) chloroplasts, (3) nucleus, (4) terminal branches.

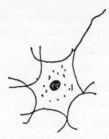

6. A synapse is found between the (1) cyton and dendrites of neurons, (2) cyton and axon of neurons, (3) dendrites and terminal branches of neurons, (4) cyton and terminal branches of neurons.

7. The involuntary beating of man's heart is controlled by (1) the autonomic nervous system, (2) a pair of fused ganglia, (3) the nerve net, (4) the spinal cord.

8. The nerve impulse may best be considered as (1) movement of the stimulus toward the spinal cord, (2) an electric current, (3) a wave of negativity, (4) a response to a stimulus.

9. In man, the stimulation of one neuron by an adjacent neuron is associated with the (1) production of electrical discharges, (2) secretion of neurohumors, (3) secretion of ATP, (4) rapid oxidation of glucose.

10. The sense of balance in man is regulated by the (1) spinal cord, (2) cerebrum, (3) cerebellum, (4) medulla.

11. In the development of which part of the central nervous system has man most clearly surpassed other animals? (1) spinal cord, (2) cerebrum, (3) medulla, (4) cerebellum.

12. Which of the following is characteristic of the nervous system of man? (1) dorsal nerve cord, (2) nerve net, (3) ventral nerve cord, (4) brain composed of a pair of fused ganglia.

13. A person under the increasing influence of alcohol loses, first, his ability to talk; second, his ability to walk straight; third, his ability to breathe normally. From this evidence, in which order may it be reasoned that alcohol affects the central nervous system? (1) medulla, cerebellum, cerebrum; (2) cerebrum, cerebellum, medulla; (3) cerebellum, medulla, cerebrum; (4) cerebellum, cerebrum, medulla.

14. Which of the following is a neurohumor? (1) secretin, (2) noradrenaline, (3) insulin, (4) catalase.

15. Loewi's experiment on frog hearts led to the discovery of (1) pacemaker, (2) acetylcholine, (3) accelerator nerve, (4) blood types.

Regulation in the body is achieved by the endocrine system and the nervous system working together. In the previous chapter, we considered the nervous system; now we will consider the endocrine system.

A | THE ENDOCRINE SYSTEM

The endocrine system of man consists of about ten ductless glands that produce hormones. Many of these endocrine glands affect one another so as to form an interrelated system. The endocrine system helps:

1. to coordinate the life processes of the body
2. to maintain a stable balance in the internal environment of the organism
3. to permit the animal to respond to its external environment.

Hormones

A *hormone* is a chemical substance that is carried by the blood to other parts of the body where it has its effect. Hormones are often called *chemical messengers*. They can either stimulate or inhibit the "target organ" to which they carry a message. They affect many activities in animals including growth patterns and rates, metamorphosis in some animals, and periodic behavior (for example, mating activity). Most hormones are produced by endocrine (ductless) glands. (See table, page 198.)

The precise biochemical mechanism by which hormone molecules have their effect is not known. It is suspected that they function as components of enzyme systems. For example, it has long been known that the hormone insulin helps the cells of the body to use glucose. Now it is known that this is accomplished by the effect of insulin on an enzyme that participates in the oxidation of glucose.

Hormones have been found in some invertebrates; for example, various insects, crustaceans, and mollusks. These invertebrates exhibit some variations in the structures they have for secreting hormones. Insects have organized endocrine glands that produce secretions which regulate growth and metamorphosis. In other invertebrates, it is probable that scattered groups of cells produce chemicals that help to regulate body processes. In animals without blood systems, the hormones probably diffuse from cell to cell.

The hormones in vertebrates will be discussed later in this chapter. Plants, too, produce hormones known as *auxins*. These will be considered in Chapter 16.

Glands

Glands may be divided into two types: (1) duct glands and (2) ductless glands. These are compared in the table on the following page.

The location of the main endocrine glands of man is indicated in Fig. 11–1.

COMPARISON OF DUCT AND DUCTLESS GLANDS

Duct Glands	Ductless Glands
They liberate their secretions through a duct.	They liberate their secretions into the capillaries of the bloodstream.
Their secretions are juices such as gastric juice, saliva, sweat, bile, and tears.	Their secretions are hormones, such as thyroxin, adrenaline, and insulin.
Their effects are produced in the region where the duct empties.	Their effects may be produced throughout the body or in special "target organs."
They are externally secreting, or **exo**crine glands.	They are internally secreting, or **endo**crine glands.

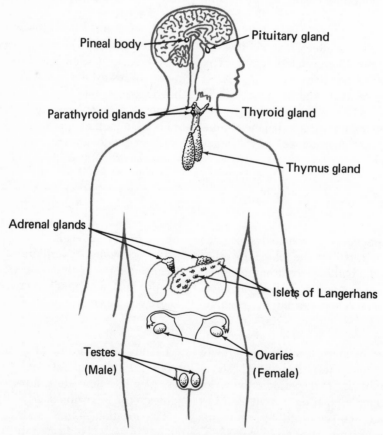

Fig. 11–1. Location of the endocrine glands.

The Nervous System and Endocrine System

Together the nervous system and the endocrine system share the job of controlling the body's activities. These two systems are compared in the table below.

 B **THE HUMAN ENDOCRINE SYSTEM**

The Pituitary Gland

The pituitary gland is often called "the Master Gland of the Body" because it produces hormones which affect many other endocrine glands. However, the title of "Master Gland" is misleading; it does not take into consideration that the other glands also produce hormones which in turn affect the pituitary gland. A more accurate concept is that many of the endocrine glands stimulate and inhibit one another in an interrelated fashion. In this way, they assist the nervous system in coordinating the functions of the body.

The pituitary gland is attached to the base of the brain by a stalk. It is about the size of a large pea and consists of two parts: (1) the anterior lobe, and (2) the posterior lobe.

The anterior lobe of the pituitary gland produces a number of hormones. Among them are growth hormone, thyrotrophic hormone, lactogenic hormone, gonadotrophic hormones, and ACTH.

The *growth hormone* regulates growth of the body by stimulating the growth of the long bones. An excess of growth hormone in childhood causes *giantism,* a condition in which the individual grows very tall, although the various parts of the body are in proper proportion. When hypersecretion (excess secretion) begins in later life, *acromegaly* results. This is a condition in which the jaw, nose, hands, and feet are enlarged out of proportion to the rest of the body. Hyposecretion (undersecretion) of growth hormone in childhood results in *dwarfism,* a condition in which the individual is small but well-proportioned and has normal intelligence. Many of the circus "midgets" are pituitary dwarfs.

COMPARISON OF THE NERVOUS SYSTEM AND ENDOCRINE SYSTEM

Nervous System	Endocrine System
Similarities	
Nerves secrete chemical substances called neurohumors.	Endocrine glands secrete chemical substances called hormones.
Nerves secrete noradrenaline	The adrenal medulla secretes noradrenaline
The nervous system helps to maintain homeostasis.	The endocrine system helps to maintain homeostasis.
Differences	
Nerve responses are rapid and of short duration.	Endocrine responses are slow but last for a long time.
Nerves transmit impulses via neurons.	Hormones are carried by the plasma of the blood.

The *thyrotrophic hormone* stimulates the thyroid gland to secrete its hormones. It is also called Thyroid Stimulating Hormone (TSH).

Prolactin is the lactogenic hormone. This hormone stimulates the mammary glands to secrete milk.

Gonadotrophic hormones affect the gonads (testes and ovaries) in a complicated cycle of interrelationships which is essential to reproduction. (See page 304.)

ACTH (Adrenocorticotrophic hormone), literally translated, means "the hormone which affects the adrenal cortex." ACTH stimulates the adrenal cortex to liberate *cortisone*. Both cortisone and ACTH are useful in treating a variety of ailments including arthritis, inflammations, allergies, and rheumatic fever. According to one theory, these hormones function by permitting the body to respond better to general conditions of stress.

The posterior lobe of the pituitary secretes two hormones. One of these, *oxytocin*, causes contraction of the smooth muscle of the uterus during the later stages of pregnancy. The second is *vasopresin* which controls water resorption in the kidneys.

The Thyroid Gland

The thyroid gland is an H-shaped structure located in the neck on both sides of the trachea. It produces the hormone *thyroxin* which contains large amounts of *iodine*. Thyroxin regulates the general rate of the body's metabolism.

Hypothyroidism is an undersecretion of thyroxin; *hyperthyroidism* is an oversecretion of thyroxin. In hypothyroidism, the rate of metabolism is lowered.

There are several forms of hypothyroidism, among them are:

1. *Cretinism.* A cretin is a person who has had a deficiency of thyroxin since birth. Such an individual may be physically and mentally retarded. He may be a dwarf whose body parts are out of proportion.

2. *Myxedema.* This disease of adults is characterized by a slowing of mental processes, lowered metabolic rate, and a puffy swelling of the face and body.

3. *Endemic goiter.* A goiter is an enlargement of the thyroid gland. In endemic goiter the swelling is caused by lack of iodine in the diet. The gland swells in the "attempt" to produce a sufficient amount of thyroxin. This type of goiter is prevalent in regions of the world where the drinking water lacks sufficient iodine. Endemic goiter can be prevented by using "iodized salt" (table salt to which an iodine compound has been added).

In hyperthyroidism, the individual's rate of metabolism is increased; he is energetic and restless. Hyperthyroidism may result in *exophthalmic goiter;* the thyroid gland swells and the eyeballs protrude. This disease may be treated by surgical removal of a portion of the thyroid gland or by various new drugs. Where the goiter is cancerous, it may be treated with radioactive iodine, I-131. The thyroid gland picks up the radioactive iodine. The radiations are more destructive to the cancerous cells than to the normal cells of the thyroid.

To determine whether an individual is suffering from abnormal thyroxin production, a doctor may direct that he be given a *basal metabolism test.* This test measures the rate at which the person takes in oxygen and gives off carbon dioxide while he is completely at rest, and compares his rate with that of average individuals for his age and size. The basal metabolism test is being replaced by another test — a blood test for the amount of *protein-bound iodine*.

The Parathyroids

The parathyroids are a group of four minute glands located on the lobes of the thyroid gland. The hormone they secrete, *parathormone*, regulates the amount of calcium in the blood and the rate at which calcium is deposited in the bones. A shortage of parathormone results in transfer of calcium from blood to bones. The lowered calcium in the blood produces violent involuntary muscular contractions and convulsions, called *tetany*. The proper metabolism of calcium is also needed for nerve functions, and blood clotting, as well as for the proper growth of bones and teeth.

The Pancreas

The pancreas has a portion that secretes digestive enzymes, and special endocrine cells that secrete hormones. Thus, the pancreas is both a duct gland and a ductless gland. We noted in an earlier chapter the action of the pancreas as a duct gland. Now we are ready to study its function as a ductless gland.

In 1869, the German physician Paul Langerhans examined the microscopic structure of the pancreas. He noted certain groups of cells that appeared different from the others adjacent to them. They stood out as little "islands" in the mass of other tissue. These cells were named the *Islands of Langerhans*.

We now know that these endocrine cells secrete the hormones insulin and glucagon.

The hormone *insulin* is actually indirectly named after the Islands of Langerhans, just as the island possessions of the United States are called the "insular possessions." This hormone was first extracted in usable form by Sir Frederick Banting and his co-workers. For this they were awarded the Nobel Prize in 1923.

Insulin is one of the hormones that regulates carbohydrate metabolism in the body. It helps remove glucose from the blood by (1) causing glucose to be converted to glycogen which is then stored in the liver and in muscle, and (2) controlling the oxidation of glucose in the cells. When there is not enough insulin, the concentration of glucose in the blood increases, a condition called hyperglycemia.

The disease that comes as a result is called *diabetes mellitus*. Since Banting's extraction of insulin, the lives of many diabetics have been saved. As stated on page 174, the presence of glucose in the urine is an indication that a person has diabetes. The proper treatment of diabetes includes (1) administration of insulin by injection or orally by mouth, and (2) careful control of carbohydrates in the diet.

Lining Cells in Stomach and Small Intestine

Gastrin and secretin have already been discussed. (See Chapter 5.) *Gastrin,* secreted by cells in the lining of the stomach, stimulates the flow of gastric juice. *Secretin,* secreted by lining cells of the small intestine, stimulates the flow of pancreatic juice and bile. Secretin was the first hormone to be discovered.

The Adrenal Glands

There are two adrenal glands, one on top of each kidney (*ad-renal* means "on the kidney"). Each gland consists of two parts: (1) the medulla or inner part, and (2) the cortex or outer layer.

One of the hormones produced by the adrenal medulla is called *adrenaline* (or epinephrine). This gland is often called "the Gland of Combat." The metabolic effects produced by adrenaline are appropriate for meeting emergency situations. In general, adrenaline helps to bring glucose and oxygen to the cells so that cellular respiration can release extra energy in an emergency.

Specifically, adrenaline accomplishes the following:

1. *The heart beats more vigorously and more rapidly.*
2. *The rate of breathing increases.*
3. *The bronchioles of the lungs dilate.* Physicians administer adrenaline in attacks of asthma, an allergic condition in which the bronchioles have become constricted.
4. *The muscles in the walls of arterioles constrict in some sections of our body or relax in other parts in such a manner that blood is diverted from the skin and digestive tract to the large muscles of the body which are used in locomotion. Peristalis decreases.* Mealtimes should be pleasant occasions in restful surroundings, for noise, distractions, and minor arguments cause the release of small amounts of adrenaline, which interferes with digestion.
5. *Glycogen of the liver is changed to glucose which is carried to the cells by the bloodstream.*
6. *The blood clots more rapidly.*

Another hormone secreted by the adrenal medulla is noradrenaline (or norepinephrine); they are similar (but not identical) in function.

The adrenal cortex secretes "cortin" which is composed of at least six active hormones, one of which is *cortisone*. The functions of the adrenal cortex hormones are:

1. *They increase the amount of glucose in the blood by assisting in the change of glycogen, proteins, and fat to glucose.*
2. *They regulate the concentration of mineral ions in the body fluids and therefore affect the water content of the tissues.*
3. *Some of them have effects which are similar to those of male sex hormones.* For example, tumors of the adrenal cortex can give rise to masculine characteristics, as in the bearded lady of the circus.
4. *They are used by physicians to prevent inflammation and to treat allergy and arthritis.* In the case of arthritis, cortisone is necessary for the proper composition and thereby functioning of cartilage in joints at the end of bones.

The Gonads

The gonads are the two testes (in the male) and the two ovaries (in the female). In addition to producing sperm or ova, these glands also produce *sex hormones*. These hormones play an important part in the sexual cycle and will be discussed more fully in Chapter 18.

The ovaries produce *estrogen* and *progesterone*. Estrogen is responsible for the development of secondary sex characteristics such as growth of breasts and widening of the hips and the development of the uterus. Progesterone, along with estrogen, controls the production of ova by the ovaries and the associated changes in the uterus that result in the phenomenon called the menstrual cycle. (See Chapter 18.)

The male sex hormones collectively are called androgens. *Testosterone* is the principal male sex hormone secreted by the testes. This hormone produces the male secondary sex characteristics such as a deep voice, larger size, masculine body form, and a beard.

The Hypothalmus

The hypothalmus is a region in the lower portion of the cerebrum which produces hormones. The hormones produced by nerve cells are called *neurohormones*. Interactions between the hormones of the hypothalmus and of the anterior pituitary gland control a number of body functions including the coordination of reproductive activity.

The Pineal Gland

The pineal gland, about the size of a small pea, is located at the base of the brain. The pineal gland may possibly have an endocrine function but this has not been conclusively proved.

The Thymus Gland

The thymus gland is located in the upper chest cavity. It was formerly believed to have an endocrine function because it increases in size in early childhood but practically disappears in adults. However, its function is now believed to be the production of antibodies in the immune response.

(c) COOPERATION BETWEEN HORMONES

Endocrine glands often work together to control body functions. We will consider two examples — glucose metabolism and the pituitary-thyroid relationship.

Glucose metabolism. The concentration of glucose in the blood remains fairly constant although some body processes remove, and others add it to the blood. A state of equilibrium (homeostasis) is maintained by the combined action of insulin and adrenaline. As you will recall, insulin is secreted by the pancreas and adrenaline by the adrenal medulla. The interrelationships are shown in Fig. 11–2.

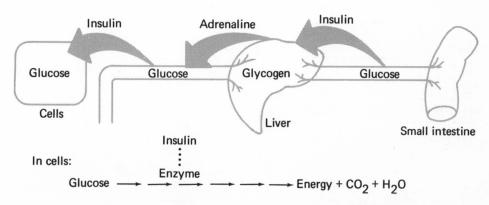

Fig. 11–2. Cooperation between insulin and adrenaline in glucose metabolism.

As previously noted, insulin regulates the removal of excess glucose from the blood in several ways:

1. It acts to change glucose to glycogen which is stored in the liver and in muscle cells.

2. It stimulates the passage of glucose through cell membranes so that it may be used by cells or stored by them as glycogen.

3. Insulin aids in oxidation of glucose within the cells.

Adrenaline serves to counterbalance the removal of glucose by adding it to the blood. Let us see how this happens. You will recall that glycogen is an *insoluble* compound that cannot pass through cell membranes whereas glucose is *soluble* and readily passes into the bloodstream. Adrenaline stimulates the formation of glucose from glycogen. Thus, the soluble glucose passes through cell membranes and is added to the blood. (*Note: Glucagon,* secreted by the pancreas, also helps to change glycogen to glucose. Thus, the pancreas secretes two hormones which have op-posite effects upon the concentration of glucose in the blood; insulin which *decreases* the level of glucose in the blood, and glucagon which *raises* the level.)

Feedback. The speed of a steam engine is regulated by a governor. When the speed exceeds a certain amount, the turning of the engine itself shuts off the power and decreases the engine's speed. This is an example of *feedback.* Another example is the thermostat in the home. In this device, the temperature in the room is used to feed information to the heating plant so that the temperature of the room remains stable. Feedback control mechanisms are an important aspect of modern automated factories.

Feedback also plays a significant role in the homeostatic ("same state") regulation of the internal environment of the body. This is shown in the relationship between glucose and insulin: the greater the level of glucose present in the bloodstream the greater is the amount of insulin secreted by the pancreas to reduce the level of glucose in the blood. The mechanism by which

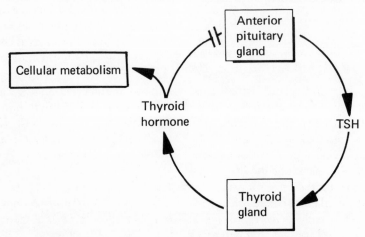

Fig. 11–3. Feedback in pituitary-thyroid relationship.

glucose influences the secretion of insulin is not known.

Pituitary-thyroid relationship. The relationship between the anterior pituitary gland and the thyroid gland is shown in Fig. 11–3. In this diagram a line with an arrow head means production or stimulation; a line with a roadblock (|||) means inhibition, or the blockage of action.

Now locate the anterior pituitary gland in the diagram. This gland secretes a hormone called TSH (Thyroid Stimulating Hormone). TSH stimulates the thyroid gland to liberate the thyroid hormone (thyroxin). The thyroid hormone increases the rate of cellular metabolism; however, a high level of thyroid hormone in the blood *inhibits the pituitary* from producing TSH. With less TSH now present in the blood, the thyroid decreases its output of thyroid hormone. In this way, there is a feedback mechanism so that an excessive amount of thyroid hormone, operating through the pituitary, reduces its own level.

Nervous System and Endocrine System

Not only do the nervous system and the endocrine system cooperate generally to regulate the body, but there are also examples of a *direct* relationship between the two systems:

1. The pituitary gland is attached by a stalk to a portion of the brain that is known as the hypothalmus. Since the hypothalmus is part of the brain, it is actually part of the nervous system. Nerve tracts that run from the hypothalmus to the posterior lobe of the pituitary gland secrete chemicals ("hormones") called *neurohormones.* Blood vessels are present that can carry these neurohormones to the anterior lobe of pituitary gland. Here, therefore, is a situation in which the nervous system (hypothalmus) secretes a *chemical* that directly affects an endocrine gland.

2. During the development of the embryo, two of the endocrine glands develop from cells of the nervous system. These two glands are the adrenal medulla and the posterior lobe of the pituitary gland.

3. Neurons secrete noradrenaline at synapses, and the adrenal medulla also secretes this substance.

HORMONES AND EVOLUTION

We have already discussed hormone production in invertebrates. Now we will consider hormone production in the vertebrates. As one might expect, the animals more closely related to man have the same hormones as man. Accordingly, man can use insulin produced by the pig and sex hormones excreted in the urine of pregnant mares.

In some cases, the same hormone has a different effect in a lower vertebrate than it has in man. Consider the action of thyroxin in the frog and man. Injections of thyroxin cause tadpoles to metamorphose more quickly into frogs, and the removal of the thyroids of the tadpole prevents metamorphosis entirely, so that the animal remains permanently a tadpole. Metamorphosis is unlike anything in human development, but in both frogs and man *metabolism* is in a sense increased by an increase in thyroxin. The thyroid gland has been traced back to a grooved structure in certain early chordates.

In man and in other mammals, there is a hormone, *prolactin,* which stimulates the production of milk in the mother when the young are born. In birds, however, this identical hormone has a different function. In pigeons it stimulates the production of "pigeon's milk." This is not milk but a secretion in the crop portion of the digestive system that is regurgitated and fed to the young. In hens, prolactin leads to broodiness — a desire to sit and hatch the eggs. It is assumed that prolactin evolved before milk did and that it took over a new function in higher animals such as man. Nevertheless, it should be noted that the production of milk in mammals, the production of pigeon's milk, and broodiness are all forms of maternal behavior.

As new data is acquired concerning endocrine function in the animal kingdom, evolutionary relationships become more apparent.

COMPLETION QUESTIONS

A 1. Coordination of body function is accomplished by the endocrine system and the . . . . system.

2. The "chemical messengers" of the body are the . . . .

3. The organ which is affected by a hormone is called the . . . . organ.

4. Hormones are carried to the various parts of the body by the . . . .

5. Another name for a ductless gland is a (an) . . . .

6. In insects, hormones have been found that regulate . . . .

7. Glands that secrete directly into the capillaries are called . . . . or . . . . glands.

8. Gastric juice and saliva are both secreted by . . . . glands.

9. Hormones are secreted by . . . . glands.

10. Compared with nerve responses, most endocrine controlled responses last a . . . . . . time.

B 11. The hormone that controls growth of the long bones of the body is . . . .

12. Giantism is caused by an excess of hormone from the . . . . gland.

13. Hypersecretion of growth hormone in later life leads to the condition of . . . .

14. The pituitary hormone which stimulates the adrenal cortex is . . . .

15. An H-shaped endocrine gland located on both sides of the trachea is the . . . .

16. Hyposecretion of thyroxin in later life results in the disease . . . .

17. Tetany results from a lack of the hormone . . . .

18. A gland which is both duct and ductless is the . . . .

19. The gland which contains the Islands of Langerhans is the . . . .

20. Frederick Banting is known for the isolation of . . . .

21. Diabetes is caused by improper functioning of the . . . .

22. A hormone secreted by glands of the small intestine is . . . .

23. The "Glands of Combat" are the . . . .

24. Severe attacks of asthma are treated by physicians with an injection of . . . .

25. A hormone produced by the ovaries is . . . .

(c) 26. A hormone which lowers the level of blood sugar is . . . .

27. Glycogen in the liver is changed to glucose by the action of . . . . and . . . .

28. A hormone which aids in the cellular oxidation of glucose is . . . .

29. Coordination between parts of the body to reduce an excess supply of a substance is known as . . . .

30. Hormones produced by nerve tissue are called . . . .

CHAPTER TEST

1. Ductless glands secrete their substances directly into (1) special tubes, (2) the bloodstream, (3) digestive juices, (4) receiving organs.

2. Removing all of the thyroid gland from immature experimental animals is most likely to produce (1) a type of dwarfism, (2) a goiter, (3) giant-sized laboratory animals, (4) an increase in the iodine content of the blood.

3. Insulin and glucagon are both produced in the (1) liver, (2) pancreas, (3) muscle cells, (4) adrenal glands.

4. Which may be used as a tracer in studying the thyroid function of the human body? (1) carbon-14, (2) hydrogen, (3) iodine-131 (4) strontium-90.

5. Which hormone does *not* participate directly in controlling the amount of sugar in the blood? (1) parathormone, (2) glucagon, (3) insulin, (4) adrenaline.

6. Undersecretion by the Islands of Langerhans usually results in an increased (1) blood sugar concentration, (2) oxidation of glucose, (3) energy output of the organism, (4) rate of growth.

7. Which hormone stimulates the pancreas to produce pancreatic juice? (1) adrenaline, (2) insulin, (3) ACTH, (4) secretin.

8. Which is characteristic of endocrine systems but *not* characteristic of nervous systems? (1) Their secretions are carried by the blood transport system. (2) They are found in vertebrates and invertebrates. (3) They secrete chemical messengers. (4) They regulate growth.

9. The glucose content of the blood is regulated by (1) insulin alone, (2) adrenaline alone, (3) combined action of insulin and adrenaline, (4) nerve stimulation.

10. The emergency gland of combat produces (1) adrenaline, (2) thyroxin, (3) estrogen, (4) parathormone.

11. Hormones from the adrenal cortex have been found useful in treating certain forms of (1) anemia, (2) arthritis, (3) diabetes, (4) myxedema.

12. A hyperthyroid condition is usually associated with (1) mental retardation, (2) high blood sugar, (3) increased rate of metabolism, (4) low blood pressure.

13. Cretinism results from (1) an excess of adrenaline, (2) a deficiency of adrenaline, (3) an excess of thyroxin, (4) a deficiency of thyroxin.

14. Which is true of both the nervous system and the endocrine system? (1) They bring about rapid responses to stimuli. (2) They assist in the maintenance of homeostasis. (3) They send messages by impulses. (4) They are under the direct control of the cerebellum.

In questions 15–17 write the number preceding the word or expression that best completes the statement or answers the question. Base your answers on the experiment below and your knowledge of biology.

An experimenter removed the pancreas of a living embryonic rat and observed the symptoms that resulted from the operation. From several dead embryonic rats, he removed the pancreases, which he ground with a small quantity of saline solution in order to extract the hormone present. He then injected this hormone preparation into the live rat whose pancreas had been removed.

15. By removing the pancreas of the live rat, the experimenter was trying to produce the symptoms of (1) goiter, (2) diabetes, (3) cretinism, (4) acromegaly.

16. Of which hormone was the experimenter attempting to prepare an extract? (1) insulin, (2) thyroxin, (3) estrogen, (4) cortisone.

17. The scientists who originally isolated the hormone that the experimenter was seeking were (1) Florey and Chain, (2) Banting and Best, (3) Sabin and Salk, (4) Bell and Langerhans.

Increases, Decreases, Remains the Same

18. As the amount of glycogen in the liver decreases, the amount of glucose in the blood usually . . . .

19. As the amount of adrenaline in the blood decreases, the blood pressure . . . .

20. As the amount of insulin in the blood increases, the amount of sugar in the blood . . . .

CHAPTER 12 / Synthesis and Locomotion

A tiger may capture and devour an animal with body parts quite similar to its own (bones, muscles, liver, kidneys, nerves, and glands). Yet, by the time the tiger's food has been digested and carried by the blood to all of its cells, it has been broken down to relatively simple molecules (glucose, amino acids, fatty acids, and glycerol). The tiger must then reassemble these simple unit molecules into the complex compounds and structures which are characteristic of it.

Of course, the heredity of the tiger will determine which compounds are formed, for tigers do not make the identical kind of molecules that lambs make. Furthermore, each tiger is different from every other tiger. How the tiger's heredity handles this task is another story and another chapter. In this chapter, we will consider how the chemical compounds of an organism are assembled. This process is *synthesis*. We will also consider how animals move from place to place, or *locomotion*.

A SYNTHESIS

As stated in Chapter 1, *synthesis* is the process by which simple materials are put together to form more complex materials.

A common method of synthesis in animal cells, already described in Chapter 2, is dehydration synthesis. In this process, molecules are joined to form larger molecules in enzyme-controlled

reactions whereby a molecule of water is split off. For example,

glucose + glucose ⟶ maltose + water

Glucose is a monosaccharide and maltose is a disaccharide. Further synthesis results in polysaccharides. For example, glycogen is a polysaccharide found in animals. It is sometimes called "animal starch," for it resembles the starch present in plants, also a polysaccharide. Glycogen is a storage product containing the potential chemical energy of many C–C and C–H bonds. Present in animal liver and muscle, it can be formed as the end product of dehydration synthesis from numerous glucose units.

The dehydration synthesis of amino acids into progressively larger and more complex peptides and polypeptides may continue as the numerous kinds of protein found in the organism are formed.

Synthesis and Hydrolysis

As pointed out in Chapter 2, the opposite of dehydration synthesis is *hydrolysis*. In hydrolysis, large molecules are enzymatically broken down to smaller units by the addition of water. The digestion of food is largely a matter of chemical hydrolysis.

The two opposite processes of synthesis and hydrolysis go on simultaneously, their rates depending on how much of each chemical is present in the organism. An example of this relationship is shown in Fig. 12–1.

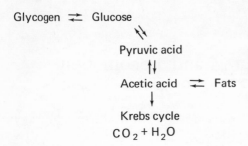

Fig. 12–1. Feedback relations of synthesis and hydrolysis.

Locate acetic acid in the diagram. If there is more acetic acid than is used up in the Krebs cycle, the excess is turned into fats, or some may return to glucose to be stored as glycogen. If glucose is in short supply, the reactions go the other way, which decreases the amount of glycogen and fat.

Enzymes are needed for every step of all these reactions. Enzymes determine the rate of the reaction, in whichever direction it proceeds, but do not determine the direction of a reversible reaction.

The Products of Synthesis

The substances synthesized include (1) secretions, which are used for maintaining life activities, and (2) structural compounds, which form the various structures of the organism.

Secretions. A *secretion* is a substance produced in one part of an organism and in most cases used elsewhere in the organism. Many secretions result from the activity of specialized groups of cells which are organized into *glands*. Some important secretions include enzymes, hormones, and neurohumors.

Enzymes are organic catalysts. The multicellular animal produces numerous enzymes that function within the cells (for example, in cellular respiration). In addition, it has glands that secrete enzymes into the alimentary canal for digesting food. Each enzyme is highly specific for the reaction that it catalyzes. All enzymes are proteins, and each is synthesized from a different arrangement of amino acids. Thousands of different enzymes are produced in a multicellular animal.

Hormones, the secretions of ductless glands, regulate a wide variety of physiological processes. As shown in Chapter 11, they generally function by accelerating or retarding activity. One of the hormones, insulin, is a protein. It is the first protein whose arrangement of amino acids was deciphered by scientists. Hormones vary in their chemical composition, and many of them are not proteins. All of them, however, must ultimately be synthesized from simpler substances.

Neurohumors, such as acetylcholine and noradrenaline (see Chapter 10), are secreted by the ends of nerve cells and stimulate nearby nerve cells. They can be compared to hormones.

Other specialized secretions found in animals include *poisons, mucus, oils* and *waxes*, and *hydrochloric acid*.

Among the animals that secrete poisons are the coral snake, the scorpion, the tarantula, the gila monster, digger wasp, sea anemone, jellyfish, termite, duck-billed platypus, and sea cucumber. In most animals, the chambers that store deadly poisons are lined with a special membrane that seems to insulate the animal from its own poison.

Mucus is largely a protein substance secreted by cells in the lining of the digestive system. This slippery secretion coats the lining of the digestive system.

A lipid which is a solid at room temperature is referred to as a *fat;* one which is liquid at this temperature is an *oil.* Both are synthesized from fatty acids and glycerol. *Waxes* differ slightly in basic structure from fats in that they are synthesized from a long-chain alcohol instead of from glycerol. Beeswax is an example of an animal wax.

Hydrochloric acid is secreted in concentrated form by specialized glandular cells of the stomach. It is diluted to one-tenth its original concentration in gastric juice, which has a pH of about 2.0. This pH is needed for the action of enzymes in the stomach, and it also helps to kill many microorganisms which might cause disease and the decay of food.

Structural compounds. Many of the compounds which form the structures of animal cells are proteins. The cell membrane is a three-layered structure resembling a sandwich of protein-fat-protein. This triple layer is also found in the membranes of the mitochondrion. All the structures which give the animal cell, and indeed the entire multicellular animal, its support and form are the result of synthesis from the basic end products of food intake and digestion.

Synthesis of Proteins

The most decisive feature that distinguishes the chemistry of one cell from that of another is its proteins. Red blood cells have hemoglobin; muscle cells, contractile fibers; some gland cells have digestive enzymes; some epithelial cells have waving protein cilia. All cells have hundreds — perhaps thousands — of different proteins, mostly serving as enzymes. The proteins of each species have characteristic differences from those of other species, and even the proteins of two individuals of the same family are not identical (except perhaps in the case of identical twins). Approximately twenty amino acids make up natural proteins, and they can be assembled in an almost infinite variety of ways.

Proteins are manufactured at the ribosomes of cells. Detailed instructions for their production are carried by molecules known as RNA, which come out of the nucleus.

The growth and metabolism of the cell are controlled by enzymes produced at the ribosomes. Instructions for the process come to the ribosomes from the nucleus in the form of RNA molecules, which were formed, in their turn, under instructions from the DNA. RNA and DNA are discussed in later chapters.

Outcome of Synthesis

As a result of synthesis, the animal can (1) participate in growth and repair, and (2) produce food reserves.

Growth and repair. A cell synthesizes structures such as endoplasmic reticulum, Golgi bodies, mitochondria, nuclear membranes, chromosomes, and cell membranes. At the same time, it synthesizes other components of the cell. For example, the cell liquid contains intermediate products that enter into the makeup of proteins, fats, and chromosomes. As a result of this increase in materials, the individual cells become larger, or *grow.*

Growth of the entire organism results from the division of cells and their subsequent enlargement. The pattern of growth and development is determined by the organism's heredity and environment.

Repair is the growth of new cells

to replace lost, dead, or diseased cells. These new cells fit into the organism's body pattern. Thus, a starfish that grows back a new arm and a boy who replaces the skin scraped from his knee are engaging in repair.

Production of food reserves. Many food materials which are not immediately used by the animal are stored. The most common storage products in animals are fat and glycogen.

Limitations of Synthesis

Man cannot synthesize some substances from available raw materials. These include vitamins and certain amino acids. As a consequence, these substances must be present in the diet. Since a well-balanced diet will include needed substances, food supplements are seldom required.

A number of plants and animals synthesize the amino acids and vitamins that cannot be synthesized by man.

REASONING EXERCISES

1. What is synthesis? How does the body obtain materials for synthesis?
2. What is glycogen? How is it formed in the body?
3. Distinguish between dehydration synthesis and hydrolysis. How are these two processes interrelated in the human body?
4. How are tissues and secretions produced in the body?
5. What do fats and oils have in common? How do they differ?
6. What is metabolism? How do synthesis and hydrolysis both function as a part of metabolism? What chemical compound controls the metabolism of the cell? Where is this compound found in the cell?
7. Distinguish between growth and repair.
8. How are the food reserves of the body produced? Why are vitamins and proteins an essential part of every person's diet?

B   LOCOMOTION

A sponge, a coral, and a barnacle are firmly fastened to one spot and cannot go searching after food. Such organisms which do not move are *sessile;* others which do move are *motile.* Locomotion, or the ability to move from place to place, has provided animals with advantages in the struggle for survival.

Advantages of Locomotion

Procuring food. Sessile animals, such as the sponge, the coral, and the barnacle, are found only in shallow regions of the sea where microscopic plants and animals are plentiful. These sessile animals require a constant and abundant source of food in their watery surroundings. Motile animals — both plant eaters like the cow and meat eaters like the wolf — are not so limited and can range into new environments in search for food. Thus, locomotion offers an advantage in procuring food.

Ability to seek shelter. When the tide goes out, the sessile barnacle closes its shell to avoid drying out. But its motile cousin, the crab, can follow the receding tide to seek the shelter of the water. For the crab, the sea is as much of a shelter as the cave was for primitive man.

Movement away from toxic wastes. If a sea anemone is placed in a saltwater aquarium, it seems quite content where it comes to rest. It looks like a relatively sessile organism (Fig. 12–2). By the next day, however, it will be quite a distance from where it was placed. The sea anemone moved to avoid taking in water containing its own poisonous wastes.

Fig. 12–2. The sea anemone is a close relative of *Hydra*.

Escape from predators. The lobster slowly and cautiously advances on its legs, but when danger threatens, it gives a forward flip of its powerful tail to retreat quickly from a possible enemy. The mole, the rabbit, and the deer owe their survival in a world of meat eaters to their ability to escape from their enemies.

Mating. If the sperm of a fish were released haphazardly into the sea, they would have only a slight chance of finding ova. Many kinds of fish swim near each other when they release their reproductive cells. Locomotion increases the likelihood of contact between individuals capable of mating with each other.

Locomotion in Protozoa

Protists and animals have a wide variety of adaptations for locomotion. The adaptations among the protozoa include the use of flagella, pseudopods, and cilia.

Use of flagella. *Euglena* is an example of a protist which moves by means of whiplike motions of a flagellum. As the flagellum whips, the organism moves forward. (See Fig. 12–3.)

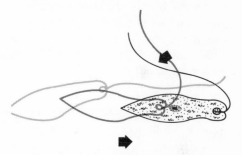

Fig. 12–3. Locomotion in *Euglena*.

Use of pseudopods. *Amoeba* crawls over solid surfaces by means of its pseudopods, or "false feet." The entire organism carries on locomotion by a flowing of the cell contents, and the cytoplasm appears to push the gradually receding cell membrane. The pseudopods may extend upward and downward as well as horizontally. This type of motion is called *amoeboid movement* and is also found in white blood cells. The energy for this movement is supplied by ATP.

The formation of an ever-advancing new outside layer of *Amoeba* is explained by sol-gel relationships. The interior portion of the protoplasm, or *endoplasm*, is in the rather fluid state called a *sol*. The exterior portion, or *ectoplasm*, is in the more firm state called a *gel*. (For example, when jello hardens, a sol is changed to a gel.) At the tip of the advancing pseudopod, the sol changes to a firmer gel as it encounters the water environment. Near the posterior region of *Amoeba*, the gel is changed to a liquid sol which is added to the flowing stream of protoplasm. The change from liquid sol to firm gel is accompanied by a contraction.

Use of cilia. *Paramecium* is an example of a protozoan which swims through the water by the beating of numerous minute cilia. Cilia are living protoplasmic structures which are present on some microscopic organisms and should not be confused with hairs, which are nonliving and much larger structures found in mammals.

Locomotion in *Hydra*

Hydra is essentially a sessile organism. Instead of pursuing food, it takes in minute organisms which accidentally make contact with its waving tentacles. However, it does have primitive muscle cells and can make three types of movement: (1) It can somersault, base over tentacles, (2) it can bend over, attach its tentacles to the bottom, and then bring the base closer, in a kind of "inching-along" fashion, and (3) it can glide along on its base. This motion is accomplished with the aid of amoeboid cells at the base.

Locomotion in the Earthworm

The earthworm has longitudinal muscles which extend along the length of the animal and circular muscles which go around it. When the longitudinal muscles contract, the worm becomes short and fat. When the circular muscles contract, it becomes long and thin. On its lower body wall, the earthworm has *setae* (bristles) which can dig into the soil. First the earthworm becomes long and thin; then the setae in the front dig into the soil and anchor the front end while the rear is brought up. After this, the setae in the

rear portion dig in and serve as anchors while the front end is protruded. The earthworm does not have a skeleton to which the muscles are attached.

Exoskeletons and Endoskeletons

The skeleton of an animal may be on the outside or on the inside of the body.

An *exoskeleton* (outside skeleton) is present in invertebrates such as crabs, lobsters, spiders, millipedes, and insects. The exoskeleton is composed of *chitin*, a nonliving material that may be thick and heavy (as on the lobster's claws) or thin and light (as at the grasshopper's joints or on its wings).

One disadvantage of an exoskeleton is that it is not living; therefore, it does not increase in size as the young animal grows. The exoskeleton must be periodically shed, a process known as *molting*. During this time a crab, for example, is virtually unprotected until the new soft skeleton is hardened.

An endoskeleton (inside skeleton) is present in chordates such as the fish, frog, reptile, bird, and mammal. The skeleton of some fish is composed of flexible cartilage, but other fish and other chordates have a skeleton composed mainly of living bone. The vertebrate endoskeleton is alive and grows with the animal.

Locomotion in the Grasshopper

The grasshopper uses its six legs and four wings to move much faster than the earthworm. One of the advantages that the grasshopper has over the earthworm is that it has a skeleton (an exoskeleton). Let us see why having a skeleton is an advantage.

The appendages of the grasshopper can be moved rapidly because they act as levers which are pulled by the grasshopper's muscles. As you may recall, a lever is a simple machine that can multiply force or distance. A pull exerted through a short distance on one portion of a lever can make another part of the lever move a greater distance. The grasshopper's appendages have a stiff supporting skeleton; therefore, they can act as levers to be pulled by muscles.

The bones of your arm are surrounded by the muscles that move them, but the hollow skeleton of the grasshopper's leg has the muscles on the inside. The grasshopper's muscles work in *opposing pairs* so that if one set of muscles raises a leg another set of muscles lowers the leg.

Locomotion in Man

Bone, which forms the internal skeleton of man, is a living tissue. It contains bone cells that are arranged in a pattern around minute branches of the circulatory system which nourish these cells (Fig. 12–4). The material between the cells is the *matrix,* which

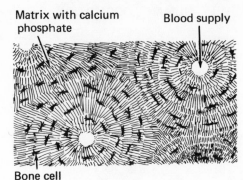

Fig. 12–4. Bone tissue.

contains calcium phosphate that helps to make the tissue hard and strong. To have strong bones, you must include adequate amounts of calcium and phosphorus in your diet. You must also have vitamin D to assist in the proper utilization of calcium.

The muscles of man, like those of the grasshopper, interact by working in opposing pairs. Also, as in the grasshopper, the muscles move parts of the skeleton which act as levers.

REASONING EXERCISES

1. How is locomotion an advantage to the survival of an individual animal?
2. How does locomotion aid a species to perpetuate itself?
3. What is amoeboid movement? Name three species which have adaptations for amoeboid movement.
4. Distinguish between an exoskeleton and an endoskeleton.
5. Why is a grasshopper able to move more rapidly than an earthworm?
6. What is bone? How is bone produced in the body? What are the dietary requirements for healthy bones?
7. What does the skeleton of man have in common with the skeleton of a grasshopper? How do their skeletons differ?
8. Discuss the various possibilities for movement available to man and to the grasshopper.

(c) HUMAN LOCOMOTION

The skeleton of man (Fig. 12–5) consists of the main axis and the appendages (arms and legs). The vertical axis includes the skull, the column of vertebrae through which passes the spinal cord, and the rib cage. The arms are attached to a group of bones which form the *pectoral girdle* and the legs are attached to the *pelvic girdle*.

The numerous small bones that constitute the skull are held together in an immovable manner. However, many other bones are connected at movable *joints* by ligaments. *Ligaments* connect bones to bones. This tissue possesses many strong, somewhat elastic fibers. Another kind of tough connecting tissue strengthened by fibers is *tendon*, which connects muscles to bones. When muscles pull on tendons, the skeleton is bent at the movable joints.

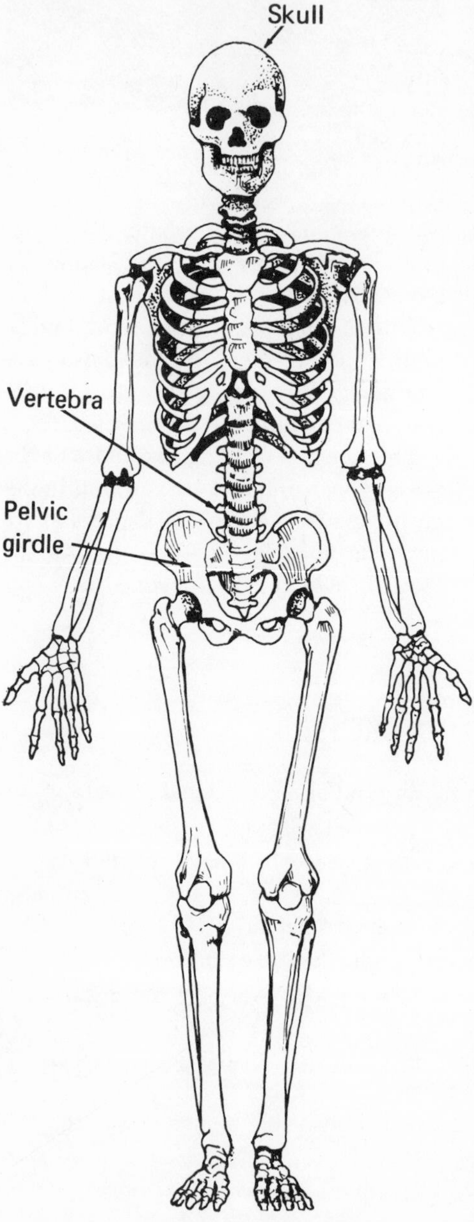

Skull

Vertebra

Pelvic girdle

Fig. 12–5. The human skeleton.

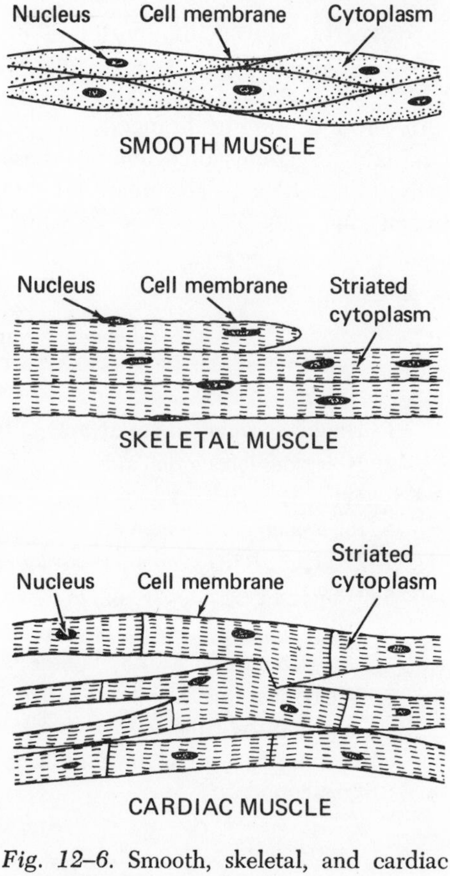

Nucleus   Cell membrane   Cytoplasm

SMOOTH MUSCLE

Nucleus   Cell membrane   Striated cytoplasm

SKELETAL MUSCLE

Nucleus   Cell membrane   Striated cytoplasm

CARDIAC MUSCLE

Fig. 12–6. Smooth, skeletal, and cardiac muscle as seen under the microscope.

The Muscles of Man

Man has three types of muscle: (1) smooth muscle, (2) skeletal muscle, and (3) cardiac muscle (Fig. 12–6).

Smooth muscle consists of distinct cells, each with its own nucleus. These spindle-shaped cells fit together in such a manner that the tissue appears "smooth" under the microscope. There are no cross stripings, or striations. Smooth muscle tissue cannot be consciously controlled ordinarily and is thus *involuntary*. It is present in the walls of the alimentary canal, respiratory passages, arteries, veins, iris of the eye, and the diaphragm.

Skeletal muscle consists of long fibers with many nuclei arranged around the periphery of the fibers. Cross stripings, or *striations*, are visible under the light microscope. Of the three types of muscle, skeletal muscle contracts the most rapidly. Because skeletal muscle can be consciously controlled, it is known as *voluntary muscle*. It is skeletal muscle which is attached to the bones and moves the bones and is thus responsible for locomotion in man.

From the viewpoint of their function, skeletal muscles may be classified into two types: flexors and extensors (Fig. 12–7). A *flexor muscle* acts to bend a joint toward the body, and an *extensor muscle* straightens it out again. These muscles work in *opposable pairs* and move the skeleton by coordinated interaction.

Cardiac muscle is a special type of muscle found only in the heart. It resembles smooth muscle in that the nuclei are distributed within the cytoplasm instead of being arranged around the outside of the fiber as in skeletal muscle, and it functions like smooth muscle in being involuntary. Cardiac muscle resembles skeletal muscle in being *striated* and in having several nuclei for each branched cell.

TYPES OF MUSCLE CELLS

Smooth	Skeletal	Cardiac
involuntary not striated	voluntary striated	involuntary striated

Functions of Bones

The functions of bones include the following: (1) support of body structures, (2) anchors for muscle action, (3) levers for body movement, and (4) protection for delicate internal organs.

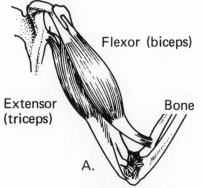

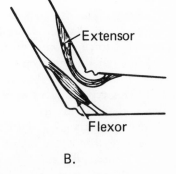

Fig. 12–7. Flexor and extensor muscles. A. Man (endoskeleton).
B. Grasshopper (exoskeleton).

A.

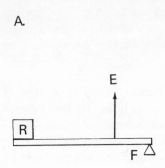

B.

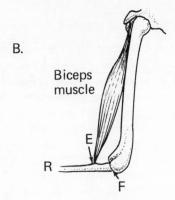

Fig. 12–8. Bones as levers. A. Lever of the third class. B. Biceps muscle bending the arm at the elbow.

Support. A frail jellyfish floating in the sea has all of its organs buoyed up by the water. However, the organs of man would collapse into a heap if there were no solid upright bony structure for supporting and suspending the organs in their proper positions.

Anchorage. Some of man's muscles are not connected to bones; for example, the smooth muscles which cause contractions in the alimentary canal. However, the skeletal muscles responsible for locomotion are attached to a bone at each end of the muscle. One bone serves as the anchor for the muscle, and the other bone is the one which is moved when the muscle contracts.

Leverage. The ordinary seesaw is a kind of lever that merely changes a downward force at one end to an upward force at the other end. The lever shown in Fig. 12–8A has a different effect. This type of lever increases distance and speed in an action. Thus, a pull exerted through a short distance, as represented by the effort (E), moves the resistance (R) a longer distance, making the resistance move faster during any period of time. The action of the biceps muscle at the elbow (Fig. 12–8B)

has a similar effect on a bone of the lower arm. This allows you to throw a ball fast.

Protection. The bones of the skull protect the brain, the internal ear, and parts of the eye. In addition to serving as a backbone to support the body, the vertebrae enclose and protect the vital spinal cord (Fig. 12–9). The ribs protect the heart and lungs.

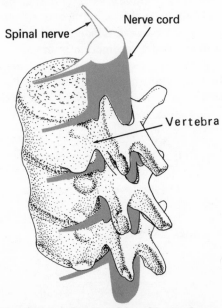

Fig. 12–9. The vertebrae fit together to form the spinal column or backbone.

COMPLETION QUESTIONS

[A] 1. Glucose, amino acids, and other simple molecules are built up into more complex molecules within the body by the chemical process of ....

2. Maltose results from the dehydration synthesis of molecules of ....
3. "Animal starch" is known as ....
4. Glycogen is stored in liver and in ....
5. The basic units for the synthesis of molecules of proteins are molecules of ....
6. The breaking down of large molecules by an enzyme-controlled reaction involving the addition of water molecules is known as ....
7. Secretions are produced in the body by the chemical process of ....
8. The secretions of ductless glands are known as ....
9. Hormone-like substances secreted by nerve cells are the ....
10. A lipid that is liquid at room temperature is a (an) ....
11. The central layer of the cell membrane is composed of ....
12. Proteins are produced within the cell at the ....
13. Instructions for the synthesis of proteins are carried from the nucleus to the cytoplasm by .... molecules.
14. The replacement of lost or wornout tissues is called ....
15. The most common storage products in animals are glycogen and ....

[B] 16. Sessile animals are those which do not have the ability to ....
17. A life function which enables some animals to escape from predators is ....
18. The structure used for locomotion by *Euglena* is the ....
19. Bristles used in locomotion by the earthworm are known as ....
20. The grasshopper's appendages are levers which are pulled by the grasshopper's .....
21. The grasshopper's skeleton is composed largely of ....
22. The shedding of the young grasshopper's skeleton is known as ....
23. The type of skeleton which grows as the animal grows is the ....
24. Muscles interact with each other in .... pairs.
25. The matrix of human bone is hardened by the presence of the compound ....

(C) 26. The arms of man are attached to the .... girdle.
27. Bones are attached to bones by ....
28. Muscles are attached to bones by ....
29. Voluntary muscle is also called .... muscle.
30. A muscle which bends a joint is called a (an) ....
31. A muscle which straightens out a joint is called a (an) ....
32. Two types of involuntary muscle are cardiac and ....
33. Two types of striated muscle are cardiac and ....
34. Bones serve the functions of protection, leverage, and ....
35. The vertebrae enclose and protect the ....

MULTIPLE-CHOICE QUESTIONS

[A] 1. The mechanism in the body by which the short supply of a chemical causes its replacement is called (1) relevance, (2) rejection, (3) repair, (4) feedback.

2. The principal reserves of the body are (1) proteins and glucose, (2) glycogen and fat, (3) vitamins and enzymes, (4) glycogen and vitamins.

3. The instructions for the synthesis of proteins are carried from the nucleus to the ribosomes by (1) DNA, (2) vitamins, (3) RNA, (4) neurohumors.

B 4. An example of an animal that is both sessile and motile is (1) the sponge, (2) *Euglena*, (3) *Hydra*, (4) the earthworm.
5. The skeleton of a grasshopper consists of (1) chitin, (2) living tissue, (3) bone, (4) cartilage.
6. A material *not* found in the human skeleton is (1) calcium, (2) matrix, (3) cellulose, (4) living tissue.

C 7. A tough tissue that connects parts of the human skeleton is (1) cartilage, (2) tendon, (3) epithelium, (4) chitin.
8. The muscles in the human stomach are (1) cardiac, (2) skeletal, (3) smooth, (4) striated.
9. The spinal cord is protected chiefly by (1) the bones of the skull, (2) the ribs, (3) the blood, (4) the vertebrae.
10. In the human body the levers for movement are (1) muscles, (2) bones, (3) tendons, (4) ligaments.

CHAPTER TEST

1. Enzymes are synthesized from (1) amino acids, (2) fatty acids, (3) glucose, (4) glycogen.
2. A characteristic of enzymes is that they are (1) unaffected by heat, (2) inorganic catalysts, (3) highly specific, (4) carbohydrates.
3. Which of the following least resembles the others in chemical composition? (1) fat, (2) wax, (3) starch, (4) oil.
4. The pH of gastric juice is closest to (1) 2, (2) 7, (3) 9, (4) 14.
5. The cell membrane is composed of layers of (1) fat, (2) protein, (3) fat-protein-fat, (4) protein-fat-protein.
6. *Paramecium* moves by means of (1) pseudopods, (2) hairs, (3) flagella, (4) cilia.
7. Somersaulting is a method of locomotion in the (1) *Amoeba*, (2) *Paramecium*, (3) *Hydra*, (4) earthworm.
8. Which moves without the use of skeletal levers? (1) grasshopper, (2) man, (3) earthworm, (4) frog.
9. An exoskeleton is present in the (1) *Hydra*, (2) grasshopper, (3) fish, (4) bird.
10. Muscles which are voluntary are (1) skeletal and cardiac, (2) smooth and cardiac, (3) smooth only, (4) skeletal only.

UNIT THREE

PLANT MAINTENANCE

This unit deals with the functions for maintenance found in plants. All plants maintain themselves by nutrition, transport, respiration, excretion, synthesis, and regulation. These processes are similar to the same processes in animals. However, there are important differences, as we shall see during the study of this unit.

CHAPTER 13 | Nutrition: Photosynthesis

All animals including man are heterotrophs. Among the plant-like protists, examples of heterotrophs are nongreen organisms such as yeast, bacteria, and mushrooms. In *heterotrophic nutrition,* the organism takes in and uses preformed organic molecules containing much energy in their bonds. By contrast, *autotrophs* synthesize their food from relatively simple substances in their environment. This making of complex food molecules from simple substances requires the addition of energy. Two kinds of autotrophic nutrition are *photosynthesis* (synthesis by the use of light) and *chemosynthesis* (synthesis by chemical means). In this chapter, we will consider photosynthesis.

A | PHOTOSYNTHESIS

Photosynthesis is a process by which green plants use carbon dioxide and water, with the energy from light (usually sunlight), to produce carbohydrates and oxygen. During this process light energy is converted into the chemical-bond energy of organic compounds.

Most of the photosynthesis (80%) which occurs on this planet is carried on by algae ("one-celled green plants") which live in the oceans and fresh water. It is also carried on by multicellular green plants—the familiar low plants, shrubs, and trees of the land. The chemical process of photosynthesis is essentially the same in both the one-celled protists and the multicellular plants.

A simplified summary equation for photosynthesis is:

$$\text{carbon dioxide} + \text{water} + \text{energy} \xrightarrow[\text{enzymes}]{\text{chlorophyll}} \text{glucose} + \text{oxygen}$$

As a chemical equation, this simplified summary may be written as follows:

$$6CO_2 \quad + \quad 6H_2O \quad + \quad energy \xrightarrow[enzymes]{chlorophyll}$$

(6 molecules of carbon dioxide)   (6 molecules of water)   (from light)

$$C_6H_{12}O_6 \quad + \quad 6O_2$$

(1 molecule of glucose)   (6 molecules of oxygen)

The chlorophyll acts as a *catalyst* — a substance which promotes (or hinders) a chemical reaction but does not itself become changed as a result of the reaction. The light energy becomes converted to the chemical bond energy of the glucose molecule (one of the products of photosynthesis). The glucose formed may be used by the plant as a source of energy in cellular respiration, or it may be converted into other compounds such as starch, proteins, or fats.

In this way, the energy of the sun becomes the source of energy for all plants, and indirectly for all animals. When man obtains useful energy by burning wood, coal, and petroleum products, he is using the sun's energy which was trapped long ago during photosynthesis.

Note from the equation that oxygen is also formed during photosynthesis. This is important, for photosynthesis is the source of almost all of the oxygen in the air.

Laboratory Experiments on Photosynthesis

A number of demonstrations are commonly performed to show the importance of some of the components of the equation for photosynthesis.

1. To determine if light is needed for photosynthesis

This experiment can be performed with a geranium plant. Select one leaf, but do not remove it from the plant. Cover the upper and lower surfaces of the same *portion* of the leaf so as to prevent light from reaching the cells of that portion of the leaf. Use small pieces of cork or black paper held in place by pins or a paper clip to achieve the desired separation. See Fig. (13–1A). The part of the green leaf capable of receiving sunlight will serve as a control. Let the plant stand in the dark overnight and, the following day, expose the plant to sunlight for several hours.

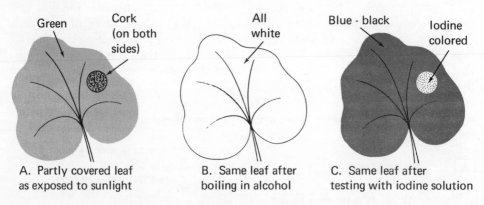

Green   Cork (on both sides)

All white

Blue - black   Iodine colored

A. Partly covered leaf as exposed to sunlight

B. Same leaf after boiling in alcohol

C. Same leaf after testing with iodine solution

Fig. 13–1. Is light needed for photosynthesis?

Remove the leaf from the plant and take off the cork or black paper. Boil the leaf in water to soften it and then boil the leaf in alcohol to extract the chlorophyll. (**Caution:** The alcohol vapors are highly flammable! Use an electric hot plate rather than an open flame.) When the leaf appears white (Fig. 13–1B), wash off the alcohol with water and place the leaf in a shallow dish. Flood the leaf with iodine solution and look for any color change in the leaf.

Upon examining the leaf, we find that the part of the leaf that was exposed to light turned blue-black, showing the presence of starch. However, the part of the leaf that did not receive light remained the color of the iodine stain, indicating the absence of starch. Knowing that the sugar which is made is changed to starch, we conclude that *light is needed for this green plant to carry on photosynthesis.*

In conducting a scientific experiment, we always establish a *control* as a regular part of the experiment. This control serves as a basis for comparison. The control repeats every part in the main portion of the experiment except for one factor—the condition which is being tested. Let us see what this means.

In the present experiment, the covered portion of the leaf is the control to the extent that it repeats every condition in the main portion of the experiment with the exception of one factor (the presence of light). Thus, light is the variable factor. Between the two parts of the experiment, there should be only a single variable. In this experiment, are you satisfied that there is a single variable? Why, or why not? Why was the plant allowed to stand overnight before starting the experiment?

2. To determine if chlorophyll is needed for photosynthesis

For this experiment, we need a leaf that is partly green and partly colorless. If we were to remove the chlorophyll from part of the leaf by boiling it in alcohol, we would kill the leaf. Fortunately, there are plants whose leaves are variegated—partly green and partly white. Examples are coleus and tradescantia (Fig. 13–2).

The steps in this experiment are as follows: Remove the leaf from a coleus plant that has been in the light for several hours. Boil the leaf in water. Then extract the chlorophyll in boiling

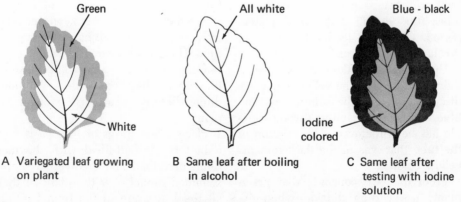

A Variegated leaf growing on plant — Green — White

B Same leaf after boiling in alcohol — All white

C Same leaf after testing with iodine solution — Blue - black — Iodine colored

Fig. 13–2. Is chlorophyll needed for photosynthesis?

alcohol as above (**caution!**). Since the chlorophyll has been removed from the green portion of the leaf, the leaf is now all white. (See Fig. 13–2B.) Place the leaf in a shallow dish, and flood the leaf with iodine solution.

Upon examining the leaf, we find that the areas which contained chlorophyll have turned blue-black in color, while the other parts, lacking chlorophyll, have remained iodine colored. We can, therefore, conclude that chlorophyll is necessary for photosynthesis.

3. **To determine if plants use carbon dioxide when exposed to light.**

In conducting this experiment, we use brom-thymol blue, a chemical *indicator*. When brom-thymol blue is dissolved in water, it forms a blue solution, which will turn yellow if carbon dioxide is bubbled into it. If carbon dioxide is removed from the yellow solution, the solution turns blue again.

Place about 50 ml of a pale blue solution of brom-thymol blue into a beaker. Introduce carbon dioxide into the solution by blowing through a straw or glass tube until the solution turns yellow. Divide the yellow solution between two test tubes. Now put a sprig of *Anacharis* into each test tube and cork the tubes. Expose one tube to the light for several hours and keep the other tube in the dark for the same period. (See Fig. 13–3.) The tube in the dark is the control.

Upon examining the tube exposed to light, we find that the yellow color has disappeared, and the solution is now blue, indicating that the carbon dioxide has been removed. The solution in the tube that was in the dark remains yellow colored.

From this, we conclude that green plants use carbon dioxide when exposed to light.

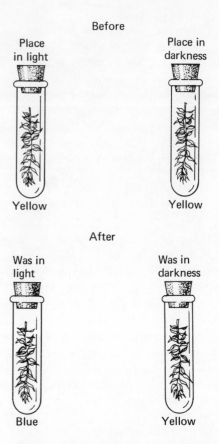

Fig. 13–3. Do green plants use carbon dioxide in the sunlight?

4. **To determine if green plants in the light give off oxygen**

To perform this experiment successfully, we use healthy sprigs of *Anacharis*, which have recently been removed from a pond or well-balanced aquarium. The *Anacharis* may be kept temporarily in water to which potassium bicarbonate has been added. This medium will provide carbon dioxide for the plant.

Place a few sprigs of *Anacharis* into a battery jar half-filled with the medium described above. Invert a short-stemmed funnel over the plant so that all the leaves are in the funnel. (Fig. 13–4). *Fill* a test tube with water and

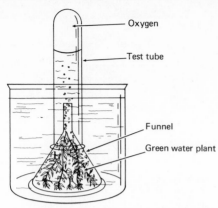

Fig. 13–4. Do green plants in the light give off oxygen?

invert it over the stem of the funnel so that the mouth of the tube is in the water and the test tube remains *filled* with water. Set the apparatus in bright sunlight. If the experiment is continued overnight, supply continued illumination with a fluorescent light. Why?

Bubbles of gas emerge from the plant and accumulate at the top of the test tube. When enough gas has accumulated, remove the test tube and insert a glowing splint into it as a test for oxygen. The splint bursts into flames, indicating the presence of oxygen.

As a control, you may use either an identical set-up which is placed in the dark or you may use a set-up without a plant. Better yet, you may use both. What gas will accumulate in the test tube when the plant is kept in the dark?

The Role of Light in Photosynthesis

Light is a form of energy. Some of its properties can be accounted for by the assumption that it is an electromagnetic wave. It differs in wave length from other electromagnetic waves such as cosmic rays, X-rays, and radio waves. Visible light is that portion of the electromagnetic spectrum with a wave length ranging from about 400 mμ (millimicrons) to 700 mμ.

White light is a mixture of varying wave lengths. When white light is passed through a glass prism, it is *dispersed* (separated) into varying wave lengths as shown in Fig. 13–5.

The "color" of light is the sensation produced in the brain when light of a certain combination of wave lengths falls upon the retina of the eye. Man does not receive a visual sensation from infra-red or ultraviolet rays. Many insects are not receptive to light of the wave length which we call "yellow." Therefore, in the summer we use yellow "bug lights" which provide light for us but which do not attract these insects.

The color of an object depends upon the wave lengths of the light that it reflects or transmits to our eyes. When white light falls upon a "green" opaque object, that object has the property of absorbing all of the wave lengths *except* those that produce a sensation of green

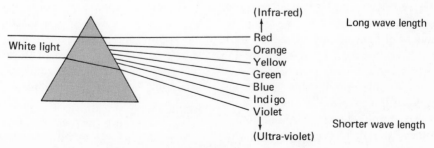

Fig. 13–5. Dispersion of white light into the visible spectrum.

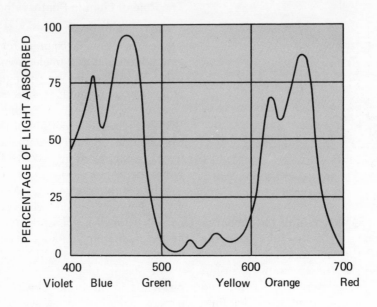

WAVELENGTH OF LIGHT (IN Mµ)

Fig. 13–6. Absorption spectrum of chlorophyll in alcohol.

when reflected to our eyes. Accordingly, if blue light falls upon this "green" object, the wave lengths for blue are absorbed. Since no light is reflected, we would see the object as black. Thus, a green object in a blue light appears black. If green light falls upon the object, this green light is not absorbed but is reflected. A green object in a green light appears green to us.

Chlorophyll is usually seen as a green pigment because it absorbs most of the other wave lengths and reflects the green to our eyes. The *spectrophotometer* is an instrument that measures the amount of light of various wave lengths which is absorbed by a solution. Fig. 13–6 is a graph of the absorption spectrum of chlorophyll; this graph was made with a spectrophotometer.

The graph shows that chlorophyll absorbs mainly red-orange and blue-violet light. *Chlorophyll absorbs very little green light, and these rays are not used in photosynthesis.* Chlorophyll reflects green light to our eyes. Since the wave lengths for green are reflected, they are not used for photosynthesis. Most plants use mainly the red-orange and blue-violet portions of the visible spectrum for photosynthesis.

To demonstrate the absorption of various wave lengths by chlorophyll, pass white light from an intense source (such as a slide projector) through a glass prism. As indicated before in Fig. 13–5, the complete spectrum appears on a screen. Now insert a solution of chlorophyll in the path of the beam, as shown in Fig. 13–7. Most of the colors disappear or become faint, and only the green appears well on the screen.

Chloroplasts

Chloroplasts are small green bodies within the cytoplasm of green plant

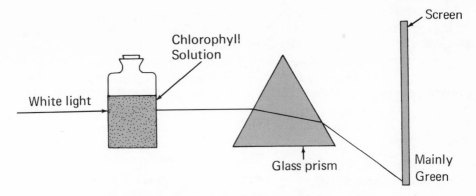

Fig. 13–7. Absorption of wave lengths by chlorophyll.
Which wave lengths are absorbed the least?

cells. In most plants, the complete chemical reactions of photosynthesis occur within the chloroplasts.

A chloroplast has a number of flat double membranes, called *lamellae* (sing. – lamella), arranged in layers. There are also small oval-shaped structures called *grana* (sing. – granum) which appear to consist of thickened regions of the lamellae. (See Fig. 13–8.) The chlorophyll is contained within the grana where the chlorophyll molecules are arranged in an orderly manner between layers of protein and lipid.

Chlorophyll

Chlorophyll, the green pigment of plants, is a highly complex molecule that has more than 100 atoms. Its structure is very similar to that of the non-protein part of the molecule of *hemoglobin,* the red oxygen-carrying pigment found in the blood of many animals. This fact has evolutionary significance because it indicates chemical relationship. One difference is that each chlorophyll molecule contains an atom of *magnesium* but hemoglobin contains an atom of iron. The arrangement of the atoms in the chlorophyll molecule permits it to trap certain wave lengths of light energy and pass this energy on to the chemical reactions involved in the process of photosynthesis.

There are at least six different kinds of chlorophyll in plants. Two of these are shown below:

Chlorophyll *a* $C_{55}H_{72}O_5N_4Mg$

Chlorophyll *b* $C_{55}H_{70}O_6N_4Mg$

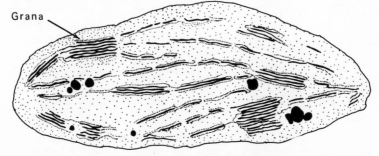

Fig. 13–8. The chloroplast as seen under the electron microscope.

Plants may contain other pigments, such as carotenoids (yellow) and anthocyans (red, blue). Sometimes these pigments mask the green color of chlorophyll. If green leaves are crushed and dissolved in acetone, many of the pigments present can be separated from one another by special techniques of chromatography.

What Is the Source of the Oxygen Liberated by a Green Plant?

By the end of the 19th century, a balanced chemical equation for the process of photosynthesis could be written as follows:

(1) $6CO_2 + 6H_2O + \text{light} \xrightarrow{\text{chlorophyll}}$
energy
$C_6H_{12}O_6 + 6O_2$

In accordance with the Law of Conservation of Matter (matter is neither created nor destroyed), the number of atoms of each kind on each side of the arrow are equal.

However, scientists were eager to explore deeper into this important process which is the source of food for living things. Although they knew what went into the process and what came out, the process itself was one big mystery, like the box shown in Fig. 13–9.

Fig. 13–9. The riddle of photosynthesis.

One important question was: "Is it the carbon dioxide or the water which is the source of the oxygen that is liberated?" Put another way, which of the following two equations summarizes the formation of carbohydrate? [(CH_2O) is the abbreviation for carbohydrate.]

$$C|O_2 + |H_2O| \longrightarrow (CH_2O) + O$$
O from CO_2

or

$$CO_2 + H_2|O \longrightarrow (CH_2O) + O$$
from H_2O?

Samuel Rubin and Martin D. Kamen of the University of California investigated this problem by supplying plants with "tagged" oxygen as a tracer. They used the stable isotope of oxygen, O^{18}. (The more common form is O^{16}.) When the oxygen of the CO_2 was thus labeled, these investigators found that the *sugar* which was produced contained the heavy or "tagged" oxygen. But when the oxygen of the H_2O was labeled, they found that the *oxygen* which was liberated contained the tagged atoms. By this method, they showed that *the source of oxygen which is liberated in photosynthesis is water,* not carbon dioxide. This was an important first step in obtaining a deeper knowledge of the chemistry of photosynthesis. It showed that, *essentially, carbon dioxide combines with hydrogen.*

Since O^{18} is not radioactive, its presence could not be determined by the use of a Geiger counter or photographic plate. Instead, Rubin and Kamen used a mass spectrometer — an instrument that identifies tagged atoms by their mass, or "weight."

When an asterisk is used to indicate the tagged atoms, equation (1) above becomes modified as follows:

(2) $6CO_2 + 6H_2O^* \longrightarrow C_6H_{12}O_6 + 6O_2^*$

However, this equation is not balanced because there are only 6 *tagged* oxygen atoms on the left side, but there are 12 *tagged* oxygen atoms on the right side. Accordingly, we can start with $12H_2O^*$ instead of $6H_2O^*$ and write:

$$6CO_2 + 12H_2O^* \longrightarrow C_6H_{12}O_6 + 6O_2^* + 6H_2O$$

The equation shows that the 12 tagged oxygens $(6O_2^*)$ come from water. An overall equation for photosynthesis may thus be written as follows:

$$\text{Energy} + 6CO_2 + 12H_2O \xrightarrow[\text{enzymes}]{\text{chlorophyll}}$$

$$C_6H_{12}O_6 + 6O_2 + 6H_2O$$

Carbon Fixation

During the first stage of photosynthesis, the H_2O, a raw material for photosynthesis, is broken down into hydrogen and oxygen, and the oxygen is given off. In essence, the remainder of the process consists of the union of CO_2 with hydrogen to form $C_6H_{12}O_6$. This attachment of the carbon atoms to produce the 6-carbon glucose is called *carbon fixation*. (When you "fix" something you attach its parts.) Biologists have been able to trace the steps of carbon fixation by use of the radioactive form of carbon — carbon-14.

By photosynthesis, light energy is converted into the chemical-bond energy of organic compounds. The energy is stored in the form of carbon-carbon and carbon-hydrogen bonds.

The rate of photosynthesis is affected by a number of factors including temperature, carbon dioxide concentration, wave length of light, chlorophyll concentration, and availability of certain minerals.

REASONING EXERCISES

1. If all of the green plants in the world were to die, how would this affect other living things?
2. A green plant by a window can be observed to incline its leaves toward the light. In what way is this behavior a valuable adaptation?
3. In the experiment to determine if chlorophyll is necessary for photosynthesis, why do we use a variegated leaf?
4. In the fall of the year, the leaves of some trees change from green to red. What reason can you give for this color change?
5. At which time of the year do you think photosynthetic activity is the greatest — the summer or the fall? Explain your answer.
6. A green plant produces both oxygen and carbon dioxide. How do you account for this?
7. How might a *sharp* drop in temperature affect the rate of photosynthesis?
8. In the experiment which shows that light is needed for photosynthesis, can you improve the controls?

(B)   **THE CHEMISTRY
OF PHOTOSYNTHESIS**

When photosynthesis is studied as a series of chemical reactions, the process can be divided into two general phases: (1) the light reactions, and (2) the dark reactions. The *light reactions* may be defined as that group of chemical reactions which require light as a source of energy; thus, they are *photo*chemical reactions. The *dark reactions* may be defined as that group of chemical reactions which do not use light as a source of energy. Normally, both of these phases occur simultaneously while the plant is in the sunlight. It is only when the chemist traces the pathway of photosynthesis that he divides its reactions into those that require light energy and

those, immediately following, that do not.

A general overview of the light and dark reactions is shown in Fig. 13–10.

The Light Reactions (Photolysis)

As you probably know, water can be decomposed into hydrogen and oxygen by electrical energy (electrolysis). *Photolysis* refers to a breakdown of water by light energy. The oxygen is released as one of the products of photosynthesis, and the hydrogen is passed on to the dark reactions. Numerous enzymes, as well as the coenzyme NADP, participate in these steps.

NADP (nicotinamide-adenine dinucleotide phosphate, formerly known as TPN) is an important coenzyme that participates in both phases of photo-

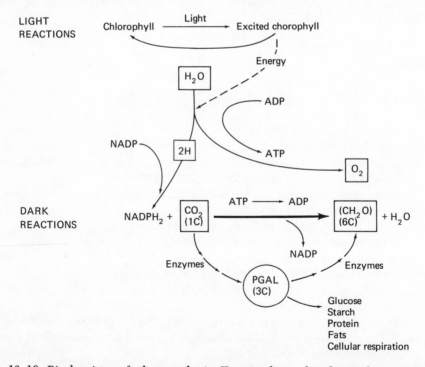

Fig. 13–10. Biochemistry of photosynthesis. For simplicity the chemical reactions are not shown as balanced equations. The 1C and 6C refer to the number of carbon atoms in the molecules of carbon dioxide and carbohydrate (CH_2O).

synthesis. This molecule carries energy and hydrogen atoms from the light reactions to the dark reactions. Acting as a hydrogen acceptor in the light reactions, it takes on two hydrogen atoms to yield the reduced form, $NADPH_2$. In the dark reactions, it is a hydrogen donor and releases two atoms of hydrogen. The reaction is reversible.

$$NADP + 2H^+ \rightleftarrows NADPH_2$$

We have noted that light energy is used to decompose water. How is this possible, since sunlight falling upon water merely evaporates it? Chlorophyll, by some little understood property of the arrangement of its atoms, traps light energy and becomes "excited." This means that some of the electrons of its atoms travel in orbits that are further removed than usual from the nuclei of these atoms. High-energy electrons from the chlorophyll are then passed by a series of enzymes to the reaction that breaks down water. (See Fig. 13–10.) As the water decomposes, some of the energy (originally from the light) is incorporated along with hydrogen into $NADPH_2$; some of the energy is used to change ADP to ATP.

The main pigment that traps light energy and passes it on to the light reactions is chlorophyll *a*. Other pigments present in the leaf may also trap light energy and pass high-energy electrons to chlorophyll *a*. After chlorophyll *a* discharges high-energy electrons, it reverts to its former unexcited state.

The results of the light phase are: (1) the energy of sunlight has been stored in $NADPH_2$ and ATP, (2) hydrogen has been taken up by NADP to form $NADPH_2$, and (3) oxygen has been released.

The Dark Reactions

As noted, hydrogen is made available during the light reactions. The dark reactions consist of the joining of hydrogen with carbon dioxide to form a carbohydrate (CH_2O).

$$CO_2 + H \xrightarrow{\text{energy}} (CH_2O)$$

These events occur by carbon fixation, and the dark phase is often called "carbon assimilation."

The hydrogen for these reactions is supplied by the $NADPH_2$ which was formed during the light reactions. The energy is supplied by $NADPH_2$ and by ATP, also formed during the light reactions.

PGAL. What are the steps by which the one-carbon molecules of CO_2 are joined to form the 6-carbon glucose? In order to determine this, Melvin Calvin and his associates at the University of California used a "lollipop" apparatus. They supplied CO_2 containing radioactive carbon-14 to algae (*Chlorella*) in a lollipop-shaped flask which were in the dark. (See Fig. 13–11.)

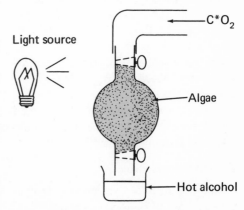

Fig. 13–11. The lollipop experiment to determine the intermediate compounds formed during carbon assimilation.

The light was turned on for a few seconds to permit photosynthesis to take place. Then the algae were run into a container of hot alcohol which killed the algae and stopped their biochemical reactions. Any tagged carbon compounds now present in the alcohol must have been produced from the C^*O_2 during the plants' brief exposure to light. The compounds produced were then identified by the technique of chromatography. When the exposure to light was as brief as two seconds, a 3-carbon tagged compound was formed. This was *PGAL* (phosphoglyceraldehyde).

PGAL is the first chemically stable compound formed during carbon assimilation. Most biologists now consider PGAL, rather than glucose, to be the product of photosynthesis. PGAL can be used by the cell as the starting point for the synthesis of glucose, starch, proteins, and oils. It can also serve as an energy-rich compound for cellular respiration.

Calvin showed that the formation of 6-carbon sugar does not proceed in a single step. Many intermediate reactions are involved, each catalyzed by its own specific enzyme.

It has been demonstrated that all of the known steps of photosynthesis can occur within the chloroplasts. Biologists discovered these steps by the use of the electron microscope, the Geiger counter, the mass spectrometer, tagged atoms (such as oxygen-18 and carbon-14), and chromatography.

Rate of Photosynthesis

The rate of photosynthesis is affected by a number of factors:

Temperature. (See Fig. 13–12.) When sufficient light is available, an increase in temperature from 0° C to 30° C results in an increase in the rate of photosynthesis. However, a further increase in temperature, from 30° C to 40° C, results in a decrease in rate. Temperature affects the rate of enzyme-controlled chemical reactions, particularly those of the dark phase of photosynthesis. This effect of temperature is shown in the top curve of Fig. 13-12. Note, however, that when the light intensity is low, as shown in the bottom curve of the figure, an increase in temperature does not increase the rate of photosynthesis. In this case, the absence of light is a *limiting* (or controlling) factor. It acts as a "bottleneck" by preventing the light phase from providing sufficient products for the dark phase.

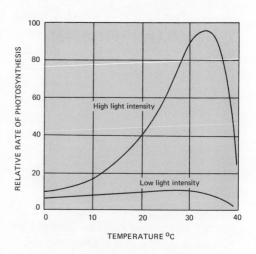

Fig. 13–12. The effect of temperature upon the rate of photosynthesis.

Carbon dioxide concentration. Under normal field conditions, the rate of photosynthesis corresponds directly to the concentration of carbon dioxide available. The normal concentration of CO_2 in the air is about 0.04 per cent. If in a greenhouse this concentration is doubled, the rate of photosynthesis will correspondingly be doubled. This, of course, assumes that there are no other variables.

Intensity of light. All other factors being equal, the rate of photosynthesis in leaves tends to be approximately proportional to the light intensity, up to a maximum of about one-third of full normal sunlight. Beyond this, there is a saturation point, and further increase in light intensity does not increase the rate of photosynthesis.

Light is needed for the light phase of photosynthesis and for the synthesis of chlorophylls. It is not only the intensity ("strength") of light that is important in photosynthesis but, as mentioned previously, also its wave length. The wave lengths of the reds and blues are most effective.

Availability of minerals. Plants in the field do not carry on photosynthesis well when they suffer from *chlorosis*. This is a condition in which the leaves are abnormally pale because of the failure of the cells to synthesize sufficient chlorophyll. To prevent chlorosis, plants require nitrogen and magnesium. Atoms of these elements are present in the chlorophyll molecule. Chlorosis also re-sults from a deficiency of iron, zinc, copper, or manganese.

Review Facts About Photosynthesis

1. Chlorophyll appears green because it reflects the wave lengths of green light and absorbs the other wave lengths of the visible spectrum.

2. Chlorophyll is present within the chloroplasts in structures called grana.

3. Chlorophyll traps light energy and makes it available for photolysis.

4. Photolysis is the breaking up of water, with the aid of light, into hydrogen and oxygen.

5. The source of the oxygen liberated is the water, not the carbon dioxide.

6. ATP carries energy from the light reaction to the dark reaction.

7. $NADPH_2$ carries hydrogen from the light reactions to the dark reactions.

8. In the dark reactions, hydrogen is fixed to carbon dioxide to form PGAL and glucose. The dark phase is also called carbon fixation.

9. PGAL is a 3-carbon product of photosynthesis.

10. Photosynthesis stores the energy of light in the form of carbon-carbon (C–C) and carbon-hydrogen (C–H) bonds.

COMPLETION QUESTIONS

[A] 1. The type of nutrition whereby organisms take in pre-formed organic molecules is . . . .

2. The type of nutrition in the green plant is . . . .

3. Autotrophs that obtain their energy for synthesis by chemical means carry on . . . . .

4. A substance that acts as a catalyst during photosynthesis is . . . .

5. During photosynthesis light energy is converted into . . . . energy.

6. Most of the oxygen in the atmosphere results from the process of . . . .

7. Chlorophyll may be removed from a leaf by boiling it in . . . .

8. The portion of an experiment that serves as a basis for comparison is the . . . .

9. A chemical that identifies different conditions by a change in color is known as a (an) ....

10. When white light falls upon a green object, the wave lengths for green are ....

11. The principal colors of light used in photosynthesis are .... and ....

12. Oval-shaped structures within the chloroplasts are the ....

13. A metallic element present in the chlorophyll molecule is ....

14. The raw material that supplies the oxygen liberated during photosynthesis is .....

15. A radioactive form of carbon that has been used to trace the steps of carbon fixation is ....

(B) 16. The breakdown of water by light energy into hydrogen and oxygen is known as ....

17. The coenzyme that carries hydrogen from the light reactions to the dark reactions is ....

18. A molecule having electrons traveling in orbits further removed from the nuclei than their normal orbits is said to be ....

19. Oxygen is released during the .... reactions of photosynthesis.

20. Carbon dioxide is used during the .... reactions of photosynthesis.

21. The energy for carbon fixation is directly supplied by .... and ....

22. Intermediate products formed during photosynthesis were identified by the technique of ....

23. The first chemically stable compound formed during photosynthesis is ....

24. ADP + P + energy yields ....

25. As a result of photosynthesis, the energy of light is stored in .... and .... bonds in glucose.

MULTIPLE-CHOICE QUESTIONS

(A) 1. The raw materials for photosynthesis are (1) CO_2 and O_2, (2) H_2O and O_2, (3) CO_2 and H_2O, (4) sunlight and chlorophyll.

2. A substance absorbs all wave lengths of visible light except that of red. When viewed in red light, this substance would appear to you as (1) red, (2) white, (3) green, (4) black.

3. O-18, employed for tracing certain reactions in photosynthesis, may be identified by use of a (an) (1) Geiger counter, (2) mass spectrophotometer, (3) mass spectrometer, (4) electron microscope.

4. Which of the following is the best equation to represent the overall process of photosynthesis?
(1) Energy + CO_2 + $H_2O \longrightarrow C_6H_{12}O_6$ + O_2
(2) Energy + $6CO_2$ + $6H_2O \longrightarrow C_6H_{12}O_6$ + $6O_2$
(3) Energy + $6CO_2$ + $12H_2O \longrightarrow C_6H_{12}O_6$ + $6O_2$ + $6H_2O$
(4) Energy + $12CO_2$ + $6H_2O \longrightarrow C_6H_{12}O_6$ + $6O_2$ + $6H_2O$

5. Which factor *least* influences the rate at which photosynthesis occurs? (1) atmospheric concentration of carbon dioxide, (2) time of day, (3) concentration of chlorophyll, (4) concentration of nitrogen in the air.

(B) 6. Which occurs during the light reactions of photosynthesis? (1) Water molecules are split, releasing oxygen. (2) Carbon dioxide molecules are split, releasing oxygen. (3) Carbon dioxide combines with hydrogen, forming carbohydrates. (4) Carbon combines with water, forming carbohydrates.

7. Which occurs during the dark reaction of photosynthesis? (1) Water is split into hydrogen and oxygen. (2) Photolysis. (3) Carbon dioxide is united with hydrogen. (4) ATP is produced.
8. The dark reactions of photosynthesis can *not* occur in the absence of (1) light, (2) chlorophyll, (3) O_2, (4) CO_2.
9. The dark phase of photosynthesis (1) consists of a single enzyme-controlled step, (2) consists of a series of enzyme-controlled steps, (3) uses energy supplied by activated chlorophyll, (4) uses energy supplied by ADP.
10. The production of glucose, a 6-carbon sugar, is most closely associated with the chemical reactions that occur during (1) the light phase of photosynthesis, (2) the dark phase of photosynthesis, (3) aerobic respiration, (4) anaerobic respiration.

CHAPTER TEST

1. The most important aspect of the photosynthetic process is considered to be the (1) production of CO_2, (2) production of H_2O, (3) production of chlorophyll, (4) conversion of radiant energy to chemical energy.
2. Scientists discovered the source of the oxygen liberated by algae primarily through the use of (1) isotopes, (2) microdissection instruments, (3) the phase-contrast microscope, (4) the ultracentrifuge.
3. Which portion of the light spectrum is *least* effective as a source of energy in photosynthesis? (1) orange, (2) red, (3) green, (4) blue.
4. Which is *not* directly required for photosynthesis? (1) water, (2) chlorophyll, (3) carbon dioxide, (4) oxygen.
5. A plant cell that lacks chloroplasts does *not* (1) give off O_2, (2) give off CO_2, (3) take in water, (4) take in food.
6. An enzyme that carries hydrogen atoms from the light reactions to the dark reactions is (1) ATP, (2) chlorophyll, (3) $NADPH_2$, (4) PGAL.
7. Which environmental change is most likely to increase the rate of photosynthesis in a bean plant? (1) a drop in temperature to 15°C, (2) an increase in the intensity of green light, (3) a rise in the oxygen concentration in the air, (4) a rise in the carbon dioxide concentration in the air.
8. A scientist observes that blue and red light are absorbed by a chlorophyll solution. This supports the concept that (1) chlorophyll can be blue or red in color, (2) chlorophyll can liberate oxygen, (3) blue and red light play a role in photosynthesis, (4) green light is necessary for photosynthesis.
9. Which has recently been found useful in adding to our knowledge of the chemical reactions that take place during photosynthesis? (1) electron microscope, (2) oxygen-16, (3) uranium-238, (4) carbon-14.
10. Which of the following is true of the dark reactions of photosynthesis? (1) $NADPH_2$ is formed, (2) carbon dioxide molecules are joined to hydrogen atoms, (3) the reactions proceed without the use of enzymes, (4) oxygen is liberated.
11. Energy is carried from the light reactions to the dark reactions by (1) ATP, (2) ADP, (3) CO_2, (4) activated chlorophyll.
12. If an organism uses carbon dioxide as a hydrogen acceptor, it is most likely a (an) (1) yeast, (2) alga, (3) protozoan, (4) virus.

Base your answers to questions 13 through 17 on the graph below of various wave lengths of light passing through a chlorophyll solution. Light that is not transmitted by the solution is absorbed.

13. The percentage of light absorbed at a wave length of 600 mμ is approximately (1) 15%, (2) 45%, (3) 90%, (4) 600%.
14. The color of light which is absorbed most by this solution is (1) blue, (2) green, (3) orange, (4) red.

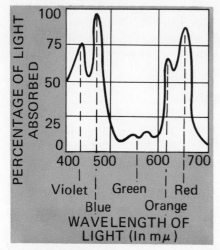

15. The color of light which is transmitted most by this solution is (1) blue, (2) green, (3) orange, (4) red.
16. The visible light with the longest wave length is (1) violet, (2) blue, (3) green, (4) red.
17. The wave length of light which is *least* effective during the process of photosynthesis is (1) 400 mμ, (2) 475 mμ, (3) 550 mμ, (4) 625 mμ.

Base your answers to questions 18 through 21 on the equations below and your knowledge of biology.

Equations

$$(1) \quad 6H_2O + 6CO_2 \xrightarrow[\text{light}]{\text{chlorophyll}} C_6H_{12}O_6 + 6O_2$$

$$(2) \quad C_6H_{12}O_6 + 6O_2 \longrightarrow 6H_2O + 6CO_2 + \text{energy}$$

18. Equation 1 takes place in all (1) plants, (2) animals, (3) plants and animals, (4) green plants.
19. The common name of the compound $C_6H_{12}O_6$ formed in equation 1 is (1) simple sugar, (2) protein, (3) glycerol, (4) amino acid.
20. Equation 2 represents a process by which energy is (1) stored, (2) released, (3) created, (4) increased.
21. Equation 2 takes place in (1) plants only, (2) animals only, (3) green plants only, (4) plants and animals.

For each of the questions 22 through 25 write the letter of the item, chosen from the list below, to which that question refers.

Item

a. Photosynthesis (but not respiration)   c. Both photosynthesis and respiration
b. Respiration (but not photosynthesis)   d. Neither photosynthesis nor respiration

22. Can take place in organelles called chloroplasts.
23. Is regulated by a series of enzymes.
24. Liberates free nitrogen.
25. Results directly in the building up of complex organic molecules.

CHAPTER 14

Plant Nutrition: Adaptations and Modes

If we study a large land plant (such as an elm tree), we soon become conscious of its needs and requirements. Most of its green leaves (containing chlorophyll) are far from the earth, and are supported on a firm trunk (stem), which is rooted in the earth. We can easily see that such a large multicellular organism has different requirements from either an animal or a simple water-dwelling plant.

As plants became many-celled and moved to the land, they had the advantages of specialization; but, they also faced certain problems. Many-celled plants are larger in size than single-celled plants and require special means for transporting essential materials to the interior. In addition, they must have suitable structures and processes to protect them while living on land where conditions are dry. Since they cannot depend upon a water environment to buoy them up, they must have adequate supporting structures. A many-celled land plant must be able to:

1. Use efficiently its limited surface area to provide for the needs of cells in the interior.

2. Transport materials (such as water and minerals) between the exterior cells and the interior cells, and transport food to all parts of the plant.

3. Obtain, transport, and conserve water.

4. Coordinate the activities of a many-celled organism.

5. Support the leaves in the air to obtain sufficient carbon dioxide and light. (Carbon dioxide and carbonates are plentiful in the ocean, but only 0.04% of the air is carbon dioxide. Green plants must hold their leaves aloft in such a manner that they are not shielded by one another from the sunlight for significant periods of time.)

6. Transfer reproductive cells on land.

7. Adapt to extreme variations in the environment. (By contrast, the physical environment of the sea is relatively constant.)

Multicellular plants have evolved many adaptations (special structures and kinds of behavior) to solve these problems.

A | THE LEAF

The specialized structure for photosynthesis in most plants is the leaf. The main regions of the leaf of a land plant are shown below. (See Fig. 14–1.)

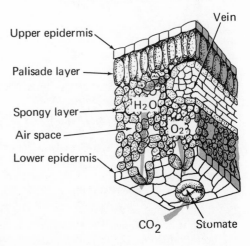

Fig. 14–1. Cross section of a leaf.

Structure of the Leaf

Upper epidermis. On the upper surface of the leaf is a single layer of colorless cells, which forms a tough waterproof covering. This is the *upper epidermis*. It protects the underlying cells and helps to prevent the loss of water. In many plants, the upper epidermis is further covered by a waxy *cuticle*, a layer that is practically impervious to water vapor.

Palisade layer. Below the upper epidermis is a layer of long, closely packed cells. This is the *palisade layer*. The cells of this layer contain numerous chloroplasts. Note that the position of the palisade layer is such that its cells are well exposed to the light.

Spongy layer. Below the palisade layer is the *spongy layer*. This is a region of loosely packed cells with many *air spaces* between them. Cells of the spongy layer also contain chloroplasts.

The spongy layer and the palisade layer carry on most of the photosynthesis of the leaf. The air spaces in the spongy layer increase enormously the surface area available to gases diffusing into and out of the cells that carry on photosynthesis. The air spaces are continuous with pores in the lower surface of the leaf.

Veins are also present in the spongy layer. These are tube-like bundles of vascular (conducting) cells. They are continuous with tubes in the stem which carry water upward from the root and others which carry food materials downward.

Lower epidermis. The lower epidermis consists of colorless cells tightly arranged together except for the presence of pores called *stomates* (Latin—*stomata*). Each stomate (Latin—*stoma*) is an opening formed by two guard cells. (See Fig. 14–2.)

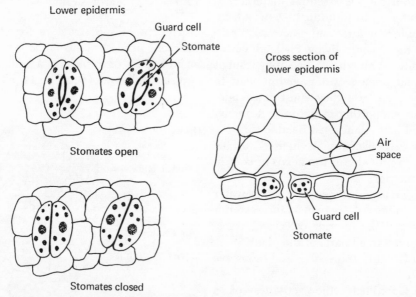

Fig. 14–2. Stomates. At the left are views of the lower epidermis, as seen from below. (The top drawing shows the guard cells in a turgid state so that the stomates are open; the bottom drawing shows the stomates closed.) At the right is a view of the stomate as seen in a cross section of the leaf.

Exchange of Gases in the Leaf

As you will recall, water is one of the raw materials of photosynthesis. Water is transported through the leaf by the veins (Fig. 14–1) and passes into the air spaces. From the air spaces, water vapor readily diffuses into the chlorophyll-bearing cells of the spongy layer and palisade layer. Carbon dioxide, another raw material of photosynthesis, enters the air spaces through the stomates in the under surface of the leaf. (Some leaves also have stomates in the upper surface.) The moist cell membranes of the cells that carry on photosynthesis permit carbon dioxide to diffuse into them. Oxygen, a by-product of photosynthesis, likewise diffuses out of the moist cell membranes into the air spaces. It then passes out of the leaf via the stomates. Water vapor also passes out of the stomates.

Transpiration. The process by which the leaf gives off water vapor is called *transpiration.* The amount of water given off by a plant during a day is considerable. For example, in a single day, a maple tree may lose 50 gallons of water. Transpiration results in the loss of a substance that is precious to land plants, but the loss is unavoidable since the moist cells inside the leaf are in contact with the air circulating in the spongy tissue. As we shall see later, transpiration also serves a most important function for the plant.

Action of the stomates. Much of the loss of water through transpiration is prevented by the remarkable capacity of the stomates to open and close. The stomates open during the day when carbon dioxide must enter for photosynthesis and they close during the night, which minimizes the loss of water by transpiration. The mechanism by which stomates open and close is explained as follows:

The stomates are regulated by guard cells—the only cells in the lower epidermis which contain chloroplasts. When light strikes the guard cells, they carry on photosynthesis, thereby producing sugars. The high concentration of sugar in the guard cells causes osmosis of water into these cells. As a result they swell, a condition known as *turgor.* However, the inner walls of the guard cells—those adjacent to the stomate opening—are thicker than the walls of the outer surface (Fig. 14–2). As a result, in swelling, the curvature of the guard cells is increased so that they assume the shape of a curved sausage. The space between the guard cells (the stomate) consequently becomes enlarged. At night, when photosynthesis stops, this swelling process is reversed. The guard cells lose their turgor and the openings close.

The shape of the guard cells is dependent not only upon their exposure to light but also upon their content of CO_2. The acidity of the guard cells, affected by their CO_2 content, also plays a part in the opening and closing of the stomates.

The action of the stomates is an example of conservation of water by land plants, for it helps to keep loss of water by transpiration within bounds. It is also an example of homeostasis, for it helps the plant to maintain a stable internal environment.

Complementarity of Structure and Function

When one thing complements another, it is said to complete it; thus, the two together make an entity. In this way, special structures in an organism complement special functions. This relationship of structure to function is characteristic of living organisms. The adaptation (special design) of a structure in order to carry on a special function is called *complementarity of structure and function*. In what way is this phenomenon discernible in the leaf?

A complex process such as photosynthesis has many functional requirements. For example, light, chlorophyll, water, and carbon dioxide must be brought together in adequate quantities. The leaf has structures that are adapted for these and other functional requirements. Adaptations of the leaf for photosynthesis include the following:

1. The thick walls of the epidermal cells prevent the loss of water.

2. Since the leaf is held aloft in the dry air, it must have adaptations for keeping certain cell membranes moist. The air spaces provide a region where cells of a leaf can have moist membranes for the exchange of gases with the external atmosphere.

3. The air spaces provide a relatively large surface for the exchange of gases between the air and the cells of the spongy layer. However, the openings of the air spaces to the outside, the stomates, are relatively small. Recall that there is a similar adaptation in man where the large moist surface of the lungs is serviced by a small exterior opening.

4. The guard cells are specialized cells of the epidermis which contain chloroplasts and which have a thick inner wall. These adaptations permit the guard cells to swell in such a manner as to open the stomate for the entrance of carbon dioxide when light is present.

REASONING EXERCISES

1. What problems did plants face as they moved from water to land?
2. What use do green plants make of carbon dioxide?
3. What adaptations of the leaf help a land plant to obtain carbon dioxide in sufficient quantities?
4. What adaptations of a green land plant help it to expose chlorophyll to the light?
5. In what ways is water a precious commodity for the green land plant?
6. How is the green leaf adapted to prevent loss of water?

B AUTOTROPHIC BACTERIA

The green plants are autotrophs, for they make their own food from inorganic materials instead of taking in pre-formed organic molecules. *Auto-troph* literally means "self-feeder," and *heterotroph* means "other-feeder." Most of the bacteria are heterotrophs. These are the ones that cause decay, fermentation, and disease. However, some bacteria are autotrophs.

The autotrophic bacteria require special consideration now, for their modes of nutrition differ significantly from that of the green plants.

Photosynthesis

The bacteria that carry on photosynthesis are the green sulfur bacteria and the purple sulfur bacteria. These bacteria contain a chlorophyll-like pigment that is dispersed along the inside of the cell membrane instead of being localized in chloroplasts. They use light as the source of energy, but do not use water as a raw material. Instead, they may use hydrogen sulfide (H_2S). They liberate sulfur instead of oxygen.

Chemosynthesis

Certain bacteria form carbohydrates without using energy from the sun. Instead, they use the energy produced by inorganic reactions within the cell. The formation of carbohydrates by the use of the energy of chemical reactions (rather than that of light) is called *chemosynthesis*.

The chemosynthetic bacteria include the iron bacteria, sulfur bacteria, hydrogen bacteria, and nitrifying bacteria. The iron bacteria use energy formed by oxidizing iron compounds, and the sulfur bacteria, by oxidizing sulfur compounds. Hydrogen bacteria oxidize hydrogen to obtain energy.

The *nitrifying bacteria* oxidize ammonia and nitrites as their source of energy, meanwhile forming nitrates. They are thus an important part of the nitrogen cycle whereby green plants are supplied with nitrates which they need in their metabolism. The nitrogen cycle is discussed in detail in Chapter 22.

C | HETEROTROPHIC NUTRITION

Heterotrophic plants lack chlorophyll and are unable to synthesize their own complex organic compounds. These plants obtain pre-formed organic molecules from other organisms and their products. In most cases, they carry on extracellular digestion. Bacteria, yeast, molds, and other fungi are heterotrophic plants.

Examples of Heterotrophic Plants

Bacteria. Some bacteria are autotrophs, but most bacteria are heterotrophs, taking in pre-formed high-energy organic molecules. They secrete digestive enzymes that pass out of the cell, where digestion of the food material occurs. Digested foods then pass through the cell wall and cell membrane into the bacterial cell. Here they are used as sources of energy.

Yeasts. Yeasts are tiny one-celled plants which do not have chlorophyll. A package of dried yeast purchased in a grocery store is composed of millions of yeast cells that renew their activity when placed in a sugar solution. Since the yeasts take in pre-formed organic molecules such as sugar, it is evident that their nutrition is heterotrophic. Yeasts obtain energy by anaerobic respiration (or fermentation). They release carbon dioxide and alcohol.

Molds. It is easy to grow bread mold for examination under the microscope. Place a small piece (about ¼ slice) of bread on a dish and moisten it. Keep the moist bread in the open for a short time so that mold spores in the air and on dust can fall on it.

Cover the dish with a glass tumbler in order to prevent drying and keep it in a warm dark place for a few days. Soon a mass of white threads becomes

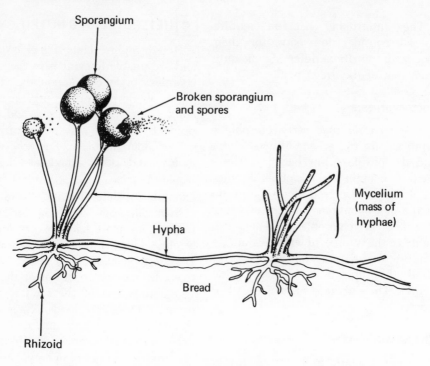

Fig. 14–3. Bread mold.

Slime Mold

visible, and then a mass of tiny dark spheres. The mass of threads is called the *mycelium,* and each individual thread is called a *hypha* (plural—*hyphae*).

Using needles, transfer a tiny portion of the mycelium to a drop of water on a microscope slide. Tease the threads apart. Cover the slide with a coverslip and examine under the microscope. Since the organization of the mold will be disturbed by the manipulation, examine several areas of the preparation. This should reveal the structure of the mold as shown in Fig. 14–3.

Special hyphae which penetrate the bread are called *rhizoids.* These have two functions: (1) attachment of the mycelium to the bread, and (2) digestion. Digestion by plants is discussed later in this chapter.

Slime Mold

The slime mold is a unique type of organism with both plant and animal characteristics. We described the three stages of the slime mold in Chapter 4, page 74.

During the plasmodium stage, the slime mold may consist of a mass of strands which creeps over the surface of logs and leaves. It sends out "arms" that engulf and digest bacteria.

The streaming of the slime mold plasmodium can readily be observed under the microscope. You can determine its rate of travel and its reaction to stimuli such as vinegar and oatmeal.

Economic Importance of Fungi

Some of the activities of molds are useful to man. Molds give flavor to cheese, decay dead trees, and produce antibiotics such as penicillin.

On the other hand, some molds are .harmful to man. Molds cause human diseases such as athlete's foot and ringworm. Plant diseases such as the rusts and smuts take a great toll on the wheat, rice, and corn crops. The potato blight, which wiped out the Irish potato in 1845, was caused by a mold. Wet towels left in a laundry hamper will get mildew, a form of mold. Leather shoes left in a damp cellar may develop a luxuriant mold growth in the form of a fuzz.

Mushrooms are a food delicacy, but mushrooms collected in the field should not be eaten because some species contain a deadly poison.

Saprophytes and Parasites

A *saprophyte* is a heterotrophic plant that obtains food from dead organisms. The bread mold is a saphrophyte because the bread that serves as its food originated from a *dead* wheat plant.

A *parasite* is an organism (plant or animal) that obtains its food from living organisms. Some of the relatives of the bread mold are classed as parasites, for they obtain their food from *living* plants. Examples are wheat rust, which causes great economic loss, and bracket fungi.

DIGESTION IN PLANTS

Test a geranium leaf during the daytime and it will show the presence of starch. On the following morning, however, no starch is present in the leaf. What has happened to it? The starch has been digested to simple sugar which can pass from cell to cell, and the sugar has been transported to other parts of the plant.

Plants do not have specialized digestive systems such as occur in animals, but they do carry on digestion. Chemically, digestion in plants consists of hydrolysis (as in animals) and is controlled by enzymes.

Intracellular Digestion in Plants

Starches, lipids, and proteins are stored in plant cells. These large molecules are converted to simpler usable forms by enzymes that function within the cells. Since this digestion occurs within the cells, it is called *intracellular digestion.*

The principal carbohydrate reserve in plants is starch. It is found in large amounts in roots, stems, and seeds. *Diastase* is a plant enzyme that hydrolyzes starch to the disaccharide maltose. Once digestion is completed, the end products may be used in the cell, or transported to other parts of the plant for use or storage.

The white potato, which is an underground stem, contains large amounts of starch. When the shoot begins to grow from it, the starch is changed to sugar, which is then available to the new plant.

The cotyledons of a bean seed are its food storage organs. Use iodine to test the cotyledons for starch, and boil cotyledons with Benedict's solution to test for sugar. Also try these tests on the embryo plant within the seed. Only the test of the cotyledons for starch will be positive. Now perform these tests on a young seedling which has just developed from a bean seed.

Extracellular Digestion in Plants

In bacteria, yeasts, and molds, digestion is *extracellular*—outside the cell. The general pattern of extracellular digestion in plants is as follows:

Enzymes secreted by the cell diffuse outward through the cell membrane into the external environment. Here they digest the food, producing such simple, soluble end products as amino acids, simple sugars, fatty acids, and glycerol. These then diffuse back into the cell. Here they are used for energy and growth. Insectivorous (insect-eating) plants such as the pitcher plant and the Venus's fly trap also carry on extracellular digestion.

COMPLETION QUESTIONS

A 1. The percentage of carbon dioxide in the air is approximately . . . . %.

2. Vascular tissue is specialized for the . . . . of water and its dissolved materials.

3. Loss of water from the upper epidermis is prevented by the presence of a waxy . . . .

4. An adaptation of the leaf that increases the surface area for diffusion of gases to the cells is the presence of . . . .

5. Most of the photosynthesis in the leaf occurs in the . . . . and the . . . . layers.

6. Water is carried into the leaf by the . . . .

7. Gases enter and depart from the leaf through the . . . .

8. The process by which leaves give off water is known as . . . .

9. A stomate opens when the guard cells become . . . .

10. Cells in the epidermis which contain chloroplasts are the . . . .

B 11. The purple sulfur bacteria use . . . . . . instead of H_2O in the process of photosynthesis.

12. Iron bacteria are autotrophs that obtain the energy for food making by the process of . . . .

13. Bacteria which oxidize ammonia to nitrates are known as . . . . bacteria.

C 14. Most bacteria carry on . . . . nutrition.

15. An individual thread of a mold is called a (an) . . . .

16. The structures of the bread mold that digest food are the . . . .

17. The stage of the slime mold that travels by a streaming motion is called the . . . . . .

18. Plants that obtain their food from dead organisms are known as . . . .

19. Plants that obtain their food from living hosts are . . . .

20. A plant enzyme that digests starch to maltose is . . . .

CHAPTER TEST

1. An example of an autotroph is the (1) bread mold, (2) moss, (3) grasshopper, (4) earthworm.

2. A plant that carries on very little transpiration in its natural environment would most likely have a (1) thick leaf cuticle and few stomates, (2) thick leaf cuticle and many stomates, (3) thin leaf cuticle and few stomates, (4) thin leaf cuticle and many stomates.

3. Bread mold gets its food by means of (1) pinocytic vesicles, (2) aerobic respiration, (3) intracellular digestion, (4) extracellular hydrolysis.

4. Many simple demonstrations rely on the presence of starch as proof that photosynthesis has occurred. This is possible because (1) starches are made from sugar, (2) sugar and starch are really the same substance, (3) sugar does not exist in living tissues, (4) Fehling's solution will not indicate starch.

5. Which substances enter a green plant chiefly through its leaves? (1) oxygen and minerals, (2) water and minerals, (3) carbon dioxide and oxygen, (4) carbon dioxide and water.

6. Bread mold absorbs nutrients through (1) modified hyphae, (2) roots, (3) specialized root hairs, (4) xylem.

7. An organism which manufactures its own food from inorganic materials is known as a (an) (1) parasite, (2) heterotroph, (3) autotroph, (4) saprophyte.

8. The fact that most bacteria are saprophytes means that they live by (1) feeding on other living things, (2) making food by photosynthesis, (3) changing carbon dioxide to food, (4) feeding on dead organic matter.

9. A response of some plants to exposure to light is a sudden enlargement of the stomates. This makes possible an increased (1) intake of carbon dioxide, (2) intake of water, (3) intake of oxygen, (4) discharge of carbon dioxide.

10. Other environmental factors being the same, the *least* growth would most likely occur in bean plants grown under (1) orange light, (2) green light, (3) red light, (4) violet light.

11. The concentration of heterotrophic bacteria around a spirogyra (alga) colony exposed to red light is greater than is the concentration around the same plant when it is exposed to green light. This phenomenon occurs because (1) green light affects enzyme action in bacteria, (2) red light supplies the energy needed by the spirogyra to liberate oxygen, (3) red light kills the spirogyra plant, thus producing more food for the saprophytic bacteria, (4) the bacteria can utilize the energy from red light but not from green light.

12. During photosynthesis the energy of light is transformed mainly into (1) heat energy, (2) chemical bond energy, (3) kinetic energy, (4) nuclear energy.

13. The fact that, under certain conditions, stomatal openings become smaller enables the plant to avoid excessive loss of (1) carbon dioxide, (2) water, (3) essential plant hormones, (4) glucose.

14. Which process adds oxygen to the earth's atmosphere? (1) nitrogen fixation, (2) fermentation, (3) respiration, (4) photosynthesis.

15. A molasses solution in which yeast has been grown is boiled, and the vapor is condensed. Which substance may be found in the condensed vapor? (1) sugar, (2) alcohol, (3) starch, (4) glycogen.

16. Which of the following functions are performed by the rhizoids of bread mold? (1) anchorage and digestion, (2) digestion and photosynthesis, (3) anchorage and photosynthesis, (4) digestion and reproduction.

17. Which of the following cells of the leaf do *not* contain chloroplasts? (1) cells of the upper epidermis, (2) palisade cells, (3) guard cells, (4) cells of the spongy layer.

18. When guard cells carry on photosynthesis the stomates (1) open because water enters the guard cells, (2) open because carbon dioxide departs from the guard cells, (3) close because water departs from the guard cells, (4) close because sugar is made in the guard cells.

19. In a green plant, nitrates are used in the production of (1) proteins, (2) starch, (3) carbon, (4) carbohydrates.

20. A plant that lacks chloroplasts does *not* (1) give off O_2, (2) give off CO_2, (3) take in water, (4) take in food.

CHAPTER 15 | Plant Transport

Complex plants are similar to animals in that they have a system for transporting materials among the various parts. *Vascular tissues* are those groups of cells that function in transport. The stem, the root, and the leaf have significant roles in this important function.

A   THE STEM

Stems of plants are classified as either herbaceous or woody. *Herbaceous stems* (such as that of the bean plant, tomato, milkweed, and sunflower) are generally soft, thin and green. *Woody stems* (such as that of the maple tree) are tough, have an outer layer of cork, and live for a number of years. When a biologist speaks of a "stem," he means anything from the slim twig of a tree to the trunk of a giant sequoia.

Structure of the Herbaceous Stem

A cross-section of a herbaceous stem reveals some circles arranged in a ring.

(See Fig. 15–1A.) These are *vascular (conducting) bundles* which run up and down the stem. Outside the ring of vascular bundles is the *cortex*. Inside the ring of vascular bundles is the *pith*. The stem as a whole is enclosed by a layer of epidermal cells.

The structure of a vascular bundle is shown in Fig. 15–1B. Inside each vascular bundle are xylem, phloem, and cambium.

Xylem cells in the inner region of the vascular bundle possess thick walls. Xylem cells provide support, and they conduct materials upward through the stem.

Cambium cells are present in the vascular bundle between the xylem and the phloem. Cambium cells have the ability to divide and give rise to specialized plant tissues. As cambium cells divide, they form either xylem or phloem.

Phloem cells are thin-walled cells that primarily transport food downward in the stem.

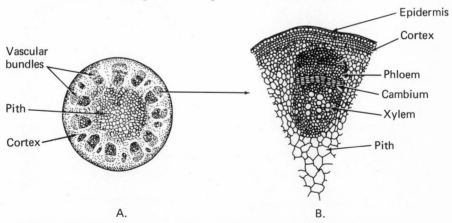

Vascular bundles

Pith

Cortex

Epidermis

Cortex

Phloem

Cambium

Xylem

Pith

A.

B.

Fig. 15–1. Cross section of herbaceous stem.

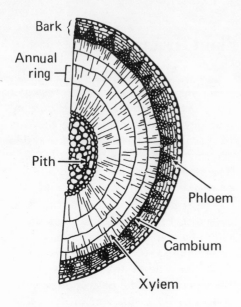

Fig. 15–2. Cross-section of a woody stem.

Structure of the Woody Stem

The general structure of the woody stem is shown in Fig. 15–2.

A comparison of this diagram with that of the herbaceous stem reveals a difference in the way the two types of stem are organized. The herbaceous stem has separate bundles of conducting tissue; the woody stem is organized in concentric rings. From the center to the outside of a woody stem, these rings are: the pith, xylem, cambium, phloem, cortex, and the outer bark.

The *xylem* constitutes the major region of the stem. In time, xylem cells develop heavy walls and die; thus, the inner region of the xylem consists of dead cells only. This is the "heartwood" portion of the stem that is used for lumber. The outer region of living xylem cells constitutes the "sapwood."

The pith and cortex are composed of large cells whose primary function is storage.

Annual Rings. Surrounding the xylem are the soft growing cells of the cambium. As the cambium cells divide, they form xylem cells toward the inner part of the stem and phloem cells toward the outer part. The xylem cells formed in the spring are larger than those formed later in the summer. This variation in size results in the appearance of annual rings, which are seen in cross-sections of tree trunks. The age of a tree can be determined by simply counting its annual rings. Note, however, that some plants have two seasons for growth and therefore two rings are formed each year. One must be familiar with the growth pattern of the plant being studied to determine age by counting the rings.

The climate of a particular year in the distant past can be estimated from a careful study of its annual rings. Deducing the climate of a particular year in the distant past is a complex matter, and involves comparison of the annual rings of many specimens from different time periods, as well as other studies.

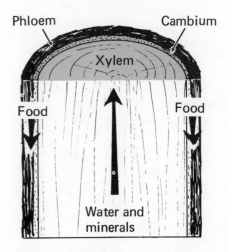

Fig. 15–3. Direction of transport in a woody stem.

Transport in the Woody Stem

The vascular tissue of the woody stem is the cylinder of xylem, cambium, and phloem. As illustrated in Fig. 15–3, the primary function of the xylem is to carry water and dissolved minerals *upward* to the leaves. The phloem carries food *downward* to the roots. However, the phloem also carries food materials upward in the spring in the form of sap.

Girdling a tree by cutting a band through the bark completely around the trunk cuts all the phloem tubes. As a result, the roots die after they have exhausted their supply of food, and the tree as a whole dies.

Adaptations of Vascular Tissue for Transport

The xylem region contains two kinds of cells for transport: (1) *tracheids,* and (2) *vessel* cells. (See Fig. 15–4A.) As previously stated, the cells in the xylem region develop thick walls and die. Then, they lose their cytoplasm, and become hollow. The thick walls of tracheids develop thin areas (pits) and become freely permeable to water and dissolved materials. During the development of vessels, holes form in the ends of adjoining cells so that, in effect, the vessels are long, hollow tubes extending up and down the stem. Thus, the water and dissolved materials can pass upward without interruption. In addition to these adaptations for transport, the xylem region contains fiber cells for support.

The phloem region contains (1) *sieve tube cells,* and (2) *companion cells.* (See Fig. 15–4B.) The cytoplasm of the sieve tube cells remains alive. Perforations in the end walls permit the cytoplasm to be continuous from cell

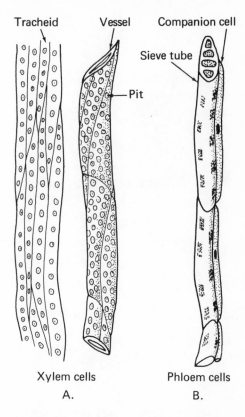

Xylem cells
A.

Phloem cells
B.

Fig. 15–4. Xylem and phloem cells.

to cell. Although sieve tube cells retain their cytoplasm, they lose their nuclei. Accordingly, the metabolism of sieve tube cells is probably regulated by the smaller adjacent companion cells.

THE ROOT

The root is continuous with the stem. The root has the following functions: (1) anchorage of the plant, (2) absorption of water and soluble salts from the soil, (3) conduction of materials to the stem, and (4) storage of food. Reproduction is a function of the root in some plants.

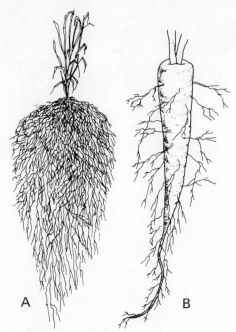

Fig. 15–5. Fibrous roots and taproots.

Kinds of Roots

Roots may be divided into three main classes: fibrous roots, taproots, and storage roots.

Fibrous roots. These roots usually consist of a network of branches that spreads out over a large area. (See Fig. 15–5A.) Grasses have fibrous roots.

Taproots. These are long main roots that go deep into the ground. The root of the carrot is an example. (See Fig. 15–5B.)

Storage roots. When the growing taproot has become fleshy with stored food, it is a thick storage root. Carrots, turnips, beets, and radishes have storage roots.

Structure of the Root

As seen in cross-section, the root has three main regions: the *central cylinder,* the *cortex,* and *epidermis.* (See Fig. 15–6A.)

The central cylinder contains the conducting tissue (xylem and phloem). Note from the diagram that the xylem in cross-section resembles a wheel and its spokes. Phloem is present between the arms of the xylem. Cambium, when present, is between the xylem and the phloem.

The *cortex* is the cylindrical region surrounding the central cylinder. The cells of this region are thin walled and adapted for storage of food.

The *epidermis* is the outer covering.

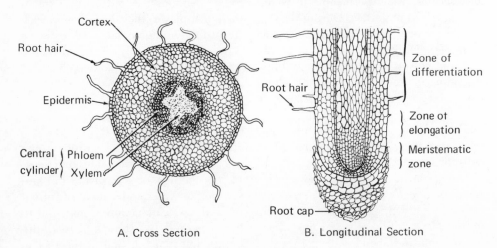

A. Cross Section   B. Longitudinal Section

Fig. 15–6. Structure of the root.

It consists of a single layer of small cells which are used for absorption. Epidermis is found only in the tiny, young ends of the roots.

The root hairs. Root hairs are located slightly behind the growing region of the root. (See Fig. 15–6B.) Root hairs are specialized cells of the epidermis which form delicate, thin-walled projections. (See Fig. 15–7.)

Root hairs absorb water and minerals from the soil. Because they protrude from the surface of the root, the

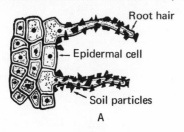

A

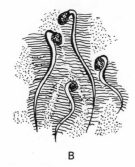

B

Fig. 15–7. A. Root hairs. B. Germinating seedlings showing root hairs.

root hairs provide greatly increased surface for absorption. A single rye plant has 14 billion root hairs with a total area of over 4000 square feet.

Substances that diffuse through the semipermeable membranes of root hairs include water, gases (such as oxygen), and ions of soluble minerals. The minerals absorbed include sulfates, phosphates, nitrates, calcium compounds, iron, and trace amounts of copper, zinc, boron, manganese, magnesium and molybdenum. Active transport also plays a part in absorption. The absorbed materials pass from cell to cell toward the region of the central cylinder. Here the xylem transports materials upward toward the stem and leaves.

THE LEAF

As previously noted, the "veins" in leaves are vascular bundles containing xylem and phloem. Because they also contain fibers, they are known as *fibrovascular bundles*. Xylem and phloem tissue extends from the roots to the very edges of leaves.

Although biologists still do not have a complete explanation of how liquids rise in stems, this process involves capillary action, root pressure, atmospheric pressure, and transpiration pull.

REASONING EXERCISES

1. What does a biologist mean by a stem?
2. Distinguish between a herbaceous stem and a woody stem.
3. How can you tell the age of a woody stem?
4. How do xylem and phloem function in transport in a woody stem? Why will girdling kill a woody stem?
5. How does the structure of a stem complement its function?
6. How is the root adapted for its functions?

B THEORIES FOR UPWARD TRANSPORT

Any theory to explain the rise of liquids in stems must account for upward transport to the heights reached by redwood trees, which are over 100 meters tall. Upward transport is probably the result of a combination of the following factors:

Capillary action. Liquids rise in tubes of small diameter by a process called *capillary action*. This rise is due to attraction: (1) between the water molecules and the walls of the tube (adhesion), and (2) of the water molecules for one another (cohesion). The narrower the tube, the higher the water will rise. It is capillary action that lifts water into a blotter.

The vessels and tracheids of the xylem are dead cells that form long tubes of microscopic diameter. These tubes can raise liquids by capillary action; however, this action will not raise water more than a few inches in height. Moreover, the rise of water by capillary action is not fast enough to explain the rapid rise of water in stems. Capillary action by itself cannot explain the rise of water in tall plants.

Root pressure. When a plant stem is cut off near the ground, sap comes out of the portion of the stem attached to the ground. This flow is due to upward pressure caused by the osmosis of water (osmotic pressure) into the root hairs. Root pressure undoubtedly contributes to the rise of liquids in the xylem, but is not sufficient to account for its rise to great heights.

Atmospheric pressure. When water is "sucked up" through a straw, the water moves upward because certain upward and downward forces have become unbalanced. Let us see why this is so.

The pressure of the air (atmospheric pressure equal to 14.7 pounds per square inch) is transmitted through the water in the container in such a manner as to exert an upward force upon the water in the straw. (See Fig. 15–8A.) Before a person sucks on the

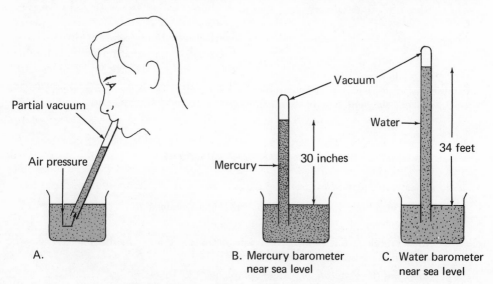

Partial vacuum

Air pressure

A.

Vacuum

Mercury —— 30 inches

B. Mercury barometer near sea level

Water ——

34 feet

C. Water barometer near sea level

Fig. 15–8. Examples of atmospheric pressure.

straw, the downward pressure caused by the weight of the water serves to counterbalance the upward pressure. However, the mouth creates a partial vacuum above the water in the tube, which reduces this downward pressure. When the atmospheric pressure pushing upward in the tube exceeds the pressure of the weight of the air pushing downward, the water rises in the tube.

A mercury barometer (Fig. 15–8B) shows the height that atmospheric pressure raises a column of mercury. Note that there is a vacuum above the mercury. The greater the atmospheric pressure, the higher the mercury rises in the tube. At sea level, atmospheric pressure raises a column of mercury to a height of about 30 inches (760 mm). Because water is much lighter than mercury, atmospheric pressure can raise water to a height of about 34 feet (Fig. 15–8C). Pumps that raise water by creating a partial vacuum above the column of water cannot raise water higher than 34 feet. It is the difference in air pressure inside and outside this column that raises the water, not any mysterious "suction."

As water is used by the leaves of the tree, a partial vacuum is probably formed at the tops of the xylem tubes. Atmospheric pressure might raise the water; however, this force could explain the rise of water only to a height of about 10 meters (30 feet). Our tall trees reach a height of 100 meters (300 feet). Atmospheric pressure alone cannot explain the rise of water to the tops of tall trees.

Transpiration pull. According to this theory, water is *pulled up by the leaves*, rather than *pushed up* by root pressure, capillary action, or atmospheric pressure. In the process of transpiration,

water evaporates from the air spaces in the leaf and passes out of the stomates. As the amount of water in the air spaces decreases, water moves toward them by osmosis. In this way, water moves from cell to cell and is drawn away from the tops of the xylem tubes. Accordingly, a "pull" is exerted on the water in the xylem tubes. Would this pull be sufficient to lift a column of water which is 300 feet in height? Let us explore this question briefly.

Because of the force of cohesion that water molecules have for one another, the thin, unbroken columns of water that fill the xylem tubes are very strong. It can be shown that the cohesive force of water in tubes as narrow as those in the xylem is sufficient to withstand a pull to a height of 4000 feet. Thus, the *combination of transpiration pull and cohesion-tension* of water is probably sufficient to raise water to the tops of tall trees. This theory has the widest acceptance among biologists. However, it is recognized that atmospheric pressure, root pressure, and capillarity probably also play a part, particularly for low-growing plants without well-developed vascular systems.

DOWNWARD TRANSPORT

The downward transport of materials in stems has not been satisfactorily explained. It is known that the sieve tubes of the phloem carry soluble materials such as sugars down to the roots where they are used for energy or are stored. (The sieve tubes also carry food upward in the spring.)

Sieve tubes present a different physical problem from that of the tracheids and vessels of the xylem, which are dead and hollow. As previously noted, sieve tubes contain living cytoplasm, which is continuous from cell to cell. The transport of materials through this cytoplasm is much too rapid to be explained by simple diffusion. In most cases, the direction of transport from cell to cell is opposite to the concentration gradient. Active transport and the use of energy seem to be involved in downward transport.

COMPLETION QUESTIONS

[A] 1. Groups of cells that participate in the transport of liquids in plants are called .... tissue.

2. The vascular bundles of a herbaceous stem contain phloem, cambium, and ....

3. The primary function of pith and cortex is ....

4. All of the cells outside of the xylem of a woody stem constitute the ....

5. Water is carried upward in the .... region of woody stems.

6. Food is carried downward in woody stems by the .... region.

7. Two kinds of cells present in xylem are tracheids and ....

8. Two kinds of cells present in phloem are companion cells and ....

9. The region of a root which contains conducting tissue is the ....

10. An adaptation of roots which provides increased surface for absorption is the presence of ....

(B) 11. Any acceptable theory of upward transport in the xylem must explain a rise to a height of .... feet.

12. The phenomenon which explains the rise of liquid in a blotter is known as ....

13. The flow of sap from a stem cut near the ground is due to upward pressure caused by the .... of water.

14. The greatest height to which atmospheric pressure could raise water in plant stems is ....

15. Probably the best explanation for the rise of liquids to the tops of tall trees is a combination of transpiration pull and ....

CHAPTER TEST

1. The process of osmosis is best illustrated by the entrance of (1) water into root hairs, (2) water through stomates, (3) oxygen into lung capillaries, (4) white corpuscles into intercellular spaces.

2. Food manufactured in the leaves of plants is transported to the roots chiefly through the (1) xylem, (2) cambium, (3) phloem, (4) pores in the epidermis of the stem.

3. In multicellular green plants an example of an adaptation that facilitates the entry of mineral salts is the (1) cambium, (2) stomates, (3) vascular bundles, (4) root hairs.

4. The major factors causing the rise of water to the top of tall trees are (1) root pressure and cohesion, (2) capillary action and adhesion, (3) cohesion-tension and transpiration pull, (4) atmospheric pressure and root pressure.

5. In the xylem are found (1) vessels, (2) sieve tubes, (3) palisade cells, (4) companion cells.

6. If the cut end of the stem of a white carnation is placed in red dye, the petals will eventually become red. This is evidence that (1) some plants can grow without roots, (2) the cambium layer is present in the stem, (3) xylem tissue is located throughout the length of the cut plant, (4) root hairs are adapted for absorption.

7. In a woody stem, the xylem and phloem tubes are separated from each other by a layer of cells known as (1) cambium cells, (2) spongy cells, (3) palisade cells, (4) epidermal cells.

8. By counting the annual rings of a woody tree, one can (1) determine the climate during each stage of growth, (2) predict the climate, (3) determine the age of the tree and the climate for each year of its growth, (4) determine the age of the tree only.

9. A plant's ability to raise water up its stem is *least* affected by the (1) thickness of its bark or epidermis, (2) amount of leaf surface area, (3) number of its root hairs, (4) diameter of its xylem tubes.

10. That mosses lack xylem accounts in part for the fact that mosses (1) often grow best in mountainous regions, (2) can survive in a very dry region, (3) are unable to absorb and store water, (4) seldom grow more than a few inches tall.

11. The surface area of a root is increased by the presence of (1) cilia, (2) root hairs, (3) flagella, (4) villi.

12. Most of the water carried to the leaves of plants travels through the (1) cambium, (2) companion cells, (3) xylem cells, (4) phloem cells.

Directions (13-17): Select the *letter* of the expression which matches the question most closely.

A. upper epidermis
B. root hairs
C. stomate
D. tracheids

E. palisade layer
F. cambium
G. sieve tubes

13. Food travels downward in stems.

14. Increase in diameter of stem.

15. Provides increased surface for osmosis.

16. Regulates rate of transpiration.

17. Tissue where photosynthesis occurs.

CHAPTER 16 / Respiration, Growth, Regulation

During the school year, if you place lima bean seeds in a jar containing moist blotting paper, you may see the seeds germinate and young roots and leaves emerge. If this happens, transfer one of the seedlings to soil in a flower pot and observe it develop into a young plant with numerous leaves. When spring comes, transplant the plant outdoors. As the summer advances, the bean plant will grow rapidly and may achieve a height of 15 feet or more as it twines around nearby supporting structures. Colorful flowers appear and, late in summer, a crop of pods may be harvested. As you open a pod, you should find a half dozen or so lima bean seeds. By this time, you will have witnessed a complete cycle in the life of a bean plant — from seed to seed.

As a representative of the autotrophic way of life, the bean plant illustrates two of the unifying themes of biology:

1. **Complementarity of structure and function.** Living things develop special structures to carry on special functions. For example, as the bean plant grows, you can observe some of its adaptations for support ("twining") and for photosynthesis (numerous leaves).

2. **Diversity of type but unity of pattern.** Although living things may seem very different, there is a great underlying similarity in their ways of solving problems. For example, animals and plants have structures that provide large surface areas for absorption. Moreover, in some cases (villi, root hairs), the solutions to the problem are very similar.

The theme of diversity of type but unity of pattern has numerous other applications. Plants and animals have many similarities in their ways of (1) obtaining energy from food, and (2) secreting chemicals to coordinate body parts. We will note some of these similarities in this chapter as we study respiration, excretion, synthesis, growth, and regulation. There are, of course, also many differences. These differences show how organisms adapted to varying living conditions during their evolution.

A | RESPIRATION

Respiration occurs 24 hours a day in plants (as in animals). Accordingly, respiration is a continuous process.

The chemical reactions of respiration in plants are very similar to those in animals. (See Chapter 7.) Much of the chemical energy of complex food molecules is transferred to ATP which then makes energy available for reactions in the cell.

You will recall that respiration processes are classified by whether or not they require oxygen. In anaerobic respiration, oxidation occurs without the use of molecular oxygen. In aerobic respiration, by contrast, oxygen is required. Anaerobic respiration produces less ATP and less energy from a molecule of glucose than does aerobic respiration.

Anaerobic Respiration

Anaerobic respiration in simple organisms is also known as *fermentation*.

In the continued absence of oxygen, ethyl alcohol or lactic acid may be produced by fermentation. Accordingly, there are two kinds of fermentation — *alcoholic fermentation* and *lactic acid fermentation*.

Alcoholic fermentation is carried on by yeasts and some bacteria. The products of the process are alcohol and carbon dioxide. A summary equation for alcoholic fermentation is:

$$\text{glucose} \xrightarrow{\text{enzymes}} \text{alcohol} + CO_2 + \text{energy}$$

Lactic acid fermentation is carried on by certain molds and some bacteria. It results in the formation of lactic acid.

$$\text{glucose} \xrightarrow{\text{enzymes}} \text{lactic acid} + \text{energy}$$

In fermentation, the products formed depend on the kinds of enzymes that are present in the plant cells.

The production of gases during the fermentation of yeast may be demonstrated in the laboratory with a fermentation tube. (See Fig. 16–1.)

Gas is given off in both arms of the tube. However, since one end of the tube is closed, and since the gas takes up space, the liquid rises in the other arm.

Although the amount of energy released in anaerobic respiration is relatively small, it is sufficient to maintain lower forms of life, including yeasts and bacteria.

Aerobic Respiration

In aerobic respiration, the plant cell uses molecular oxygen. Accordingly, the aerobic respiration of a molecule of glucose produces about 20 times as many molecules of ATP as anaerobic respiration produces. Both forms of respiration take place in a series of enzyme-controlled biochemical reactions. (See Chapter 7.)

A summary equation for aerobic respiration is:

$$\text{glucose} + \text{oxygen} \xrightarrow{\text{enzymes}} \text{carbon dioxide} + \text{water} + \text{energy}$$

Multicellular green plants (such as the bean plant) carry on aerobic respiration. The green plant obtains oxygen through stomates in the leaf and through *lenticels* (small openings) in the stem. Oxygen in the soil diffuses through the moist membranes of the root hairs. The oxygen taken in is then available for use in aerobic respiration.

The chemical reactions associated with energy release occur in the mitochondria. Carbon dioxide and water vapor, the gaseous products of aerobic respiration, are given off through the stomates and the lenticels.

Respiration and photosynthesis compared. The multicellular green plant carries on aerobic respiration and photosynthesis. Carbon dioxide produced in the leaf during respiration may be used

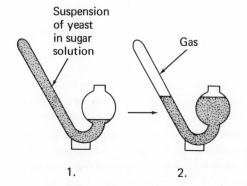

Suspension of yeast in sugar solution

Gas

1.   2.

Fig. 16–1. Fermentation tube.

RESPIRATION AND PHOTOSYNTHESIS COMPARED

	Respiration	Photosynthesis
Energy-rich compounds	1. Food *broken down* (oxidized) 2. ATP produced	1. Food accumulated 2. PGAL and glucose produced
Energy	1. Source: glucose, etc. 2. *Released* by oxidation	1. Source: sunlight 2. Stored in glucose, etc.
Carbon dioxide, oxygen, and water	1. Carbon dioxide and water vapor given off 2. Oxygen taken in	1. Carbon dioxide and water taken in 2. Oxygen given off
Occurrence	1. Day and night 2. In all living cells	1. *Only* in light 2. In presence of chlorophyll

immediately for photosynthesis. Oxygen produced during photosynthesis may be used for respiration. However, it should be borne in mind that photosynthesis occurs only when light energy is available.

Both photosynthesis and respiration proceed during the daytime. However, the net result of the daylight activity is that the plant utilizes carbon dioxide from the environment and adds oxygen to the environment. This disparity occurs because photosynthesis proceeds to a greater extent than the reverse process of respiration. Overall, green plants release far more oxygen than they take in. Although the stomates are partially closed at night, the amount of oxygen taken in through stomates, lenticels, and root hairs is sufficient for the relatively small amount of respiration occurring.

EXCRETION

Excretion is the removal of the waste products of metabolism. Plants do not have specialized organs for excretion as the higher animals do. They are able to re-use many of the products of metabolism, giving off only those products no longer useful to them.

If the carbon dioxide and water produced in respiration are not used in photosynthesis, they diffuse into the air spaces and pass out through the stomates and lenticels. Thus, stomates and lenticels may be considered as excretory structures as well as respiratory structures.

Some of the products of plant metabolism, such as organic acids, might be toxic (poisonous) to the plant. These are stored in vacuoles where, in effect, they are "sealed off" and cause no injury to the plant.

Some of the breakdown processes of plant metabolism result in the production of ammonia and other nitrogen compounds. Ordinarily, these are used by the plant. They are combined with nitrates absorbed by the plant for the synthesis of amino acids and other important compounds.

SYNTHESIS

By *synthesis,* the plant forms a wide variety of chemical substances, including auxins, poisons, waxes, fibers, drugs, and flavorings. In addition, storage products (for example, starch and oils) are synthesized, as well as the materials needed for growth and repair of the cell. (See Chapter 12.)

GROWTH

Growth is increase in size by the synthesis and organization of new material. Growth in animals occurs throughout the organism, but growth in plants takes place in specific regions. These growing regions include the tips of stems and roots. In plants there is growth (1) in length, at the tips of stems and roots, and (2) in width of the stem and root. The processes of growth include increase in number and size of cells and differentiation. Let us see what this means.

The Processes of Growth

Increase in number of cells. Division of a cell increases the number of cells. However, since two daughter cells occupy the same volume as the original cell, there is no increase in size.

Increase in size of cells. After cell division is completed, the newly formed cells increase in size (elongate).

Differentiation. The multicellular organism is not merely a mass of similar cells. Early during development, cells acquire special structural characteristics which adapt them to carry on their special functions. *Differentiation* is the formation of different kinds of cells. Accordingly, as an unspecialized cell divides, two cells that are specialized for particular functions may be formed by differentiation. The differentiated cells originate from undifferentiated cells in the plant called *meristem*.

One type of meristem (called *cambium*) undergoes differentiation, yet retains cells that remain undifferentiated. When cambium cells divide, one daughter cell may become differentiated — for example, into a xylem cell — whereas the other remains as undifferentiated cambium. The cause of differentiation is not well understood, but it is known that chemicals released from nearby cells play a significant role. Direct contact with the environment also plays a part, as when the cambium tissue of a cut stem begins to grow into roots when it is placed in moist soil.

The Regions of Growth

Growth at the root tip. The root tip is covered by a thimble-shaped group of cells, the *root cap*. The function of the root cap is to protect the delicate cells behind it as the root advances through the soil.

Behind the root cap are three regions of growth which illustrate the three processes of growth we have just discussed. These three regions are shown in the longitudinal section of a growing root on page 250 (Fig. 15–6B).

Immediately behind the root cap is the *meristematic zone*. As the name indicates, the cells in this zone are undifferentiated. These cells are actively dividing. Each cell is small, and the cells appear to be packed closely together. This zone is sometimes called the region of cell division.

Then follows a *zone of elongation*, where a marked increase in size takes place. Here the cells are long, with large vacuoles occupying much of the cell volume.

The *zone of differentiation* is a region where different kinds of cells are forming. Here are found root hairs and the beginnings of xylem and phloem.

Growth at the tip of a stem. Growth at the tip of a stem (Fig. 16–2) is similar to growth at the root tip. The same three regions are present: (1) meristem at the apex or tip, (2) zone of elongation, and (3) zone of differentiation. The cells in the region of elongation become enlarged as much as 20 times the length of the meristematic cells. The increase in length at the tip of a growing stem results mainly from this elongation.

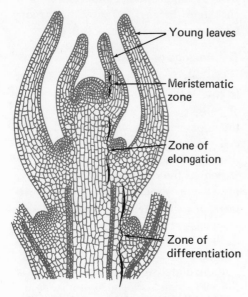

Fig. 16–2. Longitudinal section through the tip of a growing stem.

Increase in diameter of the stem. The increase in diameter of a stem results from cell division by the cambium. In the woody stem, cells produced from the cambium toward the inner region of the stem become differentiated as *xylem.* Those formed on the outer side result in the *phloem.* This addition of cells results in an increase in the diameter of the stem. As the outer circumference expands, the bark of many trees becomes cracked, furrowed, or peeled in distinctive patterns.

REGULATION

In this section, we will consider growth hormones and their role in regulation. Two types of growth hormones are (1) auxins, and (2) gibberellins.

Auxins

Auxins are chemical substances produced in one part of the plant and transmitted to another part of the plant where they have their effect. In this respect, plant hormones resemble animal hormones. Like other hormones, auxins are effective in small quantities. Auxins are produced at the growing tips of stems and roots and are also produced by young buds, young leaves, and developing flowers. They can either stimulate growth or inhibit growth, depending upon the concentration of the auxin and the kind of cell affected.

Auxins increase growth in two ways:

1. Increase in number of cells. Auxins can start cell division and increase the rate of cell division in stems, roots, buds, and cambium.

2. Elongation of cells. When auxins reach the zone of elongation of young stems and roots, they help these cells to elongate. The cell walls become more plastic, permitting them to absorb water, which is needed for elongation.

Effect of concentration. We have noted that auxins can *stimulate* growth (cause its rate to increase). Auxins can also *inhibit* growth (cause its rate to decrease). Which action will occur depends on the concentration of the auxin. A concentration that causes increased rate of growth (stimulation) in stems may result in decreased rate of growth (inhibition) in roots. This relationship is shown in Fig 16–3.

The naturally occurring auxin, IAA (indole acetic acid), has been identified. Compounds that are chemically related to IAA have been synthesized by artificial means. These artificial auxins include NAA (naphthaleneacetic acid) and 2,4-D.

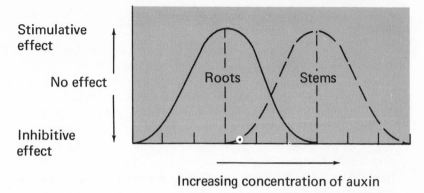

Increasing concentration of auxin

Fig. 16–3. The effect of the concentration of auxin upon growth of stems and roots. In general, to stimulate growth, roots require a lesser concentration than do stems. The concentration that has the greatest stimulative effect upon stems can have the greatest inhibitive effect upon roots.

Gibberellins

Gibberellins are growth-promoting hormones that cause a remarkable increase in the length of stems. They affect the whole stem, not just the region behind the growing tip. In some unknown manner, the gibberellins cause an increase in the rate of cell division. They do not cause a bending of seedlings. One of the gibberellins is gibberellic acid.

Kinins

Kinins (also known as cytokinins) are cellular stimulating substances that promote cell division in leaves and lateral buds. One of the best known is *kinetin.*

The kinins are not always considered to be hormones, since they affect the cell in which they are formed instead of other cells away from the place of production.

Growth Hormones a⊙ Regulators

Plants do not have a nervous system, but they do produce chemical regulators. Growth hormones, such as auxins and gibberellins, regulate the growth of portions of the plant so that the pattern of the plant as a whole is coordinated. Auxins affect: (1) the fall of leaves, (2) the development of the fruit, (3) the pattern with which flower buds open, (4) tropisms (see page 262), and (5) photoperiodism (see page 265).

Uses of Plant Hormones

Inexpensively produced synthetic plant hormones are now used in the following ways:

1. *To stimulate the formation of roots when plant cuttings are made.*

2. *To produce seedless fruit.* When IAA or NAA is applied, the ovaries of some flowers develop into fruits without pollination having occurred.

3. *To inhibit the sprouting of stored potatoes.* NAA is often used for this.

4. *As weed killers.* When proper concentrations of the synthetic plant hormone, 2,4-D, are applied to certain plants, their growth becomes abnormal and the plant dies. However, this effect is much greater upon dicot plants than upon the monocots. Inasmuch as most weeds are dicots, but grass is a monocot, 2,4-D is effective in killing weeds without damaging lawns.

REASONING EXERCISES

1. Define respiration. Compare anaerobic respiration and aerobic respiration from the standpoint of oxygen requirements and energy yield.
2. If you hammer a nail into a young tree at the height of your shoulders, and then return ten years later, how high will the nail be? Why?
3. What gases are given off by the green plant during the daytime? By what processes?
4. Why does the green plant need oxygen? Why do green plants release far more oxygen than they take in?
5. How are synthesis and growth related? Before trying to answer the question, review Chapter 12.
6. In what region of the root tip do cells increase in size? What causes this increase in size?
7. What is differentiation? What kinds of differentiated cells are found in the growing region of the root?
8. What are auxins? How does the concentration of auxins affect growth?
9. What activities of the plant do auxins regulate?
10. Distinguish between auxins and gibberellins.

B   TROPISMS

The bending of a stem toward the light is often said to be "caused" by tropisms. But what is a tropism? A *tropism* is a growth movement of a part of a plant toward or away from a stimulus. (The response of a plant to light is called *phototropism* and the response of a plant to gravity is called *geotropism*.)

To say that the bending of a stem toward the light is "caused by phototropism" is merely circular reasoning and a play on words that gets us nowhere. The word "tropism" merely describes a situation without explaining it.

In recent years, however, much progress has been made in explaining how some tropisms occur. This was made possible by experiments with tropisms leading to the discovery of auxins. Let us now review some of these experiments.

Development of Knowledge About Auxins

Charles Darwin of England experimented on the well-known curving of grass seedlings toward a light source. He covered various parts of the developing leaf and found that light must strike the *tip* for bending to occur. He realized that it is *the tips of the plant which receive the stimulus.*

Boysen-Jensen of Denmark extended Darwin's work by performing experiments with the coleoptiles of oat plants. The *coleoptile* is the first part of a grass seedling to appear as it emerges above ground. Actually, the coleoptile is a sheath that surrounds the young leaves.

Boysen-Jensen placed a gelatin between the coleoptiles of oat plants and the remainder of the plant. Thus, the gelatin separated the tip, which receives the stimulus, and the region of curvature. Under these conditions, bending still occurred. How was the stimulus carried to the part of the plant where the response occurred? Perhaps a chemical was involved.

Boysen-Jensen took a layer of impervious mica (through which chemicals could not diffuse) and placed it between the coleoptile and the region of curvature. Under these conditions, no bending occurred. He then reasoned that a chemical was produced by the tip, passed through the gelatin, and stimulated the cells in the region of curvature.

Boysen-Jensen next placed the mica strips under the coleoptiles only part of the way through the seedlings. In Fig. 16–4, two such seedlings are shown — one with the mica facing *away* from the light source, and one with the mica facing *toward* it. Note that the latter seedling bends toward the side with the strip of mica. Since the mica is impervious to the passage of chemicals what did these results mean? Boysen-Jensen reasoned that the chemical spread down the side with no mica and caused growth on that side. Accordingly, the stem bends toward the side of the stem where there is less growth. Later, when

we study phototropism, we will gain a fuller understanding of how this takes place.

As a result of Boysen-Jensen's work, it was clear that *a chemical substance was involved and that it caused growth.* However, many questions remained. What was this unknown growth-promoting chemical? Could it be isolated? Could its strength be measured?

Frits Went, an American, helped to answer these questions. Went cut off the coleoptiles of many oat seedlings and placed them on a thin layer of solidified agar. The unknown growth-promoting chemical diffused from the coleoptiles into the agar.

Then, Went took the coleoptiles off the agar, and cut the agar into small blocks. He transferred these blocks to the tips of oat seedlings whose coleoptiles had been removed. When he placed the agar on the left side of the seedling tip, the seedling curved to the right. This showed that the agar block contained the growth-promoting substance.

Next, Went placed twice as many coleoptiles on agar, and repeated each step of the previous procedure. When he placed one of the agar blocks on a seedling tip, the angle of bending was greater than before. Went had found a method of measuring and comparing the amount of growth-promoting chemical present.

We now know that the growth-promoting chemical is auxin. Went had shown that it can be isolated and its strength measured. Later investigators determined the chemical composition of auxin, and its effect on various regions of the plant. They made synthetic auxins and found uses for auxins. Now, investigators are trying to find out just what auxin does to the cells in order to

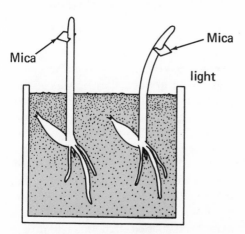

Fig. 16–4. Boysen-Jensen experiment. Bending is toward the side with the mica.

promote their elongation and cell division.

As previously stated, the concentration of auxin affects the response of the plant stem (page 260). You can test this yourself. Obtain two potted bean plants about a foot tall. Dip a glass rod into commercially prepared auxin paste and then rub the material on one side of a stem about a few inches from the tip. Place both plants in the dark. Within an hour, notice that the treated plant bends in the region where the paste was applied. Try various concentrations of auxin and measure the angle of bending with each concentration. See if bending occurs toward the auxin at any concentration.

Tropisms are referred to as *positive* when the growth is toward the stimulus, and *negative* when the growth is away from the stimulus. Organisms that are motile (capable of moving from place to place by their own energy) may respond to a stimulus by moving toward or away from the stimulus. Such an oriented movement is called a *taxis* (plural — *taxes*) and differs from a tropism by being a movement rather than a growth response. In current usage, the term "tropism" is reserved for plant responses.

Phototropism

In *phototropism*, light is the stimulus that causes the leaves and the stem of the plant to respond by bending toward the light. This is a positive phototropism. (*Heliotropism* is the response only to sunlight.)

You can demonstrate positive phototropism by using a potted plant whose leaves face toward a window. Then turn the plant around so the leaves face away from the window. Examine the plant in a day or two and notice how

the leaves have turned and are again facing the light.

Positive phototropism is of advantage to a green plant because it allows maximum reception of light. This does not imply that the plant bends toward the light because it "feels a need" for it.

The mechanism of bending is illustrated in Fig. 16–5.

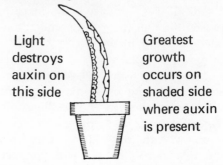

Light destroys auxin on this side

Greatest growth occurs on shaded side where auxin is present

Fig. 16–5. Positive phototropism of stem. Note that bending is toward the light.

Cells at the tip of the stem produce auxin that diffuses downward toward the region of elongation. However, light (coming from one side) *destroys the auxin* on that side. The auxin present on the other side increases growth on that side. Bending is toward the light.

Geotropism

Roots grow in the direction of the pull of gravity; they are positively geotrophic. Stems and leaves grow up, or away from gravity; they are negatively geotrophic. You can demonstrate both kinds of geotropism with a single plant.

Obtain a young seedling plant about 3 to 4 inches tall. Place the potted plant on its side in a dark place where it will not be exposed to light. Be sure that the seedling plant is in a horizontal position. In a few days, examine the plant and notice that the stem and leaves have grown upward. This is a negative geotropism. Remove some of the soil from the roots. Notice that the roots have turned and grown downward in response to the pull of gravity. This is a positive geotropism.

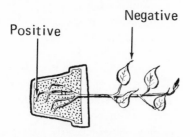

Fig. 16–6. A plant responding to gravity — geotropism.

Positive Geotropism in Roots

How do auxins cause roots to grow downward? Fig. 16–7 shows the tip of a root that is growing parallel to the ground. It may have emerged from the seed in this direction, or it may have been diverted by an obstruction.

We note the three stages of the action shown in Fig. 16–7. (1) The tip of the root produces auxin that diffuses toward the region of elongation. (2) Gravity pulls the auxin toward the lower portion of the root. (3) We observe that the tip bends downward. We then reason that if the tip bends downward,

the greatest growth must have been in the upper portion of the root. Therefore, the auxin that diffused downward must have *inhibited* the growth of cells in the lower portion of the root. The auxin has an effect where it is, not where it is not. This reasoning supports the statement made earlier that in proper concentration auxins can inhibit growth as well as stimulate growth.

Other Tropisms

Other kinds of tropism found in plants are: (1) *hydrotropism* — response to water, (2) *chemotropism* — response to chemicals, (3) *thigmotropism* — response to contact, as when a vine twines around an object, (4) *rheotropism* — response to currents of air or water, (5) *thermotropism* — response to temperature. There are many more. Why not discover some for yourself. Investigate electricity, magnetic fields, pressure, etc.

PHOTOPERIODISM

Plants and animals engage in certain activities at definite times of the day, month, or year. *Photoperiodism* is the

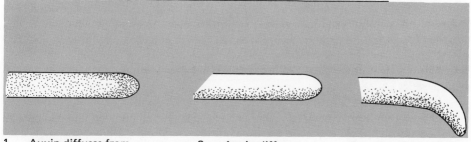

1. Auxin diffuses from the root tip.
2. Auxin diffuses downward
3. Bending occurs

Fig. 16–7. A root grows toward the gravitational pull.

response of plants to the number of hours of daylight and darkness during each 24-hour day. Photoperiodism was originally discovered in connection with the onset of flowering in plants. Recently, scientists have found that it is the length of the period of *uninterrupted darkness* that is responsible for flowering. In general, 12–14 hours of light is the critical light period for many species of plants.

Plants are classified as long-day, short-day, and day-neutral plants, depending upon the number of hours of daylight required for them to produce flowers. Long-day plants (spinach, radish, iris, clover) flower in late spring or early summer when the days are becoming longer. Short-day plants (poinsettias, dahlias, cocklebur, goldenrod)

flower in late summer or early fall when the days are becoming shorter. Day-neutral plants (beans, tomatoes, dandelions, roses) seem to be insensitive to the length of day.

Long-day plants which normally do not bloom until late spring may be made to produce flowers earlier, during the short-day winter months. This is accomplished by the use of artificial lights in the greenhouse. Commercial flower growers make use of these principles to have flowers ready for special holiday seasons.

A newly discovered plant pigment, called *phytochrome*, has been found to be important in photoperiodism. This pigment is affected by "far red" light, which has a longer wave length than the red light which we ordinarily see.

COMPLETION QUESTIONS

A 1. Cellular respiration which occurs without the use of molecular oxygen is known as ....

2. Two products produced in fermentation by yeast are .... and ....

3. Openings in stems for the passage of gases are the ....

4. The part of a 24-hour period during which a green plant gives off oxygen is the ....

5. The .... and .... may be considered as excretory structures of plants.

6. Toxic products of plant metabolism are stored in ....

7. Ammonia produced during plant metabolism may be re-used by the plant to synthesize ....

8. For growth to occur in a region of a plant, cell division must be followed by cell ....

9. The formation of different kinds of cells for specific functions is called ....

10. A general name for unspecialized plant cells is ....

11. When cambium cells in a woody stem divide, those cells which are formed toward the outer part of the stem are called ....

12. Two general types of plant growth hormones are .... and ....

13. Plants produce chemical regulators called ....

14. An auxin can stimulate or .... growth, depending on its concentration.

15. An example of an artificial plant hormone used as a weed killer is ....

B 16. The bending of a part of a plant toward or away from a stimulus is known as a (an) ....

17. The sheath surrounding the young leaves of grasses is the ....

18. A scientist who developed a method for measuring the amount of auxin produced by the tips of oat plants is . . . .
19. A movement of an animal toward or away from a stimulus is known as a (an) . . . . .
20. The bending of a plant toward the light is described as a positive . . . .
21. Light coming from the right side of a plant stem destroys auxin on the . . . . side of the stem.
22. When a root tip is growing below ground in a horizontal direction, a difference in the concentration of auxin in the upper and lower regions of the root is caused by the action of . . . .
23. The regulation of plant activities by the length of periods of uninterrupted darkness is known as . . . .
24. Plants which seem to be insensitive to the length of the day are called . . . . plants.
25. A plant pigment which is sensitive to far-red light is . . . .

MULTIPLE-CHOICE QUESTIONS

A 1. In respiration much of the chemical energy of organic molecules is transferred to become part of molecules of (1) ATP, (2) ADP, (3) carbon dioxide, (4) water.
2. Respiration in yeast (1) is aerobic, (2) is anaerobic, (3) results in production of lactic acid, (4) does not produce ATP.
3. Stomates are most often present in (1) root hairs, (2) xylem, (3) leaves, (4) phloem.
4. Most of the enzymes of aerobic respiration are located on the (1) Golgi bodies, (2) mitochondria, (3) ribosomes, (4) cell membranes.
5. Which substances enter a green plant chiefly through its leaves? (1) carbon dioxide and oxygen, (2) carbon dioxide and water, (3) water and minerals, (4) oxygen and minerals.
6. The greatest growth in the tip of a root is seen in the (1) root cap, (2) zone of cell division, (3) zone of elongation, (4) zone of differentiation.
7. Stems of some plants increase in diameter chiefly as a result of the cell division in the (1) cambium, (2) xylem, (3) root tip, (4) phloem.
8. Plant cells increase in size by (1) respiration and excretion, (2) respiration and cell division, (3) synthesis and elongation, (4) cell division and differentiation.
9. Which of the following statements is *not* true? (1) Growth in plants may be inhibited or stimulated by auxin. (2) Auxin can start cell division. (3) Auxin can increase the rate at which cells divide. (4) Auxin acts on the whole stem, rather than a part of the stem.
10. Naturally occurring auxin is (1) NAA, (2) DNA, (3) IAA, (4) 2,4-D.

B 11. The scientist who reasoned that a chemical substance produced in the tip of the root causes growth was (1) Frits Went, (2) Boysen-Jensen, (3) Charles Darwin, (4) Melvin Calvin.
12. If a root tip should start to grow horizontally, as in the diagram at the right, it would bend downward because auxin (1) stimulates elongation of

Ground line
Root   B

cells in region A, (2) stimulates elongation of cells in region B, (3) inhibits elongation of cells in region A, (4) inhibits elongation of cells in region B.

13. A plant responding to contact provides an example of (1) chemotropism, (2) thigmotropism, (3) phototropism, (4) geotropism.

14. The critical light period for classifying a plant as long-day or short-day is (1) 4–6 hours, (2) 8–10 hours, (3) 12–14 hours, (4) 16–18 hours.

15. Florists can obtain flowers earlier in the spring by the (1) addition of water, (2) circulation of air, (3) use of artificial lights, (4) spacing plants far away from each other.

CHAPTER TEST

1. An undifferentiated region of a green plant is the (1) meristem, (2) phloem, (3) root hairs, (4) epidermis.

2. One difference between the bean plant and man is that of the two, only the bean plant (1) produces DNA, (2) has its growth limited to special regions, (3) oxidizes carbohydrates to release energy, (4) utilizes digestive enzymes.

3. The bending of the tip of a stem toward light is the result of (1) geotropism, (2) unequal auxin distribution, (3) cambium stimulation, (4) transpiration pull.

4. Respiration in bread mold (1) is anaerobic, (2) is aerobic, (3) results in the formation of carbon dioxide and alcohol, (4) does not produce waste products.

5. In aerobic respiration the final hydrogen acceptor is (1) carbon dioxide, (2) oxygen, (3) ADP, (4) water.

Questions 6–9 refer to the following two equations:

$$\text{(A)} \quad 6H_2O + 6CO_2 \xrightarrow[\text{light}]{\text{chlorophyll}} C_6H_{12}O_6 + 6O_2$$

$$\text{(B)} \quad C_6H_{12}O_6 + 6O_2 \longrightarrow 6H_2O + 6CO_2$$

6. Equation A takes place in all (1) plants, (2) animals, (3) green plants, (4) plants and animals.

7. A common name for the compound $C_6H_{12}O_6$ formed in equation A is (1) simple sugar, (2) protein, (3) glycerol, (4) amino acid.

8. Equation B represents a process by which energy is (1) stored, (2) released, (3) created, (4) increased.

9. Equation B takes place in (1) plants only, (2) animals only, (3) green plants only, (4) plants and animals.

10. An energy-releasing process which does not use free oxygen is (1) aerobic respiration, (2) fermentation, (3) burning, (4) photosynthesis.

REPRODUCTION AND DEVELOPMENT

Reproduction is the process by which organisms produce new individuals of their kind. A new individual may be reproduced from a single parent or from two parents, depending on the nature of the species and the circumstances under which it occurs. Reproduction is concerned with the preservation of the *species;* the other life processes are concerned with the survival of the *individual.* The essential feature of reproduction is the separation from the parent organism of a representative portion of living substance that serves to form the next generation.

CHAPTER 17 | Asexual Reproduction

Methods by which organisms form new members of their kind can be grouped into two categories — sexual reproduction and asexual reproduction.

In *sexual reproduction,* new individuals are produced by the fusion of the nuclei of two cells, usually from two parents. The cell that fuses (unites) with another cell is a *gamete.* In *asexual reproduction,* there is no fusion of nuclei. Most examples of sexual reproduction involve two parents (some may involve a single parent), whereas in asexual reproduction only a single parent is involved. To understand asexual reproduction, it is essential that we have an understanding of mitosis.

A   MITOSIS

The division of a cell involves two distinct processes:

1. *Mitosis* — the exact duplication of the nucleus to form two identical nuclei. This involves a *doubling* and subsequent separation of nuclear material.

2. *Cytoplasmic division* — the *division* of the cytoplasm into two approximately equal portions.

These two steps are separate and distinct processes which normally occur at the same time. Although, strictly speaking, "mitosis" applies only to the changes in the nucleus, the term is often used to include the cytoplasmic division as well.

Mitosis may be defined as the orderly process by which the hereditary material in the nucleus of a cell is precisely and equally distributed to two daughter nuclei.

Chromosomes

We saw in Chapter 12 that the production of proteins in the cell is controlled by the material in the nucleus known as DNA. Within the nucleus, the DNA is bound to large protein molecules to form material known as *nucleoprotein* (nucleic acid plus protein). This is visible under the microscope because it accepts certain stains and develops a dark color. It is present in the non-dividing nucleus as long, thin, twisted strands called *chromosomes* (*chromo,* color + *soma,* body). However, the chromosomes are clearly visible only when the cell is undergoing division. These chromosomes contain DNA molecules with thousands of subunits arranged in a special sequence. The chromosomes of the two daughter cells produced in mitosis are identical in composition to the chromosomes of the parent cell (except for occasional accidents).

The normal functioning of any cell depends absolutely on its having the proper pattern of DNA in its chromosomes. This is of utmost importance, for DNA controls the metabolism of the cell, its reproduction, and the passage of hereditary traits. The chemistry of DNA is discussed in a later chapter; here it is necessary to stress its relevance to mitosis. When a cell divides, it is absolutely essential that each of the two daughter cells receives exact copies of the DNA of the mother cell. This, in fact, is the essence of mitosis.

Mitosis in an Animal Cell

Mitosis is a smoothly continuing process, but for convenience, a number of stages have been identified by biologists. These stages are indicated below by diagram and description.

Step 1. The resting stage or interphase. The resting cell carries on all the metabolic activities of the cell. It is "resting" only insofar as the visible

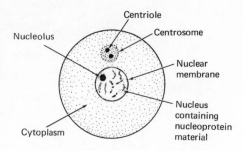

Fig. 17–1A. Resting animal cell.

steps of cell division are concerned. The resting stage (Fig. 17–1) occurs between nuclear divisions.

During early interphase, an exact duplicate is made of each chromosome. This process of self-duplication is called *replication.* When a chromosome replicates, the molecules of DNA replicate within it.

At the beginning of the visible steps of mitosis, each chromosome is already a doubled chromosome. Each portion of the doubled chromosome is called a *chromatid.* The two chromatids are held together by a minute structure, the *centromere* (Fig. 17–1B).

The *centrioles* are complex structures, visible under the compound microscope as tiny dots. They are present in a region called the *centrosome.* (See Fig. 17–1A.)

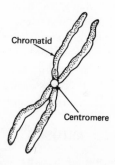

Fig. 17–1B. A replicated chromosome, consisting of a pair of chromatids.

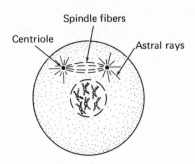

Fig. 17–2A. Step 2.

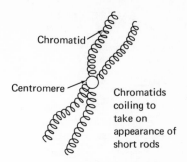

Fig. 17–2B. A double chromosome consisting of two coiled chromatids held together at the centromere.

Step 2. The prophase. The centrioles separate (Fig. 17–2A). As the two centrioles move apart, a system of gel-like protein fibers, called *spindle fibers,* forms between them. All these fibers together make up the *spindle. Astral rays* radiate out of the centrioles.

Changes occur in the nucleus also: the nucleolus disappears and the nuclear membrane begins to disappear. The thin, long chromosomes coil (Fig. 17–2B) and begin to take on the appearance of short, thick, deeply staining rods which are easily seen separately and can be counted. Each visible chromosome at this stage still has two chromatids. Fig. 17–2A represents a cell containing six chromosomes.

Step 3. The metaphase. The centrioles move to opposite poles of the cell, and the astral rays reach the cell membrane. The spindle fibers, seemingly connected to the centrioles, form a double-coned structure, the *spindle.* (See Fig. 17–3A.)

During this stage, the nuclear membrane has completely disappeared. The doubled chromosomes are arranged along the equatorial plate of the cell. The centromere of each doubled chromosome is attached to a spindle fiber. The doubled chromosomes are so short, thick, and tightly coiled that each dou-

bled chromosome appears under the usual compound microscope as a single short rod. The doubled chromosomes of Fig. 17–3A, therefore, appear as six heavily stained chromosomes. A double-stranded chromosome may be represented as shown in Fig. 17–3B.

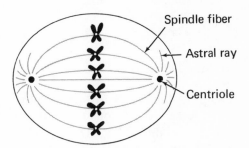

Fig. 17–3A. Step 3. Six doubled chromosomes are present along the central region of the cell—the equatorial plate.

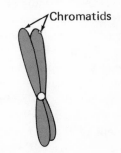

Fig. 17–3B. A double stranded chromosome.

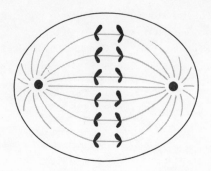

Fig. 17–4A. Step 4. The six doubled chromosomes separate to form twelve chromosomes.

Constriction of
cell membrane

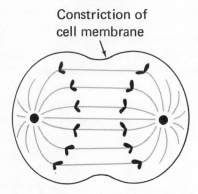

Fig. 17–4B. A later stage of step 4.

Step 4. The anaphase. The centromeres of the chromosomes divide and seem to be pulled by the shortening spindle fibers toward the centrioles. The chromatids of each doubled chromosome have separated. Each chromatid may now be called a chromosome.

In Fig. 17–4A, the number of chromosomes is now 12. According to an older concept, the chromosomes "split" at this stage. We now know that at this stage there is a separation of the parts of a doubled chromosome which had been replicated during the resting stage.

Fig. 17–4B represents a later stage in the movement of the chromosomes toward opposite ends of the cell. The mechanism by which the chromosomes move apart is not known; however,

ATP seems to be involved. Cytoplasmic division of an animal cell may begin at this point as a pinching in, or furrowing, of the cell membrane, probably due to some sort of action of the astral rays.

Step 5. The telophase. The chromosomes move to opposite ends of the cell, and they uncoil and become long and thin again. The nucleolus reappears, and the nuclear membrane forms. (See Fig. 17–5.)

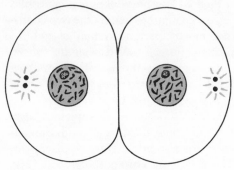

Fig. 17–5. Step 5.

During this stage, the spindle and astral rays dissolve away. The centrioles divide in preparation for the next cell division. The cytoplasmic division becomes complete, resulting in two new daughter cells.

Cytoplasmic Division in Animal Cells

The cytoplasm of the parent cell divides into two approximately equal portions. In the animal cell, this division is accomplished by a pinching of the cell in half.

Mitosis in Plant Cells

The typical plant cell does not have any centrioles and does not form any astral rays. The chromosomes coil, line up, and separate just as in animal cells.

Cytoplasmic division. After the chromosomes have moved apart, a thick *cell*

plate forms in the center of the spindle. This grows outward to the edges of the cell, dividing it in half.

Result of Mitosis. In plant and animal cells, the overall result is the same. Each daughter cell receives an exact copy (replica) of each chromosome that was present in the parent cell. In the series of diagrams (pages 270–272), a cell with a chromosome number of six forms two cells, each of which has a chromosome number of six. The replication of each chromosome to form a pair of identical chromatids took place during the resting stage, before the visible signs of mitosis began.

Chromosome Number

All of the cells of a particular organism (with the exception of certain reproductive cells) have a specific number of chromosomes that is characteristic of that organism. For example, the characteristic chromosome number of the fruit fly is 8; that of man is 46.

Chromosomes occur in pairs; man usually has 23 pairs of chromosomes.

Most organisms have a chromosome number that falls in the range between 10 and 50. However, some organisms may have as few as one, and others as many as 1600!

Biologists hope that as understanding of the normal process of cell division increases, light will be shed on abnormal cell division such as occurs in cancer. All aspects of the process that "pulls" the chromosomes apart are being carefully studied. Biologists also want to know what starts the division of a cell.

General Scheme of Mitosis

As previously stated, each daughter cell produced by mitosis has the same *number* of chromosomes as the parent cell. Now it should be noted that each daughter cell receives the same *kind* of chromosomes.

The general scheme of mitosis is

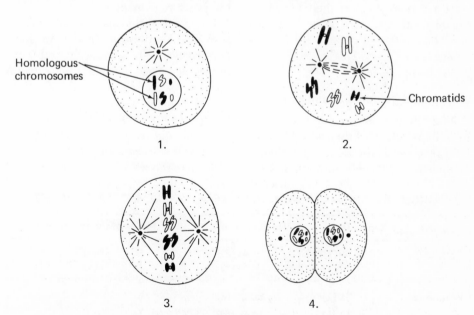

Fig. 17–6. Schematic diagram of mitosis.

shown in Fig. 17–6. The diagram emphasizes the distribution of the chromosomes. It should be understood that even before any doubling (replication) occurs, the chromosomes in the nucleus occur in pairs. The members of a pair are called *homologous* chromosomes. In the diagram, members of a pair are drawn so they have the same shape, but the members are shaded differently.

Six chromosomes (3 pairs) are shown in the parent cell. In the second drawing, each chromosome has replicated, and there are 12 chromatids. In the third drawing, the replicated chromosomes seem to form a single line along the equator. The chromatids of each doubled chromosome then separate as two *identical* chromosomes. Each daughter cell receives 6 chromosomes (3 pairs) — the same *number* and the same *kind* of chromosomes as in the parent cell.

Meiosis

At a certain stage of their life cycle, organisms that reproduce sexually may carry on a different type of cell division known as *meiosis*. This process is described on page 288.

Surface and Volume Relationships

As a cell increases in size, it increases in volume and surface area. Do they both increase at the same rate? Simple mathematics provides an answer. Consider a sphere (not a circle) whose radius is 1. What happens as the radius increases to 2? The area of the surface of a sphere is given by the formula $S = 4\pi r^2$. The volume of a sphere is given by the formula $V = \frac{4}{3}\pi r^3$. Note how these formulas are applied in the table below to obtain an answer to the question raised.

The column at the right of the table shows that as the radius of a sphere increases from 1 to 2, the surface increases 4 times. However, the volume increases 8 times. Other numbers may be substituted for the radii of these spheres. In each case, it will be found that *the volume increases more rapidly than the surface.*

If a cell continued to increase in size without dividing, the volume would soon be too large to be adequately serviced by the surface. The surface would be insufficient to transport enough materials into and out of the cell. It would be difficult for materials to diffuse to the inner regions of the larger cell.

When a cell that has an unfavorable surface-volume ratio divides to form two smaller cells, a more favorable ratio results. Cell division results in an increase in surface area. To illustrate this fact, place two books side by side to represent a single cell. Now separate the "cell" into two "cells" and note the additional surface area provided.

What is the stimulus for cell division?

CHANGES IN SURFACE AND VOLUME AS A SPHERE INCREASES IN SIZE

Sphere	Radius (r) $= 1$	Radius (r) $= 2$	Ratios of the two sizes
Surface (S)	$S_1 = 4\pi(1)^2$ $= 4\pi(1)$	$S_2 = 4\pi(2)^2$ $= 4\pi(4)$	$\dfrac{S_2}{S_1} = \dfrac{4\pi(4)}{4\pi(1)} = 4$
Volume (V)	$V_1 = \frac{4}{3}\pi(1)^3$ $= \frac{4}{3}\pi(1)$	$V_2 = \frac{4}{3}\pi(2)^3$ $= \frac{4}{3}\pi(8)$	$\dfrac{V_2}{V_1} = \dfrac{\frac{4}{3}\pi(8)}{\frac{4}{3}\pi(1)} = 8$

What is the stimulus for cell division? According to one hypothesis, the un-favorable surface-volume relationship serves as a trigger that initiates cell division. A more recent hypothesis is concerned with the ratio between the respective volumes of the *nucleus* and the *cytoplasm*. According to this hypothesis, cell division results when the nucleus loses the capacity for maintaining the large cytoplasm in a stable condition. However, the immediate stimulus that initiates cell division is still not known.

REASONING EXERCISES

1. Distinguish between sexual reproduction and asexual reproduction.
2. Distinguish between mitosis and cytoplasmic division.
3. What is nucleoprotein? From the standpoint of composition, what is the relation between nucleoprotein and chromosomes?
4. What is the essence of mitosis? Why is each chromosome a doubled chromosome at the beginning of the visible steps of mitosis?
5. What is a chromatid? At what point does a chromatid become a chromosome?
6. Why does each daughter cell produced in mitosis have the same *number* of chromosomes as the parent cell?
7. Mitosis is frequently called "nuclear division." Why?
8. How does mitosis in a plant cell differ from mitosis in an animal cell?
9. What is meant by chromosome number?
10. What are homologous chromosomes? Why does each daughter cell produced in mitosis have the same *kind* of chromosomes as the parent cell? How many pairs of homologous chromosomes do you have in one of the cells from your arm?

B   ASEXUAL REPRODUCTION

As stated previously, in the process of asexual reproduction, new individuals are produced from a single parent. There are various methods of asexual reproduction, including binary fission, budding, spore formation, regeneration, and vegetative propagation.

Binary Fission

Almost all single-celled plants and animals reproduce by mitotic division into two more or less equal parts, a process called *binary fission*. If the organism is motile, the daughter cells swim apart. If it is non-motile, the daughter cells may adhere to each other; further binary fission may result in a colony consisting of cells that all originated from the same parent.

Binary Fission in Amoeba. An *Amoeba* that reaches a certain size reproduces by binary fission. The nucleus duplicates itself exactly by mitosis, and each nucleus becomes oriented at opposite ends of the cell. (See Fig. 17–7.) The *Amoeba* then divides by constricting in the middle and pulling apart. The cytoplasm is divided between the two new cells in approximately equal portions. In this way, two new *Amoebas* (daughter cells) are produced from the parent cell. The "mother cell," of course, no longer exists.

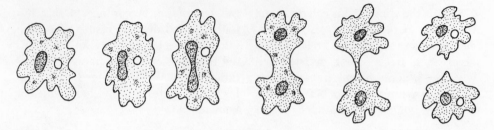

Fig. 17–7. Binary fission in *Amoeba*. The parent cell divides into approximately equal parts.

Binary fission in *Paramecium*. *Paramecium*, which is more complicated than *Amoeba*, has an oral groove, two contractile vacuoles, and two nuclei (the macronucleus and the micronucleus). The macronucleus seems to regulate the general metabolism of the cell, and the micronucleus is concerned with reproduction. Only the micronucleus undergoes mitosis; the macronucleus merely pinches in half.

A simplified presentation of binary fission in *Paramecium* is shown in Fig. 17–8. Note that the oral groove gradually disappears as the nuclei divide and the cytoplasm begins to pinch approximately in half. In each of the daughter cells, a new oral groove is formed and an additional contractile vacuole appears. When two complete cells have been formed, they pull apart and swim away.

Paramecium also undergoes conjugation, a form of sexual reproduction. (See page 287.)

Budding

Budding is a type of asexual reproduction in which the new individual is produced as an outgrowth of an older one. The new individual is called a *bud*.

Budding in one-celled organisms resembles binary fission in that the nucleus is divided into equal portions by

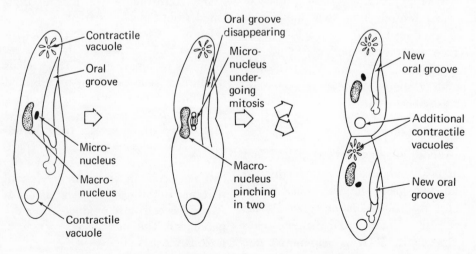

Fig. 17–8. Binary fission in *Paramecium*.

   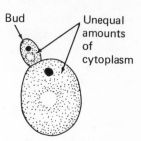

Fig. 17–9. Budding in yeast.

mitosis. Budding differs from binary fission in that the cytoplasm is *not* divided into equal portions. The bud has a smaller amount of cytoplasm.

Budding in yeast. A yeast cell is a one-celled plant that reproduces by budding. In yeast cells, the nucleus divides by mitosis, forming two identical nuclei. While this is occurring, the cell wall has softened and a bud has begun to form by pushing outward from the parent. One of the newly formed nuclei moves into the bud; the other remains in the parent. A new cell wall forms between the bud and the parent, thus forming two new cells. (See Fig. 17–9.)

The bud now increases in size and may eventually break away. Sometimes the bud remains attached to the parent cell and develops a bud of its own.

Budding in *Hydra*. Budding occurs in some of the simpler multicellular animals, for example, *Hydra*. When *Hydra*

reproduces by budding, a protrusion grows out from the side of the parent organism, and develops a mouth and tentacles. (See Fig. 17–10.) Each bud then breaks away from the parent to live an·independent life. Under proper conditions, a culture of well-fed *Hydra* may contain many organisms with two or more buds. *Hydra* also reproduces sexually.

Spore Formation (Sporulation)

Spore formation is a method of asexual reproduction in which a parent forms spores which produce new organisms. A *spore* is a cell with a hard, protective covering that permits it to survive unfavorable environmental conditions such as freezing and drying.

Some spores are formed by the rapid division of a nucleus into many nuclei. A cell wall develops around each nucleus and includes some cytoplasm. After a spore is released from the par-

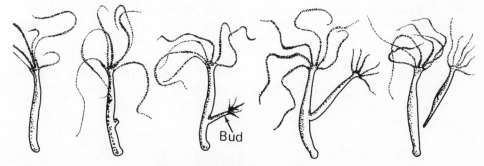

Fig. 17–10. Budding in *Hydra*.

ent, it is capable of developing into a new organism when environmental conditions are favorable.

Under certain conditions a single yeast cell may produce four spores.

Spore formation in bread mold. The structure and development of bread mold have been described in Chapter 14. Specialized hyphae develop upward into the air. At the tip of these hyphae, the *spore cases* form. Within the spore cases, as many as 70,000 asexual spores are produced. Each spore consists of a bit of cytoplasm and several nuclei. When the spores are mature, the spore cases break open and air currents scatter the spores.

Under favorable conditions of food, warmth, and moisture, each spore germinates to produce a new mycelium (mass of hyphae). Bread mold also reproduces sexually by conjugation. (See page 286.)

Regeneration

When cut into two or more parts, certain animals have the ability to grow back the missing parts. This process is called *regeneration.* By regeneration a missing part may be replaced by the animal, or actual reproduction may occur. Reproduction occurs when parts of the same animal grow into complete animals.

When a starfish is cut into several pieces, each piece that has a portion of the central disk develops into a new individual. Lower invertebrates such as *Planaria* (a flatworm) have great powers of regeneration. A planarian, if cut properly, can produce three or more complete worms by regeneration. (See Fig. 17–11.) However, if the cut is made too far to the rear, the small rear portion will not grow a new front

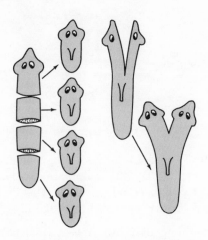

Fig. 17–11. Regeneration in *Planaria.*

portion. Two-headed planaria can also be produced.

Vegetative Propagation

The major structures of a plant are the root, stem, leaf, and flower. Plants that have flowers carry on sexual reproduction, and reproduction is the special function of the flower. As shown in Fig. 17–12, the major structures are classified as reproductive (flower) or vegetative (root, stem, and leaf).

The vegetative structures have functions that pertain to growth. Sometimes, however, the vegetative structures of a plant can produce additional plants. Accordingly, this process is called *vegetative propagation.* It may be defined as the production of new plants from roots, stems, or leaves.

Vegetative propagation occurs directly in nature. It also may be artificially induced by man to hasten the formation of new plants.

Natural Vegetative Propagation

Most of the examples of planting with which we are familiar involve

STRUCTURE   MAIN FUNCTION

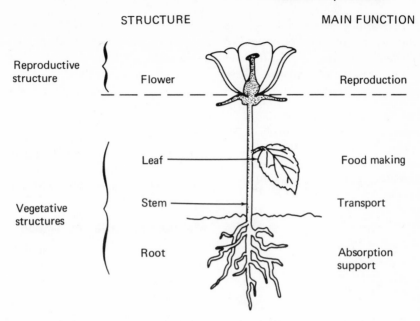

Reproductive structure   Flower   Reproduction

Leaf   Food making

Vegetative structures   Stem   Transport

Root   Absorption support

Fig. 17–12. Structures of a flowering plant and their functions.

seeds, produced by sexual reproduction. Since the seed is not part of the root, stem, or leaf, it is evident that this is *not* vegetative propagation. On the other hand, in some cases a new plant may reproduce itself by natural vegetative propagation.

Bulbs. A bulb is a short underground stem with thick, fleshy leaves containing stored food. (See Fig. 17–13.)

When a bulb is placed in the earth, its fleshy leaves supply food for the growth of the plant until it can carry on photosynthesis. Onions, tulips, daffodils, and lilies are examples of plants which survive the winter in this manner. Healthy bulbs branch to produce *bulblets,* which grow into new plants in the spring.

Tubers. A tuber is a fleshy portion of an underground *stem* which has buds. The white potato plant produces many tubers (potatoes), which store food. The "eyes" of the potato are actually tiny buds of the stem. (See Fig. 17–14.)

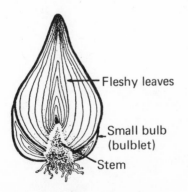

Fleshy leaves

Small bulb (bulblet)

Stem

Fig. 17–13. Tulip bulb.

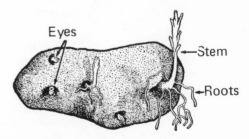

Eyes

Stem

Roots

Fig. 17–14. A white potato.

When a farmer plants potatoes, he cuts the tuber into pieces, each of which has at least one "eye." The eyes sprout and produce new plants. These new plants later produce underground stems with the next crop of tubers.

Runners. A *runner* is an overground stem. The branchlike stem grows or "runs" along the surface of the ground and produces new roots and upright stems at various points. (See Fig. 17–15.)

Fig. 17–15. Runners of strawberry.

When the interconnecting runner dies, the result is a number of new independent plants. As many as a dozen runners may grow from a single strawberry plant in one season. Each runner may be as long as 10 feet, and produce a number of new plants.

Lawn grasses spread rapidly by means of runners. In addition, many garden weeds, such as chickgrass, reproduce quickly by runners.

Rhizomes. A *rhizome* is a thickened woody underground stem. Nodes (swellings) develop along the rhizome. From the upper surface of the nodes come buds which grow into leaf-bearing branches. From the lower surface of the nodes come the roots.

Separate plants arise as the older parts of the rhizome die. Plants which reproduce by rhizomes include ferns, iris, quack grass, poison ivy, and cattails.

Artificial Vegetative Propagation

Portions of roots, stems, and leaves are used by man to produce new plants. Techniques include *cuttings, layering,* and *grafting.*

Cuttings (slips). A piece of a stem, called a slip or cutting, may be removed from some plants and placed in water or moist sand. Soon the cutting develops roots and leaves to form a new independent plant. It may then be transplanted to soil. (See Fig. 17–16.) Plants that reproduce in this manner include the geranium, hydrangea, rose, ivy, and grape.

Auxins, applied to the bottom portion of the cut stem before it is planted, assist in the formation of roots. For-

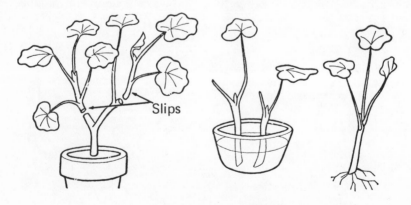

Fig. 17–16. Geranium cuttings.

merly, plants such as yew and holly could rarely be propagated by cuttings. Now they can be propagated effectively by treatment of the cut stem with auxins.

Leaf cutting. When a leaf of the *Bryophyllum* plant is placed on moist soil, tiny plants, complete with roots, stem, and leaves, develop at the notches of the leaves. (See Fig. 17–17.) As the old leaf dies, several new independent plants are formed. Begonia may also be propagated by leaf cuttings.

Fig. 17–17. Leaf cutting in *Bryophyllum.*

Layering. When the stems of certain plants bend over so that the tip is covered by soil, they take root at the covered area. Soon a complete new plant is formed. These new plants may then be cut from the parent plant. Although this method occurs naturally, man hastens the process by bending the stem underground and fastening it in place.

Layering occurs in many plants, including the blackberry and rambler rose.

Grafting. In *grafting* a stem is cut from one plant and attached to the stem of another plant. The portion of stem which is attached to the other plant is the *scion*. The stem which

remains attached to its roots is the *stock*. In order for a graft to "take," the cambium layers of the scion and stock must be in contact, at least in a small region. Close contact of the two cambiums is desirable, but this is rarely achieved in actual practice. The grafting procedure is as follows:

The scion and stock are selected. Then they are cut in such a manner that the cambium layers will make contact and the two pieces will fit snugly. After the scion is placed into position, the two sections are fastened with cord. Wax is then applied to prevent infection and prevent the sap from flowing out. All parts that grow above the graft, on the scion (leaves, flowers, and fruit), will have the hereditary characteristics of the scion. *There is no mixing or blending of hereditary characteristics in grafting.* For a graft to be successful, the two plants must be closely related.

Grafting permits the farmer to combine the desirable characteristics of two plants in one plant. For example, the stock may have roots which are resistant to soil pests, whereas the scion may bear a desirable kind of fruit. Grafting is also used to propagate plants which have seedless fruit and to hasten the production of crop-bearing trees.

For ornamental purposes, the home gardener may purchase an apple tree bearing scions of many different types of apples. In this manner, Macintosh, Greening, Baldwin, and Delicious apples can all be grown on one tree.

Advantages of Vegetative Propagation

No variations. When plants are propagated by seeds, the undesirable characteristics of remote ancestors may appear. However, in vegetative methods there are no ancestral throwbacks. All

no change in heredity from one generation of cells to the next.

Speed. When a scion is grafted to a stock which possesses a well-developed root system and a thick stem, it takes less time for a mature, fruit-bearing tree to develop than when seeds are used. Also, a stem cutting which is two feet long may represent several years' growth of a seedling.

Certainty. The stored food in a bulb, tuber, or rhizome gives the new plant a good start. In comparison, a large percentage of seeds may fail to germinate.

Seedless fruits. Flowering plants that do not produce seeds can be reproduced only by vegetative propagation. Examples are plants that produce seedless oranges, bananas, and seedless grapes.

Combination of desirable characteristics. Grafting permits the combination in one plant of the desirable qualities of the root (stock) and the desirable qualities of the stem, leaf, flower, and fruit (scion). Remember, in grafting, there is no "mixing" of the heredity of the scion and stock.

The Role of Undifferentiated Cells

How is it possible for an entire plant to develop from the tip of a runner?

The answer lies in the presence in the plant of undifferentiated cells called *meristem* (for example, cambium). As you will recall, specialized cells (xylem, phloem, epidermis, cortex, etc.) develop from meristem. The characteristics of the specialized cells that develop depend upon the influence of physical and chemical factors. In one environment, meristem cells can develop into a pattern that is characteristic of a growing stem; in other environments, they can develop into patterns that are characteristic of a growing root, leaf, flower, or fruit. Undifferentiated meristem cells are present in most regions of the plant, which accounts for the amazing ability of plants to carry on vegetative propagation.

Animals also possess undifferentiated (or embryonic) cells. Undifferentiated cells are much more extensive in the invertebrates than in the vertebrates. Budding occurs in *Hydra* and regeneration occurs in *Planaria*. Even higher invertebrates such as the lobster can regenerate a new claw. However, the more complex vertebrates have much greater specialization of their cells, although some embryonic tissue does remain. Man can regenerate tissue lost in a cut, scrape, or burn, but he cannot grow back the tip of a finger.

COMPLETION QUESTIONS

A 1. A cell which fuses with another cell during sexual reproduction is a (an) . . . .

2. Reproduction from a single parent is generally by . . . . reproduction.

3. The exact duplication and subsequent division of nuclear material occurs during the process of . . . .

4. Structures visible under the compound microscope and which carry DNA are the . . . .

5. The formation of an exact copy of a chromosome is known as . . . .

6. Replication of the chromosomes occurs during the .... stage of mitosis.
7. The structure which holds together two chromatids is the ....
8. In mitosis of the animal cell, the group of fibers passing from one centriole to the other is called the ....
9. A spindle fiber is attached to a chromosome at the ....
10. Energy for movement of the chromosomes is supplied by ....
11. In the animal cell, cytoplasmic division is accomplished by a .... of the cell in half.
12. In the plant cell, cytoplasmic division is accomplished by the formation of the .....
13. The members of a pair of chromosomes are called .... chromosomes.
14. A cell having eight chromosomes divides by mitosis to form two daughter cells, each of which has .... chromosomes.
15. As a cell increases in size, the ratio of its surface to its volume becomes .....

B 16. Equal division of the nucleus accompanied by unequal division of the cytoplasm occurs during the process of ....
17. Bread mold reproduces asexually by the process of ....
18. The process by which a starfish grows back a missing arm is called ....
19. The production of a new plant by use of roots, stems, or leaves is known as ....
20. The eyes of a potato are actually ....
21. A thickened woody underground stem is called a (an) ....
22. Stem cuttings grow better if .... is applied to the portion put into the ground.
23. The rooted portion of a graft is the ....
24. The portion of a plant that grows above a graft has the same heredity as the ......
25. Undifferentiated plant cells are called ....

MULTIPLE-CHOICE QUESTIONS

A 1. Which life function is not needed for survival of the organism but is essential for the survival of the species? (1) assimilation, (2) reproduction, (3) digestion, (4) circulation.

2. The process of mitosis insures equal distribution to the new cells of the (1) vacuoles, (2) cytoplasm, (3) chloroplasts, (4) nucleoprotein.

3. Of the following, the most significant event in mitosis is (1) the disappearance of the nuclear membrane, (2) the separation of the centrosomes, (3) the replication of chromosomes, (4) the appearance of the spindle.

4. Of the following, the best way to determine if a cell is a plant or animal cell is to determine the way it (1) replicates its chromosomes, (2) divides its cytoplasm during cell division, (3) replicates its DNA, (4) releases energy from organic compounds.

5. It is believed that cells will divide when they grow to a certain size because (1) the surface area increases faster than the volume, (2) the volume increases faster than the surface area, (3) the surface area is a multiple of the volume, (4) both surface area and volume increase proportionately.

B 6. In plants, budding and fission are alike in that (1) both processes involve mitosis, (2) both processes involve centrosomes, (3) two equal daughter cells are produced in both processes, (4) spindle fibers are not formed.

7. Yeast reproduces by (1) regeneration, (2) budding, (3) binary fission, (4) vegetative propagation.

8. A tuber is a (1) stem, (2) root, (3) flower, (4) leaf.

9. A farmer found that one tree in his pear orchard bore especially delicious fruit. Which method would most quickly provide a large crop of these pears? (1) planting the seeds of the pears from this tree, (2) crossing this tree with another tree in the orchard, (3) providing the tree with special fertilizer, (4) grafting from this tree to other healthy pear trees in the orchard.

10. Vegetative propagation is possible in plants because of the widespread presence of (1) auxins, (2) undifferentiated tissue, (3) conducting tissue, (4) gametes.

CHAPTER TEST

Directions for Questions 1–5:
   Write *B* if the statement is true of *both* plant and animal cells.
   Write *A* if the statement is true of *animal* cells but *not* of plant cells.
   Write *P* if the statement is true of *plant* cells but *not* of animal cells.
   Write *F* if the statement is false.

1. Replication occurs during the resting stage.

2. Cytoplasmic division occurs by formation of a cell plate.

3. Centriole is usually present.

4. The chromosomes at times are wound up into tight coils.

5. Between mitotic divisions the cell carries on no metabolic activities.

6. The drawing at the right shows a nuclear structure subsequent to replication. The part labeled *X* is a (1) centromere, (2) chromatid, (3) centriole, (4) chromosome.

7. A red rose scion growing on a white rose stock normally produces roses that are (1) red only, (2) white only, (3) pink, (4) red and white.

8. A cell having 20 chromosomes would, during mitosis, give rise to two cells, each of which had a chromosome number of (1) 10, (2) 20, (3) 40, (4) 80.

Directions for questions 9 and 10: Base your answers to these questions on the procedure described below:

   A leaf of the *Bryophyllum* plant was pinned to moist sand contained in a box. Several weeks later, when the leaf was examined, it was observed that young plants were growing from the edges of the leaf.

9. Which method of reproduction is demonstrated by this procedure? (1) grafting, (2) cross-pollination, (3) fertilization, (4) vegetative propagation.

10. The chromosomes of the shoots growing from the leaf are (1) identical with those of the leaf, (2) often different from those of the leaf, (3) half the number of those in the leaf, (4) double the number of those in the leaf.

CHAPTER 18 / Sexual Reproduction in Animals

In asexual reproduction, heredity remains constant from generation to generation. By contrast, in sexual reproduction, where two parent cells contribute hereditary material, there can be much variation from parent to offspring. This hereditary variation permits organisms to adapt to changing environmental conditions, and is a major factor in evolution.

In sexual reproduction there is a fusion of the nuclei of two cells. The *gametes* are the cells which unite in sexual reproduction; the *zygote* is the cell which is formed as the result of the union of two gametes. In some species of plants and animals, the two gametes are produced by the same individual, but in most cases two parents are involved in sexual reproduction.

A | INTRODUCTION TO SEXUAL REPRODUCTION

Sexual reproduction may be divided into two types, depending upon whether the gametes are similar or dissimilar in appearance. Similar gametes are called *isogametes* (*iso,* equal); dissimilar gametes are called *heterogametes* (*hetero,* different).

Conjugation is sexual reproduction by the union of similar gametes. Examples of organisms carrying on conjugation are *Spirogyra,* bread mold, and *Paramecium.*

Fertilization is sexual reproduction by the union of dissimilar gametes. Examples of this type of sexual reproduction are found in all phyla of plants and animals. This is the method present in the flowering plants and in man.

Conjugation

The process of conjugation is found only in certain of the simpler organisms which carry on sexual reproduction. Conjugation in plants and animals is considered in this section.

Conjugation in *Spirogyra*. *Spirogyra* is a green alga which may be found in ponds or gently flowing streams. It looks like a mass of green threads. Each thread, or filament, consists of a long row of cells that divide by binary fission. In autumn, as unfavorable conditions for growth develop, conjugation occurs between cells lying in adjacent filaments (Fig. 18–1).

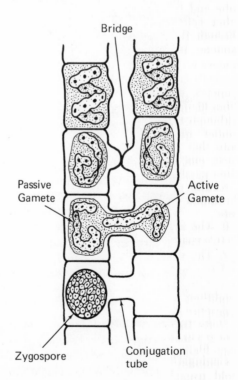

Fig. 18–1. Conjugation in *Spirogyra.*

In the diagram, successive stages of conjugation are shown one below the other. The steps are as follows:

1. The cells of two adjacent filaments send out small projections toward each other.

2. These projections meet and form a continuous tube (the conjugation tube). Because many cells in the adjacent filaments form tubes, a ladder effect is produced.

3. The contents of each cell lose water and the cell becomes more rounded, forming a *gamete.*

4. One gamete passes through the tube and fuses with the gamete of the other cell. The gamete which passes through the tube is called the *active gamete;* the one which remains stationary is called the *passive gamete.* All the cells of one filament produce active gametes, whereas all the cells of the other filament produce passive gametes. Although the gametes of *Spirogyra* are similar in appearance (isogametes), note that there is a functional difference, one kind being active and the other passive.

5. The cell formed as the result of the union of the two gametes is a *zygote.*

6. The zygote develops a thick protective coating, and is now a *zygospore.*

7. The empty cell walls decompose, and the zygospores fall to the bottom of the body of water. When favorable conditions return in the spring, each zygospore develops into a new filament.

Note: In *Spirogyra* conjugation may also occur between cells lying in the same filament.

Conjugation in bread mold. Bread mold usually reproduce asexually by spore formation. However, when conditions are unfavorable, conjugation takes place (Fig. 18–2).

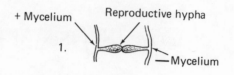
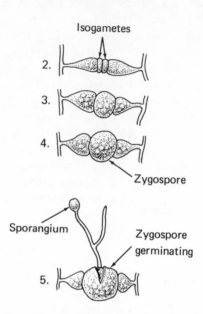

Fig. 18–2. Conjugation in bread mold.

The steps are as follows:

1. Two mycelia each send out *reproductive hyphae.* These grow toward each other.

2. At the end of each reproductive hypha a *gamete* is formed.

3. The gametes fuse to form a *zygote.*

4. The zygote develops a protective coat and becomes a *zygospore.*

5. Under favorable conditions, the zygospore germinates to form a new mycelium.

Although the two gametes of bread mold appear similar in structure, there is a difference in function between them. There are at least two different kinds of mycelia, called *plus* and *minus* strains. Gametes can only conjugate with those of the other strain.

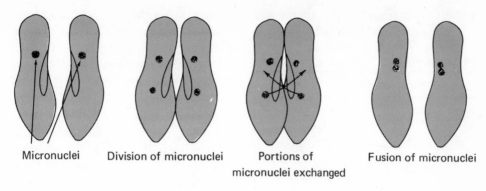

| Micronuclei | Division of micronuclei | Portions of micronuclei exchanged | Fusion of micronuclei |

Fig. 18–3. Conjugation in *Paramecium.*

Conjugation in *Paramecium.* After reproducing a limited number of times asexually by binary fission, *Paramecium* may carry on conjugation. Two distinct strains, or mating types, exist.

During conjugation, two paramecia of opposite strains join at the region of the oral groove. (See Fig. 18–3.) Their cell membranes become thin at the point of union so that a bridge of protoplasm is formed between the conjugating pair. A complex series of divisions takes place in the micronuclei of each *Paramecium.* Some of the material from the micronucleus of each cell moves across the protoplasmic bridge and fuses with the micronuclear material of the other cell.

Each organism now contains some of its original micronucleus and some of the micronucleus from the other *Paramecium.* After conjugation, the cells separate.

Conjugation in *Paramecium* is not actually a form of reproduction. No zygote is produced from which a new individual develops. However, this process does introduce hereditary variation because each individual now has a hereditary makeup different from before. Conjugation in *Paramecium* is more of a rejuvenation or revitalizing process

than a reproductive process. After conjugating, each of the paramecia can again reproduce asexually by binary fission.

Conjugation in bacteria. Recent experimental studies have shown that sexual reproduction occurs in some bacteria, and may occur in all bacteria. Two mating types, designated as male and female, have been found in *E. coli.* This is a kind of bacterium which lives in the human large intestine. A cytoplasmic bridge joins the male and female bacteria and the male injects its chromosome into the female. Soon afterward, the male bacterium dies because it lacks the DNA needed to control its cell metabolism.

Fertilization

We have noted that the gametes (isogametes) participating in conjugation are similar in structure — with slight differences in function; for example, *Spirogyra* has active and passive gametes, and bread mold has plus and minus strains. In fertilization, however, the gametes (heterogametes) differ greatly in both structure and function. In animals, these heterogametes are called the *male gamete* (or sperm cell) and the *female gamete* (or ovum). A general comparison of the sperm and ovum of animals is given below.

COMPARISON OF SPERM AND OVUM

	Sperm	Ovum
Generalized diagram	Head ↓ Tail ↓ Middle piece	Yolk
Size	Small	Much larger
Motility	Motile	Non-motile
Number	Many	Few
Food	None	Contains yolk
Where produced	In testes of male	In ovaries of female

The *gonads* are organs that produce the gametes. The *testes* are the gonads of the male, and the *ovaries* are the gonads of the female.

MEIOSIS

In this section we will study a type of cell division called *meiosis,* a very important process that occurs in the gonads (testes and ovaries).

Maintenance of the Chromosome Number

As previously stated, each organism has a characteristic number of chromosomes in its body cells. In man, this number is 46. The chromosomes occur in pairs; in man there are 23 pairs of homologous (similar) chromosomes. At fertilization, the nuclei of the sperm and the ovum fuse, so that the zygote (fertilized egg) has double the number of chromosomes present in each of the gametes. From the zygote, all the cells of the body are produced by mitosis. This introduces a puzzling situation, for it would seem that the cells of the offspring should have double the number of chromosomes (92) present in their parents. This doubling does not occur because of the process of meiosis. Let us see why this is so.

Meiosis is a kind of cell division in which the number of chromosomes is reduced to half. In man the 46 chromosomes in the cells of the testes and the cells of the ovaries are reduced in number to 23 chromosomes in the sperm and 23 chromosomes in the ova. However, these gametes do not receive just any 23 chromosomes. They usually obtain a whole set consisting of one member of each of the 23 pairs.

The *monoploid* number is the number of chromosomes in a single set of homologues. This is the number found in the sperm or egg. The *diploid number* is the number of chromosomes in the double set, found in all body cells. For example, in man the monoploid number is 23, and the diploid number is 46. In the fruit fly, whose body cells possess 4 pairs of homologous chromosomes, the monoploid number is 4 and the diploid number is 8. The symbol n is used to represent the monoploid number and $2n$ to represent the diploid number. (The monoploid number is sometimes called the *haploid* number.)

Since fertilization doubles the chromosome number and meiosis reduces it to half, it is evident that the processes of fertilization and meiosis together keep the chromosome number constant from generation to generation. (See Fig. 18–4.)

Development of Gametes (Maturation)

Gametogenesis is the process by which gametes are produced. This process occurs when the organism is old

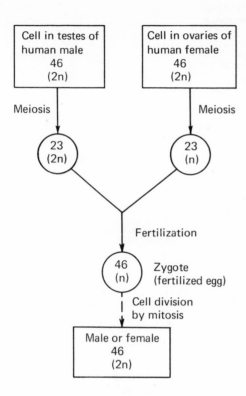

Before any division occurs, the long, thin chromosomes of the resting stage come together in homologous pairs, called *tetrads*, a process called *synapsis*. *Tetra* means "four"; the name comes from the fact that each chromosome consists of two chromatids, so when a pair unites it forms a structure made of four chromatids. These twist around each other and, in the process, the chromatids break at one or two places, and bits of one join onto another. This is called *crossing over*. In crossing over, it is always homologous parts that are exchanged between chromatids, so that each of the four chromatids is still complete.

The tetrads now coil to form short, fat rods. They line up at the equator of the cell, and each tetrad is pulled in half as the cell divides. This division results in two cells, each with three half-tetrads of two chromatids each. Then, instead of uncoiling and going into the resting phase (such as occurs in mitosis), the half-tetrads immediately line up for a *second meiotic* division. This division of two cells results in four cells, each with a single chromatid representing each homologous pair of the primary sperm cell. These now uncoil and form the monoploid nucleus that soon becomes part of a sperm. The cells lose most of their cytoplasm, and the centrosome becomes located in the "neck" region from which the long flagellum of the sperm cell grows.

Fig. 18–4. Maintenance of the chromosome number.

enough to produce gametes. Gametogenesis in the male is known as *spermatogenesis* — the production of sperm. Gametogenesis in the female is called *oogenesis* — the production of ova (or egg cells).

Spermatogenesis. Fig. 18–5 shows how spermatogenesis occurs in the formation of sperm in a testis. The primary sperm cell is in the resting stage; it is diploid; it contains three pairs of homologous chromosomes. Locate them in the diagram. Since replication has already occurred, each chromosome contains two chromatids.

In meiosis, a single diploid cell divides *twice* to form four monoploid cells. Trace the action in the diagram.

To understand the change from the diploid number ($2n$) to the monoploid number (n), bear in mind that a monoploid number of tetrads is established with synapsis. After this, it is largely a matter of each tetrad being divided twice to form four sets of homologues, each with the monoploid number. The number of single-stranded chromosomes

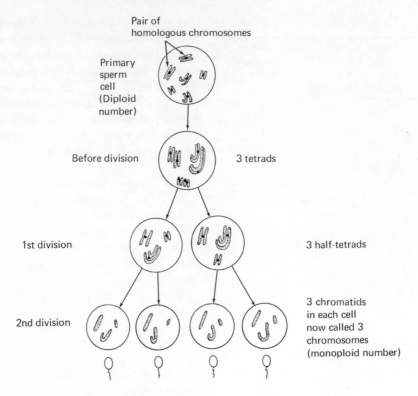

Fig. 18–5. Schematic diagram of spermatogenesis.

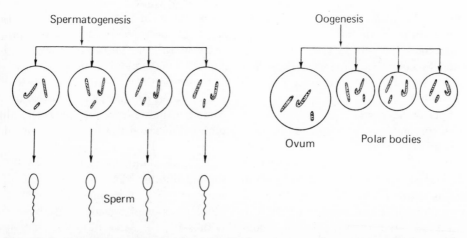

Fig. 18–6. Comparison of the end products of two types of gametogenesis (spermatogenesis and oogenesis). Assume that the primary sperm cell (spermatocyte) and the primary egg cell (oocyte) each had a diploid number of 6. The mechanism of meiosis is the same in each case.

in a mature sperm cell (three in this diagram) is exactly half as many as existed in the primary sperm cell before replication occurred (at that time, six single-stranded chromosomes).

Oogenesis. The chromosome changes during oogenesis parallel those occurring during spermatogenesis. However, there is a difference in the size of the cells that are formed. (See Fig. 18–6).

As shown in the diagram, each primary egg cell, as it matures in the ovary, finally produces one large ovum and three small *polar bodies.* The polar bodies disintegrate. Because of the unequal divisions of the cytoplasm during the two divisions, the single large ovum (egg) has much cytoplasm and may have a supply of yolk granules stored as food. Both the ovum and the polar bodies have the monoploid (*n*) chromosome number.

Comparison of Mitosis and Meiosis

The processes of mitosis and meiosis are compared in Fig. 18–7. For conve-

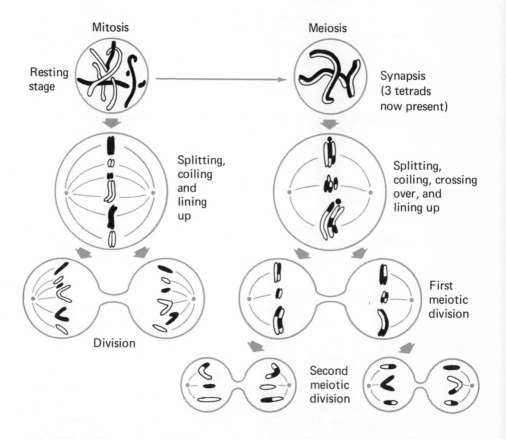

Fig. 18–7. Mitosis and meiosis compared. Note that the tetrads in the synapsis stage of meiosis have been simplified for ease of study; the chromatids are not shown.

nience of comparison, the two processes are shown proceeding from the resting stage of the same cell. In this cell, two homologous sets of chromosomes are shown, one set in black and the other in white. They are all long and tangled around each other.

The main events of the two processes are compared in the table below.

Mitosis	Meiosis
Chromosomes coil to form short, fat rods, with 2 chromatids each.	Chromosomes synapse to form tetrads, with 4 chromatids each.
	Tetrads exchange pieces and coil to form short, fat rods.
Chromatids separate to opposite sides.	Tetrads divide and the halves separate. Each half now has 2 chromatids.
Cell divides to form 2 diploid cells.	Half-tetrads divide again and move to opposite sides.
	The diploid cell has divided twice to form 4 monoploid cells.

Fertilization in Animals

Fertilization is the process by which a sperm nucleus unites with an egg nucleus to form a zygote. As the nuclei fuse, there is a union of hereditary material. Sperm require liquid for swimming to the eggs. This liquid may be the water in which the animals live, or it may be a liquid in the oviduct — the tube that carries the eggs of the female to the outside. *External fertilization* occurs outside the body of the female. *Internal fertilization* occurs inside the body of the female.

In many kinds of fish, the female deposits eggs in the water and the male swims over the same region and deposits sperm in the water — at the same place and at about the same time. Fish and amphibia generally reproduce by external fertilization.

Animals which live on land (reptiles, birds, and mammals) reproduce by internal fertilization. In this method, the sperm are introduced into the moist environment of the oviduct where they unite with the eggs.

Eggs which are deposited in the water have less chance of being fertilized than eggs which are fertilized internally. Animals which practice external fertilization produce more eggs than those which carry on internal fertilization. The production of a large number of eggs helps a species to survive.

Parthenogenesis

Parthenogenesis is the development of an egg into a new individual without its being fertilized by a sperm cell. Inasmuch as there is no fusion of gametes, this process may be considered to be a form of asexual reproduction that occurs in animals capable of sexual reproduction.

Natural parthenogenesis. The queen honeybee receives only one insemination with sperm. She seems to store that sperm for several years. As she lays eggs, the queen bee may or may not allow them to be fertilized. Fertilized eggs develop into females (workers or queens), whereas unfertilized eggs develop parthenogenetically into males (drones). The females are diploid and the males are monoploid. Natural parthenogenesis occurs in many insects, including wasps and aphids, and in microscopic animals known as rotifers. In some organisms, the parthenogenetically produced individual is monoploid, and in others it is diploid.

Artificial parthenogenesis. A variety of artificial treatments have been devised by scientists to trigger the development of an unfertilized egg into an embryo. These methods include the use of inorganic salts, organic solvents and acids, temperature change, mechanical stimulation, and electric shock. Complete new frogs, rabbits, and sea urchins have been produced in this manner. Great success has been achieved in the production of adult turkeys by parthenogenesis.

Recently, a research worker removed the nucleus from a frog's ovum and replaced it with the 2n nucleus from a cell of the intestine from another frog. This ovum developed into a complete frog. This showed that the nucleus of the specialized intestinal cell possessed all the hereditary characteristics for a complete organism.

REASONING EXERCISES

1. What is a zygote?
2. Distinguish between isogametes and heterogametes. In what respect are the two kinds of isogametes produced by *Spirogyra* dissimilar? Distinguish between the active gamete of *Spirogyra* and a sperm cell.
3. Distinguish between conjugation and fertilization. Why are they both considered to be sexual reproduction?
4. Why does the male bacterium die after sexual reproduction?
5. What is meiosis? In what kind of cells does it occur? Why is meiosis necessary to organisms that reproduce sexually?
6. How many cells result from a single cell as a result of meiosis? What happens to the chromosome number in this process? How are the chromosomes distributed?
7. What is synapsis? What exchange takes place during synapsis? Why does the reduction of the chromosome number to half require two nuclear divisions?
8. What is gametogenesis? Distinguish between the results of oogenesis and spermatogenesis. Explain how the chromosome number is kept constant from generation to generation.
9. Distinguish between mitosis and meiosis.
10. Distinguish between internal fertilization and external fertilization. Which type is more likely to be carried on by organisms that live on land? Why?

(B) DEVELOPMENT

As the fertilized egg develops, the embryo which results must receive food, have a proper environment, and be protected against danger.

External Development in Water

The eggs of amphibians and of most fishes are fertilized externally, and they develop externally in the water. The food for the embryo is the yolk stored in the egg. Generally, there is little or no parental care in these forms, but notable exceptions include mouth-breeding fish, and fish which build nests.

The large number of eggs produced by fish and amphibians serves to maintain the species despite the losses caused by predators.

External Development on Land

The eggs of reptiles and birds have a shell. Fertilization takes place internally before the shell is applied, but the fertilized eggs usually develop externally. Except for some snakes which carry their eggs internally, all reptiles and birds lay eggs on land. Birds and many kinds of snakes protect their eggs and their young in nests. Birds and reptiles produce fewer eggs than the fish and amphibians, which are characterized by external fertilization and external development. The food for the developing reptile and bird embryo is the yolk which is present in the egg.

Although there are numerous exceptions, the reproduction of vertebrates may be tabulated in a general way as shown below.

Adaptations of the Bird Embryo

The problems arising from evolution to a land environment are met in the developing reptile and bird egg by a number of adaptations.

Shell. The tough rubbery shell of the reptile egg and the calcareous shell of the bird egg provide protection. These shells are porous enough to permit the passage of gases in and out of the egg.

Embryonic membranes. The developing bird embryo has three membranes which help to provide a favorable environment for its development. These are the yolk sac, the amnion, and the allantois. (See Fig. 18–8.)

The *yolk sac* protrudes from the digestive tract of the embryo and encloses the yolk. Blood vessels transport the stored food from the yolk sac to the developing embryo.

The *amnion* encloses the embryo with a watery fluid, the *amniotic fluid*. This fluid provides a watery environment for the embryo and protects it against mechanical shock.

The *allantois* is a saclike structure which protrudes from the digestive tract and lies against the outer shell. Well-supplied with blood vessels, the allantois carries on an exchange of gases with the external environment through the outer shell. It serves the embryo as an organ for respiration and excretion. Nitrogenous wastes are excreted from the embryo into the allantois in the form of insoluble deposits of uric acids.

REPRODUCTION IN VERTEBRATES

Vertebrate	Fertilization	Development	Number of Eggs	Parental Care
Fish	External	External	Many	None
Amphibia	External	External	Many	None
Reptiles	Internal	External	Few	Little
Birds	Internal	External	Few	Much
Mammals	Internal	Internal	Few	Much

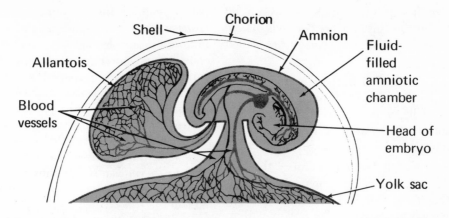

Fig. 18–8. Developing chick embryo.

Internal Development

Most mammal embryos develop inside the mother in a structure called the *uterus*. This is an enlarged portion of the oviduct. The embryo is attached to the uterus by means of the *placenta*, a structure from which it obtains nourishment.

Although external development is generally characteristic of fish and reptiles, some species develop internally, within the oviduct of the female. For example, persons who raise tropical fish are familiar with "live breeders" whose young emerge from the oviduct as young fish rather than as eggs. However, these vertebrate embryos obtain nourishment from the yolk of the eggs, and are not attached to the mother. There is no placenta such as occurs in mammals.

The duckbill (*Platypus*) is a primitive mammal which lays reptile-like eggs. It is a *non-placental* mammal. The marsupials, such as the kangaroo and the opossum, have poorly developed placentas. Their eggs have some yolk. The tiny young are born in an immature state and crawl into a pouch (*marsupium*) of the mother

where they receive shelter, warmth, and milk.

Most mammal embryos obtain food by diffusion from the bloodstream of the mother. Their eggs lack yolk and are microscopic in size. By contrast, the fish, frogs, reptiles, and birds have relatively large amounts of yolk in their eggs. The egg of the elephant (a mammal) is microscopic in size, whereas the egg of a frog is several millimeters in diameter.

Embryological Development

Embryology is the study of the development of the fertilized egg into a fully formed embryo. There are differences in this process among the various vertebrates largely because of varying amounts and varying distribution of the yolk in the egg. The main events in the development of a primitive vertebrate embryo may be summarized as follows:

1. The fertilized egg cell divides by mitotic division into two cells. Repeated cell divisions, in different planes, result in four cells, then eight cells, and so on until a solid ball of cells is formed. As the cells divide, they do not grow in

size. It is as if a meat cleaver were used to cut the egg into smaller cells. These early stages of embryology are called *cleavage.*

2. The solid-ball-of-cells stage is called the *morula* ("bunch of grapes"). Actually, this ball is not completely solid because a space develops in the interior.

3. As cell division continues, the cells exert pressure upon each other so that they form a single layer of cells enclosing a central cavity. This hollow-ball-of-cells stage is the *blastula.*

4. Take an old rubber ball which has lost much of its air and push in with a finger on one side. This indented

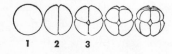

1. Zygote
2. Two-cell stage
3. Four-cell stage

ball resembles the beginning of the *gastrula* stage. The inpushing continues until there results an inner layer of cells and an outer layer of cells. The early gastrula is a cup-shaped stage consisting of an outer *ectoderm* layer and an inner *endoderm* layer. In the late gastrula, an additional layer of cells develops between the ectoderm and the endoderm. This is the *mesoderm.* Three *primary germ layers* are thus formed, the ectoderm, the endoderm, and the mesoderm.

Can the inpushing of the blastula to form the gastrula be explained on mechanical principles? One hypothesis is based upon the slower division of cells which contain more yolk. As shown in Fig. 18–10, the cells which contain more yolk are in the lower region of the blastula. Since this group of cells divides more slowly than the others, it is rea-

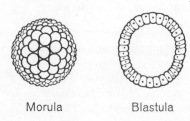

Morula   Blastula

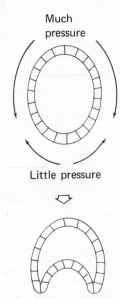

Much pressure

Little pressure

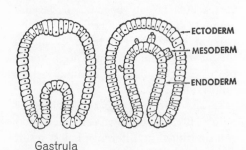

Gastrula

Fig. 18–9. Series of diagrams showing embryology.

Fig. 18–10. Formation of gastrula because of the pressure of rapidly dividing cells. Motion picture studies taken through a microscope show cells rapidly migrating into the opening of the gastrula.

soned that the cells at the bottom exert less pressure than the cells which are dividing rapidly. The cells exerting less pressure are pushed out of the way and form either an outpocketing or an inpocketing. The normal situation is for an inpocketing to occur. (See Fig. 18–10.)

Under experimental conditions, outpocketings have occurred. These resulted in deformed embryos which soon died.

Let us now review our definitions of cleavage and differentiation. *Cleavage* is the division of the fertilized egg into many small cells. *Differentiation* is the formation of special groups of cells with special structures and functions. Differentiation begins with the formation of the two-layered gastrula.

The Fate of the Germ Layers

All the organs and systems of the human body develop from the three primary germ layers; for example, the nervous system develops from ectoderm. To see how this occurs, let us first trace the development of the neural tube from the ectoderm, as shown in Fig. 18–11.

A pair of ridges develops along the length of the dorsal surface. These ridges move together and then join, forming the hollow neural tube. An expansion of the neural tube in the anterior portion of the embryo becomes the brain. All of the nerves develop from the neural tube and from additional portions of ectoderm.

Examine Fig. 18–10 again to note the relative positions of the ectoderm and endoderm. The endoderm develops into the alimentary canal. At first, the opening of the gastrula is the only opening for the alimentary canal. Later, a breakthrough occurs in the posterior region of the embryo, resulting in a continuous tube with openings for the mouth and anus. The lungs develop as an outpocketing from the endoderm of the alimentary canal.

The differentiation of the three primary germ layers into the body systems is indicated in the table below.

Explanations of Development

Embryologists have become skilled in tracing the development of the fertilized egg into the adult animal. But what causes these changes? From the time of Aristotle, two general theories have alternately received favor in explaining the mechanism of development:

Preformation. *Preformation* is the idea that the egg (or, as some said, the sperm) actually contains the new individual already formed in miniature. Development consists merely of the unfolding and growth of this individual. Some early microscopists, with their crude lenses, thought they could actually detect the minute individual (a

DIFFERENTIATION OF PRIMARY GERM LAYERS

Ectoderm	Mesoderm	Endoderm
Nervous system	Skeleton	Digestive tract
Epidermis of skin	Muscles	Respiratory system
	Circulatory system	Liver, pancreas
	Excretory system	
	Gonads	

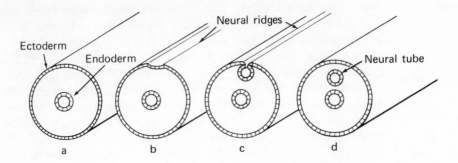

Fig. 18–11. Formation of the neural tube from ectoderm.

homunculus) within the sperm or egg.

Epigenesis. *Epigenesis* is the idea that the individual is not preformed but that it develops from material in the egg.

Within the past 75 years many studies have been performed by experimental embryologists to determine how development occurs. Some of these researches tended to support the general concept of preformation; others tended to support the idea of epigenesis. Many of the experiments were done on the embryos of sea urchins and salamanders.

In one experiment, tissue from one part of a salamander embryo was transplanted to another region of the embryo. The scientists wanted to see whether this tissue would retain its original destiny or change to something else because of its new environment in its new location. Experiments of this type have led to the discovery of embryonic induction. *Embryonic induction* is the ability of a structure which is already differentiated to induce changes in nearby embryonic tissue which has not yet been differentiated.

Abnormal Development—Cancer

Cancer is not a single disease. Cancer is rather a name which is applied to a large variety of disturbances in which cells continue to divide without restraint. Normal cells only divide sufficiently to conform to the general body pattern. Cancer cells, by their uncontrolled growth, deprive nearby tissues of food and upset their metabolic balance. Cancer is not merely a matter of rapidity of growth and division, for some cancerous cells do not divide more rapidly than non-cancerous ones. Rather, it is a matter of unceasing cell division.

After differentiation has occurred and the animal is fully grown, some cells have lost the ability to divide. During development they may even change their chromosome number (by losses or duplications), retaining only those parts of chromosomes needed to perform their specialized functions. But even in the adult, some cells retain the full diploid set of chromosomes, and they can divide to replace worn-out cells or to repair injuries. Something unknown limits their division to just the amount needed. Cancer cells have lost this limitation.

Viruses have recently been implicated as a cause of some kinds of cancer. Researchers believe that the nucleic acid (DNA) of the virus becomes intimately involved in the cell's own genetic machinery and gets passed along during cell division for many generations as a hidden "provirus." Some outside change then causes the viral DNA to produce virus again and to affect the cell. Studies of viruses in cancer cells are helping to unravel the mysteries of growth and heredity. Prospects for finding a cure for cancer have become much brighter as biologists have developed a better understanding of normal heredity, growth, and development.

HUMAN REPRODUCTION

The sea provided a relatively stable environment for primitive animals. It was also favorable for reproduction in that water permitted the sperm to swim to the egg. As animals moved to the land, they evolved adaptations for reproduction in this new environment. The sequence by which these specializations probably evolved may be traced as follows:

1. Internal fertilization, which increased the chances of the sperm meeting the egg.

2. Reduction in the number of ova produced.

3. Internal development of the embryo within the female to provide greater protection for the embryo.

4. Development of the placenta to provide the embryo with food from the mother's blood supply.

5. Reduction in the size of the egg.

6. Development of mammary glands to provide the newborn young with milk as a source of nourishment.

In perpetuating his species, man added to these characteristics of his fellow mammals such human characteristics as greatly extended parental care of the young, love, and the high devel-

opment of the family as a social unit.

Male Reproductive Structures

The *testes* of man are paired structures which lie in an outpocketing of the body wall, called the *scrotum*. Temperature in the scrotum is 2 to 4 degrees lower than body temperature, and this slightly lower temperature is advantageous for the production and storage of sperm. The sperm are produced in a series of highly coiled tubes within the testes, and are transported by a series of ducts to the outside of the body. The *seminal vesicles* and *prostate gland* supply a fluid and the mixture of fluid and sperm is known as *semen*. The sperm are produced in great numbers and at mating as many as 300 million sperm are transferred to the female. The liquid transport medium for the sperm is an adaptation for life on land.

The testes also produce *male sex hormones* which regulate such *secondary sexual characteristics* as the pitch of the voice and the growth of hair on the face. The production of sperm in the male, and ova in the female, are primary sexual characteristics.

Female Reproductive Structures

The ovaries (Fig. 18–12) are paired structures located in the lower portion of the abdominal cavity. The ovaries produce eggs in tiny cavities called *follicles*. As a follicle matures, it becomes filled with fluid and then discharges the ripe egg and fluid from the surface of the ovary. *Ovulation* is the discharge of the egg. In the adult human female, an ovum is produced by an ovary of the female approximately once every 28 days. Usually, the two ovaries alternate in this, with one producing an ovum and then the other producing an ovum. Occasionally, two ova may be liberated at the same time, in which case fraternal twins may result. (See page 301.)

All of the eggs that a female will ovulate during her lifetime are present in her ovaries when she is born. As previously noted, the ovaries produce ova as a primary sexual characteristic. Ovaries also produce female sex hormones which regulate female secondary sex characteristics. These include the development of the breasts, the lack of a beard, and the broadened pelvis.

After it leaves the ovary, the egg is drawn to the opening of the oviduct as shown in Fig. 18–12. The *oviducts,* or *Fallopian tubes,* transport the egg to the uterus. The *vagina* is the opening of the female reproductive tract. If sperm have been introduced into the vagina, they may swim up both branches of the oviducts and may meet an egg if one is present. Fertilization of the egg occurs in the oviduct. After the egg has been fertilized, it begins a series of mitotic cell divisions (*cleavage*). It is already a

tiny embryo with a fringe of outpocketings when it reaches the uterus about four days later.

The uterus is a thick-walled muscular sac in which the embryo develops. Under the influence of hormones, the inner lining of the uterus becomes prepared for the attachment of the embryo. The lining of the uterus becomes thick, soft, and moist with fluid. Its blood supply increases greatly. If the egg has been fertilized, the young embryo becomes attached to the lining of the uterus and continues to develop further.

If the egg has not been fertilized, the inner lining of the uterus breaks down and there is a discharge from the vagina of cellular debris and a slight to moderate amount of blood. This discharge takes place during the process of *menstruation.* The remains of the unfertilized egg pass out at the same time. After menstruation, the uterine wall again builds up in preparation for the next embryo. In this way a periodic menstrual cycle of about 28 days is maintained. If an embryo is implanted

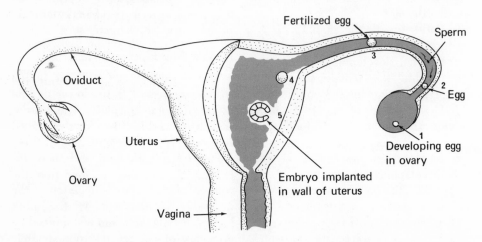

Fig. 18–12. Structure of the human female reproductive system, showing the fertilization of the ovum in the oviduct and the implantation of the embryo in the wall of the uterus.

in the uterus, hormones produced by the placenta cause the menstrual cycle to stop until after the baby is born.

Multiple Births

As previously indicated, *fraternal twins* may result when two ova are produced simultaneously. Each ovum is fertilized by a different sperm cell, and the heredity of the two offspring is different. Since fraternal twins originate from separate fertilized eggs, they do not necessarily look alike any more than other brothers and sisters. They need not be of the same sex.

Identical twins result when a single ovum fertilized by a single sperm cell later results in two embryos. In the human female, this separation into two embryos probably occurs after a

small mass of cells has developed. Since identical twins possess identical DNA from the same zygote, the twins resemble each other greatly in physical characteristics. Identical twins are always of the same sex.

Multiple births of the fraternal or identical type may yield triplets, quadruplets, and quintuplets. The offspring of mammals such as the cat and dog, which are born in litters, are actually fraternal twins. The armadillo is a mammal which normally produces identical quadruplets.

Adaptation for Internal Development

As the embryo develops in the uterus, a number of special membranes and organs appear. (See Fig. 18–13.) These are homologous with embryonic mem-

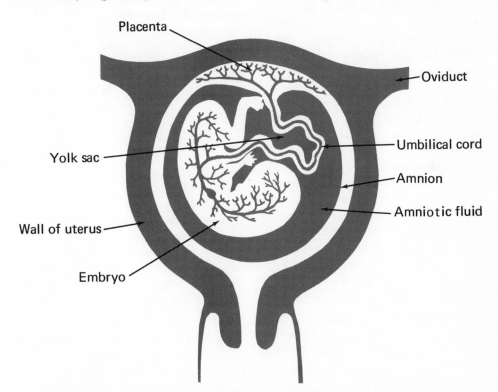

Fig. 18–13. Embryo in uterus, showing embryonic membranes.

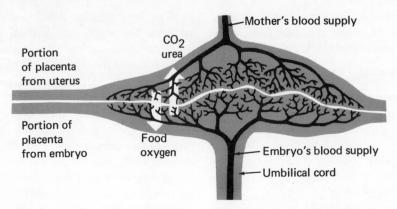

Fig. 18–14. Diagrammatic representation of exchange between mother's blood and embryo's blood at the placenta.

branes found in the fertilized egg of the reptile and bird.

The placenta. Fingerlike projections from an outer membrane of the embryo grow into the soft tissues of the lining of the uterus to form the double-walled *placenta.* One portion of this wall is derived from the embryo and the other portion is derived from the uterus. The placenta connects the embryo to the wall of the uterus and provides for the exchange of materials between the blood of the embryo and the blood of the mother. The two blood systems are largely separate — little or no blood of the mother flows into the blood system of the embryo. However, the two blood systems are close enough so that substances can diffuse from the capillaries of one system into the capillaries of the other. As shown in Fig. 18–14, the exchange is as follows:

To the embryo: nutrients, oxygen

To the mother: waste products, such as carbon dioxide and nitrogenous wastes.

The placenta thus serves the embryo as an organ for nutrition, respiration, and excretion, as well as for attachment.

The umbilical cord. The umbilical cord attaches the embryo to the placenta. It contains arteries and a vein which carry blood to and from the placenta. The *umbilicus* or *navel* is the remains of the umbilical cord on the baby after birth.

The amnion. The amnion is a membrane which surrounds the embryo. It encloses the *amniotic fluid* which supplies a watery environment for a land embryo. It also protects the embryo from mechanical shock.

Pregnancy

The period of time during which the mother protects and nourishes the developing embryo is known as the period of *gestation* (pregnancy). After the first six weeks, when the embryo is now recognizable as human, it is called a *fetus.* The length of gestation varies in different mammals. Some approximate times are shown in the table on p. 303.

Birth. At birth, hormones cause repeated contractions of the muscles of the uterus ("labor pains"). These contractions push the fetus through the vagina. The doctor ties off the umbilical

PERIODS OF GESTATION

Man	280 days	Cat	60 days
Mouse	21 days	Horse	350 days
Rabbit	32 days	Whale	360 days
Dog	60 days	Elephant	over 1½ years

cord and cuts it. Shortly thereafter further contractions of the uterus expel the remainder of the umbilical cord and placenta as the "afterbirth." During pregnancy changes occur in the mammary glands so that soon after birth these glands are ready to secrete milk for the nourishment of the baby.

Medical care during pregnancy. During pregnancy the mother's kidneys, lungs, heart, and digestive system supply the demands of two individuals. The mother's diet must be carefully controlled to insure her own good health and the proper development of her baby. An insufficient supply of calcium in the mother's diet may interfere with proper bone development in the embryo. As a result, calcium may be removed from her own bones and teeth to supply the demands of the embryo within. On the other hand, if the mother eats too much, excessive size of the fetus may cause the delivery of the child to be difficult. The possibility of "an Rh baby" (see page 136) must also be considered.

As a result of wise educational campaigns, it has become standard practice in the United States for husband and wife to visit the family physician at the earliest sign of pregnancy in order to obtain expert medical advice on the management of the pregnancy. As a result, the number of cases of prenatal and infant death has greatly decreased in the United States and whole generations of babies have started their lives in good health. However, recent evidence shows that many defects of intelligence and personality are caused by poor nutrition before birth.

The Menstrual Cycle

The *menstrual cycle* is a cycle of changes in the ovary and uterus. Hormones from the pituitary gland, ovaries, and uterus regulate the cycle. The stages are as follows:

1. Menstruation. In the absence of an embryo in the uterus, the lining of the uterus sloughs off with a slight loss of blood and the discharge of cellular debris. Menstruation occurs when there is a proper balance of hormones produced in the ovary.

2. Follicle stage. During this stage an egg is developing in an enlarged follicle (cavity) in the ovary. The follicle also produces a fluid containing the hormone *estrogen*.

3. Ovulation. The follicle ruptures at the surface of the ovary and ejects a mature egg.

4. Corpus luteum stage. The follicle heals over with a yellow-colored tissue called the corpus luteum ("yellow body"). The corpus luteum acts as an endocrine gland and produces the hormone *progesterone*.

In the human female, this cycle takes about 28 days although there is considerable variation among normal individuals. Ovulation occurs on or about the 14th day after the onset of menstruation.

The menstrual cycle continues from the time of *puberty*, when the ovaries mature, until *menopause* ("change of life") when the cycle ceases. Human females in the United States usually start to menstruate at about the age of 13 or 14. However, there is much variation, and normal individuals may start to menstruate much earlier or later. The menstrual cycle is interrupted by pregnancy and may be suspended during illness.

Estrous Cycle

Among mammals other than man and the anthropoid apes, the females have a periodic rhythm of sexual activity called the *estrous cycle*. This is not the same as the menstrual cycle of the human female although it is also controlled by hormones. In animals such as dogs, cats, and cows, the female will receive the male for mating only at the time of the year when she is "in heat." This is at the time of estrus when ovulation occurs. Thus fertilization of the egg is virtually assured.

Hormonal Control of the Menstrual Cycle

The menstrual cycle is an example of the regulation of body activity by the endocrine system. The hormones which play a part in the menstrual cycle are produced by both the pituitary gland and by the ovaries. Not only do the pituitary hormones affect the ovary, but the ovarian hormones also affect the pituitary gland.

A mechanism of feedback is involved as shown in Fig. 18–15. For simplicity, letters are used in place of the hormones.

As shown in the diagram, the pituitary gland secretes hormone A which stimulates the ovary to liberate hormone B. Hormone B has an effect upon the lining of the uterus. Also, a sufficient concentration of hormone B acts upon the pituitary to inhibit the production of hormone A. In this manner, the supply of hormone B is reduced at the proper time.

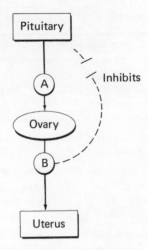

Fig. 18–15. Feedback.

The hormones produced by the pituitary and the ovaries are listed in the table on page 305.

Regulation of the reproductive cycle is accomplished not only by the presence or absence of hormones but also by their changing relative concentrations. For example, ovulation occurs when one of the hormones from the pituitary gland is increasing in concentration and another of these hormones is decreasing in concentration. Ovulation occurs when there is a proper balance between the declining level of FSH in the blood and the increasing level of LH. Menstruation occurs when the corpus luteum degenerates and the level of progesterone decreases.

Hormones in Pregnancy

Why does the menstrual cycle stop when a young embryo is implanted in the uterus? The chorion membrane of the embryo produces a hormone which functions like LTH and stimulates the corpus luteum to secrete progesterone. This maintains the uterine lining.

HORMONES AFFECTING MENSTRUATION

Pituitary Hormones	Ovarian Hormones
FSH (Follicle Stimulating Hormone). Acts on the follicle in the ovary to promote its growth. LH (Luteinizing Hormone). Acts on the follicle after it has released an egg. Promotes the formation of the corpus luteum.	Estrogen. Secreted by the follicle as it develops in the ovary. Causes preliminary development of the uterine lining by increasing cell division, increasing the blood supply, and by increasing the accumulation of fluids. (Also causes the breasts to enlarge.)
LTH (Luteotrophic Hormone, also called prolactin). Stimulates the corpus luteum to produce progesterone. (Also causes the secretion of milk.)	Progesterone. Secreted by the corpus luteum in the ovary. Causes thickening and final preparation of the uterine lining so that it forms a thick, rich, spongy layer. (Also causes milk glands in the breasts to develop.)

When the placenta is formed, this organ also secretes progesterone and estrogen in such concentrations as to maintain the condition of the uterine lining. Progesterone and estrogen also prepare the breasts to secrete milk.

LTH (or prolactin) causes the breasts to secrete milk. This hormone is secreted in small amounts during pregnancy by the pituitary gland, and it is needed to maintain the functioning of the corpus luteum. After the baby is born, prolactin production increases greatly, and milk forms in the breasts.

Another pituitary hormone (pitocin) has two effects: it produces the contractions of labor that push the baby into the world; after the baby is born, this hormone is responsible for the "letdown reflex." This reflex starts with the stimulation of the mother's nipples by the baby. Nervous pathways into the brain cause the hypothalmus to secrete a hormone that goes directly to the pituitary. Here, pitocin is released and, when it gets to the breast, milk flows.

Male Sex Hormones

The pattern of sex hormones in the male seems to be considerably less complicated than that in the female, since the male hormones are not concerned with the developing embryo. The hormones produced by the testes are known under the general name of *androgens*. The most important of the androgens is *testosterone*. This is responsible for the development of male secondary sex characteristics. Pituitary hormones are involved in the production of sperm.

COMPLETION QUESTIONS

[A] 1. Cells which unite in sexual reproduction are called ....

2. Reproduction by isogametes is called ....

3. A zygote with a heavy wall is a (an) ....

4. The motile gamete is the ....

5. The organs in animals that produce gametes are the ....

6. A pair of similar chromosomes are .... chromosomes.

7. The number of chromosomes in a single set of homologues is the .... number.

8. Maintenance of the chromosome number from generation to generation is accomplished by the processes of .... and ....

9. The process by which sperm are produced is ....

10. The pairing of homologous chromosomes during meiosis is called ....

11. From each primary egg cell the number of ova produced is ....

12. In an organism with a diploid number of 18, the number of chromosomes in a polar body is ....

13. If the monoploid number is 10, the number of chromosomes in a kidney cell is .....

14. Animals which live on land generally reproduce by .... fertilization.

15. The development of an embryo from an unfertilized egg is called ....

(B) 16. Birds have .... fertilization and .... development.

17. A watery environment for the embryos of mammals is provided by the ....

18. The embryo of the cow develops in the ....

19. A mammal which lays eggs is the ....

20. The hollow-ball-of-cells stage in embryology is the ....

21. The cup-shaped stage in the development of an embryo is called the ....

22. The primary germ layer which develops into the nervous system is the ....

23. The cavities within the ovary in which the mammalian eggs form are the ....

24. The release of the mature ovum from the follicle is known as ....

25. The structure in which fertilization occurs in mammals is the ....

26. The time required for a mammal embryo to develop is known as the period of ....

27. The cycle of change in the human uterus is known as the .... cycle.

28. The yellow tissue which fills the space of the follicle is the ....

29. In addition to producing sperm or eggs, the gonads also secrete ....

30. A hormone which leads to the formation of male secondary sexual characteristics is ....

MULTIPLE-CHOICE QUESTIONS

A 1. Fertilization is accomplished when (1) the sperm has entered the egg, (2) the egg nucleus and the sperm nucleus have fused, (3) a mature egg meets a mature sperm, (4) a fertilization membrane has formed around the egg.

2. In the embryonic development of higher animals, which process occurs immediately after fertilization? (1) mitosis, (2) meiosis, (3) growth in weight, (4) nuclear fusion.

3. If gametes were formed solely as a result of mitosis, each time that fertilization occurred the species chromosome number would be (1) reduced by half, (2) constant (unchanged), (3) doubled, (4) quadrupled.

4. The head of a sperm cell is of primary importance in the formation of a zygote because it provides (1) cytoplasm, (2) stored food, (3) the diploid number of chromosomes, (4) the monoploid number of chromosomes.

5. The diploid number of chromosomes is restored by (1) maturation, (2) spermatogenesis, (3) reduction division, (4) fertilization.

6. Meiosis occurs during (1) oogenesis, (2) embryo formation, (3) budding, (4) replication.

7. As compared with the sperm cell, the human egg cell contains more (1) chromosomes, (2) mitochondria, (3) centrosomes, (4) cytoplasm.

8. Replication occurs in (1) fertilization, (2) mitosis only, (3) meiosis and mitosis, (4) meiosis only.

9. Identical DNA composition in organisms is maintained from generation to generation by means of (1) sexual reproduction, (2) asexual reproduction, (3) external fertilization, (4) internal fertilization.

10. A major advantage of sexual reproduction over asexual reproduction is that sexual reproduction results in (1) more rapid production of offspring, (2) a greater number of offspring, (3) larger offspring, (4) greater variety of offspring.

B 11. Generally, animals that carry on external fertilization (1) produce more eggs than animals carrying on internal fertilization, (2) have much parental care of the young, (3) have little yolk in the eggs, (4) live on land.

12. Eggs containing very little yolk are usually associated with (1) external fertilization, (2) parthenogenesis, (3) internal fertilization and external development, (4) internal fertilization and internal development.

13. Meiosis is a process of cell division which occurs in (1) a rapidly cleaving zygote, (2) an embryonic plant tissue, (3) a vegetative structure, (4) a reproductive organ.

14. Which is never found in any of the stages of the life cycle of an amphibian such as the frog? (1) zygote, (2) ovary, (3) amnion, (4) yolk.

15. When a cell divides during the process of cleavage, each resulting cell, compared with the parent cell, has (1) more cytoplasm, (2) less cytoplasm, (3) more chromosomes, (4) fewer chromosomes.

16. Which structure makes nutrients and oxygen directly available to the human embryo? (1) uterus, (2) oviduct, (3) placenta, (4) ovary.

17. Which will a developing human embryo normally *not* receive from the mother? (1) oxygen, (2) glucose, (3) calcium, (4) red blood cells.

18. If a placenta were found in an animal, the animal could be (1) a cat, (2) an eagle, (3) a turtle, (4) a platypus.

19. The amnion of a human embryo is (1) the principal organ of nutrition and respiration, (2) a sac containing the fluid in which the developing embryo is suspended, (3) a sac which stores embryonic waste products until birth, (4) a cord which attaches the embryo to the mother.

20. Which is *not* a stage of the menstrual cycle? (1) follicle stage, (2) menstruation, (3) fertilization, (4) ovulation.

CHAPTER TEST

1. During cleavage, increase in cell number (1) is usually accomplished by an increase in cell size, (2) is accomplished by successive meiotic divisions, (3) results in the formation of a blastula, (4) results in the formation of a zygote.

2. If a sperm with a monoploid chromosome number fertilized an egg with a monoploid chromosome number, the resulting zygote would be (1) diploid, (2) haploid, (3) abnormal in chromosome number, (4) unable to undergo cleavage.

3. The ectoderm of the gastrula gives rise to the (1) blood and blood vessels, (2) muscles, (3) digestive glands, (4) brain and nervous system.

4. The development of internal fertilization accompanied the evolution of (1) water animals, (2) land animals, (3) invertebrates, (4) monoploid animals.

5. The species number of an organism is maintained by (1) mitosis and meiosis, (2) fertilization and replication, (3) fertilization and mitosis, (4) fertilization and meiosis.

6. Artificial parthenogenesis is ordinarily accomplished by stimulating (1) a zygote, (2) a polar body, (3) an unfertilized egg, (4) a sperm cell.

7. Which is classified as sexual reproduction? (1) conjugation, (2) budding, (3) fragmentation, (4) sporulation.

8. Of the following, the smallest eggs are produced by the (1) sparrow, (2) codfish, (3) elephant, (4) ostrich.

9. The survival of the human embryo, which is developed from an egg with little yolk, is in part due to the fact that it (1) is milk fed, (2) needs little food, (3) has a short period of development, (4) is attached to a placenta.

10. A membrane used for respiration and excretion by an embryo bird is the (1) allantois, (2) amnion, (3) yolk sac, (4) placenta.

CHAPTER 19 / Sexual Reproduction in Plants

Fertilization in plants, as in animals, is the union of sperm and egg nuclei. In plants such as mosses and ferns, the sperm swim through water to the eggs. Fertilization occurs only when water is available. In flowering plants, the situation is different. Here the transfer of the male gamete is accomplished in the absence of a watery environment.

The evolution of plants from water to land parallels the evolution of animals. The higher forms of both are adapted for reproduction in the absence of water.

Conjugation (sexual reproduction by isogametes) in *Spirogyra* and bread mold was described in Chapter 18. This chapter is devoted to sexual reproduction in plants by heterogametes.

A | ALTERNATION OF GENERATIONS

In the life cycle of algae, mosses, and ferns, there exist two separate kinds of individuals:

1. The diploid *sporophyte generation.* This is an asexual stage that produces monoploid spores. The spores give rise to

2. The monoploid *gametophyte generation.* This is a *sexual* stage that reproduces by male and female gametes.

Fertilization by the union of the male and female gametes restores the diploid sporophyte generation. This method, by which an organism reproduces asexually in one generation and sexually in the succeeding generation, is called *alternation of generations.* Many variations exist in the details of the life cycle, but a typical example of alternation of generations is shown in Fig. 19–1. Trace the main events of the cycle in the diagram.

In mosses, the gametophyte generation is the conspicuous green plant. Accordingly, it is the sexual generation which is commonly recognized as the moss plant. The sporophyte generation, on the other hand, is insignificant in appearance.

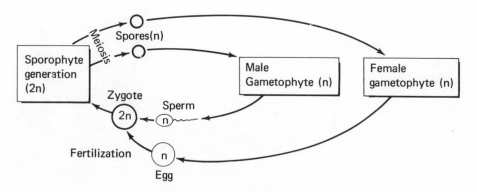

Fig. 19–1. Alternation of generations in a moss.

As higher plants evolved, however, there was a progressive reversal of the prominence of these two generations. The sporophyte became progressively more prominent, whereas the gametophyte became reduced. Most people never see the inconspicuous gametophyte generation of the ferns; the sporophyte generation is the conspicuous one.

The familiar fern (Fig. 19–2) is the sporophyte generation. This is the asexual generation, and it reproduces by spores. The spores are produced in small brown dots (sori) which are composed of spore cases. In many common ferns, these are on the underside of the fronds (leaves).

When the spores are mature, the spore cases burst open and spores are released. If a spore falls in a favorable environment, it germinates into a gametophyte plant (*n*). This plant is the sexual phase of the fern's life cycle. It is a small, heart-shaped plant which contains the female sex organs that produce the eggs and which also contains the male sex organs that produce the sperm (*n*).

When the eggs and sperm mature and there is sufficient water present, the sperm leave the male organs and swim to the female organs. The sperm enter the female organs and an egg is fertilized, forming a zygote (*2n*). The zygote then develops into a new sporophyte (the familiar fern), and the cycle is repeated. Note that ferns can carry on sexual reproduction only if there is a watery environment available for fertilization.

In plants which produce flowers, the evolutionary trend for the rise of the sporophyte generation and the suppression of the gametophyte generation is carried still further than in the ferns. The conspicuous plant which is seen, whether it be an oak tree or a bean plant, is the diploid sporophyte generation. The monoploid gametophyte generation has been so greatly reduced that it does not lead an independent existence. It exists only for a short time, as a structure composed of few cells, within the flower. The flower nourishes and supports within it, for a brief time, the simple male gametophyte and the female gametophyte which produce the sperm and eggs. The flower is the reproductive structure of the sporophyte generation.

One difference in how higher plants and animals carry on reproduction

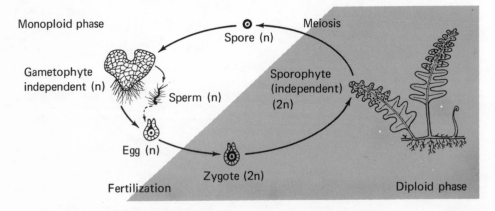

Monoploid phase

Meiosis

Spore (n)

Gametophyte independent (n)

Sperm (n)

Sporophyte (independent) (2n)

Egg (n)

Zygote (2n)

Fertilization

Diploid phase

Fig. 19–2. Alternation of generations in the fern plant.

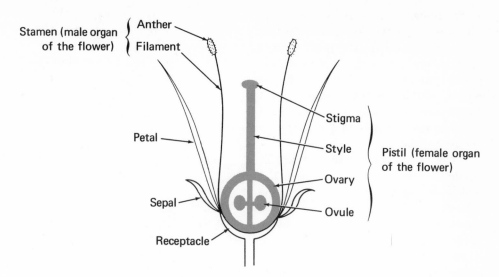

Fig. 19–3. Diagram of a generalized flower.

should now be noted. In higher animals, meiosis occurs within the gonad as a part of the development of the mature sperm or egg. In higher plants, however, meiosis occurs as a part of spore formation. The spore then produces a gametophyte, which produces the gametes.

THE FLOWER

A flower is a modified stem with highly specialized leaves for carrying on reproduction. Meiosis and fertilization both occur within the flower.

Structure of the Flower

Fig. 19–3 is a diagram of a generalized flower. There are many variations on this general plan.

The parts of the flower are:

Receptacle. The receptacle, or base, is the enlarged part of the stem of the flower. It supports the entire flower structure.

Sepals. The sepals are flower leaves that form an outer circle at the base. All of the sepals together make up the *calyx*. The sepals protect the flower before it opens, that is, when it is a bud. Typically, the sepals are green, but they may be as brightly colored as the petals.

Petals. The petals are the inner circle of flower leaves. All the petals together constitute the *corolla*. The petals of many flowers have bright colors and odors that attract insects.

Stamens. Located inside the corolla is a ring of stamens. The *stamens* may be considered as the male organs of the flower because they ultimately give rise to the male gametophyte. A knob-like sac at the top of the stamen is the *anther*, which produces pollen. The *filament* is a threadlike stalk which supports the anther.

Pistil. Inside the circle of stamens is the pistil. The *pistil* may be considered as the female organ of the flower because it ultimately gives rise to the female gametophyte. The pistil consists of three parts: (1) The *stigma,* an expanded sticky top which receives the pollen; (2) the tubelike *style,* which connects the stigma with the ovary; and (3) the enlarged base of the pistil, the *ovary.*

A flower may have several pistils. Inside the ovary are one or more *ovules* which are attached to the ovary. Some flowers may have as many as several hundred ovules. The ovules within the ovary later become seeds.

The *essential organs* of a flower are the stamens and pistils. The *accessory organs* are the petals and sepals, which help indirectly in reproduction.

Types of Flowers

A flower that possesses all four parts — calyx, corolla, stamens, and pistil — is a *complete* flower. If any one of these parts is missing, the flower is *incomplete.* Incomplete flowers may be further classified as follows:

Perfect — possessing both pistils and stamens

Imperfect — lacking pistils or stamens

Staminate — having stamens only

Pistillate — having pistils only.

A plant bearing only staminate flowers is called a male plant. A plant bearing only pistillate flowers is called a female plant. In some species, such as corn, a single plant may carry both staminate and pistillate flowers.

Development of the Pollen Grain

Pollen grains develop within the anther. Let us see how this occurs. First, spores are produced by meiosis. In this way, the diploid number of chromosomes which is found in the parent plant is reduced to the monoploid num-

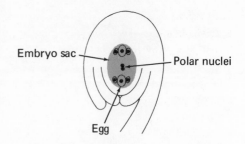

Fig. 19–4. The ovule at time of fertilization.

ber. Then the pollen grains develop from these spores. Each mature pollen is an incipient gametophyte. It consists of a cell wall, cytoplasm, and usually two monoploid nuclei. When the pollen is ripe, the anther splits open, releasing the pollen.

Development of the Ovule

The ovules develop in the ovary early in the formation of the flower. In some flowers, the ovary contains only one ovule; in others, the ovary may contain hundreds. Each ovule is attached to the ovary wall by a thin stalk. The ovule is surrounded by a wall composed of two layers of cells. The *micropyle* is a small opening in the wall which leads to the interior of the ovule.

Within the ovule, one cell divides by meiosis, reducing the chromosomes to the monoploid number as four spores are produced. Three of these spores degenerate; the other spore grows into a tiny female gametophyte, called the *embryo sac,* an oval-shaped sac containing eight nuclei.

In many plants, some of these nuclei degenerate. However, at least three nuclei remain. These are two *polar nuclei* and one *egg nucleus.* Each of these has the monoploid (n) chromosome number. (See Fig. 19–4). The egg nucleus is the female gamete.

Pollination

Pollination is the transfer of pollen from an anther to a stigma. When the anthers burst, the pollen is released. Some of the pollen grains may be carried by the wind to the stigmas of other flowers. This type of pollination is called *wind pollination.*

Sometimes insects that enter the flower in search of nectar pick up pollen on their bodies and carry the pollen to other flowers. This is known as *insect pollination.* Many farmers keep beehives in their orchards to insure pollination of the fruit trees.

When the pollen from an anther falls on the stigma of a flower on the same plant, the process is called *self-pollination.* When the pollen is transferred to the stigma of another plant of the same kind, the process is called *cross-pollination.* The production of seeds in flowering plants occurs only after pollination has taken place.

Pollination may be accomplished by wind, water, insects, wandering animals, and birds. In *artificial pollination,* man uses a brush or the anther itself to transfer pollen from an anther to a stigma.

The thick wall of the pollen grain prevents dehydration of its contents during its transfer to the female reproductive organ. In this way, flowering plants solve the problem of transferring sperm in a dry external environment.

Growth of the Pollen Tube

Pollen grains that fall on the stigma are held there by a sticky, sugary secretion. The pollen grain absorbs food and water from the stigma and germinates (starts growing) into a male gametophyte called the *pollen tube.* A chemical from the pistil stimulates the pollen grain to develop into the pollen tube.

The pollen tube digests the tissues of the pistil as it grows down through the stigma, style, and ovary to the ovule. Within the pollen tube are present a tube nucleus and two sperm nuclei. (See Fig. 19–5.) Each of these has the monoploid chromosome number.

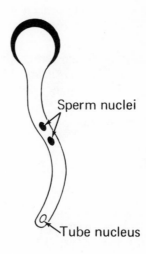

Fig. 19–5. Pollen tube.

Pollen may be germinated in the laboratory by use of the proper concentration of sugar solution and plant hormones.

Fertilization

The pollen tube enters the ovule through the micropyle and discharges its two sperm nuclei into the embryo sac. Note that the sperm do not swim through water to reach the egg. Two fertilizations result:

1. One sperm nucleus (n) unites with the egg nucleus (n). The diploid zygote resulting from this fertilization begins a series of mitotic cell divisions that ultimately result in the embryo of the new plant. This is the beginning of the sporophyte generation.

2. The other sperm nucleus (n) unites with the two polar nuclei to form a triploid ($3n$) *endosperm* nucleus. Cell divisions involving this nucleus result in the formation of the endosperm. This tissue provides food for the developing embryo. In many plants the endosperm disappears and its function of storing and supplying food is taken over by a structure called the *cotyledon*, or seed leaf.

Fertilization in flowering plants is known as *double fertilization* because one sperm nucleus unites with the egg nucleus and the second sperm nucleus unites with the two polar nuclei.

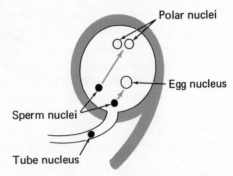

Fig. 19–6. Double fertilization. One sperm nucleus unites with the egg nucleus; the other sperm nucleus unites with the two polar nuclei.

REASONING EXERCISES

1. Distinguish between the sporophyte and gametophyte generations. Which is monoploid and which is diploid?

2. What is "alternation of generations?"

3. In mosses, how does the sporophyte generation arise? How does the gametophyte arise?

4. What function does meiosis serve for both plants and animals? State a basic difference in the way gametes are formed in higher plants and higher animals.

5. Which part of the flowering plant is the male gametophyte? Which part is the female gametophyte? Which part is the sporophyte?

6. Compare the gametophyte of a fern and the gametophyte of a flowering plant from the standpoint of size and water requirements.

7. Distinguish between pollination and fertilization. If you were conducting experiments in cross-pollination and you wanted to prevent self-pollination of some of the flowers, which structure would you remove from these flowers?

8. What is double fertilization?

(B) **THE SEED AND FRUIT**

The fertilized egg goes through a series of mitotic divisions and becomes a multicellular embryo. The *embryo* is the young plant. The endosperm nucleus develops into a mass of tissue, the endosperm, which helps to nourish the embryo. The outer covering of the ovule develops into the seed coat. This hard protective coat insures survival of the embryo during unfavorable conditions of dryness. A *seed* is a ripened ovule containing an embryo, stored food, and one or more protective coats.

While the ovule is ripening into the seed, changes are also occurring in the ovary. Generally, the ovary gets larger, develops a thick wall, and stores food and water. The changes which occur depend upon the type of plant. The ovaries of different plants develop in different ways to provide protection and dispersal of the seeds within. As a result of these changes, the ovary develops into a fruit.

A *fruit* is an enlarged ovary with its seeds and any parts of the flower which remain attached. An apple, an orange, or a watermelon are familiar examples of fruits, but a fruit need not be edible. For example, the "polly-noses," which develop on maple trees in the springtime are fruits, as are the "stickers" of the cocklebur plant, which catch on clothing in the autumn. Moreover, biologists do not recognize the distinction between "fruits and vegetables" made in the food market. A fruit is the ripened ovary of a flower; consequently, tomatoes and cucumbers are fruits. The major functions of the fruit are to aid in the dispersal of the seeds and to protect and nourish the seeds.

Structure of the Seed

A bean seed is shown in Fig. 19–7. On one edge of the external surface is a long scar called the *hilum*. This marks the place where the ovule was attached to the wall of the ovary. The micropyle of the ovule is still visible as a tiny opening near the hilum.

After a bean seed has been soaked in water, it is easy to remove the seed coat

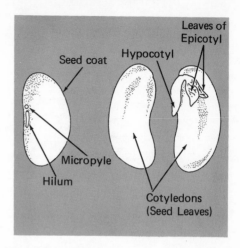

Fig. 19–7. External and internal view of bean seed.

(or *testa*) and disclose the embryo. The two large portions of the embryo are cotyledons, or seed leaves. They supply food to the rest of the embryo as it develops into a young plant. Between the cotyledons are the *epicotyl* and *hypocotyl*.

The upper portion is the epicotyl, which will develop into the leaves and upper portion of the stem. The lower portion is the hypocotyl, which will develop into the lower portion of the stem and the roots.

Seed Dispersal

Many fruits have special adaptations which insure dispersal of the seeds. Maple trees produce fruits with a winglike structure, thus enabling the fruit to float along on the wind. The cocklebur and burdock have burrs which catch on clothing or in animal fur. The poppy and the touch-me-not shoot their seeds a short distance. Many fruits (cherry, peach, apple) are eaten by animals which spread the seeds.

Although pollination and seed dis-

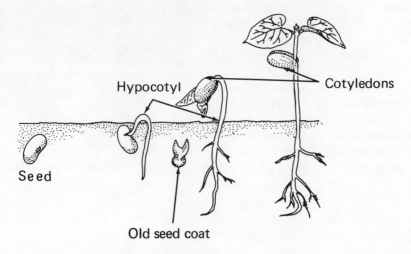

Hypocotyl

Cotyledons

Seed

Old seed coat

Fig. 19–8. Germination of a bean seed.

persal are both forms of dispersal, a careful distinction should be made between these two processes. *Pollination* is a method of transferring the male gamete to the region of the female gamete before fertilization takes place. *Seed dispersal* is a method of spreading the embryos which result from this fertilization.

Germination of Seeds

After a seed is formed, it enters a period of dormancy (resting stage). The length of this period varies, depending on external conditions and on the type of seed. The seed can withstand unfavorable conditions of dryness and temperature. Favorable conditions cause dormancy to end and active life to begin again.

Germination is the growth of a new plant from the seed. (See Fig. 19-8.) Some of the requirements of germination are (1) sufficient moisture, (2) proper temperature for enzyme activity, and (3) oxygen.

During the early growth of the plant, food is obtained from the endosperm or from the cotyledons. After green leaves are produced, the young plant makes its own food. In the case of the bean seedling, the cotyledons are carried above ground and are present for a time at the lower portion of the stem. When their food supply is exhausted, they shrivel and drop off.

The transformation of a seed into a mature plant which is capable of reproduction involves cell division, differentiation, and growth. (See pages 259 and 260.)

COMPLETION QUESTIONS

A 1. Cells that unite during sexual reproduction are called . . . .
2. When alternation of generations occurs, the stage that carries on sexual reproduction is the . . . .
3. The variation between a sexual stage and an asexual stage in the life cycle of an organism is known as . . . .
4. In lower plants the sperm reach the eggs by . . . .
5. All the petals of a flower together make up the . . . .
6. The organ of the flower that produces the male gametophyte is the . . . .
7. The stigma is connected to the ovary by the . . . .
8. The essential organs of the flower are the . . . . and the . . . .
9. A flower lacking pistils or stamens is said to be . . . .
10. The pollen-bearing structure at the top of the stamen is the . . . .
11. The gamete contained within the ovule is the . . . .
12. In pollination pollen is transferred to a . . . . of a flower.
13. In the pollen tube are formed the . . . . nuclei and a . . . . nucleus.
14. The embryo of a plant results from the union of a sperm nucleus and the . . . . nucleus.
15. The union of a sperm nucleus and the two polar nuclei results in the formation of the . . . ., which has a chromosome number of . . . .

B 16. The young plant in a seed is called the . . . .
17. The outer protective coat of the seed is called the . . . .
18. A seed is a ripened . . . .
19. A fruit is a ripened . . . .
20. The part of the plant embryo which develops into leaves is the . . . .
21. The part of the plant embryo which develops into the roots is the . . . .
22. Plant embryos are spread by the process of . . . .
23. The resting stage of a seed is called the period of . . . .
24. The growth of a new plant from a seed is the process of . . . .
25. Before green leaves are produced, the germinating bean seed obtains its food from the . . . .

MULTIPLE-CHOICE QUESTIONS

A *Questions 1 to 5 refer to structures in the diagram below.*

1. Which two structures produce cells that carry on the process of meiosis? (1) 1 and 2, (2) 2 and 5, (3) 6 and 7, (4) 4 and 6.
2. Which structure contains a female gamete? (1) 1, (2) 2, (3) 6, (4) 4.
3. Which structure may develop into a seed? (1) 1, (2) 6, (3) 7, (4) 4.
4. On which structure does a pollen grain normally begin to germinate? (1) 1, (2) 7, (3) 3, (4) 4.
5. Which structure may develop into a fruit? (1) 1, (2) 5, (3) 3, (4) 4.

(B) 6. After the sperm nucleus and the egg nucleus fuse, the resulting embryo develops by (1) mitosis, (2) fertilization, (3) gametogenesis, (4) parthenogenesis.
7. Botanists refer to a bean pod as a (1) vegetable, (2) fruit, (3) seed, (4) cotyledon.
8. Which would *not* be found in a mature bean seed? (1) digestive and respiratory enzymes, (2) an embryo, (3) cells with monoploid nuclei, (4) starch.

Questions 9 and 10 refer to structures in the diagram below.

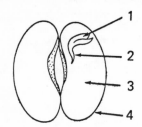

9. Which part would most probably turn blue-black if treated with iodine solution? (1) 1, (2) 2, (3) 3, (4) 4.
10. Which part develops into the true leaves and the upper part of the stem? (1) 1, (2) 2, (3) 3, (4) 4.

CHAPTER TEST

1. Fertilization in higher plants takes place in the (1) ovule, (2) pollen tube, (3) stigma, (4) style.
2. In higher plants the embryo develops inside the (1) stigma, (2) stamen, (3) ovule, (4) pollen tube.
3. If a plant has large colorful flowers, it is probably (1) wind pollinated, (2) self-pollinated, (3) insect pollinated, (4) water pollinated.
4. The stamen is to the sperm nucleus as the pistil is to the (1) stigma, (2) egg nucleus, (3) embryo sac, (4) ovary.
5. Which structure is found in bean seeds but *not* in fern spores? (1) cell membrane, (2) embryo, (3) nucleus, (4) chromosome.
6. The fruit is to the ovary as the seed is to the (1) stigma, (2) embryo, (3) ovule, (4) pollen tube.
7. The ripened ovary of a jewelweed plant pops open when it is touched. This is an example of adaptation for (1) seed dispersal, (2) protection, (3) self-pollination, (4) cross-pollination.
8. As seeds germinate, the starch they contain is converted to sugar. This process is necessary because (1) the enzyme used aids digestion, (2) starch contains no energy, (3) only soluble materials may enter a living cell, (4) amino acids are necessary for growth.
9. Pollen tubes serve as passageways for (1) pollen grains, (2) ovules, (3) sperm nuclei, (4) egg nuclei.
10. A cell which eventually will produce a new organism by combining with another cell is called a (1) pollen grain, (2) spore, (3) zygote, (4) gamete.

UNIT FIVE

GENETICS

Even in ancient days people noticed that offspring tend to resemble their parents. However, they saw that offspring may also differ from their parents; for example, brown-eyed parents can have blue-eyed children. To the medieval noblemen trying to breed sturdier warhorses, or to the farmer trying to obtain cows which produce more milk, there seemed to be no pattern by which traits are passed on from one generation to the next. Today, knowledge about the inheritance of traits is provided by the study of genetics. *Heredity* is the transmission of characteristics or traits from one generation to the next. (An individual's heredity may also be considered as the set of "chemical instructions" received by an individual through the gametes of his parents.) *Genetics* is the branch of biology which studies the process of heredity. The science of genetics examines not only the similarities in heredity between parents and offspring but also the differences or variations.

CHAPTER 20 / Patterns of Heredity

In the middle of the 19th century, Gregor Mendel, an Austrian monk, described a number of his experiments with garden peas. Mendel's work received little attention from the outstanding scientists of his day and it was not until 35 years later that its importance was recognized. Today, Mendel is generally considered to be the founder of genetics.

Mendel viewed an organism as a collection of traits, each of which is inherited separately. Through the years, it has been shown that there are indeed discrete units of heredity in the egg and sperm, which are called *genes*. These have now been identified as sections of the DNA molecules. It is known that the genes are arranged linearly on the chromosomes.

A   HOW DNA WAS FOUND TO CARRY HEREDITY

DNA is the stuff of heredity, the material passed from parent to offspring, that controls the development of the offspring. Nucleic acid itself was discovered as a result of studies by Friedrich Miescher, a Swiss biochemist who, in 1869, extracted from the nuclei of pus cells a nonprotein material which he called "nuclein." Because of its acid properties, later workers called it nucleic acid. However, the role of one kind of nucleic acid — DNA — in heredity was not discovered until 50 years later. Let us examine some of the experiments which showed that DNA carries heredity.

Bacterial Transformation

Transformation is a change in heredity caused by a transfer of dissolved DNA from one cell to another. Knowledge of transformation began in 1928 with the experiments of Frederick Griffith, a British bacteriologist. Griffith was working with pneumococcus, the bacterium which causes pneumonia. Two types of pneumococcus have these characteristics:

Type S forms smooth colonies. These bacteria possess a capsule. They are virulent (cause disease).

Type R forms rough colonies. These bacteria do not possess a capsule. They do not cause disease.

When Griffith injected mice with living S cells, the mice developed pneumonia and died. When he injected dead S cells (killed by heat), the mice did not die. Injection of the harmless R cells did not cause disease, whether these cells were living or dead. However, when Griffith first mixed the harmless R cells with killed S cells and injected this mixture, he was surprised to find that the mice died. When the dead mice were examined they were found to contain live bacteria of the virulent S type. The living harmless R cells had been changed into living type S cells by something from the dead S cells! These new type S cells passed their characteristics on to succeeding generations.

Fifteen years later, Dr. Oswald T. Avery and his co-workers at the Rockefeller Institute for Medical Research in New York set out to discover the transforming principle that changed the harmless bacteria into the virulent type. They grew both types of bacteria in separate dishes and found that an extract from the type S cells could transform R cells into S cells. Laboriously they isolated and purified this transforming substance. It turned out to be a nucleic acid, DNA. This showed that, of all the chemical substances present in cells, it is the DNA that carries heredity. This is the material that Miescher had found in the nucleus in 1869.

Bacteria have become important experimental organisms for modern investigations in heredity because:

1. They have a rapid rate of reproduction (a new generation every 20 minutes).

2. They are present in large numbers — billions of cells can be examined for mutations.

3. Techniques have been discovered for detecting and isolating mutant forms.

4. They carry on sexual reproduction.

5. Bacteria are the host organisms for growth of *bacteriophages* — viruses that develop inside bacteria.

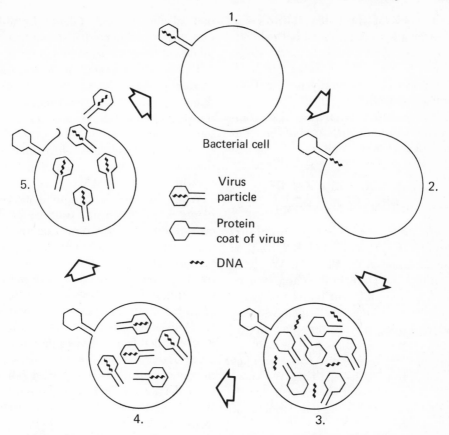

Fig. 20–1. Cycle by which one kind of bacteriophage infects bacterial cells.

Virus Studies

Studies of viruses have shown that here also DNA is the usual hereditary material. One kind of bacteriophage has a tail. The 'phage is composed of DNA and a coat made of protein. The cycle by which this bacteriophage infects bacterial cells and is then released has been discovered by use of the electron microscope and radioisotopes of sulfur and phosphorus. This cycle is shown in Fig. 20–1.

The events of the cycle are as follows:

1. The virus attaches itself to a bacterial cell by its tail. The tail is not used for locomotion but to digest the wall of the bacterial cell.

2. The DNA content of the virus is injected into the cell.

3. The virus DNA "takes over" the metabolism of the cell so that virus DNA is manufactured instead of bacterial DNA. New virus protein coats are also produced.

4. Virus DNA and virus protein coats are assembled to form complete virus particles.

5. The bacterial cell bursts and large numbers of new virus particles are released.

Two hundred new virus particles have been shown to be produced in this manner within an interval of 20 minutes. Each of these new viruses can then invade another bacterial cell. Viruses may differ greatly from the shape shown in Fig. 20–1. Viruses which attack plant cells generally possess RNA instead of DNA.

The chemistry of DNA, and its companion substance RNA, will be considered later in this unit.

THE CONTRIBUTION OF GREGOR MENDEL

We have seen how the nucleus, through the action of its DNA, controls the metabolism of the cell. We have seen that the DNA in the chromosomes is replicated and distributed to daughter cells every time a cell divides so that each resulting cell has the same supply. Furthermore, we know that in fertilization the zygote receives equal supplies of DNA from its two parents. We know that DNA is passed from parent to offspring and controls the development of the offspring.

Gregor Mendel did not know about DNA; yet, he was able to initiate the science of genetics. He performed an eight-year series of experiments on garden peas in the monastery at Brünn, now part of Czechoslovakia. Although he published his results in 1865, little attention was paid to his basic principles until 1900.

Mendel summarized his results in a set of four laws, known as *Mendel's Laws of Heredity*. While it is now known that only one of these laws is universal, all four have led to discoveries that have unlocked the deepest mysteries of the biological control of development.

Reasons for Mendel's Success

Mendel possessed unusual mathe-matical and scientific talents. In addition, his experiments had a number of qualities which brought him success:

1. Use of favorable material. The garden pea is easy to cultivate, produces a number of generations in a reasonably short time, and exists in several easily recognizable varieties. Because the plants are self-pollinating, the offspring usually have the same heredity as the parent.

When Mendel wished to mate two plants that differed in a trait, he would cross-pollinate them. For example, in mating a plant which had the trait of tallness with a plant possessing the contrasting trait of shortness, he removed the anthers from a flower of the short plant before its own pollen was mature. He then dusted onto the stigma of this flower pollen from a flower growing on a tall plant. He also cross-pollinated in the opposite direction by dusting pollen from flowers of short plants onto the stigmas of tall plants. He was careful to prevent any stray pollination.

2. Use of pure strains. Mendel started his experiments with plants whose heredity was pure for a certain trait, rather than a mixture. For example, after the flowers of tall plants self-pollinated themselves, Mendel took the resulting seeds and planted them. These seeds developed into tall plants. The flowers of this next generation of plants self-pollinated themselves and produced seeds which again developed into tall plants.

After carrying on this procedure for four generations, Mendel was reasonably sure that a tall plant's heredity was not mixed with heredity for shortness. He had a strain of pea plant that was *pure* for tallness. Similarly, Mendel obtained plants that were pure for the opposite trait of shortness. He could then perform an experiment to cross these two pure types which contrasted in the trait of height.

3. **Careful choice of traits.** It is not possible to divide people into two well-defined categories, tall and short, because there are so many variations between these two extremes. When there are more intermediates than extremes, the traits are said to *vary continuously*. Most traits in living things vary continuously.

Mendel chose to study traits that *vary discontinuously*. When traits vary discontinuously, all of the individuals of the species fall into a small number of well-defined categories on the basis of the traits. For example, all people have one of four blood types, and there are no intermediates. While a minority of traits vary discontinuously, they are the simplest to study, and a large part of Mendel's success was due to this choice. A "short" pea plant is a low herb, while a "tall" pea plant is a vine that needs a support, and it is always possible to tell one from the other. At first, Mendel dealt with the traits one at a time, which made the problem even simpler.

4. **Use of large numbers.** Instead of studying only a few offspring resulting from a single mating, Mendel made the same kind of cross hundreds of times. In one experiment, there were over 8000 offspring. Mendel thus had large numbers of offspring to study. Mendel kept careful records of all his plants.

5. **Use of mathematics.** Instead of merely noting in a general way the kind of offspring produced, Mendel counted the numbers of different kinds of offspring. He was thus able to calculate the ratios of offspring and to use these ratios in a mathematical manner to determine general principles.

The Idea of Unit Characters

Mendel viewed an organism as a collection of traits, each of which is inherited separately. Mendel concluded that each observable physical trait of the organism is represented by a "factor" (unit character) of some sort that passes to the offspring in the process of fertilization. Through the years, it has been shown that there are indeed discrete units of heredity in the egg and sperm, which are called genes. As previously stated, these have now been identified as portions of the DNA molecule. However, the only reason Mendel found a neat, direct relationship between the unit character (gene) and the trait is that he studied discontinuous traits only, and carefully chosen ones at that.

We still think of the genes as discrete units, but their relationship to the traits they produce is much more complicated than Mendel thought.

The Law of Dominance

When Mendel crossed pure tall pea plants with pure short pea plants, hundreds of seeds resulted. When planted, all (100%) of these seeds developed into tall plants. Similarly, in the six other experiments wherein he crossed plants which differed for a trait, only one of the alternative forms appeared in the next generation. For example, a cross between pure yellow-seeded plants and pure green-seeded plants resulted only in yellow-seeded plants. Accordingly, Mendel formulated the *Law of Dominance: In a cross between two pure contrasting traits only one of these traits appears in the next generation; this trait is the dominant trait and the one that does not appear is the recessive trait.*

In garden peas, the trait of tallness is dominant to shortness, the trait of yellow seeds is dominant to green seeds, and the trait of round seeds is dominant to wrinkled seeds, etc. Other dominant traits found in garden peas are shown in the second vertical column of the table below.

Today, we speak of dominant and recessive genes, not traits, and we know of many cases in which the Law of Dominance does not hold.

The Law of Segregation

When heredity is traced for several generations, the following symbols are used: P = parental generation; F_1 = first filial generation; F_2 = second filial generation. ("Filial" is from the Latin word *filius* for son and refers generally to sons and daughters.) Under the conditions of Mendel's experiments, the dominant traits appeared in the F_1 gen-

eration, the recessive traits did not.

Mendel now allowed the tall plants of the F_1 generation to self-pollinate themselves. The resulting seeds developed into the F_2 generation consisting of 787 tall plants and 277 short plants. This is a ratio of 2.84 tall to 1 short, or approximately 3 tall to 1 short. The results in all seven types of cross are shown in the table below.

In each case, the F_2 generation had a ratio of approximately 3 : 1. Mendel's results with length of stem may be summarized as follows:

P pure tall $\times$ pure short

F_1 100% tall

F_2 tall, tall, tall, short
     (75% tall : 25% short)

How could these results be explained? Mendel reasoned that since the recessive trait of shortness reap-

MENDEL'S RESULTS IN CROSSING GARDEN PEAS FOR TWO GENERATIONS

Parental Generation	F_1 Plants	F_1 Self-Pollination	F_2 Plants	F_2 Ratio
tall $\times$ short stems	all tall stems	tall $\times$ tall	787 tall stems 277 short stems	2.84 : 1
yellow $\times$ green seeds	all yellow seeds	yellow $\times$ yellow	6022 yellow seeds 2001 green seeds	3.01 : 1
round $\times$ wrinkled seeds	all round seeds	round $\times$ round	5474 round seeds 1850 wrinkled seeds	2.96 : 1
colored $\times$ white seed coats	all colored seed coats	colored $\times$ colored	705 colored seed coats 224 white seed coats	3.15 : 1
axial $\times$ terminal flowers	all axial flowers	axial $\times$ axial	651 axial flowers 207 terminal flowers	3.14 : 1
inflated $\times$ wrinkled pods	all inflated pods	inflated $\times$ inflated	882 inflated pods 229 wrinkled pods	2.95 : 1
green $\times$ yellow pods	all green pods	green $\times$ green	428 green pods 152 yellow pods	2.82 : 1

peared in the F_2 generation, some factor producing shortness must have been present, not lost, in the F_1 generation. However, since the F_1 plants were tall, they must have been carrying the heredity for tallness as well as for shortness. Using T as the symbol for tallness, and t as the symbol for shortness, the heredity of the F_1 plants could be represented as Tt.

Today, the symbol Tt implies that each cell of the organism has both kinds of genes, the dominant gene for tallness (T) and the recessive gene for shortness (t). In peas, the gene for yellow seeds is dominant over the gene for green seeds. Mendel reasoned that every plant had a pair of factors (genes) for each trait.

The alternative genes for a trait, such as the gene for tallness and the gene for shortness, are called *alleles*. An allele is dominant when its full effect is produced by a single "dose"; thus TT plants and Tt plants are both tall. A recessive gene's effects are found only if the gene is present in a "double dose"; only tt plants are short.

Recall that the chromosomes are present within the nuclei as homologous pairs. The genes on the chromosomes have been likened to a string of beads, as shown in Fig. 20–2. When F_1 organisms produce gametes by meiosis, each chromosome of a pair of homologous chromosomes separates. They carry with them the members of a pair of alleles, which also separate. Each gene of a pair of alleles thus passes to a different gamete. (Fig. 20-2.)

When Mendel removed the anthers from some of his garden pea plants, they could produce female gametes only, and acted in the capacity of a female parent. Bear this in mind as you examine Fig. 20–3, where Mendel's two

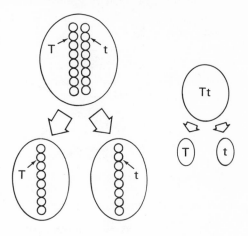

Fig. 20–2. The diagram at the left shows the separation of homologous chromosomes carrying alleles for tallness and shortness. The one at the right shows the symbols for only these genes — all other genes in the cell are disregarded. (In this simplified diagram, the chromatids are not considered.)

generations of crosses of plants differing in stem length are represented.

If 4000 F_2 zygotes are produced in a set of crosses such as that shown in Fig. 20–3, the chances are that there will be approximately 1000 of each of the following: TT, Tt, tT, and tt. Each of the zygotes then develops into a new plant by repeated mitotic divisions. Those plants whose cells carry the dominant gene T for tallness grow tall, those not carrying T develop as short plants. The result is a ratio of 3 tall : 1 short in the F_2 generation. This ratio may also be expressed as 75% tall : 25% short.

From his results in the F_2 generation, Mendel formulated the *Law of Segregation.* Mendel's law can be stated in modern terms as follows: *Genes occur in pairs which separate when gametes are formed, and only one gene of each pair goes to a gamete.* It is a tribute to

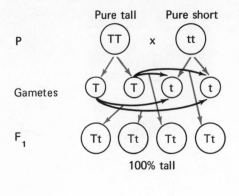

Pure tall   Pure short

P   TT   x   tt

Gametes   T   T   t   t

F₁   Tt   Tt   Tt   Tt

100% tall

T = gene for tallness (dominant)

t = gene for shortness (recessive)

Let us consider the individual at the left to be the "male parent" and the one at the right to be the "female parent."

Since all the sperm carry the gene T, and all the eggs carry t, the only possibility for the zygote is Tt. This zygote develops into a tall plant.

The male Tt plants may produce thousands of sperm; half of the sperm carry the gene T, and half carry the gene t. Similarly, half of the eggs carry T and half carry t.

The sperm which carry the gene T have an equal chance of fertilizing the T eggs and the t eggs. Similarly, the t sperm have an equal chance of fertilizing the T eggs and the t eggs.

Consequently, there are equal chances for the formation of each of these kinds of zygote: TT, Tt, tT, and tt.

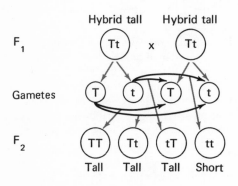

Hybrid tall   Hybrid tall

F₁   Tt   x   Tt

Gametes   T   t   T   t

F₂   TT   Tt   tT   tt

Tall   Tall   Tall   Short

Fig. 20–3. The chances for the production of various kinds of zygotes.

Mendel's genius that he formulated this law without knowing about chromosomes. It was not until 1902 that separation of chromosomes during gamete formation was observed.

This is the only one of Mendel's laws which is still considered to be completely universal in all living things, although it is occasionally violated in abnormal meiosis.

Pure and Hybrid Organisms

An organism is said to be *homozygous* (pure) for a gene if both genes of a pair of alleles are alike. Every gamete produced by such a homozygous organism carries an identical gene.

An organism is said to be *hybrid* (or *heterozygous*) for a gene if the genes of a pair of alleles are unlike. As a result of meiosis, half of its gametes carry one type of gene and half carry the other. The symbols Tt represent the genes in cells of a hybrid tall pea plant. The symbols TT represent the genes in the cells of a pure tall pea plant. Both plants are equally tall; the differing gene make-ups of the plants cannot be distinguished by their appearance. The symbols tt represent the genes in cells of a pure short pea plant.

Genotype and Phenotype

The *genotype* refers to the genetic makeup of an individual and can be ascertained by knowing its genes. The *phenotype* refers to the observable appearance of an individual. The difference in these two concepts is shown in the table below.

Genes	Genotype	Phenotype
TT	pure tall	tall
Tt	hybrid tall	tall
tt	pure short	short

Note that a pea plant whose phenotype is short must be *pure* short. If it were a hybrid, it would possess a dominant gene for tallness and consequently, would not be short.

In the F_2 generation shown in Fig. 20–3, the phenotype ratio is 3 tall : 1 short. The genotype ratio is 1 pure tall : 2 hybrid tall : 1 pure short.

Sample Problem 1

What is the expected ratio of offspring produced in crosses between hybrid tall pea plants and pure tall pea plants? (See Fig. 20–4 for the solution.)

The Punnett Square or Checkerboard

Fig. 20–4 uses lines drawn from the gametes to the F_1 generation to show the fertilizations that are possible. It is more convenient to use the *Punnett Square*, or *checkerboard method* (Fig. 20–5) to show the possible fertilizations. The method of using this checkerboard is now briefly described.

There are two kinds of sperms, depending on which gene they carry. Indicate the two kinds by the letter designating the gene at the top of the square. Similarly, indicate the two kinds of eggs at the side of the square.

Now place the letters which are at the top of the square in each of the boxes below these letters. In the same way, place the letters which are at the side of the square in each of the boxes to the right. The pairs of letters in each of the four boxes now show the various kinds of zygotes which are formed.

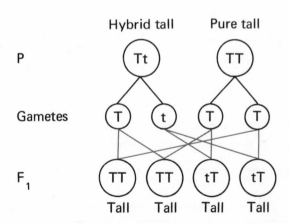

Key:
T = gene for tallness
t = gene for shortness

Answer
F_1 phenotype ratio:
100% tall

Fig. 20–4. Solution to Sample Problem 1. Unless the genotype ratio is specifically requested in the problem, only the phenotype ratio is given as the answer.

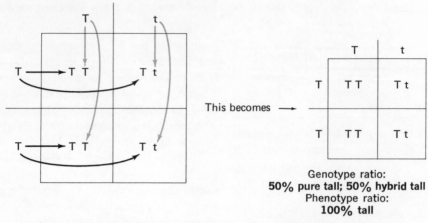

Genotype ratio:
50% pure tall; 50% hybrid tall
Phenotype ratio:
100% tall

Fig. 20–5. The checkerboard method. For convenience both ratios are given.

Sample Problem 2

What ratio can be expected in the offspring of a cross between hybrid tall pea plants and short pea plants?

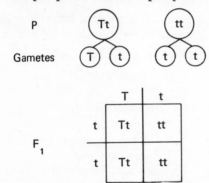

F_1 phenotype ratio: **50% tall, 50% short**

Fig. 20–6. Solution to Sample Problem 2.

Segregation Ratios

Note that the three problems which have been solved in Figs. 20–4, 20–5, and 20–6 include all of the possibilities involved in crossing hybrid plants. The phenotype ratios which were determined above are shown in Fig. 20–7.

Sample Problem 3

Mendel found that pure wrinkled-seeded plants crossed with pure round-seeded plants resulted in 100% round-seeded plants in the first generation. What ratio can be expected in crosses between pure wrinkled-seeded plants and hybrid round?

This problem could be solved by inspection merely by noting that it represents a cross between a hybrid and a

Tt × Tt	hybrid × hybrid ⟶	75% dominant : 25% recessive
Tt × TT	hybrid × pure dominant ⟶	100% dominant
Tt × tt	hybrid × pure recessive ⟶	50% dominant : 50% recessive

In addition, the following ratios are apparent:

TT × TT	pure dominant × pure dominant ⟶	100% dominant
tt × tt	pure recessive × pure recessive ⟶	100% recessive
TT × tt	pure dominant × pure recessive ⟶	100% dominant

Fig. 20–7. Summary of ratios obtained in various crosses. The above is a useful tool for solving problems quickly by inspection.

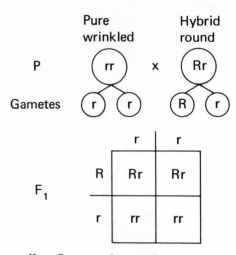

P

Gametes

F$_1$

Key: R = gene for round
    r = gene for wrinkled
Answer: 50% round, 50% wrinkled

Fig. 20–8. Solution to Sample Problem 3.

pure recessive. The ratios produced are the same as for a cross between *Tt* and *tt* as shown in Fig. 20–7.

The Need for Large Numbers

In tossing a coin, there is a 50% chance that a head will turn up. However, it is quite possible to toss a coin 10 times and obtain 8 heads and 2 tails. As the number of trials is increased, the ratios obtained are closer to the ratios predicted by the laws of chance (or the laws of probability). Which sperm and egg will unite at any mating is also a matter of chance. Consequently, the expected Mendelian ratios are approximated more closely as large numbers of offspring are produced.

Sample Problem 4

Black coat color is dominant over white coat in guinea pigs. A cross between two black guinea pigs resulted in four offspring, all of which were black. Can both parents be hybrids?

SOLUTION

The results of mating black parents can be represented by three possibilities (A, B, and C):

(A) *Bb* × *Bb* ⟶ 75% black: 25% white
(B) *BB* × *BB* ⟶ 100% black
(C) *Bb* × *BB* ⟶ 100% black

The answer is: Yes, both parents could be hybrids, as in possibility A. If the next few offspring were white, the 75 : 25 ratio would become apparent. Four offspring constitute too small a number to rule out the possibility of mating A.

The Test Cross (Back Cross)

A boy wishes to determine whether his black guinea pig is pure black or hybrid black. How can he find out?

Recall that in guinea pigs the color black is dominant and the color white is recessive. If the boy selects a white guinea pig for mating with his black one, the genotype of at least one of the parents is known — the white one is pure white. The boy can use this pure recessive to make a test. The two possible crosses are shown in Fig. 20–9.

The boy should permit a number of matings between white guinea pigs and the unknown black one which he owns. As soon as a white offspring is obtained, he knows that the unknown guinea pig is a hybrid, for there is no way that a pure black can have a white offspring. However, if no white offspring are obtained, he can never be certain, for a hybrid black could produce all black offspring just by chance. However, if he obtained a *large number* of black offspring and no white ones, he could say that the black parent is *most likely* pure. If you tossed a coin 20 times and it came out heads each time, you would surely wonder whether the coin had heads on both sides.

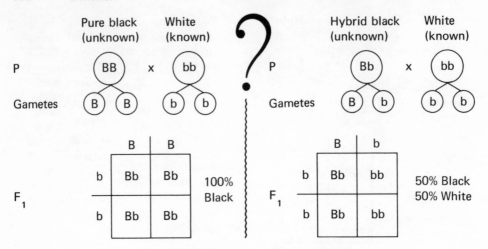

Fig. 20–9. Test cross with guinea pigs.

A *test cross,* or *back cross,* is a method for determining the genotype of an organism by mating it with a pure recessive.

The Law of Independent Assortment

A trait such as height, which has alternative aspects such as tallness and shortness, is represented by a single pair of genes (alleles). Until now, we have considered only a single such trait (pair of alleles) at a time. For example, we have considered the trait of height in pea plants and the trait of color of coat in guinea pigs.

Consider, however, a cross between pea plants that differ in two traits, such as *height* and *color of seeds.* For example, cross a pure tall plant producing yellow seeds with a pure short plant which produces green seeds. The genes for height and for color of seeds are carried on different chromosomes, as indicated in Fig. 20–10.

As shown in the diagram, when gametes are produced by a plant which is pure for both tallness and for yellow color of seeds, each gamete receives a gene for tallness and a gene for yellow seed color.

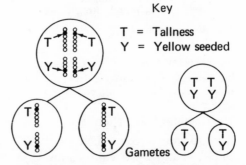

Fig. 20–10. How genes for height and seed color are carried on different chromosomes.

Now we are ready to show the cross. Accordingly, a cross between a pure tall, yellow-seeded plant and a pure short, green-seeded plant is diagrammed in Fig. 20–11.

The F_1 plants are hybrid for both height and for color. Such a plant which is hybrid for two traits is a *dihybrid.* A plant which is hybrid for only one trait is a *monohybrid.*

Consider now the effect of self-pollination in these F_1 dihybrids, as shown in Fig. 20–12.

When the homologous chromosomes line up and then separate during meiosis, two kinds of gametes formed are *TY* and *ty.* However, if the chromosomes

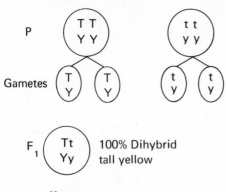

P

Gametes

F₁ 100% Dihybrid
tall yellow

Key:
T = gene for tallness
t = gene for shortness
Y = gene for yellow seeds
y = gene for green seeds

Fig. 20–11. Pure tall yellow ×
pure short green.

line up differently, there is an equal chance for the formation of the gametes *Ty* and *tY*. Consequently, a dihybrid may produce equal numbers of *four* different kinds of gametes as shown in the diagram.

To represent the possible fertilizations that can result from the gametes produced by a self-pollinating dihybrid, a Punnett Square (checkerboard) can be used. However, the Punnett Square must be enlarged to accommodate four different kinds of sperm and four different kinds of eggs, as shown in Fig. 20–13.

Four different kinds of phenotypes are produced. The relative numbers of each kind of phenotype are:

12 tall { tall yellow — 9 ⟍
tall green — 3 ⟍ 12 yellow
4 short { short yellow — 3 ⟋
short green — 1 ⟋ 4 green

A cross between two dihybrids results in a 9 : 3 : 3 : 1 ratio.

Of particular interest, however, is the fact that there are 12 tall plants to 4 short plants. The 3 : 1 ratio of tall to short appears in this dihybrid cross! Similarly, there are 12 yellow-seeded

Dihybrid tall yellow
(stamen)   ×   Dihybrid tall yellow
                                      (pistil)

F₁

Gametes

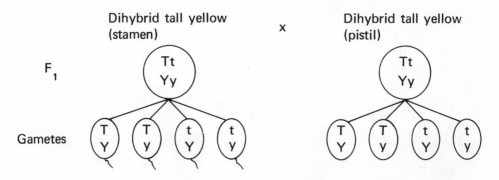

Fig. 20–12. Gametes produced in a self-pollinating dihybrid. Four kinds of gametes are produced.

	T Y	T y	t Y	t y
T Y	T T Y Y	T T y Y	t T Y Y	t T y Y
T y	T T Y y	T T y y	t T Y y	t T y y
t Y	T t Y Y	T t y Y	t t Y Y	t t y Y
t y	T t Y y	T t y y	t t Y y	t t y y

Fig. 20–13. Punnett Square for dihybrid cross.

plants to 4 green-seeded ones — again the 3 : 1 ratio! Thus, each trait appears in the same ratio as if it were alone in a monohybrid cross.

His results with dihybrid crosses led Mendel to formulate his third law, the *Law of Independent Assortment: When dihybrid plants are crossed, the factor for each trait is distributed independently of the factors for all other traits.* Alleles which are present on different chromosomes are distributed independently because the chromosomes carrying these genes segregate independently during meiosis. This law applies only if the alleles involved are on different chromosomes. It is a remarkable coincidence that each of the 7 alleles studied by Mendel was on a different

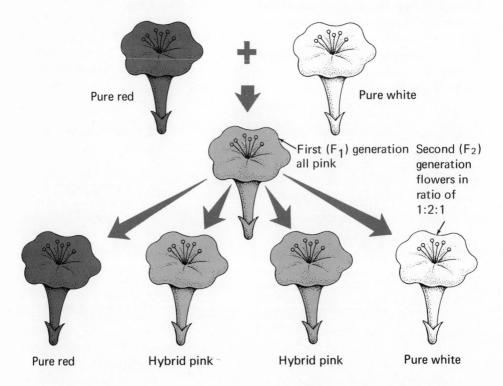

Pure red　　　　　　　Pure white

First (F₁) generation all pink

Second (F₂) generation flowers in ratio of 1:2:1

Pure red　　Hybrid pink　　Hybrid pink　　Pure white

Fig. 20–14. Incomplete dominance is illustrated by Japanese four o'clock flowers.

one of the 7 pairs of chromosomes in the pea!

Incomplete Dominance

In the Japanese four o'clock plant, red-flowered plants crossed with red-flowered plants produce red-flowered offspring. Similarly, white-flowered plants crossed with white-flowered plants produce white-flowered offspring. What is the result of crossing red with white? (One cannot tell which gene is dominant until after the cross is performed.) When this experiment is performed, surprisingly, the offspring are all pink! (See F_1 generation in Fig. 20–14 and Fig. 20–15.) Obviously, this cross represents a different type of heredity — one wherein the Law of Dominance does not operate. This type of heredity in which the hybrid is noticeably different from both purebred parent forms is known as *incomplete dominance* (also called blending inheritance).

Apparently, it takes two "doses" of the gene for red to produce the chemical changes that result in a lot of red pigment in the petals. If there is only one such gene for red, the amount of pigment produced is less and the petals are pink.

To analyze this cross, a key must first be set up as follows:

> **Key:** R — gene for red
> W — gene for white
> RW = pink (phenotype)

Since neither red nor white is dominant, a capital letter is used for each gene. Pink results from a combination of the action of the genes for red and for white. Now, using this key, we can analyze the cross as in Fig. 20–15.

Another well-known example of incomplete dominance is feather color in the Andalusian fowl. These highly

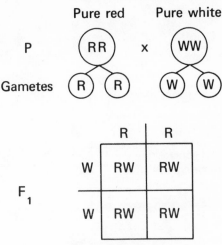

Phenotype: **100% pink**

Fig. 20–15. Red and white Japanese four o'clocks.

Sample Problem 5

What ratios may be expected in the cross between pink-flowered Japanese four o'clocks?

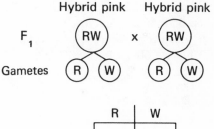

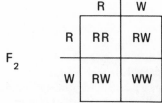

Phenotype ratio:
25% red, 50% pink, 25% white

Fig. 20–16. Solution to Sample Problem 5. (Note: The genotype ratio is the same as the phenotype ratio: 25% pure red : 50% hybrid pink : 25% pure white.)

prized roosters and chickens from the Andalusian section of Spain may be

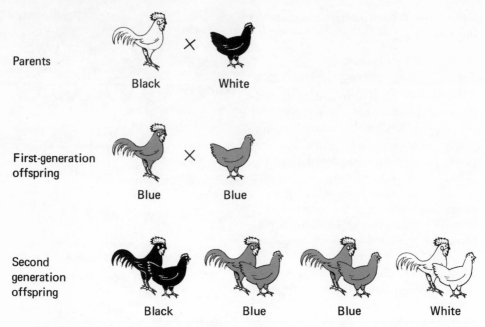

Fig. 20–17. The Andalusian fowl shows incomplete dominance of feather color.

pure black, pure white, or a speckled "blend" of black and white. This speckled blend appears *blue* from a distance. When blue Andalusian fowls are mated, their offspring occur in the ratio of 25% black : 50% blue : 25% white.

Another example of incomplete dominance is found in Shorthorn cattle. A homozygous red animal and a homozygous white animal produce offspring whose coat color has the appearance known as *roan*. When two roan animals are mated, the expected ratio among the offspring is 1 red : 2 roan : 1 white.

PROBLEMS IN GENETICS

1. Show by means of keyed and labeled diagrams the ratio of offspring expected in each of the following: (A) A cross between hybrid tall garden pea plants and pure tall garden pea plants. (B) A cross between hybrid tall and pure short. (C) A cross between hybrid tall and hybrid tall.

2. Summarize the ratios obtained in problem 1 by extending each of the following:

   $Tt \times Tt$ ⟶ $TT \times TT$ ⟶

   $Tt \times TT$ ⟶ $TT \times tt$ ⟶

   $Tt \times tt$ ⟶ $tt \times tt$ ⟶

3. Black coat color in guinea pigs is dominant over white coat color. (A) By means of keyed and labeled diagrams, show the results of crossing a hybrid black with a pure white. (B) Show the results of crossing a pure black and a hybrid black.

4. In pigs, the white color W is dominant; the black color w is recessive. Using diagrams, show the expected results of the following crosses: (A) A pure white pig is mated with a black pig. (B) A hybrid pig is mated with a hybrid pig.

5. What were the parents of the following groups of offspring from guinea pig crosses? Give all possibilities, assuming that the laws of chance operate:
   (A) One-half of the offspring are black.
   (B) Three-fourths of the offspring are black.
   (C) One-half of the offspring are hybrid black.
   (D) All of the offspring are black.
   (E) All of the offspring are white.
   (F) One-half of the offspring are white.
   (G) One-fourth of the offspring are white.

6. Explain with the aid of keyed and labeled diagrams how you would determine whether a squirrel is pure gray or hybrid gray. (Gray is dominant over black.)

7. After several matings of tan-colored birds, the following offspring resulted: 23 white, 26 brown, 53 tan. By means of keyed and labeled diagrams, show each of the following: (A) A cross that would produce 50% of all offspring, brown. (B) A cross that would produce 100% of all offspring, tan.

8. There are three types of radishes: round, oval, and long. A breeder made three crosses, and obtained the results indicated below. Using keyed and labeled diagrams, show the crosses in each.
   (A) *First cross:* Long with round gave 342 oval.
   (B) *Second cross:* Long with oval resulted in 48 long and 52 oval.
   (C) *Third cross:* Oval with round resulted in 141 oval and 137 round.

9. A black guinea pig was crossed with a white guinea pig. All of the individuals in the F_1 generation were black. These individuals were then crossed among themselves, resulting in 30 black guinea pigs and 10 white guinea pigs.
   (A) How many of the black pigs in the F_2 generation would be (1) pure dominant? (2) hybrid?
   (B) How many of the white pigs in the F_2 generation would be (1) pure recessive? (2) hybrid?

10. What ratio may be expected in a cross between a red-flowered Japanese four o'clock plant and a pink one?

11. What ratio may be expected in a cross between a white-flowered Japanese four o'clock and a pink one?

B   LINKAGE

According to the *Law of Independent Assortment*, each trait is inherited independently of all other traits. However, geneticists began to find exceptions to this rule. Some traits seemed to "stick together" with other traits. Traits that are inherited together are said to be *linked*. In the fruit fly (*Drosophila*), four large groups of traits were found to be linked. *Drosophila* has four pairs of chromosomes, corresponding to the number of linkage groups. The concept, therefore, emerged that when traits are linked it is because the genes which determine these traits are present on the same chromosome.

As previously stated, Mendel was fortunate in that the traits which he studied were on different chromosomes. He did not have to face the complexities of linkage.

Experiments with *Drosophila*

Great advances in genetics in the early 1900's were made by Thomas Hunt Morgan and his associates at Columbia University when they introduced *Drosophila melanogaster* (the fruit fly) as an experimental organism. This is a tiny fly which is often found around grapes and rotting bananas. (See Fig. 20–18.) *Drosophila* possesses the following characteristics which made it an ideal organism for experimental studies in genetics:

1. Short life cycle. Its entire life cycle is only 14 days. Thus, many generations can be bred within a short period of time.

2. Many offspring. One pair of parents may produce over 300 offspring. Thus, the large numbers required for accurate ratios are easily obtained.

3. Small size. A half-pint milk bottle can accommodate several hundred individuals.

4. Modest food requirements. Corn meal and molasses, with a little agar added, is a commonly used culture medium.

5. Few chromosomes. *Drosophila* has only four pairs of chromosomes. These

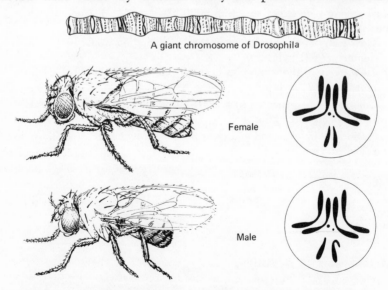

A giant chromosome of Drosophila

Female

Male

Fig. 20–18. Female and male *Drosophila* and their chromosomes.

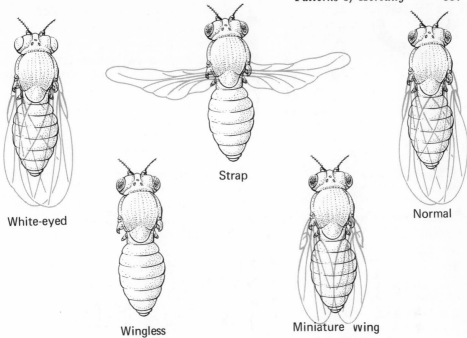

White-eyed

Strap

Normal

Wingless

Miniature wing

Fig. 20–19. Mutant forms of *Drosophila* compared with normal *Drosophila*.

are recognizable by their size and shape (Fig. 20–18).

6. Giant chromosomes. In the salivary glands of one stage of the *Drosophila* larvae were found greatly enlarged chromosomes — the *giant chromosomes* of *Drosophila* (Fig. 20–18). Light staining and dark staining bands on these chromosomes permitted the study of fine detail. Recent studies of giant chromosomes of *Drosophila* and of other organisms have shown that the DNA (genetic material) is located at the dark bands. At certain times, when swellings or "puffs" are formed, the chromosome is actively producing RNA. It is thought that these giant insect chromosomes contain as many as 1024 DNA fibers.

7. Mutant forms. Sometimes new traits appeared in *Drosophila* cultures. The new trait, if passed on to succeeding generations by heredity, is called a *mutation*. Morgan was able to locate the genes for many mutations on gene maps which he prepared of the chromosomes of *Drosophila*. Some of the mutant forms of *Drosophila* are shown in Fig. 20–19.

Although *Drosophila* remains an active subject for research, it is being supplemented by bacteria and by molds (such as *Neurospora*). These organisms are still smaller than *Drosophila*; they reproduce more rapidly, and they permit the study of the biochemical reactions caused by genes.

CROSSING–OVER

Linkage is an exception to independent assortment; crossing-over is an exception to linkage. *Crossing-over* is the breaking of linkage groups and the exchange of these linkage groups between

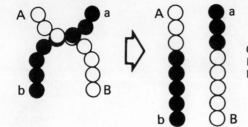

Gene A, which was linked to gene B, is now linked to gene b; and gene a, which was linked to gene b, is now linked to gene B.

Fig. 20–20. Crossing-over. A chromatid is represented as a chain of genes, like beads on a string.

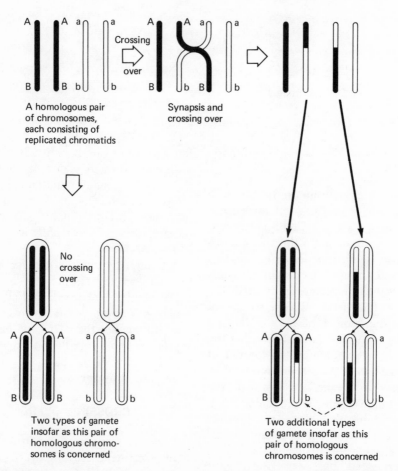

A homologous pair of chromosomes, each consisting of replicated chromatids

Synapsis and crossing over

No crossing over

Two types of gamete insofar as this pair of homologous chromosomes is concerned

Two additional types of gamete insofar as this pair of homologous chromosomes is concerned

Fig. 20–21. Crossing-over during meiosis. A single pair of homologous chromosomes is illustrated. At the left, where crossing-over does not occur, no new combination of genes results. At the right, with crossing-over, two new combinations of linkage groups result.

homologous chromosomes. This occurs when homologous chromosomes pair during synapsis and exchange fragments of chromatids. Crossing-over is illustrated in Fig. 20–20 and Fig. 20–21. It was also described in Chapter 18 (page 289).

Crossing-over does not occur in organisms that reproduce only by asexual means because these organisms do not carry on meiosis. It does, of course, occur in organisms that reproduce by sexual means. Crossing-over helps to increase the number of different kinds of gametes that take part in sexual reproduction (Fig. 20–21), thus increasing the variety of zygotes produced.

The variability of offspring produced during sexual reproduction is increased by:

1. Independent assortment of chromosomes.
2. Crossing-over of portions of chromosomes.
3. Chance mating of different kinds of sperm and eggs.

This variability of offspring produced during sexual reproduction permits a species to adapt to changing environmental conditions and is a major factor in evolution.

THEORY OF THE GENE

Thomas Hunt Morgan and his associates were largely responsible for the theory of the gene. By the late 1930's, this theory included the following concepts:

1. Genes in a chromosome are arranged in a row, like a string of beads.
2. Traits which are inherited together are controlled by genes which are linked on the same chromosome.
3. Linkage groups may be broken by crossing-over.

4. The position of genes on a chromosome may be determined and may be indicated on gene maps.

For his contributions to genetics Morgan was awarded the Nobel Prize in 1933.

SEX DETERMINATION

In the human being, the diploid chromosome number is 46; that is, there are 23 pairs of homologous chromosomes. Of these, 22 pairs do *not* determine sex (and the members of each pair are alike). These pairs are called *autosomes*. In addition, there is one pair of *sex chromosomes*. In the human female, each member of a pair of sex chromosomes resembles its homologue. These two chromosomes are designated XX. Accordingly, every cell in the body of the human female has 22 pairs of autosomes and two X chromosomes. The human male also has a pair of sex chromosomes in each body cell. However, only one member of the pair is an X chromosome. Its homologue, which is much smaller in size, is called the Y chromosome.

Since the X chromosome occurs in both the male and female and the Y chromosome occurs only in the male, it is evident that genes on the Y chromosome (in man) determine the sex of the individual.

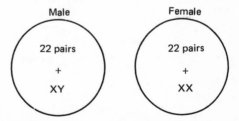

Male   Female

22 pairs   22 pairs

+   +

XY   XX

The ratio of offspring produced in matings may be determined by use of the Punnett Square (Fig. 20–22).

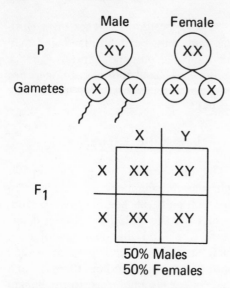

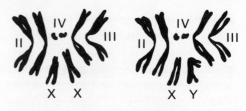

Fig. 20–24. The chromosomes of male and female *Drosophila*. The female has two X chromosomes in addition to its autosomes. The male has an X and a Y. (In the diagram, each chromosome is shown as composed of two chromatids.)

50% Males
50% Females

Fig. 20–22. Male × female. (Note what happens to the XY pair when gametes are formed.)

At any one mating, millions of sperm enter the oviducts. Half of the sperm carry the X chromosome; half carry the Y chromosome. All the ova carry the X chromosome. If an X-carrying sperm fertilizes an ovum, the fertilized egg develops into a female; if a Y-carrying sperm fertilizes the ovum, a male results. The sex of an individual is thus determined at the time of fertilization by the X or Y chromosome donated by the male. (See Fig. 20–23.)

In *Drosophila*, which has only 4 pairs of chromosomes, the sex chromosomes are relatively easy to identify (Fig. 20–24).

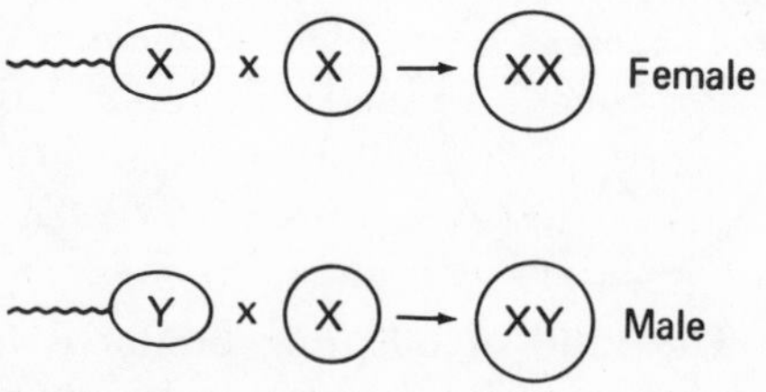

Fig. 20–23. Sex determination.

As in all cases where chance operates, the 50:50 ratio of males to females appears in large numbers of offspring. The statistics for births in the United States are very close to the ideal ratio (106 boys: 100 girls). However, when prenatal (before birth) deaths are taken into consideration, the ratio of males to females produced at conception (fertilization) is calculated to be 114 : 100. This ratio would seem to indicate an advantage for the Y-carrying sperm, and raises the possibility that factors other than chance might be operating.

The symbol ♂, representing Herme's sword and arrow, is used to designate males, and the symbol ♀, representing Aphrodite's looking glass, is used to designate females.

SEX LINKAGE

We find that some traits such as hemophilia (bleeder's disease) and red-green color blindness (inability to distinguish between the two colors) are found more frequently in one sex than in the other. Such traits are examples of *sex-linked inheritance*.

Hemophilia

Hemophilia is a hereditary condition in which the blood clots very slowly.

Prolonged bleeding can result from the slightest scratch. This disease was common in several royal families of Europe, one of which traces its lineage back to Queen Victoria of England. Today, it is found in non-royalty as well. Hemophilia is found more frequently in males than in females and the gene for the disease is carried on the sex chromosomes.

The X chromosome may be represented as a bar and the Y chromosome as a shorter bar with a hook to identify it, as in Fig. 20–25. The small white circle represents the normal gene which causes the blood to clot properly. Note that this gene is present *in that portion of the X chromosome which is absent in the shorter Y chromosome.* Sex-linked inheritance is controlled by genes which are present in that part of the X chromosome which is missing in the Y chromosome. Only one normal gene is required in order to produce the chemicals that cause the blood to clot properly. The individual represented in the diagram is a normal male.

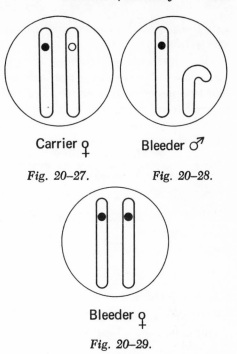

Carrier ♀   Bleeder ♂

Fig. 20–27.   *Fig. 20–28.*

Bleeder ♀

Fig. 20–29.

The black circle in Fig. 20–27, represents a defective gene which fails to lead to the production of chemicals for normal blood clotting. Because only one normal gene is needed for normal blood clotting, this female does not have hemophilia. However, because she can transmit the defective gene to her offspring, she is a *carrier*.

The male shown in Fig. 20–28 has a defective gene on the X chromosome. The Y chromosome is short and lacks the gene entirely. Not having even a single normal gene, this male is a *bleeder*, or *hemophiliac*. Note that in the male the presence of a single defective gene results in a bleeder.

The female shown in Fig. 20-29 has defective genes on both of the X chromosomes, and is a *bleeder*. Note that, in the female, two defective genes are necessary to result in a bleeder.

A single defective gene in the male results in hemophilia, whereas two

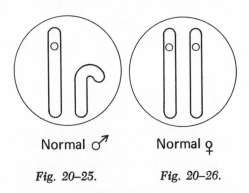

Normal ♂   Normal ♀

Fig. 20–25.   *Fig. 20–26.*

The normal female has normal genes present on both of the X chromosomes. (See Fig. 20–26.)

Compare the genes of the female in Fig. 20–27 with the normal female shown in Fig. 20–26.

genes are needed to cause this condition in the female.

The chromosome drawings shown above are awkward for use in solving genetics problems. Instead we shall use the following symbols:

X = X chromosome carrying the normal
     gene for blood clotting
X̶ = X chromosome carrying the defective
     gene for "bleeder"
Y = Y chromosome

Thus,

XX =   normal ♀   X̶X = carrier ♀

XY =   normal ♂   X̶X̶ = bleeder ♀

X̶Y = bleeder ♂

Sample Problem 6

What ratios can be expected in crosses between a bleeder male and a normal female?

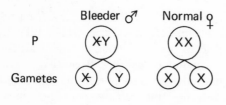

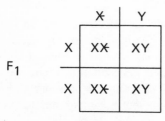

Answer
Of the males: **100% normal**
Of the females: **100% carriers**

Fig. 20–30. Solution to Sample Problem 6. (In cases of sex-linked inheritance, it is useful to give the ratios for each sex separately.)

The results of Sample Problem 6 show that all (100%) of the daughters produced in such a cross are carriers.

If these parents have five daughters, and a nonhemophiliac young man desires to marry the most charming of these daughters, he knows that she *must* be a carrier. What ratios could this young couple expect in their offspring?

Sample Problem 7

What ratios can be expected in a mating of a normal male and a carrier female?

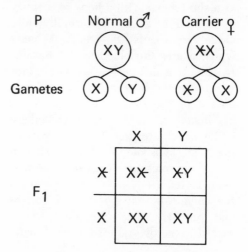

Answer
Of the males: **50% bleeders, 50% normal**
Of the females: **50% carriers, 50% normal**

Fig. 20–31. Solution to Sample Problem 7.

The results of Sample Problems 6 and 7 show that a bleeder male transmits this trait, through his daughters, to one-half of the sons of his daughters (theoretical ratio). He makes each daughter a carrier.

With care, males with hemophilia can live fairly long lives. It is rare for hemophiliac females to be produced because in relation to the entire population the chances are against a hemophiliac male mating with a carrier female.

Recently, it has been shown that the bleeder gene fails to cause the formation of certain blood proteins that are needed for clotting. Today, some hemorrhages in hemophiliacs can be controlled by administering needed blood proteins.

Other Sex-Linked Traits

White eyes in *Drosophila*. Normal *Drosophila* have red eyes. The condition of white eyes is linked and occurs more frequently in males than in females. White eyes in *Drosophila*, discovered by T.H. Morgan, was the first analysis made of sex linked inheritance.

Color blindness in man. A type of color blindness known as red-green color blindness is sex linked. This explains why color blindness is relatively rare in females.

Sample Problem 8

What heredity must the parents have in order to produce a color-blind daughter? (Note the daughter must have two defective chromosomes. Let ✗ symbolize a defective chromosome.)

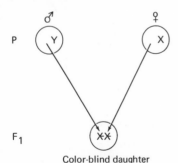

Color-blind daughter

Answer
The father must be **color blind**. The mother must be either a **carrier** or **color blind**.

Fig. 20–32. Solution to Sample Problem 8. One of the daughter's ✗ chromosomes must come from the father and one from the mother.

MULTIPLE ALLELES

We have already learned about gene pairs (called alleles). If a gene has mutated several times in the past, several varieties of this allele may exist among individuals of the population. *Multiple alleles* are several varieties of genes for a given trait, any of which may occur at the same locus (position) on a chromosome.

Blood Types

The inheritance of the major blood types in man is an example of multiple alleles. (See page 134 for a description of blood types.) On the surface of the red blood cells, there may be present two types of proteins, called *antigens:* antigen A and antigen B.

Blood type O – neither antigen is present on the red blood cells.

Blood type A – antigen A is present on the red blood cells

Blood type B – antigen B is present on the red blood cells

Blood type AB – Both antigen A and antigen B are present on the red blood cells,

Allele I^A results in the production of antigen A and allele I^B produces antigen B. When both of these genes are present as $I^A I^B$, each produces its own antigen so that the effect is a "blend," blood type AB (note that the *effect* is a blend – the genes themselves do not blend). A third allele, i, does not produce either antigen. For simplicity, the three genes, I^A, I^B and i may be designated as A, B, and O, respectively. Gene A and gene B together have the effect of blending (incomplete dominance), and each of these genes is dominant to gene O. Thus, $A = B > O$.

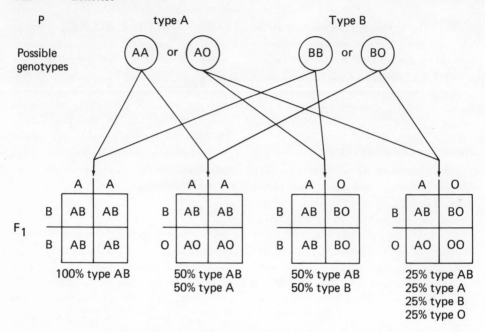

Fig. 20–33. Possible crosses between type A and type B individuals.

A cross between a type A individual and a type B individual might result in several kinds of offspring, depending upon the genotype of the parents. (See Fig. 20–33.)

The Problem of the "Switched Baby"

Occasionally, parents of a newborn baby believe that an error was made in the hospital and that the infant which they brought home is not their own. This belief might be deepened when blood tests reveal that the father is blood type A, the mother is blood type B, and the baby is blood type O. The parents could readily understand how the baby might be type A (like the father), or type B (like the mother), or even type AB (a "blend" of the two parents). But something entirely different!

Sample Problem 9

Could type A and type B parents have a type O offspring?

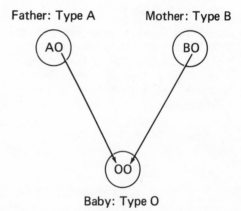

Answer
Yes, if the father is AO and the mother is BO.

Fig. 20–34. Solution to Sample Problem 9. Note that the baby with genotype OO inherits one O from the father and one O from the mother.

Paternity Suits

Courts of law sometimes have to rule on cases such as the following: A mother who is blood type A has a baby who is blood type AB. She accuses a man who is blood type O of being the child's father and asks for support of the child. Could this man be the father?

Sample Problem 10

If the mother is blood type A, could a man of blood type O be the father of a baby with blood type AB?

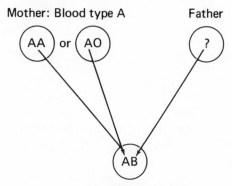

Mother: Blood type A   Father

Baby: Blood type AB

Answer

The baby obtains the gene A from its mother. The baby's B gene comes from the father. The father could be any genotype containing B: BB, BO, or AB. **He could not be type O.**

Fig. 20–35. Solution to Sample Problem 10.

The Rh Factor

Subsequent to the discovery of the A-B-O blood types, another protein antigen was found on the red blood cells of man. This protein was called the *Rh factor*, after the Rhesus monkey in which the discovery was made. The effect of this factor in causing difficulty in pregnancy was described on page 136.

An *Rh positive* (Rh+) individual is a phenotype possessing the Rh antigen on his red blood cells. An *Rh negative* (Rh−) individual either does not possess this antigen on his red blood cells or has it in such a weak form that it can be disregarded. When we say that an individual's blood is type AB+, we are therefore describing *two* of his characteristics — his A-B-O blood type and his Rh factor.

It is now believed that there are at least nine different antigens for the Rh factor and that they are governed by multiple alleles. A description of the inheritance of the Rh factor is beyond the scope of this course.

PROBLEMS IN GENETICS

1. What ratios can be expected in a cross between a bleeder male and a carrier female?

2. (A) Show by keyed and labeled diagrams the results of a cross between a normal male and a female carrier of color blindness. (B) What combination of parents could result in a color-blind female?

3. By means of keyed and labeled diagram, show how a color-blind man passes the trait of color blindness through his daughters to some of his daughters' children.

4. Nine of ten children in a family are right-handed. Right-handedness is a dominant trait. Assume that this trait is controlled by a single pair of genes and answer these questions:

   (A) What is the gene make-up (genotype) of the left-handed child?

(B) What are the possible gene make-ups of the right-handed children?

(C) What are the possible gene make-ups of the parents?

(D) Could these parents have produced only right-handed children? Explain.

(E) If these parents had produced only one child would this child have been left- or right-handed? Explain.

5. In shorthorn cattle, the hornless condition is dominant over the horned condition. Indicate by keyed and labeled diagrams the results of the following crosses: (A) Hybrid crossed with hybrid. (B) Hybrid crossed with recessive.

c | CHANGES IN HEREDITY

DNA replication and the orderly sorting of chromosomes and genes during mitosis are essentially conservative mechanisms. They preserve the hereditary *status quo* from generation to generation. However, changes in heredity also take place. These hereditary changes permit the gradual adaptation of living forms to changing environments. Without change in heredity (mutation), evolution could not occur. A *mutation* is a genetic change which can be inherited.

Some mutations affect the body cells, but not the sex cells. These mutations which affect only the body cells are called *somatic mutations*. By mitosis, somatic mutations are passed on to succeeding generations of body cells, but they do not affect the offspring. For example, a mutant gene may cause a skin cell to become abnormal. Such a change will be passed on to succeeding generations of body cells, but have no affect on offspring. Accordingly, when the individual dies, the somatic mutant genes die with him. *Only mutations which affect gamete-producing cells of the gonads may be passed on to the offspring.*

The characteristics of mutations are as follows:

1. Most mutations are disadvantageous to the organism. An organism's set of genes work fine — the organism lives. Any change is likely to "upset the apple cart."

2. Most mutations are recessive.

3. Mutations are random. The changes cannot be predicted nor, at present, can scientists direct the formation of a specific change.

Usually the term "mutation" refers to a change in a gene, but changes in the number or structure of chromosomes are also considered mutations.

Chromosome Mutations

A chromosome mutation is a change in the number of chromosomes or of the structure of chromosomes. Many of these changes can be seen under the microscope.

Polyploidy. During the process of meiosis, the sets of chromosomes may not separate properly. As a result, gametes may be formed possessing the $2n$ (diploid) number, instead of the n (monoploid) number. Fertilization of such a gamete by another $2n$ gamete results in a $4n$ (tetraploid) zygote. By mitosis, this zygote develops into an individual, all of whose body cells are 4n. Individuals may also be produced with 3n, 5n, 6n, etc., chromosomes. Such a condition, in which the cells have extra sets of chromosomes, is known as *polyploidy.*

Polyploidy in plants has been artificially induced by use of *colchicine.* This chemical interferes with the proper functioning of the spindle fibers during meiosis. Polyploid plants are usually larger than ordinary specimens. Polyploid mutants include many of the giant varieties of flowers and fruits developed for the commercial market, as well as new types of tobacco, cotton and wheat.

Nondisjunction. During the process of meiosis, the homologous chromosomes normally separate from each other after synapsis occurs. Failure of chromosomes to separate is *nondisjunction.* When a whole set of chromosomes fails to separate, the gametes may be 2n and polyploidy may result. Sometimes, however, only one pair of chromosomes fails to carry on normal disjunction. The ensuing gamete is $n + 1$ and the fertilized egg is $2n + 1$; that is, the offspring has one extra chromosome in each cell of its body.

In humans the failure of disjunction to occur in chromosome number 21 results in Down's syndrome, a severe form of mental retardation. Because of facial characteristics of children suffering from this condition, it is sometimes called Mongoloid Idiocy, but it has nothing to do with people from Asia.

Nondisjunction of the sex chromosomes can lead to the production of XYY males. Such individuals may have aggressive behavior which brings them into conflict with society and sometimes results in the individual committing criminal acts. This has raised the question of how much a person with such a genetic defect is responsible for his actions.

Changes in chromosome structure. Irregularities in chromosome activity during the early stages of the first meiotic division may result in unusual alterations in the composition of chromosomes. Some examples include:

translocations – where a portion of a chromosome is transferred to another chromosome that is not homologous with it.

inversions – where the sequence of genes in a portion of a chromosome is reversed.

deletions – where a portion of a chromosome is omitted and the corresponding genes are absent.

additions – where a portion of a chromosome is repeated and the corresponding genes are present twice.

Changes in the chromosome structure often result in marked changes in the organism because many genes are involved.

Gene Mutation

As previously stated, genes are portions of DNA molecules. Accordingly, a gene mutation is a change in the DNA molecule. The chemistry of DNA and the structural factors involved in gene mutation are considered in Chapter 21.

Mutagenic agents. A mutagenic agent is one which causes a gene change. When various forms of ionizing radiation penetrate cells, they can modify portions of the DNA molecule. In 1927, Herman J. Muller found that he could increase the rate of mutation in fruit flies by exposing them to *X-rays.* Excessive exposure to the *ultra-violet* rays of the sun or to a "sun lamp" has caused skin cancer.

Naturally occurring radioactive sub-

stances, such as radium or uranium, emit radiations. On striking DNA, these may cause gene mutations. Man-made *radioactive isotopes* are introduced into our environment by modern technology and by the atmospheric testing of nuclear weapons. These are possible causes of gene mutation.

Chemicals induce mutation by uniting with DNA. The *nitrogen mustards* are one class of chemicals which act in this manner. Some mutagenic chemicals resemble in structure the raw materials which the cell uses to make DNA. They "trick" the DNA into incorporating the "fake" chemical.

Significance of Mutations

1. Mutations are a source of the variability among living organisms that is the basis for evolution.

2. The vast majority of mutations are harmful in their effect upon living things. Even the smallest amount of radiation from radioactive isotopes or X-rays can cause harmful mutation to body cells or to the gonads. Mutations in reproductive cells may weaken or harm future generations of mankind. Above-ground nuclear bomb tests in any part of the world may produce radioactive isotopes such as *strontium–90*. These are carried world-wide by air currents and settle as *fall-out*. Taken up by plants and then by animals, strontium–90 is incorporated into bone tissue (it chemically resembles calcium) where it continues to emit radiations. Potassium–40 and caesium–137 affect the gonads.

As previously stated, the rate at which a radioactive element changes to another element by emitting particles from its nucleus is measured in a unit called the *half-life*. This is the length of time that it takes for half of the atoms in a sample to "decay." Because strontium–90 has a relatively long half-life of 20 years, it continues to emit radiation in our environment for a long time after its formation in a nuclear reaction.

The effect of exposure to X-rays builds up during an individual's lifetime. The medical use of X-rays provides the gonads of man with more exposure to radiation than all other sources combined. Physicians are well aware of this problem and are taking steps to prevent unnecessary and excess exposure to radiation. The effect of radiation in space upon living organisms is being studied by experiments conducted during space flights.

3. Biologists study the mutations of experimental organisms to increase their knowledge of genetics, development, and biochemistry.

HEREDITY AND ENVIRONMENT

The DNA from both parents which is assembled in the fertilized egg provides chemical instructions that determine the structure and function of the cells of the offspring. However, the development of cells and of the entire organism is also greatly influenced by the environment of the developing individual. The *environment* of an organism consists of all the substances, forces, and other organisms which affect it during its life.

A pea plant that has the heredity to be tall may never achieve the promise of its heredity because it lacks the proper environment (such as fertilizer). A seedling may have the heredity to produce chlorophyll rather than to be an albino (colorless). However, if it is raised in the dark, the seedling will be colorless.

Curly wings is a hereditary trait in fruit flies which is influenced by the temperature under which the larvae develop. For example, if the larvae develops at moderate temperatures of 25°C, a fruit fly may have curly wings. However, if the larvae develop at lower temperatures of 16°C, the wings of the adult will be straight, not curly.

As a result of experiments on many kinds of organisms, scientists conclude that heredity and environment are both important in determining the characteristics of an individual.

INHERITANCE IN MAN

It is difficult to study heredity in man because (1) experimental crossings cannot be made, (2) the life span of man is so long, and (3) man produces too few offspring to yield reliable ratios. However, the study of family histories (genealogies) and of records in institutions have yielded information concerning the inheritance of traits in man.

Some of the traits inherited in man are discussed elsewhere in this book. These traits, are listed below.

inheritance of sex

hemophilia

color blindness

sickle-call anemia

ABO blood types

Rh factor

Down's syndrome

Other traits inherited in man are given in the table below.

Susceptibility to Disease

Microorganisms cause many diseases, but heredity makes some individuals more susceptible than others to certain diseases, Genes, as well as the environment, are important in the development of tuberculosis. Some forms of cancer appear to run in families. Susceptibility to diabetes is inherited as a recessive trait which prevents the formation of insulin. However, environmental factors, such as the amount of

SOME TRAITS INHERITED IN MAN

Trait	Dominant	Recessive
Hair color	dark	light
Eye color	brown (several genes are involved)	blue
Hair type	wavy or curly	straight (except for the straight Mongolian hair for the Mongoloid race, which is dominant)
Skin pigmentation	normal (several "blending" genes are involved in the production of varying degrees of the pigment *melanin* in the skin)	albino
Ability to taste PTC (a substance that is bitter to some persons)	taster	non-taster
Ability to roll the tongue	roller	non-roller
Number of fingers	6-fingered (a mutation)	5-fingered

sugar in the diet, play a role in the onset of this disease. At least 50 human diseases, in which the metabolism is upset, have been described as having a hereditary basis. In many of these diseases, a gene mutation prevents the formation of a normal protein.

Intelligence

One of the difficulties in determining the heritability of intelligence is that there is no adequate definition of it. Intelligence tests provide a good measure of a person's present academic ability, but they tell nothing of what it might have been under different circumstances. Furthermore, there are other kinds of ability that these tests do not measure. Every test is standardized on some particular population, and is not valid when used with people raised in vastly different surroundings.

The only valid data concerning the inheritance of intelligence come from the study of twins. There are two types of twins. *Identical twins* result from a single fertilized egg and therefore have the same heredity. *Fraternal twins* are formed from different sperms and eggs, and are genetically different.

The studies were performed by comparing four samples: (1) identical twins raised together, (2) identical twins raised apart, (3) fraternal twins raised together, (4) fraternal twins raised apart.

It is assumed that environments are more similar when the twins are raised together than when they are raised apart. Unfortunately, samples are few, especially for the twins raised apart. With all these limitations, and with the understanding that we really do not know the biological significance of whatever intelligence tests measure, it appears that heredity and environment both play roles in determining intelligent behavior.

PLANT AND ANIMAL BREEDING

From earliest times, man has attempted to improve domesticated animals and plants. Now the knowledge from modern genetics is being applied to the problem.

Aims of Breeders

Some of the aims of breeders are as follows:

1. **Greater yield.** Examples: more wheat per acre; greater quantity of milk produced by cows; meatier livestock.

2. **Better quality.** Examples: higher butterfat content of milk; better-tasting apples; greater speed in race horses.

3. **Resistance to disease.** Examples: wheat plants resistant to wheat rust (a fungus); cattle which are immune to Texas fever.

4. **Extension of crop areas.** Examples: wheat which will grow in cold climates or in poor soil.

5. **Special traits.** Examples: ornamental varieties of plants, birds, and tropical fish; thin-shelled walnuts which may be opened easily; turkeys which will fit into the small oven of the modern American kitchen and which are suitable for small families.

The traits in which the breeder is interested nearly always vary continuously. For example, the amount of butterfat in milk is not controlled discontinuously by a single pair of genes, like the height of Mendel's peas. Probably many different genes have some influence on the butterfat content. Some of these genes will increase the butterfat content when certain feeds are used but will not do so when others are used. Many of these genes will produce effects on the cows other than the effect on butterfat. Such effects may or may not be desirable. Accordingly, the breeder cannot work with a single pair of genes. He must work with the entire genotype of the whole organism in a particular environment. His aim is to improve the combination of genes in the breed so that it will thrive in a definite environment and will be suitable for his own needs.

Methods Used by Breeders

In this section we will show some of the steps used by breeders, by considering examples of animal and plant breeding.

An example of animal breeding. Texas cattle produce good beef; however, the animals are highly susceptible to Texas fever, a disease which is caused by microorganisms and spread by the bite of ticks. A worldwide search disclosed a breed of cattle from India (Brahman cattle) that were immune to this disease. However, Brahman cattle did not possess good beef qualities. The steps in breeding were:

1. Selection. Choosing organisms to be used for breeding is called *selection*. The Texas and the Indian cattle were selected for crossing.

2. Hybridization

Texas Cattle $\times$	*Indian Cattle*
*good beef	poor beef
susceptible to Texas fever	*resistant to Texas fever

Note: The asterisk is used to designate the traits desired by the breeder.

3. Selection. Those offspring in which the desired qualities were combined were selected for further breeding.

4. Inbreeding. *Inbreeding* is the mating of closely related individuals

of the same strain. After a number of generations of inbreeding, the desired genes may be present in a homozygous state; that is, both alleles for the trait will be alike.

Inbreeding also brings together undesirable recessive genes in the homozygous state so that they appear in the phenotype. In the case of plants and domesticated animals, those undesirable genes which show up may be removed from the breeding group by destroying the specimens or by not permitting the reproduction of individuals showing the undesired characteristics.

Outbreeding, or *hybridization*, is the mating of individuals of different strains in the attempt to bring together desirable combinations of genes. Frequently, this combination of genes results in offspring that are generally more vigorous than either parent. This phenomenon is known as *hybrid vigor*. An example of hybrid vigor is provided by the mule, which is the result of a cross between a female horse and a male donkey. The large size is inherited from the horse. From the donkey the mule inherits sure-footedness, long ears, endurance, and the ability to endure hardships and to live on rough food. The mule is the result of a cross between two different species and is usually unable to reproduce — it is sterile.

Cross-bred hogs produce more offspring than their pure-bred relatives. They also grow faster and put on more weight for every hundred pounds of food eaten.

Results. Breeders were able to produce a breed of cattle which had the desirable characteristics (good beef, resistance to Texas fever) of Texas cattle and Indian cattle.

Strains resulting from crosses between the Brahman cattle of India and domestic breeds can endure the hot, humid climate of our southeastern states and the dry summer heat of the Southwest.

An example of plant breeding. Luther Burbank (1849–1926) achieved fame as a breeder of new varieties of plants. One of Burbank's experiments illustrates the methods employed by plant breeders. Burbank discovered a blackberry plant whose fruits were white in color but whose berries had a poor taste. Burbank set about developing a white blackberry with good-tasting fruit. He crossed the specimen he had found with the Lawton Blackberry:

Lawton	×	*White*
Blackberry		*Blackberry*
*good taste		poor taste
black fruit		*white fruit

The steps which Burbank employed were:

1. **Use of a mutation.** A mutation had occurred in nature, and Burbank made use of it.

2. **Selection by man.** Burbank selected the two strains which he wished to cross.

3. **Hybridization.** Burbank crossed the two strains by cross-pollination.

4. **Selection by man.** Many of the resulting plants had the undesirable traits of one or both parents. Burbank discarded these undesirable specimens but was able to find at least one plant which had the qualities which he was seeking.

5. **Vegetative propagation.** Blackberry plants may readily be propagated by layering. Accordingly, Burbank was able to obtain large numbers of the new type. Achieving this is more difficult when the breeder is working with plants that cannot be reproduced by vegetative means. In these cases, an additional series of inbreeding experiments is required.

The white blackberry did not achieve great popularity. Burbank is better known for: (1) the Burbank potato, which resists the fungus disease known as rot; (2) the spineless cactus, which is used as animal feed; (3) the Shasta daisy, a startling white flower obtained by crossing a common native daisy with a European variety; (4) the plumcot, obtained by crossing the plum with the apricot; (5) a cross between the squash and the pumpkin; (6) the thin-shelled walnut, which has tasty meat and is easy to crack open.

Mutations

Mutant forms that occur naturally are sometimes detected by man and are selected for propagation. The California navel orange appeared as a bud mutation. Grafts were then made from twigs of the branch which developed from this bud. The Ançon sheep, a short-legged variety which appeared as a dominant mutation, is useful because it cannot jump fences and twice as many can be shipped in each double-decker freight car. Naturally occurring mutations were formerly called *sports* or *freaks*.

Instead of waiting for mutations to appear, breeders can speed this process by the use of radiation and chemicals. Because mutations are of a random nature, it is impossible to predict the results of radiation. Man has not yet learned to alter the DNA molecule "to order" but scientists predict that this feat will be accomplished in the near future. *Colchicine,* applied to the shoots or buds of plants, has resulted in polyploid forms of blueberries, lilies, and cabbage. Other plants with multiple chromosome sets include cherries, grapes, strawberries, and a variety of McIntosh apple. Polyploidy generally results in greatly increased size. Radiation experiments at the Brookhaven National Laboratories of the Atomic Energy Commission on Long Island, N.Y., have resulted in flowers with interesting color patterns. Oat plants that are rust resistant were also developed by this method.

Other Examples of Breeds Improved by Man

Hybrid corn. Hybrid corn is produced by two hybridizations. Four pure line parents are combined in a "double cross." The qualities of good-tasting kernel, regular ear, sturdy stem, and disease-resistant roots are introduced into the seeds. These are purchased and planted by the farmer. The plants which grow from these seeds have all four good qualities. However, if the farmer plants the seeds from the double hybrid plants, the next generation may display inferior qualities of the four ancestors.

Nectarine — appeared as a bud mutant of the peach.

Chickens
   large number of eggs — Leghorn
   good tasting meat — Cornish
   dual purpose (egg production and meat) — Plymouth Rock
   small size, with large proportion of meat — Beltsville chicken

Beltsville turkey — small size suited for the modern kitchen oven, with larger proportion of meat and less bone.

Cattle — beef cattle — Aberdeen-Angus, Hereford, Shorthorn
   dairy cattle — rich milk: Jersey, Guernsey. Large quantity of milk: Holstein-Friesian
Swine — heavy, lard type — Poland China, Hampshire
   lean, bacon type — Yorkshire, Tamworth

PROBABILITY IN GENETICS — THE ALGEBRAIC METHOD

Problems in genetics may be solved by using an algebraic method instead of the Punnett Square. The algebraic method is based upon the laws of probability. *Probability* is the mathematical science that attempts to predict the "chances" of an event happening.

In tossing a coin, the chances that the coin will fall head may be expressed as the fraction 1/2. In rolling one die (singular of dice), what is the chance of obtaining a 5? Since there are six numbers on the die, the probability is 1/6. The probability of drawing a spade in a deck of cards is 13/52 or 1/4. With a trick coin, both of whose sides are heads, the chances of its falling head in 1000 trials would be 1000/1000, or 1. The chances of its falling tail in 1000 trials would be 0/1000, or 0. The event, tails, has a *zero possibility* — that is, it is impossible.

Now consider two events at the same time. In rolling a pair of dice, what is the probability of obtaining two 5's? The probability of obtaining a 5 on one die is 1/6; the probability of obtaining a 5 on the other die is 1/6; the probability of obtaining a 5 on *both* dice is $1/6 \times 1/6 = 1/36$. *The probability that two independent events will occur together is the product of their probabilities of occurring separately.*

In tossing a coin a large number of times, the probability of obtaining a head is 1/2. However, it is quite possible that the first eight trials all result in tails. What is the probability that the ninth trial will be a head? The answer is still 1/2. *Previous trials do not affect the results of later trials of the same event.* If parents have four male children in succession, the probability that the next child will be a boy (if no other factors are operating) is still 1/2.

In tossing two coins simultaneously, the probability of obtaining

(a) 2 heads: is $1/2 \times 1/2 = 1/4$

(b) 2 tails: is $1/2 \times 1/2 = 1/4$

What is the probability of obtaining the combination of a head and a tail? There are two ways of obtaining the combination of a head and a tail: The first coin could fall head and the second one tail, the first coin could fall tail and the second one head.

	1st coin	2nd coin	Probability
Two heads	H	H	1/4
Head and tail	H T	T H }	1/2
Two tails	T	T	1/4

This problem may also be solved by algebraic multiplication. First, state the probabilities for each coin; then multiply the probabilities.

$$
\begin{array}{ll}
1/2\,H \;+\; 1/2\,T & \text{(probability for 1st coin)} \\
\times \quad 1/2\,H \;+\; 1/2\,T & \text{(probability for 2nd coin)} \\
\hline
1/4\,HH \;+\; 1/4\,HT & \\
\quad\quad\quad 1/4\,HT \;+\; 1/4\,TT & \\
\hline
1/4\,HH \;+\; 1/2\,HT \;+\; 1/4\,TT &
\end{array}
$$

The probability of obtaining a head and a tail is 1/2.

Let us now turn to genetics. If the parent is homozygous, with the genes AA, the probable gametes are $1/2\,A + 1/2\,A = 1/1\,A$, or all A. If the parent is

homozygous *aa*, the probable gametes are $1/2\,a + 1/2\,a = 1/1\,a$, or all *a*.

If the parent is heterozygous *Aa*, the probable gametes are $1/2\,A + 1/2\,a$.

Sample Problem 11

What are the probable offspring in a cross between two heterozygous parents *Aa* × *Aa*?

SOLUTION

$$
\begin{array}{ll}
1/2\,A \;+\; 1/2\,a & \text{(gametes from one } Aa \text{ parent)} \\
\times \quad 1/2\,A \;+\; 1/2\,a & \text{(gametes from other } Aa \text{ parent)} \\
\hline
1/4\,AA \;+\; 1/4\,Aa & \\
\quad\quad\quad 1/4\,Aa \;+\; 1/4\,aa & \\
\hline
1/4\,AA \;+\; 1/2\,Aa \;+\; 1/4\,aa & \\
& \textbf{Answer}
\end{array}
$$

These results confirm those obtained by use of the Punnett Square.

Sample Problem 12

What are the probable offspring in a cross when the parents are *Aa* and *aa*?

SOLUTION

$$
\begin{array}{ll}
1/2\,A \;+\; 1/2\,a & \text{(gametes from } Aa \text{ parent)} \\
\times \quad\quad\quad 1/1\,a & \text{(gametes from } aa \text{ parent)} \\
\hline
1/2\,Aa \;+\; 1/2\,aa & \text{Answer}
\end{array}
$$

Complex problems involving dihybrids and trihybrids may be more readily solved by use of the algebraic method than by the laborious construction of gigantic Punnett Squares. For example, consider the following problem involving dihybrids:

Sample Problem 13

What are the expected results in the cross *Ttyy* × *ttYY*?

SOLUTION

In solving this problem (1) make the computation for one trait, (2) make the computation for the other trait, (3) multiply.

(1) Computation for the *T* trait.

   1/2 T + 1/2 t (gametes of Tt parent)

×   1/1 t (gametes of tt parent)

   ——————————

   1/2 Tt + 1/2 tt

(2) Computation for *Y* trait

   1/1 y (gametes of yy parent)

× 1/1 Y (gametes of YY parent)

   1/1 Yy

(3) Multiply. If the inheritance of each pair of alleles is independent of the other, we can now multiply each of the probabilities to determine the probability of their occurring together:

     1/2 Tt   + 1/2 tt

×   1/1 Yy

   ————————————

   1/2 TtYy + 1/2 ttYy

Answer. Phenotype ratio:

     50% tall yellow

     50% short yellow

PROBLEMS IN GENETICS

Sample Problem

Calculate the probable results in the cross $TtYy \times TtYy$. Your results should be: $\frac{1}{16}$ *TTYY*, $\frac{1}{8}$ *TtYY*, $\frac{1}{16}$ *ttYY*, $\frac{1}{8}$ *TTYy*, $\frac{1}{4}$ *TtYy*, $\frac{1}{8}$ *ttYy*, $\frac{1}{16}$ *TTyy*, $\frac{1}{8}$ *Ttyy*, $\frac{1}{16}$ *ttyy*. Collecting all the phenotypes results in the $9:3:3:1$ ratio previously obtained in solving this problem with Punnett Square:

Tall Yellow: $1/16 + 1/8 + 1/8 + 1/4 = 9/16$

Tall Green: $1/16 + 1/8 \qquad\quad = 3/16$

Short Yellow: $1/16 + 1/8 \qquad\quad = 3/16$

Short Green: $1/16 \qquad\qquad\qquad = 1/16$

1. Calculate the probable results in the cross $TTYy \times TtYy$.

COMPLETION QUESTIONS

A 1. The virulent type of pneumococcus is type . . . .

2. The experiments of Griffith and Avery showed that the hereditary material is . . . . .

3. Mendel mated two types of garden pea by the process of . . . .

4. Before starting an experiment, Mendel allowed each kind of pea plant to self-pollinate for several generations in order to obtain . . . . strains.

5. Pure round-seeded plants crossed with pure wrinkle-seeded plants produced only round-seeded offspring. This illustrates Mendel's Law of . . . .

6. The appearance of the recessive gene in the F_2 generation of a cross between pure parents illustrates Mendel's Law of . . . .

7. When two hybrid tall pea plants are crossed, the ratio of tall to short offspring is . . . .

8. Alternative genes for a trait are known as . . . .

9. If both alleles for a gene are alike, the organism is . . . . for that gene.

10. The set of chemical instructions received by an individual through the gametes that produced him is his . . . .

11. A ratio of 50% dominant and 50% recessive offspring is obtained in a cross between a pure recessive and a . . . .

12. The expected Mendelian ratios tend to appear when the number of offspring is ....

13. In a test cross, the unknown organism is mated with one whose genotype is ....

14. Black coat color is dominant in rabbits. If a black rabbit crossed with a white rabbit produced a white offspring, the black rabbit must have been ....

15. When 25% of a large number of guinea pig offspring are white, hybrid black parents were crossed with ....

16. The Law of Independent Assortment is shown in a cross between ....

17. When four o'clock plants bearing pink flowers are crossed with each other, the percentage of pink-flowered plants in the offspring is approximately ....

B 18. Genes inherited together on the same chromosome are said to be ....

19. The giant chromosomes of *Drosophila* are seen in the .... glands of the larval stage.

20. The exchange of linkage groups between homologous chromosomes is known as ....

21. The lining up of homologous chromosomes in twisted pairs is known as ....

22. T. H. Morgan developed the theory of the gene while working with .... as an experimental organism.

23. A breeder can determine whether a tall pea plant is pure tall or hybrid tall by crossing it with a ....

24. When a human egg is fertilized by a sperm carrying the Y chromosome, the sex of the child will be ....

25. A color-blind man and a pure normal woman will have sons of whom .... percent are normal with respect to color vision.

26. If both the father and the mother have blood type A, the baby can have blood type A or blood type ....

C 27. A genetic change which can be inherited is known as a (an) ....

28. A chemical which can be used to induce polyploidy is ....

29. The presence of one extra chromosome in all the body cells of man may result in the form of mental retardation known as ....

30. H. J. Muller caused mutations in fruit flies by exposing them to ....

31. A radioactive isotope which is readily taken up by bone tissue is ....

32. Fruit flies with the heredity for curly wings develop straight wings when the larvae are exposed to an environmental change in ....

33. Choosing animals to be used for breeding is called ....

34. The mating of closely related individuals of the same strain is known as ....

35. An animal which is an example of hybrid vigor is the ....

36. An outstanding plant breeder in the early part of this century was ....

37. The desirable characteristics of two kinds of wheat plant may be united in the offspring by ....

38. Plant mutants may be multiplied more easily than animal mutants because many plants are capable of ....

39. The first seedless oranges were the result of ....

40. A breed of cattle noted for its rich milk is the ....

MULTIPLE-CHOICE QUESTIONS

A 1. Mendel's Law of Segregation can be best illustrated by the cross (1) *AA* × *aa*, (2) *Aa* × *aa*, (3) *Aa* × *Aa*, (4) *AA* × *Aa*.

2. If a trait skips a generation, the most probable reason is that this trait (1) is dominant, (2) is recessive, (3) has mutated, (4) is linked.

3. When experimenting with the factor of height in peas, Gregor Mendel actually obtained this result from one of his series of crosses: 73.97% tall and 26.03% short. The parents were probably (1) hybrid tall peas and short peas, (2) hybrid tall peas and pure tall peas, (3) hybrid tall peas, (4) tall peas and short peas.

4. The phenotype of a mouse can best be determined by (1) test breeding, (2) crossing-over, (3) looking at it, (4) biochemical techniques.

5. In the cross *Bb* × *Bb*, approximately what percentage of offspring have the same phenotype as the parents? (1) 25%, (2) 50%, (3) 75%, (4) 100%.

6. In man, consider brown eye-color to be dominant to blue eye-color and that this trait is controlled by a single pair of alleles. If two brown-eyed parents have blue-eyed children, the genetic makeup of the parents must be (1) *Bb* and *Bb*, (2) *Bb* and *BB*, (3) *BB* and *Bb*, (4) *BB* and *bb*.

7. In garden peas, the offspring of a cross between two hybrid yellow parents would be (1) 25% yellow, (2) 50% yellow, (3) 75% yellow, (4) 100% yellow.

8. The crossing of dihybrid organisms results in the appearance of traits in the ratio of (1) 1:2:1, (2) 3:1, (3) 9:3:3:1, (4) 1:1.

9. When round squash are crossed with long squash, all offspring are oval in shape. How many genotypes are produced when round squash are crossed with oval squash? (1) 1, (2) 2, (3) 3, (4) 4.

10. A boy plans to toss a coin ten times. His first four tosses all result in heads. His chances of getting a head on the next throw are (1) 1:10, (2) 1:6, (3) 1:2, (4) 4:10.

11. Pink offspring are produced by crossing a plant bearing white flowers with a plant bearing red flowers. This is evidence of (1) dominance, (2) segregation, (3) recessiveness, (4) incomplete dominance.

12. A cross between two black guinea pigs produces some black guinea pigs and some white guinea pigs. The gene makeup of the two parents is called (1) homozygous black, (2) heterozygous black, (3) heterozygous white, (4) homozygous white.

B 13. Which is the most likely result of crossing-over in plant breeding? (1) elimination of some recessive genes, (2) weakening of the dominant gene, (3) increased number of gene mutations among offspring, (4) increased variability among offspring.

14. Genes that are found on the same chromosome might be expected to be passed on to offspring together. This is called gene linkage. However, chromosomes are frequently produced in which a new combination of the original genes appears. This is caused by a process called (1) mutation, (2) crossing-over, (3) sex linkage, (4) chromosome linkage.

15. Disregarding Rh factors, a person with blood group A who needs a transfusion is most safely given blood from a donor whose genes for blood group are (1) **AO**, (2) **BO**, (3) **AB**, (4) **BB**.

16. An embryo resulting from the mating of two albino rabbits is transplanted into the uterus of a brown rabbit. The color of the offspring can most reasonably be expected to be (1) brown with white spots, (2) white with brown spots, (3) all white, (4) all brown.

17. In a family with parents who do not have hemophilia, one son has hemophilia. He received the gene for hemophilia from the (1) sperm cell, (2) Y-chromosome, (3) autosome, (4) egg cell.

18. If a boy's father has hemophilia and his mother's cells have one gene for hemophilia, what is the chance that the boy will inherit the disease? (1) 0%, (2) 50%, (3) 75%, (4) 100%.

19. A girl inherited color blindness, which is a sex-linked, recessive trait. It is probable, therefore, that (1) both parents carried the recessive gene, (2) only one parent carried the recessive gene, (3) the presence of the recessive gene in the mother assured the presence of the trait in the girl, (4) the recessive gene was not carried by the father.

20. A colorblind man marries a woman who is neither colorblind nor carrier of this trait. Which statement best describes their probable offspring? (1) All the male children will be colorblind. (2) All the female children will be colorblind. (3) All the female children will carry the gene for colorblindness but none will be colorblind. (4) 50% of the males will carry the gene for colorblindness.

[C] 21. The most important danger to future generations from radioactive fallout is the fact that (1) radioactive substances are poisonous, (2) radiation can speed up mutation rates, (3) radioactive substances can induce leukemia, (4) some radioactive elements concentrate in bone.

22. The radioactive isotope, iodine-131, has a half-life of 8 days. After how many days would there be only 1 milligram of the isotope left from an original supply of 8 milligrams? (1) 8, (2) 16, (3) 24, (4) 32.

23. Colchicine is a chemical which has been used on plants to (1) prevent cross-pollination, (2) prevent fruit formation, (3) stimulate the formation of mutants, (4) stimulate photosynthesis.

24. The characteristics of a person are determined by (1) what he inherits, (2) his environment, (3) his education, (4) the combined effects of his inheritance and environment.

25. In sexual reproduction, the offspring arising from two hybrid parents may be completely different from either parent. The most probable explanation of this is that (1) the parents contained many pairs of genes, (2) environmental influences resulted in the mutation of some genes, (3) some pairs of genes for recessive characteristics were recombined in the offspring, (4) acquired characteristics of the hybrid parents were transmitted by somatic mutation.

CHAPTER TEST

1. A cross between red-flowered and white-flowered four-o'clock plants produces seeds, all of which give rise to pink-flowered plants. If two of these pink-flowered plants are crossed, what is the chance that white-flowered plants will be produced? (1) 0, (2) ¼, (3) ½, (4) ¾.

2. Which of the following is *not* a characteristic of mutations? (1) They are usually recessive. (2) They usually produce an improvement in the organism. (3) They are inherited. (4) They may be caused by chemicals.

3. In comparing a single zygote with a single gamete of a particular species, the ratio of numbers of genes carried would be (1) 3:1, (2) 2:1, (3) 1:1, (4) 1:2:1.

4. If dimpled cheeks are dominant over nondimpled cheeks, what proportion of offspring with dimpled cheeks may one expect if two nondimpled individuals are crossed? (1) 0%, (2) 50%, (3) 75%, (4) 100%.

5. Which statement concerning an allelic pair of genes controlling a single characteristic in man is true? (1) Both genes come from the father. (2) Both genes come from the mother. (3) One gene comes from the father and one gene comes from the mother. (4) The genes come randomly in pairs from either the father or the mother.

6. In man, brown eyes seem dominant over blue eyes. A brown-eyed man marries a blue-eyed woman and they have 8 children, all brown-eyed. The *probable* genetic makeup of father, mother and children, respectively, is (1) *BB, bb, Bb*, (2) *Bb, bb, BB*, (3) *BB, bb, BB*, (4) *Bb, bb, bb*.

7. A farmer is told that his black bull is a thoroughbred (pure black). Knowing that black color in cattle is dominant over red color, he decides to determine the purity of the strain by mating the bull with several red cows. If the bull is pure, (1) 100% of the offspring will be black, (2) 100% of the offspring will be red, (3) 75% of the offspring will be black and 25% will be red, (4) 50% of the offspring will be black and 50% will be red.

8. A series of three matings between a hybrid black guinea pig and a white one resulted in 8 black offspring. The most probable explanation is that (1) a mutation occurred, (2) the expected ratios are unlikely to appear in small numbers of offspring, (3) the white parent was also hybrid, (4) the black parent was the result of incomplete dominance.

9. In a breeding experiment in which there are 200 offspring, 50% are *WW*. The parents of the offspring were (1) *RW* and *WW*, (2) *RR* and *WW*, (3) *RW* and *RW*, (4) *RR* and *RW*.

10. Inbreeding is usually used to (1) obtain hybrids, (2) produce new types, (3) insure fertilization, (4) maintain a desired type.

Genetics today is concerned with such topics as the chemical nature of the gene and the method by which genes regulate cell activities. Much of today's knowledge centers on two kinds of nucleic acid, DNA (*deoxyribonucleic acid*) and RNA (*ribonucleic acid*). The chromosomes are composed of DNA and protein. It is the DNA portion of the chromosomes which is the main hereditary material.

A THE STRUCTURE OF DNA

Since DNA is the stuff of heredity, it is essential that we have an understanding of the chemical structure of this important nucleic acid.

Nucleotides

The fundamental unit of a nucleic acid (DNA or RNA) is a *nucleotide*. The structure of a nucleotide may be represented as shown in Fig. 21–1.

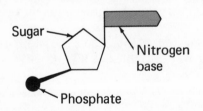

Fig. 21–1. Structure of a nucleotide.

As shown in the diagram, a nucleotide is composed of three portions: a phosphate group, a 5-carbon sugar, and a nitrogen base.

Phosphate group. A phosphate group may be represented as follows:

$$\text{HO} - \overset{\displaystyle \text{OH}}{\underset{\displaystyle \text{O}}{\overset{|}{\underset{|}{\text{P}}}}} - \text{OH}$$

The phosphate group is usually symbolized as P

Sugar. The sugar is a 5-carbon sugar rather than a 6-carbon sugar as in glucose. If the 5-carbon sugar is *ribose*, the nucleic acid produced is *ribo*nucleic acid, or RNA. If the 5-carbon sugar lacks an atom of oxygen, the sugar is *deoxy*ribose, and the nucleic acid is *deoxy*ribonucleic acid, or DNA.

Nitrogen base. There are four kinds of nitrogen bases. These are divided into two groups as follows:

Purines		Pyrimidines	
adenine	(A)	thymine	(T)
guanine	(G)	cytosine	(C)

The purine molecules are larger than the pyrimidine molecules.

The Watson-Crick Model of DNA

By means of dehydration synthesis between phosphate and the sugar, hundreds upon hundreds of nucleotides may join to form a strand of nucleic acid. DNA (Fig. 21–2) occurs as a ladder-like double strand of nucleotides. Two nitrogen bases join to form the "rungs" of the ladder and the alternating sequence of sugars and phosphates form the "uprights" of the ladder. Because of their chemical composition and shape, the bonding of the nitrogen bases on the "rungs" is always between

$$\text{A} - \text{T}$$
$$\text{G} - \text{C}$$

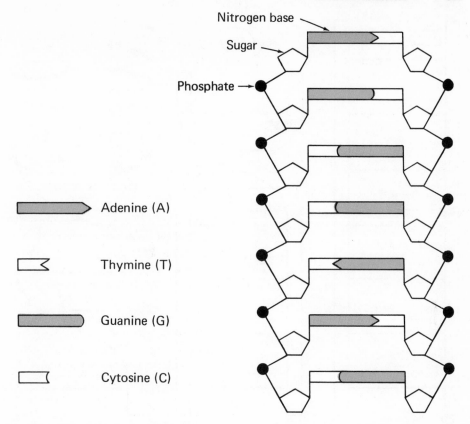

Nitrogen base

Sugar

Phosphate →

Adenine (A)

Thymine (T)

Guanine (G)

Cytosine (C)

Fig. 21–2. The ladder structure of DNA. The purines (adenine and guanine) are longer than the pyrimidines (thymine and cytosine). A combination of a purine and its complementary pyrimidine is called a "base pair." This combination of a short nitrogen base and a long nitrogen base maintains the uniform length of the "rungs" of the ladder. Weak hydrogen bonds hold together the two strands of the ladder at the union between the nitrogen bases.

If a ladder is twisted, as in Fig. 21–3, the shape assumed by each of the uprights is known as a *helix*. A "helix" is the shape assumed by a rope which is coiled around a cylinder. By contrast, a "spiral" is the flattened shape of a coiled watchspring.

The DNA molecule may be described as a double helix which resembles a twisted ladder. As shown in Fig. 21–4, each side of the ladder consists of alternating sugar (S) and phosphate (P) units, and the two sides are connected by cross bars of paired nitrogen bases.

It is the sequence of the groups of nitrogen bases which constitutes the instructions transferred to RNA.

This concept of the structure of DNA was devised by James Watson of the United States and Francis Crick of England and is known as the Watson-Crick Model of DNA. Much of the work of Watson and Crick was based upon the pioneering studies of M. H. F. Wilkins, who studied the patterns produced by X-rays after they passed through crystals of DNA. These three scientists were awarded a Nobel Prize in 1962.

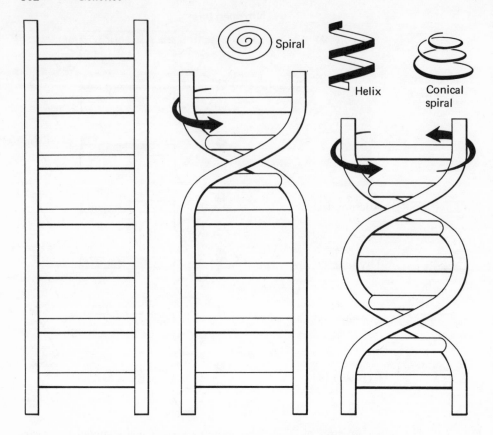

Fig. 21–3. The twisting of a ladder to form a double helix.

(B)   DNA REPLICATION

During mitosis, exact duplicates (or replicas) are made of chromosomes. The Watson-Crick model of DNA explains how this replication is accomplished. (See Fig. 21–5.)

Starting at one end of the molecule, a "zipper reaction" begins. The weak hydrogen bonds connecting the base pairs are released and the two strands come apart like a zipper. Each separation leaves an exposed purine or pyrimidine which can take on its complementary partner: an A must unite with a T and a G with a C. Within the liquid cytoplasm of the cell are present additional nucleotides. These now fall into place along the exposed complementary nitrogen base. As the double helix separates, each half thus forms a new upright to the ladder, identical to the one from which it separated.

As the "unzipping" proceeds along the molecule, a closing of new "zippers" follows behind. Where previously there was one double ladder there are now two. The two new DNA molecules are exact duplicates of the original molecule, and each double helix is part of one chromatid. Each strand of the ladder has acted as a template (pattern) for the formation of its complementary strand. This process requires energy which is supplied by high-energy phosphate bonds.

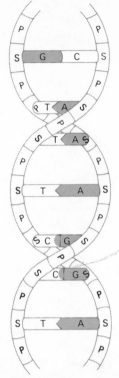

Fig. 21–4. The Watson-Crick Model of DNA.

The DNA molecule performs two functions:

1. *Replication.* The DNA molecule replicates itself so that the genetic information which it carries is passed on to the chromosomes of succeeding cells.

2. *Control of cellular activities.* The DNA molecule passes information on to the cytoplasm in such a manner as to direct the activities of the cell. In performing this function, DNA acts as a template for the formation of RNA.

RNA

An RNA molecule consists of a long chain of nucleotides. Its particular properties depend on the arrangement of the four nucleotides in the chain. The order of these nucleotides is determined by the DNA in the chromosomes. The details of how this replication is accomplished are not known.

RNA differs from DNA in several respects:

1. The sugar component of the nucleotides of RNA is ribose instead of deoxyribose.

2. RNA in most organisms is a single strand, rather than a double strand.

3. RNA does not contain thymine. Instead, it possesses uracil (U). Uracil in RNA acts as the complement for adenine. Thus, the nitrogen bases in RNA are adenine, uracil, cytosine, and guanine.

RNA and Protein Synthesis

The activities of a cell are governed by the multitude of enzymes that are present within the cytoplasm. Enzymes are proteins. As previously stated, proteins are manufactured at the ribosomes of cells. They are assembled from varying arrangements and combinations of about 20 different amino acids. How does the hereditary information present in DNA direct the precise assembly of amino acids in such a manner as to form the correct proteins at the ribosomes? Considerable evidence, much of it derived from experiments with radioactive isotopes, exists for the following sequence of events:

1. The instructions for protein synthesis are carried by the DNA in the form of a code which consists of the varying sequence of nitrogen base pairs. A strand of DNA acts as a template for the manufacture of a complementary strand of *messenger* RNA.

A crucial part of this process by which messenger RNA is formed (and

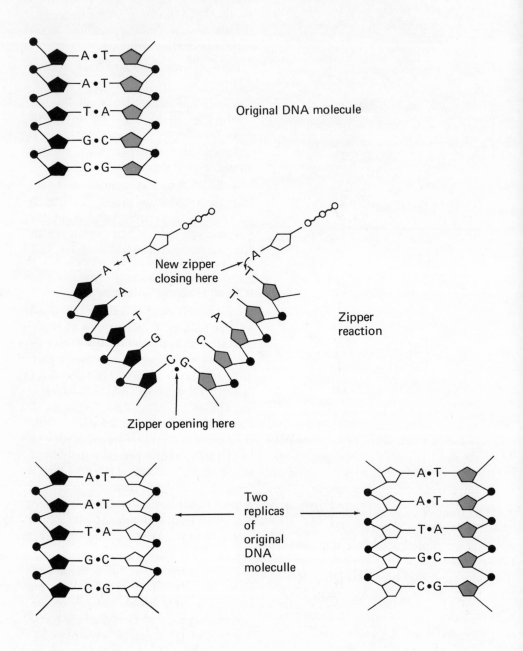

Fig. 21–5. Replication of DNA. The original DNA molecule is replaced by two replicas.

of all the reactions of DNA and RNA), lies in the phenomenon of *base pairing.* This means that the four nitrogen bases can form weak bonds with each other in a kind of molecular mating. But the bonds can form only between specific pairs:

PAIRS

guanine	bonds with	cytosine
adenine	bonds with	uracil (in RNA) or thymine (in DNA)

Thus, if the nucleotides are in a particular order in the DNA of the nucleus, the nucleotides of RNA form in a complementary order, as illustrated below.

DNA		RNA
T	.	A
A	.	U
G	.	C
C	.	G
A	.	U
T	.	A
		(RNA has U instead of T)

Fig. 21–6. RNA synthesis. Which half of the DNA molecule acts as the template for RNA? We do not as yet know.

We have seen how DNA acts as a template in the production of messenger RNA. The messenger RNA, now carrying the code, accumulates in the nucleolus and then passes to the cytoplasm where it becomes attached to a ribosome. (See Fig. 21–7.)

2. Another kind of RNA, *transfer RNA,* is present in the cytoplasm. In addition, the cytoplasm contains the 20 different kinds of amino acids. Various kinds of transfer RNA can attach to each kind of amino acid. This union is accomplished with the aid of ATP and an enzyme. Each transfer RNA mole-

cule possesses a triplet of three exposed nitrogen bases, as shown in Fig. 21–7.

3. The transfer RNA molecules, each carrying an amino acid, move to the messenger RNA which is present on a ribosome.

4. The adjacent amino acid units, one at a time, unite by dehydration synthesis to form polypeptides and then proteins, as specified by the code. The transfer RNA molecules peel off and are used again to combine with additional amino acids.

Thus, the unique sequence of nitrogen bases in the DNA of the chromosome determines the unique sequence of amino acids in the proteins which are synthesized at the ribosomes.

THE GENETIC CODE

The sequence of nitrogen bases in DNA acts as a code which ultimately designates the amino acids that are to be linked in the manufacture of protein. As an example, consider our English language. Here we have an alphabet of 26 letters which serve as the basis for spelling out words. The varying arrangements of these 26 letters are the code to spell out all the thousands of short and long words of our language. In a similar manner, the *four* nitrogen bases (A, G, C, T) of DNA act as letters to spell out the *twenty* amino acid "words." How many nitrogen base "letters" at a time would be required, in their various combinations, to spell out the twenty different amino acid "words?" Groups of two letters (AA, AG, AC, etc.) would only yield $4^2 = 16$ different words. Groups of three letters (AAA, AAG, AAC, etc.) would provide $4^3 = 64$ different words. Groups of three nitrogen bases would therefore provide more than enough combinations to en-

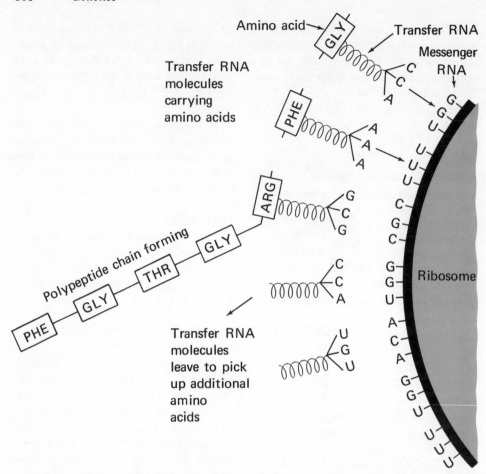

Fig. 21–7. Protein synthesis at the ribosomes. According to recent evidence, a ribosome appears to travel along the messenger RNA strand and pick up the appropriate transfer RNA and its amino acid, one at a time, in sequence. When the ribosome reaches the end of a messenger RNA section, the polypeptide chain has been completed. One messenger RNA molecule may have several ribosomes attached to it at the same time. The kind of protein produced depends not upon any specific kind of ribosome but upon the sequence of groups of nitrogen bases in the messenger RNA. The RNA also carries instructions for starting and stopping the formation of proteins, in accordance with the demands of the cell's physiology, and its homeostatic regulation.

code the twenty amino acids. The simplest code would therefore be a three-letter code, or *triplet code*. The triplet code of DNA uses the letters A, G, C, T in groups of three. This is transposed into the complementary code of RNA which uses the letters A, G, C, U (uracil replaces thymine).

What is the RNA code for each amino acid? Severo Ochoa, of New York University, prepared an artificial RNA containing only uracil nucleotides. To this he added all 20 amino acids, ATP, ribosomes, and transfer RNA's. A polypeptide was produced containing only one kind of amino acid – phenylalanine.

Thus, UUU is the RNA code for the insertion of phenylalanine into polypeptide. (The DNA code for phenylalanine is AAA, since A of DNA pairs only with U of RNA.)

By experiments of this nature, the RNA code triplets for many amino acids have been tentatively deciphered. In some cases, the precise sequence of the letters in a triplet is not yet known.

Amino Acid	RNA Code Triplet
phenylalanine	UUU
alanine	CCG, UCG
arginine	CGC, AGA, UCG
glutamic acid	GAA, UAG
tryptophan	UGG
valine	UUG
lysine	AAA, AAG, UAA

Mutations and the Genetic Code (Sickle-cell Anemia)

Knowledge of the genetic code has helped us to understand how mutations occur. For example, a mutation could be an alteration of one base pair in a section of the DNA molecule. This concept is illustrated in *sickle-cell anemia.* This disease is an inherited defect in man in which the red blood cells become distorted in the shape of sickles. They tend to form clumps which block the smaller blood vessels, and their tendency to burst brings about anemia. Victims who have this gene in the homozygous condition usually have an early death. Chemical analysis has shown that the structure of the normal hemoglobin molecule is slightly different from that of the sickle-cell hemoglobin molecule. The sequence of amino acids in a portion of each molecule is:

normal hemoglobin —
    THR, PRO, **GLU**, GLU, LYS

sickle-cell hemoglobin —
    THR, PRO, **VAL**, GLU, LYS

The amino acid valine, substituted for the amino acid glutamic acid in the hemoglobin molecule, results in this fatal disease. However, we can go a step further:

The RNA code triplet for glutamic acid is UAG.

The RNA code for valine is UUG.

Thus, a serious hereditary defect in man is caused by *the substitution of only one nitrogen base* in the DNA molecule!

THE ONE GENE–ONE ENZYME THEORY

In what manner do genes have their effects? We have seen that early studies in the field of genetics were concerned with large visible hereditary effects, such as the height of a pea plant or the color of eyes in *Drosophila*. The science of genetics has now progressed to studying the action of genes in affecting biochemical reactions within the cell. This new development in genetics is known as *biochemical genetics*.

An outstanding development in biochemical genetics was the *one gene-one enzyme theory* which provides an explanation for the mechanism by which genes have their biochemical effects. This theory grew out of the Nobel Prize-winning work of George W. Beadle and Edward L. Tatum at Stanford University, published in 1945. Beadle and Tatum used the red bread mold *Neurospora* as an experimental organism. *Neurospora* is well suited for genetic studies because:

1. It produces eight spores in a narrow sac in such manner that each spore can be identified, isolated, and grown to produce a new mycelium (mass of

threads which constitute the growing mold).

2. It can be grown on a simple culture medium (a minimal medium) containing a few simple minerals, sugar, and the vitamin biotin. *Neurospora* synthesizes all other essential compounds (such as amino acids and vitamins) in a series of biochemical reactions. Beadle and Tatum set out to study the effect of gene mutation in this series of biochemical reactions. In this way, the biochemical effect of a single gene could be investigated.

Fig. 21–8 illustrates a typical experiment performed by Beadle and Tatum. In this figure, the growth of the mycelium is indicated by lines at the top of the culture medium. The following description of this experiment is keyed to the numbers in the drawing.

1. When they grew the mycelium of the mold in a minimal medium, the mold synthesized all its needed amino acids and vitamins and grew well.

2. When the mycelium was transferred to a complete medium (containing all amino acids and vitamins) it also grew well. Here the mycelium was then *exposed to ultra-violet rays, which are known to cause mutations.*

3, 4, and 5. A single spore, carrying a possible mutation, was selected from a spore sac and transferred to a complete medium (No. 5). Here it grew and developed a mycelium. Portions of this mycelium were then transferred to a variety of other media (6a, 6b, 6c, and 6d).

6a. When the mycelium was transferred to a minimal medium, no growth occurred. This failure to grow indicated that a mutation had occurred so that the mold could no longer synthesize some essential compound. Which compound? Parts 6b, 6c, and 6d provided an answer.

6b. When the mycelium from 5 was transferred to a minimal medium to which *all the amino acids had been added,* growth still did not occur. This showed that the defect was *not* caused by the mold's failure to synthesize amino acids. Even though all the amino acids were supplied to it, the mold still could not grow. Something else was still lacking.

6c. When the mycelium from 5 was transferred to a minimal medium to which *all the vitamins had been added,* growth occurred. This indicated that the defect was caused by the mold's failure to synthesize vitamins. The mutated mold could grow only when it was supplied with the vitamins which it could not synthesize by itself. The question now remained, "Which vitamin was lacking?"

6d. A series of new tubes was now set up, each with a different vitamin added to a minimal medium. Of this new series, only tube 6d is shown. This tube contained a minimal medium plus the vitamin *thiamine.* This medium supported the growth of the mold. Since the addition of thiamine permitted the mutated mold to grow, this shows that the defect was caused by the failure of the mutated mold to synthesize this vitamin. Subsequent generations of the mold continued to grow in a minimal medium supplemented by thiamine.

In other experiments, mutations caused by ultra-violet or X-ray irradiation caused different biochemical defects to arise — it was not always a defect in the synthesis of thiamine.

The synthesis of thiamine occurs by a series of biochemical reactions, each controlled by a different enzyme, as shown in Fig. 21–9.

According to the theory of Beadle and Tatum, the mutated gene resulted in

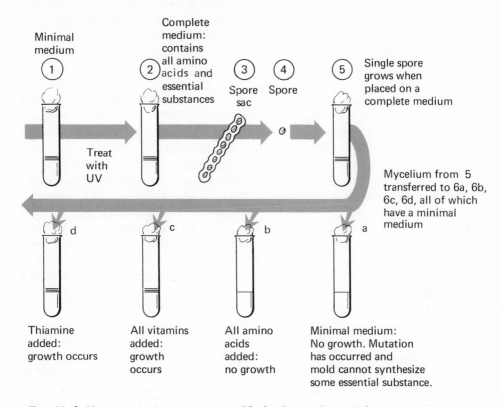

Fig. 21–8. Neurospora. A gene mutation blocks the synthesis of the vitamin, thiamine.

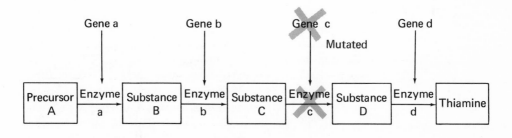

Fig. 21–9. One gene-one enzyme theory: Each gene is responsible for the production of a specified enzyme. If gene C is defective, enzyme C is not produced and substance D is not formed. However, the addition of substance D to a minimal medium permits the mold to live normally.

the inability to produce the enzyme necessary to convert substance C to substance D. When substance D is supplied, the red bread mold grows on a minimal medium. Further experiments in which the mutant strain was crossed with a normal strain showed that this failure was caused by a single gene. On the basis of many experiments of this type, Beadle and Tatum theorized that a single gene governs the synthesis of a single enzyme — the one gene-one enzyme theory. In its present form this is now known as the *one gene-one polypeptide theory*.

Phenylketonuria

Phenylketonuria (or PKU) is a hereditary disease of man in which the individual suffers lowered mentality and has pale skin coloring. The individual also excretes phenylpyruvic acid, a substance that is absent from the urine of normal people. The disease results from defective recessive mutant genes. When these genes are present in the homozygous state, an important enzyme is not produced (Fig. 21–10).

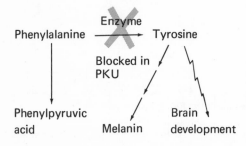

Fig. 21–10. Phenylketonuria.

The normal gene causes synthesis of the enzyme that converts the amino acid phenylalanine to a similar amino acid, tyrosine. In the absence of at least one normal gene, the enzyme is not produced, and tyrosine is not formed. Tyrosine is a precursor for the manufacture of melanin, a black pigment. In the absence of tyrosine, melanin is not synthesized, and a pale skin color results. The phenylalanine which accumulates is oxidized to phenylpyruvic acid which is excreted in the urine. Tyrosine is needed not only for the production of melanin but also, in some unknown way, for proper brain development.

In the heterozygous condition, if one normal gene is present, the enzyme is produced, and the individual appears normal. In the homozygous condition, there results a form of mental retardation known as *phenylpyruvic idiocy*.

Many states now require that the blood or urine of newborn infants be tested for indications of PKU. Placing such a baby on a diet containing almost no phenylalanine prevents the brain damage that is typical of this disease. However, this restricted diet is said to cause other damage to the body.

The study of PKU shows that genes can have their effects on the body by their production of enzymes. The presence of defective genes results in the failure to produce an important enzyme.

One Gene Can Have Many Effects

In the analysis of PKU, note how one defective gene, which participates in an early stage of a series of biochemical reactions, can block a web of succeeding stages, and therefore can have widespread effects. This situation is diagrammed in Fig. 21–11.

A lack of enzyme 5 blocks the appearance of Trait A. A lack of enzyme 3 blocks the appearance of Traits A and B. A lack of enzyme 2 blocks the appearance of Traits A, B, and C. Since each enzyme results from the action of a gene (one gene-one enzyme theory),

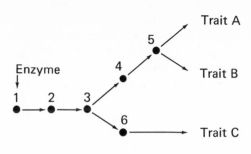

Fig. 21–11. One gene can have many effects.

some genes may appear to control several traits. A mutant gene that interferes with enough important reactions to cause death is known as a *lethal* gene. However, when a lethal gene is heterozygous with a normal gene, the individual may appear normal.

CYTOPLASMIC INHERITANCE

The importance of the nucleus and its genes in the process of heredity is well established. However, this does not rule out the possibility that substances in the *cytoplasm* can also pass on characteristics from one generation to the next.

Plasmagenes are structures in the cytoplasm which duplicate themselves and which are transmitted to succeeding cells. These include chloroplasts, mitochondria, centrioles, and structures at the base of flagella and cilia called "basal bodies." DNA has recently been found both in chloroplasts and mitochondria. Two examples that seem to show true cytoplasmic inheritance are (1) variegation in leaves and (2) kappa particles in *Paramecium*.

1. Variegation is the appearance of pale blotches in otherwise normal green leaves. The pale areas result from a mutant chloroplast. This characteristic is passed on to succeeding generations of plants in a manner that is not explained by *nuclear* genes.

2. Some paramecia, called *killers*, secrete into the water a substance that kills other paramecia, called *sensitives*. In their cytoplasm, the killers contain minute particles called *kappa particles* which are responsible for the secretion of the poison. The kappa particles contain DNA, and they can duplicate themselves. The inheritance of the killer trait is not in accordance with Mendelian laws and seems to pass from generation to generation through the cytoplasm.

The relationship of plasmagenes to nuclear genes in the overall process of heredity remains to be settled. It seems that plasmagenes are self-duplicating cytoplasmic particles which are transmitted from cell to cell, but that their replication and activity is under the control of genes in the cell nucleus.

What Then Is the Gene?

Morgan thought of the gene as the smallest portion of a chromosome which could be exchanged during crossing over. He could determine the presence of genes only when mutations resulted in a change of the phenotype from the normal kind of fruit fly. For example, when fruit flies appeared with a bar-eye instead of the normal red eye, he could map the position of the gene for bar-eye by experiments involving crossing over and linkage with known genes. How does current knowledge of DNA structure and function affect our concept of the gene? How big a section of the DNA molecule *is* a gene? Unfortunately there is no simple answer. The kind of work that a geneticist does affects his concept of the gene.

To some geneticists the gene is the

codon. This is the unit of coding, and consists of a section three nucleotides in length. To others, the gene is the *muton,* the smallest unit which can be affected by mutation, and is only one nucleotide long. This would also be the size of the *recon,* the smallest unit which can engage in recombination during crossing over.

However, for the biochemical geneticist who thinks of genes as responsible for producing enzymes that can have a functional effect upon the organism, a gene must be responsible for producing a polypeptide (one gene — one enzyme theory of Beadle and Tatum). Polypeptides consist of 300–500 amino acid residues. Thus, in accordance with the triplet code, a gene would be 900–1500 nucleotides or more in length. This unit is the *cistron.* It is the unit which most biochemical geneticists regard as the gene responsible for producing a functional protein. However, we do not know whether the "gene" which governs tallness in garden pea plants and tongue-rolling in man is the same as the cistron of the biochemical geneticist.

COMPLETION QUESTIONS

[A] 1. DNA and RNA are two kinds of .... acid.

2. The fundamental unit of structure of the DNA molecule is the ....

3. A nucleotide consists of a phosphate group, a nitrogen base, and a(an) ....

4. The sugar present in RNA is ....

5. Two classes of nitrogen bases are .... and ....

6. The "uprights" of the DNA ladder consist of sugar and ....

7. In the "rungs" of DNA, cytosine bonds to ....

8. Nitrogen bases are held together in DNA by weak .... bonds.

9. The instructions transferred from DNA to RNA depend upon the sequence of the ....

10. The modern model of the DNA molecule was devised by .... and ....

(B) 11. The production of exact copies of DNA is known as DNA ....

12. In DNA, adenine bonds to ....

13. The nitrogen base in RNA which is complementary to adenine is ....

14. The site of protein synthesis in the cell is the ....

15. Two forms of RNA are .... RNA and .... RNA.

16. Amino acids are carried to the ribosomes by .... RNA.

17. The sequence of amino acids in a protein is determined by the sequence of the .... in DNA.

18. The number of "letters" in a "word" of the genetic code is ....

19. Sickle-cell anemia is caused by a defect in a single .... in the DNA molecule.

20. A mold which is currently used in genetic research is ....

21. Beadle and Tatum developed the .... theory.

22. PKU is a hereditary disease in which a defective gene results in the failure to produce a necessary ....

23. A mutant gene which interferes with so many important reactions as to cause death is called a(an) ....

24. Structures within the cytoplasm which seem to function in heredity are called ....

25. Kappa particles in *Paramecium* provide an example of .... inheritance.

MULTIPLE-CHOICE QUESTIONS

[A] 1. DNA is most closely associated with the (1) cytoplasm, (2) nucleus, (3) centrosome, (4) cell wall.

2. Knowledge of DNA is important in understanding living plants because DNA is (1) an enzyme required for photosynthesis, (2) an essential amino acid, (3) a carrier of genetic information, (4) a spiral-shaped molecule that gives rigidity to cell walls.

3. In general, the number of adenine units in a DNA molecule is the same as the number of units of (1) thymine, (2) cytosine, (3) guanine, (4) uracil.

4. One strand of a DNA molecule is arranged in the following order: adenine—guanine—thymine—cytosine. In what order is the other strand arranged? (1) adenine—cytosine—thymine—guanine, (2) guanine—adenine—cytosine—uracil, (3) cytosine—thymine—guanine—adenine, (4) thymine—cytosine—adenine—guanine.

5. Viruses seem to consist of (1) protein crystals interspersed with RNA molecules, (2) a protein coat wrapped around either DNA or RNA, (3) a nucleus and a tail by which the viruses move, (4) a living portion and a lifeless portion.

(B) 6. The DNA code is directly dependent upon the (1) number of carbons in the sugars, (2) number of deoxyribose molecules, (3) arrangement of purine-pyrimidine pairs, (4) position of the phosphate groups.

7. The code of messenger RNA is directly dependent upon the (1) number of ribose sugars in RNA, (2) purine and pyrimidine pairs in RNA, (3) amino acid content of the cytoplasm, (4) sequence of nucleotides in the associated DNA.

8. Transfer RNA functions in the transfer of (1) fatty acids, (2) amino acids, (3) ribosomes, (4) ADP.

9. During replication, molecules of DNA are most likely to separate between the (1) cytosine and guanine, (2) phosphate and deoxyribose, (3) ribose and adenine, (4) uracil and thymine.

10. A change in the structure of the DNA molecule is called (1) replication, (2) synthesis, (3) hydrolysis, (4) mutation.

CHAPTER TEST

1. Genes control cellular activity by leading to the production of specific (1) enzymes, (2) hormones, (3) ribosomes, (4) helixes.

2. The kinds of proteins produced in a cell are directly related to (1) sequence of nucleotides in the DNA molecule, (2) number of nucleotides in the DNA mole-

cule, (3) distribution of mitochondria in the cell, (4) sequence of sugars and phosphates in the uprights.

3. The DNA Code consists of sequences of nucleotides arranged in groups of (1) variable number, (2) twos, (3) threes, (4) fours.

4. In phenylketonuria, a defective gene results in the failure to produce (1) many enzymes, (2) one enzyme, (3) many ribosomes, (4) one ribosome.

5. Chromosomes are largely composed of (1) DNA and a protein coat, (2) RNA and a protein coat, (3) RNA and DNA, (4) protein.

6. A normal pairing of bases found in a molecule of DNA might be (1) adenine — thymine, (2) cytosine — adenine, (3) guanine — thymine, (4) uracil — guanine.

7. Which sugar is obtained from the hydrolysis of RNA? (1) glucose, (2) sucrose, (3) ribose, (4) dextrose.

8. One unique property of DNA is its ability to (1) reproduce, (2) catalyze, (3) replicate, (4) metabolize.

9. A nucleotide may be composed of (1) purine and pyrimidine bases, (2) adenine and thymine, (3) pyrimidine and phosphate, (4) purine, sugar, and phosphate.

10. A bacteriophage and a yeast cell are alike in that both (1) carry on intracellular digestion, (2) can carry on aerobic respiration, (3) contain nucleic acids, (4) have a cell wall.

PLANTS AND ANIMALS IN THEIR ENVIRONMENT

In previous chapters we have emphasized the individual organism. Probably you have noticed that the very word "organism" implies that a living thing is *organized* into subsidiary units, and by this time you are undoubtedly aware of the following levels of organization: atom, molecule, cell, tissue, organ, organ system, and organism. Now it is important to become aware of other levels of organization which are higher than the organism level.

LEVELS OF BIOLOGICAL ORGANIZATION

Level	Example
atom	hydrogen atom
molecule	amino acid
cell	nerve cell
tissue	nerve tissue
organ	brain
organ system	nervous system
organism	grasshopper
population	Ecology is concerned
community	with these levels
ecosystem	of biological organization.
biosphere	

The study of biology includes every level of organization from the atom to the biosphere. As shown in the table, ecology is concerned with those levels of organization that are higher than the individual organism.

CHAPTER 22 | Ecology

Ecology is the study of the interrelationships between living things and their environment (Greek *oikos*, house + *logia*, study). The environment consists of (1) physical factors (such as soil, water, and temperature) and (2) biotic factors (other living things which directly or indirectly affect organisms).

Ecology is concerned with interacting groups of living things rather than with individual organisms. However, every group is composed of individuals, and any study of ecology must include an investigation into the ways in which each creature is suited to live in its particular environment. Every species has developed adaptations to a particular set of conditions through the process of evolution.

A PHYSICAL FACTORS IN THE ENVIRONMENT

Every living thing is affected by the factors in the environment. The physical factors considered in this section are water, soil, temperature, light, inorganic nutrients, and oxygen.

Water

As a basic constituent of the internal environment of living things, water must be in plentiful and usable supply. Where water is not plentiful, organisms evolve adaptations to secure water and prevent its loss.

Even marine organisms are faced with a water problem. Marine vertebrates tend to lose water from their bodies because of osmosis to regions of higher salt concentration. Marine fishes can thus be considered as living in an environment that is physiologically dry. These fish possess specialized urinary systems to maintain a water balance. Some fish extract fresh water from the sea and return the salt by active transport at the gills.

The horned toad, which lives in desert conditions, has a thick scaly skin which prevents loss of water. The kangaroo rat and the camel secure sufficient water as a product of the oxidation of stored food. Many desert animals limit water loss by hiding in the shade or in burrows during the day and by foraging for their food at night.

Because the temperature of water changes slowly, the environmental temperature of marine organisms is relatively stable. Water's buoyancy provides support both for delicate creatures such as the jellyfish and for heavy ones such as the whale. Water is a medium for the deposit of sperm and eggs.

As light passes through water, it is gradually absorbed; therefore, photosynthetic organisms in the sea are restricted to the upper levels of the water. Here they must be able to withstand the destructive action of waves. Protective adaptations against the action of waves in the region between the tide levels can be noticed in the kelps and rockweeds (their flexibility) and in the snails and mussels (their shells).

On the basis of their adaptations to the amount of water available, plants may be classified as hydrophytes, mesophytes, and xerophytes.

Hydrophytes live either immersed in water or with their roots covered by water. Their root systems are small and lack root hairs. Stomates may be present, but only on leaf surfaces exposed to the air. Examples include water lilies, pickerel weed, and *Anacharis*.

Mesophytes have a moderate amount of water available to the roots. Loss of water by transpiration from exposed parts is prevented by adaptations such as the following:

Adaptations to Dryness

Waxy cuticle on leaf.

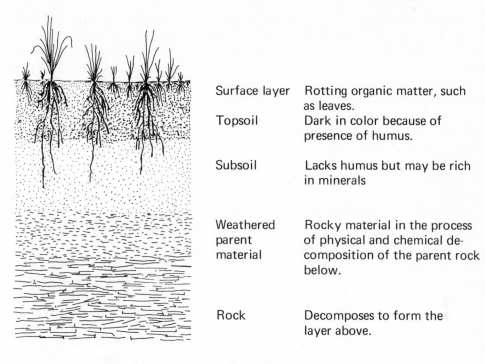

Surface layer	Rotting organic matter, such as leaves.
Topsoil	Dark in color because of presence of humus.
Subsoil	Lacks humus but may be rich in minerals
Weathered parent material	Rocky material in the process of physical and chemical decomposition of the parent rock below.
Rock	Decomposes to form the layer above.

Fig. 22–1. Vertical profile showing horizons in soil.

Stomates present only on lower surface of leaf.

Cork on the stem.

Leaves shed during winter when the soil water is frozen and unavailable.

Root systems of mesophytes may be extensive. Most of our common plants are mesophytes. Examples include the maple, the bean plant, the geranium, and the corn plant.

Xerophytes are plants that are adapted to long periods of extreme dryness. They have adaptations that help them to obtain and conserve water as shown below.

Adaptations to Extreme Dryness

Extensive root systems.

Fleshy leaves or stems for storing water.

Leaves reduced to spines.

A thick waxy cuticle.

Stomates sunken into pits, or even plugged temporarily during the dry season.

A reduced number of stomates.

Pine trees growing in moist acid soils are physiological xerophytes. Their needle-like leaves, with heavy cuticle and depressed stomates, are adaptations for the conservation of water. Some xerophytes are commonly cultivated as house plants. Examples are cacti, spurges, and beach grass.

Soil

Soil is composed of a varying mixture of rock particles, water, air, living organisms, and nonliving organic matter. Soil is formed by the slow decay of organic matter and by the weathering of rock. Some plants need an alkaline soil; others, such as azaleas and rhododendrons, require an acid soil. Clay soil has finely packed particles which may hold too much water. It may also form a layer which is impervious to water. *Humus* consists of decaying plant materials in the soil. Humus is loose, it provides minerals, and it holds water for a long period of time. Soil may also contain earthworms (which turn over and aerate the soil), nematodes, insects, protozoa, and bacteria.

A vertical profile through soil reveals five distinct horizons, as shown in Fig. 22–1.

The type of soil which is formed in a region depends upon the kind of parent rock, such as sandstone, limestone, shale, or granite, and upon the material deposited from above. Decaying pine needles and oak leaves increase the acidity of the soil. Mosses and grasses add large quantities of humus to the soil. The characteristics of the soil determine the kind of plants that can live in a region. In turn, the living organisms which are present affect the type of soil which is produced.

Temperature

With the exception of birds and mammals (both of which are warm-blooded), an organism's rate of metabolism is largely determined by the environmental temperature. Plants and animals are limited in their distribution by their ability to develop adaptations to conditions of temperature. The temperature of a region is affected by latitude, altitude, and terrain. An ascent in altitude up a mountainside has the same general effect on the habitat as a progression toward a polar region. This effect is illustrated in Fig. 22–2.

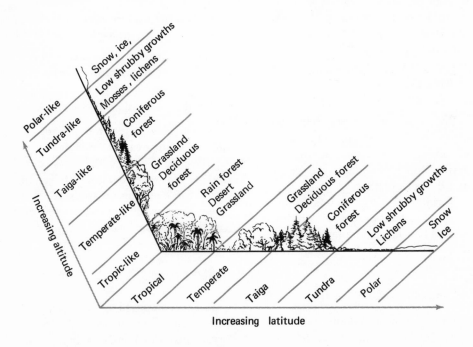

Fig. 22–2. Similarity of altitude and latitude in their effect upon habitat zones.

Ultra-violet, visible, and infra-red radiations from the sun penetrate the earth's atmosphere and are absorbed by the earth's land and water regions. These regions in turn warm the moist lower levels of the atmosphere. The temperature of the atmosphere drops on the average of 3° F for every 1000 feet of elevation. Mountain ridges and large bodies of water are elements of terrain that affect regional temperatures.

Light

Light is the ultimate source of energy for all life. However, either too much or too little of it can result in death. Light is a regulator of daily and seasonal activities for many kinds of plants and animals. Characteristics of light that affect communities of living things are its intensity, wave length, and duration.

Only about two per cent of the energy of visible sunlight reaching the earth becomes converted by photosynthesis to chemical energy. Of this amount, approximately 80 per cent of the total photosynthetic activity occurs in the sea. As light penetrates the sea, it is rapidly absorbed so that little photosynthetic activity occurs at depths greater than 300 feet. However, some photosynthetic activity does occur at depths as great as 600 feet. Accordingly, there is a zonation of marine forms:

The Photic Zone. The photic zone is the zone through which light penetrates. This is the region of primary productivity of the sea. Above the high tide mark along the coastline, blue-green algae, moistened by spray, cling to rocks. *Plankton* in the upper region of the ocean is the accumulation of floating microscopic plants and animals. Close below the surface, *Ulva* (sea lettuce) may be found attached to rocks.

Here also, and in deeper water, are rockweeds (brown algae), Irish moss (red algae), kelps (brown algae), and other red algae. Although most plants can absorb and use only the red-orange and blue-violet portions of the spectrum in manufacturing food, diatoms can use any of the wavelengths.

Sunlight is absorbed as it penetrates the water. However, all colors of light are not absorbed at the same rate. Water absorbs the long red and yellow rays faster than the short blue and green rays. Thus, deeper layers of the water are rich only in the penetrating green and blue rays.

The Aphotic Zone. At depths greater than a few hundred feet, so little light penetrates that the human eye cannot detect it. This is the *aphotic zone*, or region of no light. In this region of the ocean, animals subsist mainly on the dead and decaying organisms that descend from above. Nutrition here is heterotrophic. Many deep sea fish of the aphotic zone possess bioluminescent organs for luring prey.

Inorganic Nutrients

Living things contain carbon, nitrogen, hydrogen, oxygen, sulfur, and phosphorus. They also contain potassium, calcium, sodium, magnesium, and traces of other elements. Among the non-metallic ions needed by living things are phosphate (PO_4^{---}), chloride (Cl^-), sulfate (SO_4^{--}) and bicarbonate (HCO_3^-). These inorganic substances are incorporated into living organisms and are then returned to the environment to be used over and over again.

Oxygen

Oxygen is so plentiful in our atmosphere (about 20 percent of the atmosphere is oxygen) that we seldom think of its importance for living things. It is a necessity for all animals and must be present not only in the air but also in water and in the layers of soil which support life. Fish in a polluted stream die for lack of oxygen when the bacteria use up the available supply. In the Florida Everglades there are cypress trees whose water-covered roots cannot obtain sufficient oxygen. Accordingly, they have "knees" that are extensions of roots protruding above the water.

One should think in terms of conserving oxygen and clean air as well as in terms of controlling air pollution so that future generations will have enough of these materials to sustain life.

BIOTIC FACTORS IN THE ENVIRONMENT

Living things are interrelated with other living things as well as with their physical environment. Groups of living things are organized into (1) populations, (2) communities, (3) ecosystems, and (4) the biosphere.

1. Population

A *population* consists of organisms of the same species living together in a given location. For example, each of the following is a population: all of the *Paramecium caudatum* in a mixed culture of protozoa in the laboratory; all the snails of a particular species in a pond; all the gray birch trees in a forest.

2. Community

A *community* consists of populations of different species in a given location, interacting with each other. (The term *biotic community* is often used to distinguish between the community in the

ecological sense and the community considered from the usual viewpoint of human affairs.) A community is a self-maintaining unit in which energy and food materials circulate. Examples are: a lawn containing grasses, dandelions, earthworms, nematodes, and bacteria; a stream containing plants, worms, insects, snails, leeches, turtles, and fish.

3. Ecosystem

An ecosystem (or ecological system) consists of the (a) living community of a region and its (b) nonliving environment; these affect each other in an interrelated manner. In studying ecosystems, biologists emphasize the interrelationship of the biotic community and the abiotic (nonliving) environment. Changes in the environment affect the biotic community and changes in the biotic community alter the environment.

An ecosystem is *self-sustaining* if three conditions are met. If it (a) has a constant source of energy, (b) is a living system capable of incorporating this energy into organic compounds and of passing the energy from organism to organism, and (c) carries on a cycling of elements (such as nitrogen and carbon) between organisms and the environment.

Biologists discover interrelationships in ecosystems by careful and long observation of living organisms and by the collection of data which are interpreted by use of graphs, tables, and statistics. For example, an ecosystem under study may consist of a square meter of marshland, the edge of a stream, a tide pool, a balanced aquarium, or an entire forest, grassland, or desert. A manned self-contained space ship on a long voyage would also constitute an ecosystem.

4. The Biosphere

The biosphere is that portion of this planet in which ecosystems operate. This includes all parts of the water, soil, and air inhabited by living things. It does not include molten portions of this planet or regions high enough in the atmosphere to be devoid of life.

INTERACTIONS IN THE ECOSYSTEM

The *ecological niche* is the role that a species plays in the community — particularly its role in relation to food. The niche is not the same as the habitat. The habitat is the place where a species lives, the niche is its way of life in relation to other species. An example of a niche follows:

Woodpeckers make nesting holes in the Saguaro cactus which lives in the deserts of Arizona. Unused holes in the cactus are inhabited by two species of owl: the elf owl and the screech owl. Two adjacent holes on the same cactus plant might be occupied by an elf owl and a screech owl, and they would thus have the same habitat. However, the elf owl eats only insects, whereas the screech owl eats both insects and small rodents. Although they have the same habitat, these two species of birds fit into differing total food relationships of the community. Thus, they occupy differing niches. If insects become scarce, the screech owls could change to a diet of rodents and thus relieve the competition.

If two species having the same food requirements temporarily occupy the identical niche, competition sets in. No two species can occupy the same niche in the same location for a long period of time without one of two results ensuing. Either one of the species completely eliminates the other, or the two species each take over different parts of the niche — a process called *niche-splitting*. Two species of warblers can live in the same woods if one of them feeds on insects near the trunk of the tree and nests in the lower branches, while the other feeds in the outer limbs and nests high up. If only one of the two species is present, it may feed everywhere and nest anywhere.

Food Chains

A food chain involves the transfer of food energy through a series of organisms with repeated stages of eating and being eaten. The steps which may occur in a food chain are:

1. Producer. Producers incorporate the sun's energy in the manufacture of food from simple raw materials. Green plants are producers. At the bottom or base of all food chains is an autotroph (green plant or protist).

2. Primary consumer. Any herbivore, such as a grasshopper, cow, or minute water flea, which feeds directly upon the producers, is a primary consumer.

3. Secondary consumer. An organism such as the frog which eats the grasshopper, or man which eats the cow, is a secondary consumer. Any organism that feeds on a secondary consumer (such as the snake that feeds on the frog) could be considered a tertiary consumer. However, all consumers above the secondary level are roughly grouped as secondary consumers.

4. Decomposers. Decomposers are organisms that live upon dead things. They return inorganic materials to the environment where they are used again by other living things. Bacteria and fungi are decomposers.

The general pattern of a food chain

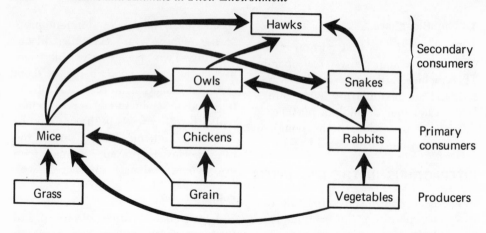

Fig. 22–3. A portion of a food web. Not shown are the decomposers which may act on each stage of the interrelated food chains. Usually the organisms at the top of the food web are bigger or heavier. For example, a mountain lion could be included near the top of this diagram.

is producer → primary consumer → secondary consumers, with decomposers acting on each stage of the chain. Some examples of food chains are as follows:

1. Diatoms → water fleas → shrimp → small fish → herring → man.
2. Shrub → June bug larva → lizard → snake → hawk.
3. Wheat plant → rabbit → coyote → mosquito → bat.

Food Webs

In a community, various food chains are interconnected in a *food web* as shown in Fig. 22–3.

If the rabbits should die off because of disease, the owls would not go without food. Because less vegetation is now consumed by rabbits, the mice have more vegetation available to them and they increase in numbers – thus substituting for the rabbits as food for the owls. However, with the attention of the owls now directed toward the mice, the few rabbits remaining have a better opportunity to increase their

numbers. The greater the number of alternative pathways in a community's food web the more stable is that community.

Sometimes the upsetting of food webs by man has unexpected results. For example, a campaign to exterminate owls because of their occasional predation upon chickens may result in an increase in other animals previously held in check by the owls – rabbits and mice. These now increase their inroads upon man's supply of grain and vegetables.

Types of Food Relationships

Methods for the transfer of food energy in a food chain fall within three categories – predation, scavenging, and symbiosis.

Predation
Scavenging
Symbiosis
  Commensalism
  Mutualism
  Parasitism

Predation. A *predator* is a consumer which kills another organism (the prey) and then eats it. The predator may kill the prey before or during the eating process. The eagle, owl, wolf, sea lion, and man are some examples of predators.

Predators are fitted for their method of food-getting by such adaptations as sensitive eyes, ears, and organs of smell and by strong jaws and long claws. On the other hand, the prey, too, has keen sense organs to escape detection and may be adapted for burrowing, or for rapid running, flying, or swimming. The squid camouflages its escape route by discharging an inky cloud into the water. The robber fly, which resembles a bumble bee, is not likely to be disturbed by an animal which has previously encountered a bee. This copying adaptation is an example of *mimicry*.

Cooperative behavior by members of a group is a behavioral adaptation which reduces the losses from attackers. For example, herds of grazing cattle or sheep arrange themselves so that the strongest are on the outside; moreover, many sets of eyes and ears are on the lookout for danger. Tiny tubifex worms bunch up in a mass that is too large for a small fish to swallow, but individual worms are readily picked off.

Insectivorous plants, which take in insects, reverse the usual pattern of predation. Such plants are predators upon animals. (See Fig. 22–4.) An insect may alight upon sensitive "trigger hairs" on the inside of the pair of leaves of the Venus's Fly Trap. The leaves close quickly by a loss of turgor (water pressure within the cells) and trap the insect.

Extracellular enzymes then digest the insect. Insectivorous plants live in acid

Fig. 22–4. Venus's Fly Trap.

bogs where the pH is too low for the growth of nitrate-forming bacteria. Insect-eating probably is an adaptation by which the plant makes up for a shortage of soil nitrates by the intake of animal proteins. Insectivorous plants are green, and they carry on photosynthesis in addition to capturing an occasional insect. Some examples of other insectivorous plants are the sundew, the pitcher plant, and the bladderwort.

Scavenging. A scavenger is an organism that "preys" upon the bodies of dead animals or plants. Vultures, jackals, and some maggots are examples of animal scavengers. Scavenging plants such as fungi and bacteria of decay, which live on *dead* things, are also known as *saprophytes.* (By contrast, an organism which lives off *living* things is a *parasite.*) Scavengers carry on the important task of returning inorganic materials into the cycles of nature. Some animals are difficult to classify as either predators or scavengers. For example, the snapping turtle is a predator most of the time, but if it finds a dead organism that is edible it will act as a scavenger.

Symbiosis. Symbiosis (*syn,* together + *bios,* life) is a nutritive relationship in which organisms of different species "live together." One organism benefits by the relationship. The other organism may be:

(*a*) unaffected — commensalism
(*b*) benefited — mutualism,
(*c*) harmed — parasitism

Commensalism. Commensalism means "at table together." In this symbiotic relationship, one organism benefits by consuming the unused food of the other organism, which is not harmed.

Some examples are: (1) The remora fish uses a suckerlike modification of the dorsal fin to attach itself temporarily to a shark. When the shark feeds, the remora picks up floating scraps. The shark does not seem to be inconvenienced. (2) Barnacles attached to whales are transported to new feeding grounds. (3) Small fish live among the tentacles of sea anemones and are immune to the anemone's stinging cells. They derive protection and probably some food from this relationship.

Mutualism. In this form of symbiosis both partners benefit. Some examples are: (1) Nitrogen-fixing bacteria, which occur in nodules on the roots of legumes, "fix" molecular nitrogen. The plant uses the nitrates and organic nitrogen compounds given off by the bacteria. (See nitrogen cycle, Fig. 22–13.) In return, the bacteria receive nourishment from the host. (2) A *lichen* is an association of a specific fungus and a specific alga. By photosynthesis the alga provides food for the heterotrophic fungus and in return receives protection from drying out. The alga and the fungus of a lichen may be separated and grown individually in the laboratory, but in nature the fungus partner does not survive alone. (3) Termites are insects which eat wood but cannot digest cellulose. This func-

tion is performed for them by symbiotic protozoa which live in the digestive tract of the termites. In return the protozoa receive food, water, and protection. (4) A certain species of crocodile opens its mouth and allows a small bird, the Egyptian Plover, to pick leeches and food particles from its gums. (See Fig. 22–5.) (5) Bacteria living in the intestine of man receive food and water, and produce vitamins which are helpful to the host. (6) Ants care for aphids and receive from them a sweet milklike secretion which the aphids liberate when stroked by the ant's antennae.

Fig. 22–5. Mutualism between the Egyptian Plover and the crocodile.

Parasitism. A parasite obtains nourishment from another living organism (the host) to which it does some degree of damage while living on or in its body. Normally the parasite does not kill the host. Some examples are: (1) Bacteria which cause diseases such as diphtheria and typhoid fever. (2) Fungi which cause wheat rust, athlete's foot, or ringworm. (3) Protozoa which cause malaria or amoebic dysentery. (4) Tapeworms and hookworms. (5) Arthropods such as lice and ticks.

The study of human disease may be considered as an ecological study involving adaptations of the parasite and the host. In the condition known as *disease,* man provides the environment for microorganisms which occupy an internal niche as parasites.

Food Pyramids

Pyramid of mass. In a pond community there are innumerable microscopic algae which are the main producers. If weighed, these algae would constitute a considerable mass. All the copepods (minute arthropods) in the pond which derive their nourishment from the algae constitute a smaller mass. The mass of all the minnows which feed upon the copepods is still less. Least of all is the mass of the minnow-eating fish (bass) which in the long run derive their food from the algae. Thus, from producers to primary consumers to secondary consumers and so on, there is a pyramid of mass in a community. This pyramid has a broad base and a decreasing mass at each level of the food web. (See Fig. 22–6.) If a 150-pound person were to eat only sheep, he would need much more than 150 pounds of sheep to sustain him.

Pyramid of numbers. The organisms

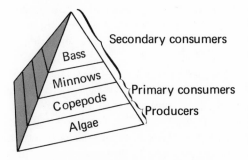

Fig. 22–6. Pyramid of mass, pyramid of numbers, and pyramid of energy in a pond community.

that are higher on the food chain are usually heavier than those lower down (compare a bass with a minnow). Accordingly, the pyramid of mass may also be interpreted as a pyramid of numbers. There are a large number of producers at the base of the pyramid and a small number of consumers at the top. There are more individual algae than bass in the pond. There are more worms than robins in a community.

Pyramid of energy. Only a small portion of the total energy incorporated by photosynthesis into the autotrophs of a community is passed on to each succeeding level of the food web. At each level, energy is lost. Loss of energy occurs because (1) each organism uses some of the chemical bond energy present in its food for carrying on its own life processes, and (2) no transfer is 100 per cent efficient — there is always some loss as heat energy. Thus, the transfer of energy in the food web may be represented as a pyramid of energy. There is more stored energy in all the producers of a community than in all the secondary consumers.

An understanding of food pyramids permits biologists to make predictions. For example, a decrease in the algae population of a lake in the spring will probably lead to a decrease in the bass population during the following year.

Growth of Populations

Under optimal conditions all populations have a tremendous capacity to increase, as shown in the typical S-shaped curve of population growth. (See Fig. 22–7.)

The increase is relatively slow at first (Phase A). However, each addition to the number of organisms already present is itself the basis for further in-

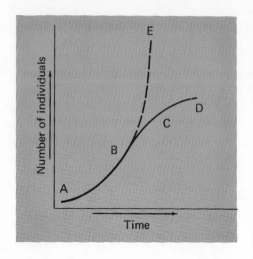

Fig. 22–7. Curve of population growth.

petition by other organisms, depletion of food supply, and the effect of predators) limit the increased rate of growth and finally the population reaches equilibrium at Phase D. Here as many individuals die as are born. The S-shaped curve is typical, *under optimal conditions*, of the populations of all living things, including man.

Cycles of Growth. Instead of following the S-shaped growth curve, populations in nature often follow a repeated up-and-down pattern or cycle. For example, the population of snowshoe hares in the Hudson Bay region of Canada follows a cycle which repeats about every ten years (Fig. 22–8). Overpopulation is followed by epidemics in which parasites reduce the numbers of hares. Populations in nature tend to be self-regulating because (1) predators and parasites reduce numbers, and (2) the capacity of the environment to supply the population is limited.

crease in number. Therefore, in Phase B there is rapid increase in population. This is known as the *phase of exponential* (or *logarithmic*) *growth*. If unchecked, the population would follow the dotted line from B to E. However, factors in the environment (such as com-

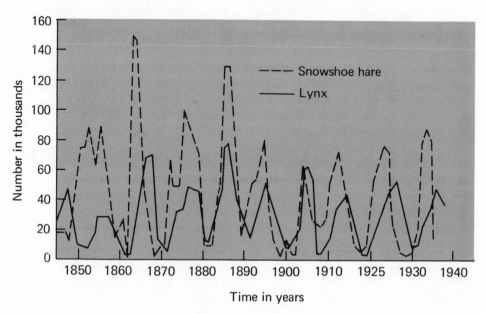

Fig. 22–8. Cycles in population of snowshoe hare and lynx.

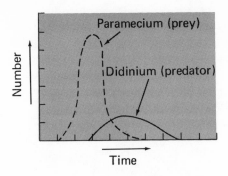 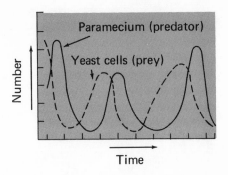

Fig. 22–9. Predator-prey relationships in laboratory cultures.

Whether one species in a community is an important source of food for another species can sometimes be determined by comparing graphs of their populations. Fig. 22–8 shows changes in the population of the lynx as well as the hare. (The lynx is a member of the cat family which is often observed to prey upon the hare.) The lynx cycle closely follows the hare cycle, but there is a slight lag so that the population peaks do not coincide. Statistical data of this kind provide biologists with large-scale evidence (but not proof) of an important ecological relationship between two species.

Other examples of predator-prey relationship are shown in Fig. 22–9.

Interfering with food chains. The Kaibib Deer. Before 1907, the Kaibib Plateau of Arizona had a herd of healthy deer with a stable population. (See Fig. 22–10.) Heavy predations by pumas, coyotes, and wolves kept the number of deer down so that the deer were amply fed by the available vegetation. In order to protect the deer from their "enemies," man killed off large numbers of the predators. As a result, within a span of 17 years the deer population increased twenty-five fold (from 4,000 to 100,000). Now the

vegetation could no longer support the deer population, and more than 50,000 suffered death by starvation within a period of two years. Thereafter the deer population slowly declined to a smaller number (10,000) of individuals who barely maintained themselves by feeding on the overgrazed land. In the original stable situation, the predators, while destroying *individuals,* had been beneficial to the *group* of prey.

Competition. Competition among organisms is of two types: intraspecific and interspecific.

Intraspecific competition (intra = within) is the competition among individuals of the same species. Plants as well as animals may compete with members of their own species for such factors as light, raw materials, food, water, and space. Some examples are: (1) Cattails compete with each other for space and soil; (2) deer compete for a limited food supply. The winners in intraspecific competition are those individuals which are best adapted to the ecological niche of the species.

Interspecific competition (inter = between) is the competition between individuals belonging to different species. Some examples are: (1) Cattails

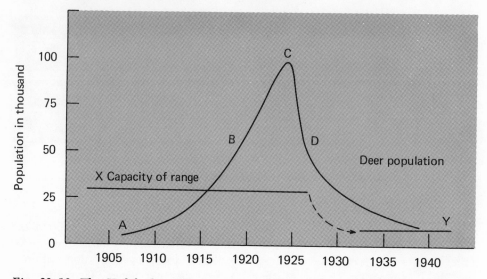

Fig. 22–10. The Kaibib deer. Changes in population show what happens when the ecological balance is upset. (A) Predators keep numbers below normal capacity of range. (B) Coyotes, pumas, and wolves were killed. (C) Overpopulation of deer. (D) Sixty percent of deer die in two years. (X) Normal capacity of range to support deer. (Y) Capacity of range reduced by overgrazing.

and duckweed compete for the same water; (2) hawks and owls compete for mice as food.

MATERIAL CYCLES IN THE ECOSYSTEM

The inorganic substances taken in by living things are returned to the environment and are used over and over again in cycles by other living things. Three of these cycles are the carbon cycle, the oxygen cycle, and the nitrogen cycle.

The Carbon Cycle

Plants obtain carbon in the form of carbon dioxide which is taken from the air or water in the process of photosynthesis. Although carbon dioxide makes up 0.03–0.04 per cent of the atmosphere, there is an additional reservoir of carbon dioxide and of bicarbo-

nate ions in the waters. Locate CO_2 in the diagram of the carbon cycle (Fig. 22–11). During photosynthesis, carbon dioxide is reduced to sugar and other organic compounds of plants. Animals incorporate carbon compounds into their bodies by eating plants. As a result of cellular respiration and decay (and the burning of plants), the carbon compounds of both plants and animals are changed to carbon dioxide.

The calcium carbonate present in limestone deposits such as coral are an additional reservoir of carbon dioxide. Coal and oil (fossil remains of plants and of protists respectively) are fuels used by man. These fuels withhold carbon from circulation until they are burned.

The Oxygen Cycle

Oxygen constitutes one-fifth of the atmosphere and is also dissolved in the waters of the earth. The oxygen given

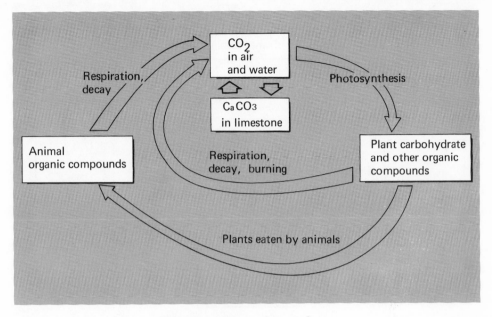

Fig. 22–11. The carbon cycle.

off during photosynthesis is split off from molecules of water. In cellular respiration and in burning, oxygen is again combined with hydrogen to form water. (See Fig. 22–12.)

The Nitrogen Cycle

Although the atmosphere contains a huge storehouse consisting of 79 per-

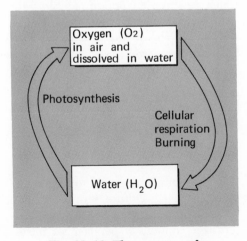

Fig. 22–12. The oxygen cycle.

cent nitrogen, this nitrogen is in the form of free (unattached) nitrogen which cannot be used by living things to produce amino acids and proteins. Free nitrogen (N_2) is *inactive*, which means that it does not readily combine with other substances. However, it is made available through the nitrogen cycle. (See Fig. 22–13.) Nitrogen is absorbed by plants from soil water in the form of *nitrates* (NO_3^-) such as potassium nitrate. Plants use the nitrogen of nitrates in synthesizing *plant proteins*. Animals eat plants and use the plant proteins to produce *animal proteins*. After plants and animals are dead, they are decomposed by *bacteria of decay*. These bacteria liberate nitrogen in the form of *ammonia* (NH_3). Some kinds of *nitrifying bacteria* obtain their energy by oxidizing ammonia to *nitrites* (NO_2^-), and other nitrifying bacteria oxidize nitrites to *nitrates* (NO_3-). This completes the major portion of the nitrogen cycle, and permits

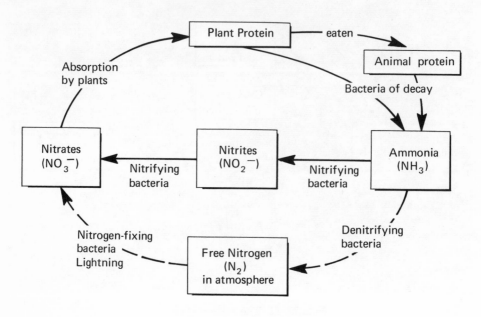

Fig. 22–13. The nitrogen cycle.

the re-use of nitrogen for living things.

Under conditions which favor the growth of *denitrifying bacteria,* however, the nitrogen which is present in ammonia is lost from the major cycle. As part of their metabolism, these bacteria convert the nitrogen of ammonia to *free nitrogen* which passes into the atmosphere or into the loose soil.

This free nitrogen may be fixed (to "affix" means to attach) in two ways: (1) By the action of lightning to form oxides of nitrogen and, (2) by the action of *nitrogen-fixing bacteria.* Some nitrogen-fixing bacteria are free-living in the soil, but many live in a mutualistic relationship in nodules on the roots of legumes (peas, beans, peanuts, alfalfa, clover). The nitrogen-fixing bacteria in nodules produce nitrogen compounds, such as nitrates, for their own use and for the use of the plant.

The American scientist, George Washington Carver, was aware of the importance of legumes in restoring the fertility of soil. Accordingly, he urged sharecroppers in the South to plant peanuts on land which had been depleted of minerals by the continual growing of cotton. This resulted in a bumper crop of peanuts for which, unfortunately, there was little market. Carver later did much research to find profitable new uses for peanuts. The planting of peanuts was a cheap way of assuring that free nitrogen would be converted to useful compounds.

Industry is also concerned with cheap methods for converting free nitrogen to useful compounds, and chemists have devised methods of nitrogen fixation which are successful on a commercial basis.

When soil is depleted of minerals by intensive agriculture, fertilizers must be added to the soil. A 5–10–5 chemical fertilizer contains 5 per cent of nitrogen, 10 per cent phosphoric acid, and

5 per cent soluble potash by weight. This fertilizer adds necessary nitrogen, phosphorus, and potassium to the soil. Homeowners add powdered limestone ($CaCO_3$) to their lawns to counteract the tendency of the soil to become too acid for proper growth of grass. In order to counteract the alkalinity introduced into the soil by concrete, they add ammonium sulfate — (NH_4)$_2SO_4$ — to beds of acid-loving magnolia, azalea, andromeda, and spruces planted alongside house foundations. Instead of using chemical fertilizers, homeowners may use organic fertilizers such as manure and compost.

CHANGES IN THE ECOSYSTEM

When biologists study biotic communities over a period of many years, they find successive stages of change in the plant and animal life. This process of change is called *ecological succession.*

Dominant Species

The *dominant species* in each of the stages of ecological succession is the one that exerts control over the other species present. For example, in one kind of forest, spruce trees are the dominant species — they create soil and light conditions which prevent the growth of ferns and of gray birch trees.

Ecological Succession Defined

Ecological succession is the orderly process by which one biotic community is replaced by another. The kinds of plants that are dominant at any one time determine the stage in a series of successions. The animal population also changes in accordance with the changes in the plants.

Changes occur because each successive community changes the environment, making it more unfavorable for itself and more favorable for the following community. Examples: (1) White pine trees produce such a heavy shade that their own seedlings cannot grow in this region. However, beech and maple seedlings grow well in the shade of pine trees and also in the shade of their own parents. Thus, a beech-maple forest may succeed a white pine forest. (2) The accumulation of humus from lake plants such as lilies and cattails forms a bog in which the lilies and cattails cannot grow. However, this bog provides good conditions for sedges and reeds.

Pioneers

The *pioneer organisms* are the first to populate a given region. For example, a lichen may be the pioneer on barren rock. Acids produced by the lichens attack the rock and help to form bits of soil. When the lichens decay, humus is produced and now conditions are right for mosses to supplant the lichens.

Climax

The *climax community* is a stable, self-perpetuating community in which populations exist in balance with each other and with the environment. In northeastern United States, climax communities include the *beech-maple, oak-hickory,* and *hemlock.* Each is named for the plant or plants which are the dominant forms. In the prairie states, where there is less rain, the climax community may be a *grassland.* The climax community may be replaced by another community as a result of major climatic, geologic, or biotic change. A forest fire may undo the natural processes of hundreds or thousands of years, and a

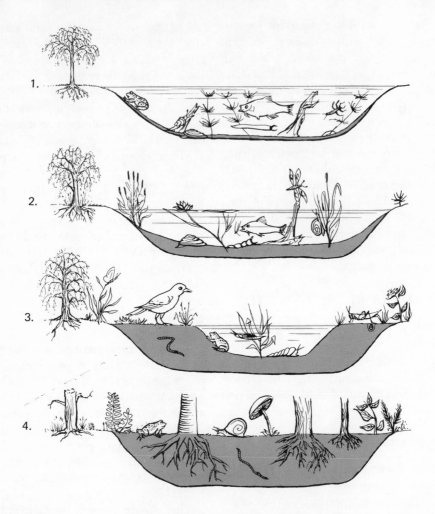

Fig. 22–14. Succession from a pond.

new succession begins with pioneer organisms.

Succession from Bare Rock

The successive stages starting from bare rock, in northeastern United States, may be:
lichens
mosses
grasses and ferns
shrubs
trees
gray birch, poplar (fast growing,
sun-loving species)
white pine
maple and beeches

Succession from a Shallow Pond

The stages by which a shallow pond changes to a woodland area may be traced by walking from the pond back into the surrounding forest:
(1) *In the pond:* Water plants such as algae, *Elodea*, pondweed, knotweed. Animals such as protozoa, invertebrates, and fish.

(2) *Shallow water:* sedges, rushes, reeds, cattails, water lilies.
(3) *Land surrounding the pond:* grasses, herbs, shrubs, low willow trees. Insects, frogs, and snakes may be present. Here silt and humus are slowly forming solid ground.
(4) *Surrounding region:* woodland.

REASONING EXERCISES

1. Plants that live in deserts and salt water marshes have some similar adaptations. Can you give any reasons for this?
2. Why are farmers concerned about the pH of soil?
3. Distinguish between the kinds of organisms found in the photosynthetic zone of the ocean and those found on the ocean bottom in the aphotic zone. What reasons can you give for the differences in kinds of organisms in each area?
4. Distinguish between a population and a community.
5. Give an example of a food chain. Show how the food chain illustrates the idea of the food pyramid.
6. Distinguish between an ecological niche and a habitat.
7. What happens when two species are in the same ecological niche?
8. How is the carbon cycle essential to the continuation of life?
9. Distinguish between nitrifying bacteria and denitrifying bacteria.
10. What is the role of the dominant species in ecological succession?

(B) WORLD BIOMES

A climax community may occupy a relatively small geographic region. However, this is not always the case. A *biome* is a climax community of plant and animal life that is typical of a broad region with one kind of climate. Not all the sites within this region need be of the climax type but, local climatic conditions permitting, there is a general uniformity of dominant plant species within the region.

The climate of a region depends upon the temperature, precipitation, sunlight, and winds. Nearness to mountains and to large bodies of water also affects the climate of a region.

Terrestrial Biomes

The oceans provide (1) a stable environment, (2) little fluctuation in temperature, (3) plenty of water, and (4) a uniform distribution of nutritive chemicals. Plants and animals which migrated to the land faced the following problems of the physical environment: (1) lack of a supporting medium for the body, (2) lack of water, (3) varying chemical composition of the soil, and (4) varying temperature. Terrestrial organisms met these problems by adapting to the conditions of restricted regional environments. Here they became organized into distinctive communities called *biomes*.

A progressive lowering of the average temperature is reflected in the sequence of the following four biomes:

1. **The Tropical Rain Forest, or Jungle.** The constant warmth of the tropical rain forest is supplemented by an abundant and continuous supply of rain water. Tropical rain forests are present in Central and South America, in Southeast Asia, and in West Africa. Vegetation is so dense in a rain forest that little light reaches the forest floor. Vines climb toward the light by twining around trees. There are numerous *epiphytes* — plants which grow on other plants without parasitizing them. They obtain their moisture from the air. Trees are tall, they do not shed their leaves, and they vary so greatly in kind that it is rare to find two members of the same species growing near each other. Many of the animals (monkeys, sloths, lizards, snakes, and birds) live in trees.

2. **The Temperate Deciduous Forest.** The average temperature is cooler in the temperate deciduous forest than it is in the tropical rain forest. The temperate deciduous forest has conditions of moderate, well-distributed rainfall, with cold winters and warm summers, such as are present in the eastern part of the United States. The trees are *deciduous;* that is, they adapt to the absence of available water in the winter by dropping their leaves. There is little variety in the trees of any one forest region. They are likely to consist mainly of beeches and maples; oaks and hickories; or willows, cottonwood and sycamore. Animals include the deer, fox, woodchuck, squirrel, and raccoon.

3. **The Taiga.** Cold temperatures, accompanied by a summer growing season, produce the coniferous forests of the taiga in Canada. Here the spruce and fir predominate. Because of the dominant position of the moose, the taiga is sometimes called the "spruce-moose" biome. Other common animals are the black bear, squirrel, wolf, and lynx. Birds migrate south in the fall from this region.

4. **The Tundra.** The northern zone surrounding the Arctic Ocean is so cold that the ground is permanently frozen a few feet below the surface. During summer thaws, however, the region is extremely wet, with many bogs and ponds. Trees are stunted; mosses (*Sphagnum*) and lichens cover vast areas. Herbs produce brilliantly colored flowers during the short growing season. The animals present include the caribou, musk ox, polar bear, and Arctic hare. Insects occur in swarms. Some birds (ptarmigan) and animals (snowshoe hare) turn white in winter.

Two other biomes found in North America are the grassland (plain) and desert. The controlling factor in these regions is water rather than temperature.

The Grassland. The major North American grassland, now largely used for growing crops, was the region east of the Rocky Mountains. This area lacks water for the following reason: As the prevailing moist winds from the Pacific rise up the westerly slopes of mountain chains, the air expands, cools, and drops its water on the west side of the mountains below. As the air continues to the east of the mountains, it is relatively dry. Here the low irregular annual rainfall prevents the growth of trees except along rivers and streams.

Grasses, adapted to survive through irregular periods of drought and downpour, are the dominant form of plant life. The plentitude of grass for fodder, and the absence of shelter from predators, provide favorable conditions for many herbivorous ungulates (hoofed mammals) which are adapted to escape from enemies by running. Bison and antelope were numerous on the level regions before the arrival of the plainsmen who planted corn and wheat. Domesticated cattle and sheep are raised in the hilly regions.

The Desert. The small amount of precipitation which does fall in deserts is likely to occur during erratic showers. The temperature may fluctuate greatly between hot days and cold nights. After a rainfall, the plants grow rapidly, bloom brilliantly, and produce seed all within a period of a few days. Desert conditions occur in some western states, such as Montana and Arizona. Xerophytes such as cactus, sagebrush, and mesquite are the dominant plants. Animals include kangaroo rats, roadrunners, lizards, insects, and arachnids. Many animals live in burrows where the humidity is greater and where there is less fluctuation in temperature.

Microclimates

The general atmospheric and weather conditions of a region constitute its climate. However, the conditions of temperature for an organism a few inches below the ground are different than for an organism on the surface. The climate for an organism which dwells on the forest floor is cooler, more humid, and less windy than for one which inhabits the tops of trees. Despite general regional climates, organisms live in their own small climates, or *microclimates*.

The Marine Biome

The two-thirds of the earth's surface covered by oceans contains more plants and animals than are found on land. Probably 90 per cent of the food-

making and oxygen-releasing on this planet occurs in the waters. Most life on our planet occurs in fresh or salt water.

The main physical factors affecting the growth of aquatic organisms are the quantity of available oxygen and carbon dioxide, the temperature, the presence of dissolved or suspended materials, and the intensity of light.

Variations in temperature are not as great in aquatic areas as on land, and the aquatic areas are thus the largest and most stable ecosystems on earth.

The oceans of the world constitute a huge continuous body of water. Some characteristics of this world ocean are:

1. It absorbs and holds large quantities of solar heat and moderates the earth's temperature.

2. It contains a relatively constant supply of nutritive materials and dissolved salts.

3. It is the most populated of all the habitats on this planet.

Environmental conditions in the oceans are far more uniform than on land, but even within the marine biome distinct zones occur (Fig. 22–15):

1. The intertidal zone. The region of land which is covered by water during high tide and which is uncovered during low tides is known as the region between high and low tides, or the *intertidal zone*. This region is occupied by marine plants and animals which can withstand periods of dryness and variations in temperature. Many of the algae and small colonial animals can hold fast to rocks, to the shells of snails, and to underwater plants. Snails and barnacles conserve water by closing their shells when the outgoing tide leaves them high and dry. Algae are the basic producers of the intertidal zone. Most of the algae are microscopic floating types, but masses of rockweed

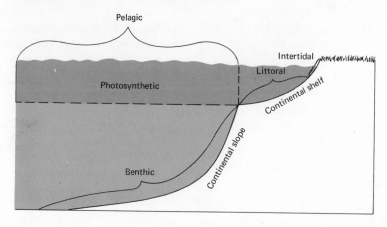

Fig. 22–15. Zones of the ocean. The *benthic zone* is the ocean bottom beyond the continental shelf. The *pelagic zone* is the open sea beyond the continental shelf. The *photosynthetic zone* is the 600-foot-deep region through which light penetrates. This zone exists close to shore and far out at sea in the pelagic zone.

and kelps are readily visible. Red and brown algae are common in the oceans.

In the intertidal zone, complex food webs may occur. Sponges, clams, mussels, and oysters filter algae from the water. Sea urchins, shrimp, copepods, and small fish feed on floating or attached plants. The carnivores include the starfish, sea anemone, octopus, and squid. Worms, crabs, and bacteria play the role of scavengers.

2. The littoral zone. This is the region of the *continental shelf.* It may extend several hundred miles from shore until it drops to a depth of about 600 feet. This is the depth to which some light penetrates.

The littoral zone along the coast is the most productive region of the oceans. Life in this region is affected by the presence of organic and inorganic materials brought down by the rivers. In these shallow waters, there may be a luxuriant growth of brown algae such as the kelps.

Plants are not abundant on unstable sandy bottoms, but crustacea, mollusks,

and annelid worms burrow into the sand. Sea cucumbers and crabs abound on muddy bottoms. On rocky bottoms, where currents are usually strong, barnacles and mussels usually attach themselves to the rocks. Fishes of the littoral zone include flounders and sting rays.

Eel grass is a seed plant which forms extensive underwater growths in shallow northern regions. In warm climates, corals, aided by the action of red algae, build reefs. Within these reefs live brilliantly colored fishes.

3. The pelagic zone. The *pelagic zone* is the region of the open sea where plankton is the basic food. *Plankton* consists of the drifting minute plants and animals on or near the surface of bodies of water. The main members of the plant plankton (*phytoplankton*) of the sea are diatoms, other kinds of microscopic algae, and dinoflagellates. The phytoplankton is the basic producer of ocean food chains.

The animal plankton (*zooplankton*) includes protozoa, copepods, small shrimp, and the larval forms of many animals. The copepods feed on algae, and the copepods in turn serve as the main food of the huge whalebone whale. From alga to copepod to whale is a short food chain. Other food chains include many kinds of carnivorous fish (e.g., sharks) and squid.

The productivity in the sea varies greatly from place to place. The presence of phosphates and nitrates seems to regulate this productivity. The continued growth of the teeming phytoplankton depends upon material washing in from the land or material rising up from below. In regions such as California and Peru, where rising currents of water bring minerals for the phytoplankton, there is a wealth of consumers. Accordingly, these areas provide good fishing territory.

The photosynthetic zone extends to a depth of 600 feet. Below this level, there is very little light for the growth of plants. The longer wavelengths of light (reds and yellows) are absorbed most rapidly. Thus the short-wave blues and violets penetrate the deepest. Green plants use mainly the reds, yellows, blues, and violets. However, some marine algae that dwell at the lower limits of light contain pigments that permit them to use all wavelengths of light.

Animals living below the photosynthetic zone depend for food on dead animals and plants that sink from above and also upon fish that swim from one level to another. Many of the fish and invertebrates of the deep dark regions have organs that glow in the dark by bioluminescence. These luminescent organs seem to serve as a lure for prey. They may also be a mark of recognition, just as color patterns function in lighted regions of the sea and land.

4. **The Ocean Bottom.** Biologists formerly believed that the tremendous pressure exerted by the water upon the ocean bottom prevented the existence of life there. However, recently many organisms have been found on the cold, dark ocean bottom. These are mainly scavengers that depend for food upon the rain of dead organisms from above. Bacteria are present in a fine ooze which is composed mainly of the skeletons of radiolarian protozoa.

Oceanography

The scientific study of the physical and biotic aspects of the seas is known as *oceanography* (or oceanology). Great national interest in oceanography has recently developed because of the following possible applications of oceanographic research:

1. **Feeding an expanding world population.** In some underdeveloped countries, the diet of the hungry is severely restricted in proteins. Children on such a diet develop the disease *kwashiokor*. One promising development for reducing this protein hunger is chemically to process whole fish, including the scales, bones, and internal organs, in such a manner as to yield a fine powder called *FPC* (fish protein concentrate). This powder is added to various foods. For example, it is being used in overpopulated countries to enrich the flour used for baking.

Oceanographers hope that our knowledge of marine ecology will increase in the future and that this will make possible a greater annual harvest of fish, lobsters, and other marine life. Obviously, such a development could be of great help in feeding an expanding world population.

2. **Desalting of sea water to provide fresh water.** Salt water is gradually invading the water table of underground

fresh water in coastal regions. This invasion makes fresh water wells useless, and also prevents the growth of deep-rooted plants. Stream pollution, the increasing demands of industry, and the needs of expanding urban populations make it imperative to obtain new sources of fresh water. Therefore, scientists are seeking methods for desalting (desalinizing) sea water to provide fresh water. One method for desalting sea water involves using atomic energy to furnish the heat for a distillation process.

3. Obtaining oil. Oil is obtained by offshore and deep-water drilling.

4. Recovery of minerals from the sea. Magnesium, bromine, boron, uranium, copper, and gold are present in the sea. Some of these are already being obtained from the sea by methods that are commercially successful.

5. Better understanding of the geology of the ocean bottom, of ocean currents, and of the biology of marine organisms.

6. Improved protection against submarine warfare.

Recent developments in oceanography include the following:

1. In 1966 Congress established the National Council on Marine Resources and Engineering Development. This agency parallels the National Aeronautics and Space Administration (NASA), and will develop a national oceanographic program. Large sums of money have been provided for oceanographic research and development.

2. Congress is considering the establishment of Sea Grant colleges to expand oceanographic research and to train needed marine scientists. Some well-established marine institutes are the Woods Hole Oceanographic Insti-

tute at Cape Cod, Massachusetts; the Scripps Institution of Oceanography at the University of California; the Institute of Marine Sciences of the University of Miami; and the Lamont Geological Laboratory of Columbia University.

3. Undersea laboratories have been developed to observe marine life and to test the adaptations of man to life at high pressure. In 1930 William Beebe descended to a depth of more than half a mile in a device called a *bathysphere*. Auguste Piccard's bathscaphe can go more than six miles deep. The Frenchman, Jacques-Yves Cousteau, developed an underwater cabin in which a colony of scientists lived and worked for several weeks at a depth of 85 feet. The United States has developed underwater cabins called *sealabs*.

4. Sonar is being used to trace movements of groups of large sea animals and of masses of plankton. *Sonar* is a method for locating underwater objects by means of the sound waves which they reflect as "echoes."

The Fresh Water Biome

The fresh water biome consists of ponds, lakes, and streams.

Ponds. Plankton occurs in ponds as well as in the oceans. The main producers in the pond plankton are green and blue-green algae, dinoflagellates, and diatoms. Consumers in the plankton include protozoa, rotifers, and minute crustacea. Most members of the plankton are drifters but some of them swim. Filamentous algae, such as *Spirogyra*, are attached to the bottom in shallow regions near the shore. *Anacharis* and *Vallisneria* are pond weeds that take root on the bottom. The food chains in a pond form a complex web that includes organisms from the nearby shore. First and second order consumers include minute crustacea, *Hydra,* colonial coelenterates, insect larvae, fish, mussels, sponges, turtles, leeches, frogs, water snakes, crayfish, herons, kingfishers, minks, and otters. In the bottom mud are decomposers such as bacteria, fungi, and tubifex worms.

Lakes. Light penetrates to the bottom of a shallow pond, but many lakes are too deep for plants at the bottom to receive light. Many of the bottom-dwelling animals depend for food upon dead organisms that sink from above. Too little oxygen diffuses from the surface of a deep lake to meet the needs of the deep-dwelling inhabitants. Every autumn, however, a massive turnover of water supplies oxygen to the deep regions.

This turnover occurs because water is denser at 4° C than at any other temperature. When the water cools slightly below this temperature, it becomes less dense and rises to the surface. When the temperature drops to 0° C, the cold water at the surface turns to ice. In this way, instead of ice forming on the bottom of lakes and ponds, it forms at the surface and permits life to continue below.

There are disadvantages to surface ice which should be noted. Since it prevents oxygen from getting into the water, many lake and pond organisms die of suffocation during the winter. Ice also hinders the passage of light.

Streams. The cold water in the upper reaches of a stream flows swiftly, and contains much oxygen. The oxygen content is high because air bubbles are trapped at rapids and waterfalls and because cold water can contain much gas in solution.

There is no floating phytoplankton in the upper reaches of a stream simply because of the turbulence of the water.

(However, algae and diatoms are attached to rocks.) Further along, in the middle parts, phytoplankton serves as the basic producer for the community. The muddy waters of the lower reaches prevent the passage of considerable light. Accordingly, the phytoplankton here is reduced.

Mosses and algae attached to rocks in the upper parts of the stream provide shelter for insect larvae that serve as food for small fish. In the middle reaches of the stream, mussels, snails, crayfish, and insects are food for secondary consumers such as catfish, turtles, and bass. The numerous types of consumers in the lower reaches include large fish, crocodiles, birds, and mammals. Turtles act as scavengers.

THE BIOSPHERE AND MAN

As a member of world ecosystems, man is not exempt from the ecological principles that apply to other living things. As we will see in this section, man faces the problem of an expanding population, as well as the problems of conserving his resources and controlling insects.

The Population Explosion

Probably the greatest social, economic, and political issues facing mankind today result from the huge increase in man's numbers that is known as the "population explosion."

The Problem. The human population of the world is in the stage of exponential growth on the S-shaped curve of population growth (Fig. 22–16). The increasing rate at which the world population doubles is shown by the graph.

It took about 1700 years from the time of the early Romans for the population to double to the figure of one-

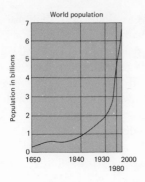

Fig. 22–16. A graph of the estimated world population shows a progressive decrease in the *doubling time*. This is the length of time it takes for the population to double itself.

half billion in 1650. Thereafter, it took only 190 years for it to double to 1 billion in 1840, and only 90 years for it to double to 2 billion in 1930. At the current rate of change, it will take only 50 years to double to 4 billion in 1980. Thus, we can see that it will take only 35 years for the present world population of 3 billion to double to 6 billion in the year 2000. This progressive decrease in the doubling time shows that not only is the population increasing but that the *rate of increase* of the population is speeding up.

In 1798, Thomas Malthus, an English clergyman, published his *Essay on Population* in which he pointed out that the increase of any population is limited by restrictions imposed by the environment (for example, food supply). True, man has devised techniques for increasing the food yield of the· environment. However, his present knowledge of the ecology of other organisms indicates that there must be some limit to the capacity of the environment. A balance between the population and the yield of the environment might be achieved by (1) a decrease in the birth rate or (2) by an increase in the death rate. As Malthus pointed out, an increase in the death rate might occur by the unpleasant methods of war, disease, and famine.

In underdeveloped countries, populations are large and are increasing — primarily because of the increasing birth rate. Man has a natural desire to perpetuate his species, and religious and social convictions have tended to support this idea. Actually, the highest rates of population growth are in the underdeveloped, already impoverished regions of the world (Latin America, Southern Asia, and Africa). Although the developed countries of the West have sufficient food and the standard of living is generally rising, two-thirds of the world's population suffers from malnutrition and hunger!

Aid programs to improve agriculture, industry, and health in underdeveloped countries have been provided. Medical advances (such as the control of malaria and of childhood diseases) have caused dramatic declines in the death rate. Although this is beneficial in itself, it should be noted that an increase in survivorship actually contributes to the overall increase in population. Accordingly, the populations of underdeveloped countries are expanding, and the standard of living has failed to improve satisfactorily.

The increase in western populations is not so much the result of an increase in the birth rate as the result of an *increase in survivorship* — the proportion of people who live to an old age. The increasing number of aged persons has introduced new social, economic, and medical problems. *Geriatrics* is an expanding branch of medical practice which treats old age and its diseases.

An underfed, continually growing population cannot raise its standard of living or develop the cultural achievements that are within the grasp of all mankind. The uncontrolled growth of populations intensifies the political, economic, and social problems of the world and makes them much more difficult to solve.

Suggested solutions. Several recommendations have been made for dealing with the problems arising from the population explosion.

We are trying to increase the yield of the environment. Accordingly, we have modernized agricultural practices and have developed industrial potential in underdeveloped countries. In addition, we have been able to provide many persons with a more balanced diet by the introduction of synthetic and processed foods rich in proteins and fats.

Within the United States, public and private agencies are disseminating information on family planning. The United States and Sweden have offered to assist other nations in programs of population control. The United Nations has also demonstrated an interest in aiding programs for population control.

Scientists are continuing research on human reproduction in an effort to devise physically and morally acceptable methods for limiting the size of the family.

CONSERVATION OF OUR NATURAL RESOURCES

We have noted that all living things affect their environment. Because of man's highly developed brain, he has been able to alter his environment to a greater extent than any other organism has. He has exploited the natural resources of the world for his own use.

Natural resources are the living and nonliving materials on earth which support man. *Renewable* natural resources include plants, animals, soil, air, and water. *Nonrenewable* natural resources include metals, minerals, coal, oil and natural gas.

Unfortunately, man's use of natural resources has often upset ecological processes — with disastrous results. Civilized man has depleted the soil, water, plants, and animals that he inherited from nature, and famine, floods, and drought have followed in the wake of these activities.

A number of man's activities have had a negative effect upon the biosphere. A large number of urban centers have been built on land formerly devoted to raising food. Overgrazing by sheep and cattle, overcropping, and overburning have destroyed the natural cover of the soil. These practices have led to dust storms and floods that removed the cover of topsoil.

Since air and water are necessary to life, the pollution of these vital resources is a serious problem. *Air pollution* is caused chiefly by wastes from burning, from industrial processes, and from automobile exhausts. The major *water pollutants* include sewage, chemicals, fertilizers, silt, and salt water. Today, when our water supply is being increasingly polluted, our water requirements are greater than ever. Water is needed in large quantities for agriculture, sewage disposal, industrial use, hydroelectric power, and recreation.

Conservation does not mean the hoarding of natural resources so that they are not used. *Conservation* is a way of using natural resources wisely for maximum production while protecting the main supply for future use. Conservation practices for air, soil, water, forests, grasslands, and wildlife are interdependent and are thus part of one broad conservation program. Since conservation problems are man-made, they can also be solved by man. Local, state, and federal groups play a part in the conservation program. Every individual also has important responsibilities in this program.

Soil Conservation

One-third of the topsoil of the United States has been lost. It takes 300 to 1000 years for nature to make one inch of topsoil. The soil is a mine and, like other mines, its mineral wealth can be exhausted. Unlike other mines, the productivity of the soil can be renewed indefinitely by wise use.

Soil is injured by erosion and by depletion. *Erosion* is the action of wind and water to remove the soil. *Depletion* is the removal of organic matter and minerals. An excess of water may carry away dissolved minerals. Continuous planting of the same crop removes valuable minerals. Soil conservation measures include the following:

Crop rotation. Since various crops differ in the minerals that they take from the soil, it is possible to prevent the depletion of minerals that occurs when the same crop is planted year after year. *Crop rotation* is a practice that prevents depletion by the growing of different crops in a sequence. For example, corn, wheat, and clover may be planted in different years. The introduction of a legume such as clover adds to the soil nitrates which were removed by the other crops.

Cover crops. Corn, beans, tobacco, and tomatoes are *row crops*. They are planted in rows and the cultivated soil between the rows is exposed to erosion by wind and water. The planting of cover crops is one method used to prevent excessive loss of soil by erosion. *Cover crops* are close-growing plants whose roots form a dense mat that holds the soil. Examples of cover crops are wheat, oats, alfalfa, and grasses.

Strip cropping. The alternation on a slope of strips of cover crops with strips of row crops is called *strip cropping*. Water that runs off the row crops is caught by the cover crops.

Contour farming. Contour plowing consists of plowing around a hill, rather than up and down. Each furrow serves as a small dam to check the flow of water (runoff) and permit it to soak into the ground.

Terracing. The breaking of a long slope into a series of short level areas is called *terracing*. Each level area is separated from the next one by a bank. Drainage ditches conduct the water around the slope. Terracing is effective for checking the eroding effect of runoff and for making efficient use of mountainous lands for farming.

Windbreaks and dams. *Windbreaks* are rows of trees planted to control wind erosion by breaking the force of the wind. Small *dams* that hold back water can be used to prevent gullies from enlarging.

Fertilizers. Fertilizers are used to restore depleted minerals to the soil. Fertilizers may be inorganic chemicals or organic materials such as manure or decomposing plants. "Green manure" is a crop (such as clover) which has been plowed under to enrich the soil.

Water Conservation

Alternate floods and droughts are chronic in many parts of the United States. Many places have suffered a decrease in the level of ground water and cannot meet the present demand of cities and farms. Pollution has made many sources of water unfit for use.

Soil conservation and water conservation are closely related. Many of the soil conservation measures are designed to get more rainfall into the ground and to control the water which runs off.

Watersheds are the mountain regions which collect precipitation and pass it to a major body of water. Where the forests of mountain regions have been depleted, they should be restored. Soil erosion should be controlled. Dams and reservoirs can be built to hold back water during flood periods and to store waters for dry periods. It is important to prevent rivers and lakes from being polluted by sewage, industrial wastes, and contamination from insecticides.

Forest Conservation

Forests protect our water supply and prevent soil erosion by controlling run-off. They provide a home for wildlife, and they provide recreation for man. In addition, forests provide many useful products such as the following:

lumber	newsprint
wood alcohol	tannic acid
paints	resin
charcoal	materials for
turpentine	plastics

The need for forest products is steadily increasing because of a growing population and new uses for wood products. To insure adequate future supplies for forest products, man must manage his forests wisely.

The major causes of forest destruction are forest fires, insects and microorganisms which parasitize trees, and unwise forestry practices. Forest conservation measures include:

1. *Reforestation.* This is simply the practice of renewing the forest cover by seeding or planting.

When forest trees are fully mature, they are harvested.

2. *The removal of dead or diseased trees.* This helps to prevent other trees in the forest which are healthy from becoming infected.
3. *The quarantine and confiscation of imported plants that are diseased.*
4. *The protection of birds in order to control harmful insects.*
5. *The organization of fire-detection and fire-fighting systems.*
6. *The education of the public to prevent fires.* A campfire should be watched carefully and extinguished thoroughly. Matches and cigarettes should be put out carefully before being discarded.

Wildlife Conservation

Many of man's practices have destroyed valuable wildlife. Breeding and feeding grounds have been destroyed in order to reclaim "waste" land for agriculture and for large residential developments. Sewage and industrial wastes have made public waters unfit for use by the biological communities that ordinarily live in those waters. Powerful insecticides have killed enormous numbers of fish.

Some species have been eliminated (or practically so) by uncontrolled hunting and fishing. Animals that have been made extinct are:

passenger pigeon	dodo
heath hen	Carolina paroquet
great aux	Labrador duck

Those threatened with extinction are:

bison (now protected in National parks)	American bald eagle
wild turkey	American crocodile
whooping crane	blue whale
grizzly bear	California condor

Wildlife conservation measures include the following:

1. *The establishment of preserves and wildlife refuges where birds and other animals may breed without being killed by man.* Animals which were near extinction and which are now increasing in numbers as a result of protection are the bison, egret, and trumpeter swan. The future of some species, such as the whooping crane, is still in doubt.

2. *The establishment of closed seasons during which fish or game may not be taken.* The closed season usually includes the period of reproduction and care of the young.

3. *Game laws to restrict the size of the fish which may be caught, the number of fish which may be caught, and the method of capture.*

4. *The artificial breeding of fish and the restocking of ponds and streams.*

5. *The prevention of stream, air, and lake pollution.*

6. *The protection of birds and wildflowers.*

7. *Permitting a natural ecological balance of predators and prey.*

MAN AND INSECTS

Insects are man's greatest competitors for domination of the earth. Insects reproduce in enormous numbers and many species produce several generations in a single season. Some insects even reproduce in their larval stage. Insects are able to adapt to changing conditions of the environment (including poisons spread by man) because of the short length of time required to produce a new generation and the large number of individuals produced. Under these conditions, mutations favorable to insect survival may occur. The large number of organisms prevents all members from being killed.

The small size of insects hinders their detection and also permits their muscle systems to be proportionately more efficient. A man who jumps six feet is jumping only the length of his body whereas a grasshopper jumping six feet may be jumping about 36 times his body length. Insects possess keen sense organs. Their organs of sight and smell are amazingly perceptive.

Camouflage is a protective coloration which permits an organism to escape detection. For example, the katydid and the walking stick resemble their background of leaves and twigs. *Mimicry* is an actual physical resemblance to another organism. For example, the viceroy butterfly resembles the unpleasant-tasting monarch butterfly; the robber fly resembles the bumble bee. Presumably camouflage and mimicry confer an advantage for survival.

Activities of Insects

Some of the activities of insects are helpful to man. Insect-pollinated fruit trees depend upon the activities of bees, butterflies, moths, and flies. If bees were to disappear, 100,000 species of flowering plants would die out.

Man obtains natural silk from the cocoon of the silkworm moth, honey from the bee's hive, shellac from the lac insect, and a red dye from the cocchineal insect. The ladybird beetle has saved orange groves by destroying the cottony cushion scale insect which feeds on the groves. Maggots and beetles help to dispose of dead animals and plants.

Many of the activities of insects are harmful to man. Harmful activities of insects include the following:

Destruction of grain, fruit, and vegetables. Insects are believed to take about one-tenth of the food which man grows. The Japanese beetle, the European corn borer, and the coddling moth (whose larvae are sometimes found in apples) are notorious examples of insects that destroy crops.

Injury to shade trees. The Gypsy moth was introduced into Massachusetts in 1868 for scientific experiments. It escaped and has spread to other eastern states. The caterpillars eat the leaves of shade trees, such as the maple and linden, and the leaves of many fruit trees.

Transmission of disease to man and animals. The female *Anopheles* mosquito transmits the malarial protozoan and has made malaria a major world health problem. Fleas transmit the bubonic plague, lice transmit typhus fever, and the house fly carries typhoid fever bacteria.

Transmission of plant disease. A beetle carries the fungus which causes Dutch Elm disease. This disease has wiped out many of the most beautiful trees in the northeast.

Destruction of buildings. Termites in rich earth gain access to wooden structures which are in contact with the ground. They then bore extensive tunnels to heights of 3–4 feet above the ground, leaving mainly a hollow shell of the timbers they have invaded.

Annoyance and injury by bites and stings. Mosquitoes, wasps, and gnats are notorious in this respect.

Effect on food. Although cockroaches are not known to carry disease, they are repugnant and may give food a bad odor.

Destruction of clothing and fibers. The clothes moth and carpet beetle cause much damage in the home by destroying clothing and fibers.

Parasitism on man and domestic animals. Fleas, lice, and the screw worm fly are examples of parasitic insects.

Control of Insects

Insect pests are controlled by three main methods: quarantine, chemical control, and biological control.

Quarantine. *Quarantine* is the detention (isolation until proven harmless) of insects or of diseased plants and animals as they are brought over territorial boundaries. The Bureau of Entomology and Plant Quarantine of the United States Department of Agriculture examines materials at ports of entry, including airports. State governments also engage in quarantine programs. Insect pests which entered this country before the days of quarantine include the European corn borer, the Japanese beetle, the Gypsy moth, the cotton boll weevil, and Hessian fly.

Chemical control. Insects may be controlled by the use of chemicals, called *insecticides,* that are poisonous to them. The kind of insecticide selected for use against a specific insect usually depends upon the insect's mouth parts.

Stomach poisons are used against insects that devour plants with their biting mouth parts. Such insects are caterpillars, beetles, and grasshoppers. Lead arsenate is an example of a stomach poison.

Contact poisons are used against insects with piercing and sucking mouth parts. These insects pierce stems of leaves and suck plant juices from them.

Accordingly, stomach poisons applied to the surface of the plant cannot harm them. Contact poisons act by poisoning the insect by absorption through its outer surface. Contact poisons are used against lice and scale insects. Nicotine sulphate, pyrethrum, and oil emulsions are examples of contact sprays.

DDT (dichloro-diphenyl-trichlorethane) acts powerfully upon the nervous system of insects. It functions both as a stomach poison and a contact poison. By mutation, flies have developed an enzyme that acts as an antidote to DDT, giving them immunity. *Chlordane* and *dieldrin* are additional examples of powerful new insecticides.

There are disadvantages to chemical control. Insects evolve an immunity to insecticides. The chemical industry must constantly strive to keep ahead of rapid evolutionary changes which permit insects to become immune to new insecticides.

Powerful insecticides may have wider ecological effects than anticipated. As Rachel Carson emphasized in her book, *The Silent Spring*, insecticides may accumulate in the bodies of fish and birds that live on insects. As a result, these natural predators of the insects die off. DDT is one of the worst offenders. Dieldrin applied to farm land drains off into rivers where minute concentrations kill large numbers of fish. Care must be taken to ensure that the foods that man eats do not contain harmful concentrations of insecticides or fungicides (chemicals which destroy fungus pests).

Biological control. This is the use of biological methods for controlling a population within an ecosystem. Examples of biological control are as follows:

1. Birds and mammals that attack crop-growing pests can be encouraged.

Farmers can do this by maintaining "living borders" of herbs and shrubs around their fields. These borders are the home for birds and small mammals which devour insects.

2. Stream and river pollution can be eliminated. This is done to provide a habitat for fish and other predators of insects, such as frogs and birds.

3. Crop rotation can be used to deprive a generation of insects of their main source of food.

4. The introduction of natural enemies of an imported pest is often an effective method of controlling the pest. For example, the Japanese beetle, which is effectively controlled by its natural enemies in Japan, proved to be a serious problem in the United States where its natural enemies were absent. The praying mantis was introduced to control the Japanese beetle.

5. Sex-attractants can be used. Gyplure, the sex-attractant of the female Gypsy moth, is used to attract males into poisoned traps. Without mates, the females are unable to reproduce.

6. Sterilization can be an effective method of biological control. The screw worm fly has been a serious pest of livestock in our southern states. Eggs laid in sores of the cattle develop into larvae that eat the tissues of the host. Two billion male flies were sterilized by gamma radiation from cobalt-60 in laboratories and then released into the fields. The females mate only once. Eggs laid by females that had mated with sterilized males did not develop. Accordingly, by sterilizing the males, man has eliminated the screw worm as a pest east of the Mississippi River.

In order to maintain the human population on earth man has devoted huge areas to the cultivation of a single species of plant, such as wheat. This rich source of food brought in primary consumers such as the Hessian fly and decomposers such as wheat rust. Because these insects and parasites competed with man for wheat, man had to use chemical controls to prevent famine within his own species. The use of chemical controls sometimes had undesirable ecological effects.

The dangers of chemical control may be alleviated by the wise use of biological controls, but even biological controls have their dangers. For example, the mongoose was introduced into the West Indies to combat rats but soon became a serious pest in its own right. Additional research in ecology is needed to provide a secure basis for biological control.

When man exploits the environment without considering the interrelationships of living things, there can be dire consequences to man himself. Ecologists have no intention of preventing man from using the animals and plants in his environment. Rather, by continued research, they seek ways of avoiding the catastrophic consequences of unplanned use of natural resources.

COMPLETION QUESTIONS

[A] 1. The study of the interrelationships between living things and their environment is called ....

2. Structures used by marine fish to return salt to the sea are their ....

3. Plants which live immersed in water are called ....

4. Plants living in a dry environment are called ....

5. Decaying plant materials in the soil constitute ....

6. An increase in altitude has the same effect upon the development of terrestrial biomes as an increase in ....

7. The accumulation of floating microscopic organisms in the upper region of bodies of water is called ....

8. All of the organisms of the same species living together in a region are known as a (an) ....

9. Populations of different species in a given location constitute a (an) ....

10. The living community of a region, together with the non-living environment constitute a (an) ....

11. All parts of this planet which are inhabited by living things make up the ....

12. The role that a species plays in the life of a community is called its ....

13. The series of organisms through which food energy in a community is transferred is called a (an) ....

14. Green plants are known as the .... of a food chain.

15. In a food chain, a cow is known as a (an) ....

16. Man plays the role of a (an) .... in a food chain.

17. Bacteria act as .... in a food chain.

18. A plant which lives on dead things is called a (an) ....

19. The symbiotic relationship between the remora fish and the shark is an example of ....

20. The form of symbiosis in which both organisms are benefited is known as ....

21. A fungus and an alga living together for each other's benefit make up a structure called a (an) ....

22. The symbiotic relationship between termites and the protozoa which inhabit their digestive system is called ....

23. There is (more, less) .... stored energy in all of the secondary consumers of a community than in all of the producers.

24. Oxygen is added to the atmosphere as a result of the process of ....

25. Ammonia is changed to nitrates by the action of .... bacteria.

26. Denitrifying bacteria change the nitrogen of ammonia to ....

27. Nitrogen fixing bacteria live in nodules on the roots of ....

28. A 5–10–5 fertilizer contains 10 percent of ....

29. A pioneer organism on barren rock is the ....

30. A beech-maple forest is the .... stage in a succession.

(B) 31. Plants that grow on other plants without parasitizing them are called ....

32. The terrestrial biome with the largest variety of plant species is the ....

33. Huge areas of mosses and lichens are found in the .... biome.

34. The ocean region above the continental shelf is the .... zone.

35. The animal portion of the plankton is called the ....

36. The organisms that serve as the major food for whalebone whale are the ....

37. Enough light for photosynthesis penetrates the ocean to a depth of about .... feet.

38. Some deep sea fishes use glowing lures which are lighted by ....

39. The most productive regions of the oceans is the .... zone.

40. The use of echoes to locate underwater objects is known as ....

41. The kind of fresh water community with the largest amount of oxygen in the water is the ....

42. According to present estimates, the population of the world in the year 2000 will be .... billion.

43. The removal of organic materials from soil is called ....

44. Plants whose roots form a dense mat which holds the soil are known as .... crops.

45. Replacing trees that have been cut down with young trees is known as ....

46. The resemblance of the viceroy butterfly to the monarch butterfly is an example of ....

47. The "worm" in an apple is the larva of the ....

48. Stomach poisons are used for insects with .... mouth parts.

49. Three examples of powerful new insecticides are ...., ..., and ....

50. The screwworm fly was controlled by .... of male flies.

MULTIPLE-CHOICE QUESTIONS

[A] 1. In general, soils containing large amounts of humus are most likely to (1) be poorly aerated, (2) be lighter in color than other soils, (3) support large communities of organisms, (4) provide little water for plant growth.

2. Green plants are unable to live at great depths in the oceans because the (1) plants have no place to anchor their roots, (2) oxygen concentration is too high, (3) carbon dioxide concentration is too high, (4) intensity of light energy is too low.

3. Which group of ecological terms is in correct order, from simplest to most complex? (1) organism — population — community — ecosystem — biosphere, (2) organism — population — community — biosphere — ecosystem, (3) population — organism — community — ecosystem — biosphere, (4) biosphere — ecosystem — organism — community — population.

4. The members of an animal community are usually similar in (1) size, (2) structure, (3) food requirements, (4) environmental requirements.

5. Which is *not* an essential condition in a self-sustaining ecosystem? (1) equal numbers of plants and animals, (2) a constant source of energy, (3) the presence of autotrophic organisms, (4) the cycling of materials between organisms and their environment.

6. During the period of time when two species occupy the same ecological niche, they are (1) dependent upon each other, (2) not affected by each other, (3) competing with each other, (4) cooperating with each other.

7. Which of the following sequences correctly illustrates a food chain? (1) algae — insect larvae — fish — man, (2) fish — insect larvae — algae — man, (3) insect larvae — algae — fish — man, (4) algae — fish — insect larvae — man.

8. If one attempts to illustrate the food relations of a habitat area by means of a sketch, the result is always a complicated weblike diagram because (1) food chains overlap at many points, (2) animal species use only one type of food, (3) consumers outnumber the producers, (4) food chains begin with green plants.

9. In a food chain consisting of photosynthetic organisms, herbivores, carnivores and organisms of decay, which two links of the chain are *not* absolutely necessary? (1) photosynthetic organisms and herbivores, (2) herbivores and carnivores, (3) carnivores and organisms of decay, (4) organisms of decay and photosynthetic organisms.

10. In a certain national park, the rangers have been killing off the wolves, coyotes, and mountain lions. Which is the most probable result of this practice? (1) There will eventually be more deer than the area can support. (2) The predators will be benefited. (3) The plant life of the area will exhibit little change. (4) The number of grasshoppers and other plant-eating insects will increase.

11. A molecule of nitrogen which you are now breathing in may have been part of a plant which lived thousands of years ago. This is an illustration of the principle that (1) dead organisms may be reincarnated, (2) molecules of cytoplasm may be replaced by inorganic salts, (3) nitrogen does not combine readily with other elements, (4) bacteria of decay return elements into circulation.

12. *Chlamydomonas* is a unicellular alga found in pond water. It serves as a source of food for many organisms. It also aids organisms in the pond by (1) removing excess amounts of carbon dioxide, (2) removing excess amounts of oxygen, (3) performing the process of conjugation, (4) bringing about the decay of organic matter.

13. An example of a common climax community in New York State is a (1) birch grove, (2) beech-maple forest, (3) sphagnum bog, (4) pasture.

14. If all the plants in an area were destroyed by fire, which would probably be the first plants to grow back? (1) annual weeds, (2) flowering shrubs, (3) maple trees, (4) pine trees.

15. The last type of plant to appear in a forest succession is the (1) algae, (2) fungi, (3) mosses, (4) seed plants.

(B) 16. A large geographical region characterized by a dominant type of vegetation is called a (1) community, (2) biome, (3) range, (4) realm.

17. The climax plants in a succession in a desert would be expected to have such adaptations as (1) broad leaves and shallow roots, (2) dull leaves and a thin epidermis, (3) a large number of stomates on the upper surface of leaves, (4) a reduced leaf surface and an extensive root system.

18. The greatest amount of food production in the world occurs in (1) tropical forests, (2) coastal waters of oceans, (3) midocean, (4) fresh water.

19. "When an increase in population exceeds food supply, the result will be war, famine, and disease." This statement was first made by (1) Hooke, (2) Darwin, (3) Avery, (4) Malthus.

20. In most populations, a period of very rapid population growth is usually followed by a period during which the population stabilizes. This stabilization is chiefly the result of (1) a decrease in the number of natural enemies, (2) an increase in the supply of available food, (3) either an increased death rate or a decreased birth rate, (4) the appearance of several new species in the region.

21. The nutritional content of the soil can best be maintained or improved by (1) gullying, (2) terracing, (3) windbreaks, (4) crop rotation.

22. A person wishing to establish a windbreak should plant several rows of (1) pine, (2) corn, (3) iris, (4) sugar cane.

23. Which animal has become extinct within the past 100 years? (1) saber-toothed tiger, (2) passenger pigeon, (3) bison, (4) wild turkey.

24. Man's chief competitors for available food are (1) fungi, (2) birds, (3) rodents, (4) insects.

25. A positive program to maintain a balance in nature would include (1) using submarginal lands for home building, (2) increasing the use of germicides to control stream pollution, (3) making use of ecological principles, (4) widespread introduction of new organisms into ecosystems.

CHAPTER TEST

1. Which statement concerning populations is *false?* (1) A population is made up of individuals of different species. (2) Populations respond to favorable conditions by increasing in number. (3) Conditions favorable for one population may be unfavorable for another. (4) Individuals of a population interact with each other.

2. That portion of our planet in which ecosystems operate is known as the (1) land, (2) ecological niche, (3) biosphere, (4) climax community.

3. A group of plants and animals living together in a suitable environment is called a (1) community, (2) succession, (3) phylum, (4) food chain.

4. The branchings and interrelationships between food chains in a biotic community is an example of (1) succession, (2) an ecosystem, (3) a food web, (4) a population.

5. Returning nutrients from dead plants and animals to the soil is chiefly the work of the (1) nitrogen-fixing bacteria, (2) pathogenic bacteria, (3) denitrifying bacteria, (4) decay bacteria.

6. In an area populated by all of the following animals, which would probably have the largest population? (1) foxes, (2) field mice, (3) grasshoppers, (4) skunks.

7. *Plankton* is the name given to the algae, protozoa and other forms of microscopic life that abound on the surface of the ocean. Which best describes the most important role in nature of plankton? (1) It serves as the basis of the food chain for oceanic life. (2) It is a source of human food. (3) It is symbiotic to land forms of microscopic life. (4) It serves directly as the basic diet of most fish.

8. Which microorganism keeps the nitrogen cycle going as a result of its metabolism of ammonia? (1) nitrogen-fixing bacteria, (2) yeast, (3) nitrifying bacteria, (4) *Paramecium.*

For each description of a biome in questions 9 through 13 write the letter preceding the name of the matching biome.

Major Biomes of the Earth
A. Desert
B. Grassland
C. Marine
D. Taiga
E. Temperate deciduous forest
F. Tropical forest
G. Tundra

9. This area receives little solar energy at any time. It has a short growing season; its precipitation is low and occurs largely in the form of snow; its soil is frozen most of the year.

10. Trees in this area have broad leaves that are shed in the fall. The weather is variable with snow seldom lasting all winter.

11. This area has 10 to 30 inches of rainfall. It supports many grazing animals. The summer growing season does not produce forests.

12. This area has great coniferous forests extending in a broad zone across Eurasia and North America. It has a multitude of ponds and lakes.

13. Over 80 inches of rainfall per year in this area are evenly distributed so that there is no well-defined dry season.

Base your answers to questions 14 and 15 on the information below.

The graph shows the relative populations of mountain lions and deer in a certain geographic area generally favorable to both animals. At the time indicated by point *A*, hunters were offered a bounty for each mountain lion killed. Later, at the time indicated by point *E*, these bounties were withdrawn and the hunting of mountain lions was discouraged.

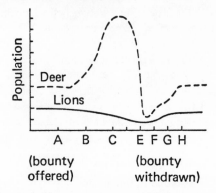

14. Which is the most probable explanation of the fact that in the graph the deer population is always higher than the mountain lion population? (1) The geographic location is more favorable for deer than for mountain lions. (2) Hunters are likely to kill more mountain lions than deer. (3) The organism serving as the food supply is normally more numerous than its predator. (4) Deer usually reproduce more abundantly than do mountain lions.

15. The graph is an illustration of the principle that (1) predators serve an important role in a balanced community, (2) man's intervention has little permanent effect on survival of species, (3) deer need the protection of man in order to survive the attacks of the more aggressive among their natural enemies, (4) mountain lions are the least of the deer's dangers in their struggle for survival.

For *each* of statements 16 through 20, write the *number* preceding the term, *chosen from the list below,* to which that statement refers.

Biological Terms
A. Commensalism
B. Mutualism
C. Parasitism
D. Saprophytism

16. Certain termites eat wood but are unable to digest the wood without the aid of certain flagellate protozoans that live within the bodies of the termites.

17. Certain young clams attach themselves to the gills of a fish. In a short time each clam becomes surrounded by a capsule formed by cells of the fish. The clams feed and grow by absorbing nutrients from the fish's body.

18. Certain luminescent bacteria living in the body of a marine fish obtain nutrition from the fish and in turn provide the fish with a lantern, a warning signal, or a recognition device.

19. Beadle and Tatum grew *Neurospora* on a medium containing water, salts, biotin (a vitamin), and several amino acids.

20. Barnacles, which have no means of locomotion, acquire mobility by attaching themselves to whales. The whales are apparently unharmed by this relationship.

Base your answers to questions 21 through 25 on the pyramid below.

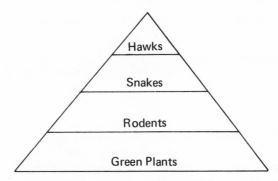

21. The greatest amount of energy present in this pyramid is found in the (1) hawks, (2) snakes, (3) rodents, (4) green plants.

22. Which represents the food chain in this pyramid? (1) hawks, snakes, green plants, rodents, (2) green plants, rodents, snakes, hawks, (3) rodents, green plants, snakes, hawks, (4) snakes, rodents, hawks, green plants.

23. This pyramid implies that, in order to live and grow, one thousand pounds of snakes would require (1) less than 1,000 pounds of green plants, (2) 1,000 pounds of rodents, (3) more than 1,000 pounds of rodents, (4) no rodents.

24. Which organism illustrated in this pyramid is an autotroph? (1) green plant, (2) rodent, (3) snake, (4) hawk.

25. Which is the primary consumer in this pyramid? (1) green plant, (2) rodent, (3) snake, (4) hawk.

EVOLUTION

Throughout this book we have found it necessary to invoke the idea of evolution. Based on facts and reasonable assumptions, this theory has provided a unifying concept which has greatly advanced the progress of biology. Without it, biology would be more like a catalogue of observations than a science dealing with broad concepts. The theory of evolution helps to direct the biologist into meaningful research, whether he is investigating the chemistry of blood proteins or the migration of birds.

Recall that, simply stated, organic evolution means living things *changed* from an ancestor which they had in common. In Chapter 3, we showed that the concept of evolution is supported by evidences acquired from the study of rocks and fossils, from comparative anatomy, from comparative embryology, from vestiges, from physiology, and from biochemistry. In the chapters on plant and animal breeding and genetics, we showed that changes in heredity can occur. In Chapter 23, we shall see that the geographic distribution of living things over the face of the earth provides another evidence for evolution. The evidences for evolution are summarized in the table below.

EVIDENCES FOR EVOLUTION

rocks and fossils	geographic distribution
comparative anatomy	physiology and biochemistry
comparative embryology	plant and animal breeding
vestiges	genetics

CHAPTER 23 / Theories of Evolution

In 1859 — a little over 100 years ago — Charles Darwin published his *On the Origin of Species by Means of Natural Selection,* a book that began a new epoch in man's understanding of life.

Several competent biologists had expressed ideas on evolution before; it fell to Darwin to gather sufficient evidence to show convincingly that evolution occurred and to present the evidence systematically with a theory that explained the mechanism of evolution.

At a fairly early age he made some conjectures which have helped us to understand one of the evidences for evolution; that is, geographic distribution.

A GEOGRAPHIC DISTRIBUTION

At the age of 22, Darwin had the good fortune to be selected for the post of naturalist on H.M.S. *Beagle* in its five-year trip to chart the islands of the Pacific and the coasts of South America. At the Galapagos Islands, off the coast of Ecuador, he was struck by the slight differences in the kinds of finches, tortoises, and snails inhabiting the various islands. For example, the finches which he found were very similar to one another, but with great differences in certain features which adapt them to their environment. The 14 species of finches have important differences in their bills, which are adapted to different methods of feeding. Darwin suggested that each species had developed on a different island, in isolation from the others, all from a single ancestor. He concluded that these birds had all developed from a single ancestral type which came from the mainland of South America. In this way, Darwin offered an explanation for the geographic distribution of the finches he had observed.

Factors in Distribution

A species tends to radiate geographically in all directions. Even a slow-moving animal such as the snail can spread widely if given enough time. An organism spreads along *highways* but it can be stopped by *barriers*. An easy route for one organism can be a barrier for another. For example, a river may be a highway for a fish, but a barrier for a land snail. Oceans are a barrier for fresh water amphibians; however, they help to spread coconut palm by carrying their fruits to other shores. A particular climate, or even a living organism, may serve as a highway or a barrier. Man has helped to distribute the rat throughout the world; the tse-tse fly is a barrier which keeps man out of vast regions of Africa.

The Distribution of Marsupials

The way the marsupials occur in different geographic regions presents an interesting problem. As you will recall, the *marsupials* are pouched mammals such as the kangaroo, opossum, and wombat. They are considered to be primitive mammals. When man arrived in Australia, he found many types of marsupials, and even more primitive mammals such as the egg-laying *Platypus*. However, there were no mammals more advanced than the marsupials. (See page 93.)

Fossils of extinct marsupials are found throughout the world, but few kinds of living marsupials occur naturally outside Australia, even though climatic conditions in other parts of the world are favorable to them. Although *fossils* of marsupials are found in the United States, only one living kind of marsupial, the opossum, exists here today. The climate of Australia can support higher mammals, and the rabbit (a rodent-like mammal) thrived when it was introduced to that continent.

How can the peculiar restriction of living marsupials to Australia be explained? The explanation is that land bridges probably provided connections between Australia, Asia, and North America during the Mesozoic era. At that time the evolution of mammals had reached the stage of marsupials, and these animals spread over these three continents and into South America. When the land areas sank or the water levels rose, the land bridges disappeared. (The tops of the mountainous bridge between North America and Asia remain as the Aleutian Islands.)

Outside of Australia, higher forms of mammals evolved. The new mammals (placentals) were possibly more efficient at obtaining food and escaping predation than the marsupials. In any event, the placentals gradually replaced the marsupials wherever the two groups existed together. Why the opossum has managed to survive is not known. Possibly survival is due to its ability to "play dead," which makes it an unappetizing prey for most carnivores.

It is assumed that the disappearance of the land bridges caused Australia to become *isolated* from the regions containing higher mammals. As a result, marsupials survived in Australia. Evolution continued to produce new types within Australia but the elimination of the brood pouch (marsupium) did not occur there.

Thus, we can see that the distribution of marsupials is explained by (1) changes in the surface of the earth, (2) evolution, (3) isolation, and (4) adaptation in the isolated region.

Distribution as Evidence

The geographic distribution of living things is considered to be an evidence for evolution because it would be difficult to explain this distribution without invoking the concept of evolution as part of the explanation. This concept, along with interpretations of rises and falls of the earth's surface, helps to provide a reasonable explanation for the distribution of living things on the surface of the earth.

HISTORY OF THEORIES OF EVOLUTION

Several biologists before Darwin had accepted the idea of evolution, and some of them offered hypotheses to explain how it might take place. Each hypothesis had to be tested against the observed properties of living things.

Lamarck's Theory of Use and Disuse

One of the first biologists to offer a theory of the mechanism of evolution was the Frenchman Jean Baptiste Lamarck. In 1809 he proposed that a change in the environment resulted in a change in an organism because of its *need* to adjust to the new environment. This hypothesis led to his two major assumptions:

1. **Use and disuse.** Lamarck's idea of use and disuse is basically simple. He believed that if an organism used an organ, that organ would become stronger and more highly developed to perform its function. On the other hand, if an organ was not used, it would tend to wither away, or atrophy.

2. **The inheritance of acquired characteristics.** Those characteristics that are developed by the organism during its lifetime are *acquired characteristics*. Lamarck believed that acquired characteristics are passed to the next generation — a belief that we do not hold today.

Let us see how Lamarck applied his two major assumptions. (1) Lamarck became interested in the absence of legs in snakes. He suggested that the ancestors of snakes, which had legs and short bodies, crawled along the ground to pass through low and narrow openings. Since their legs hampered crawling, the legs gradually disappeared. (2) He said that ducks developed webbed feet because webs were *needed* for swimming. He explained that as the feet of the ducks were used they developed tiny webs. The acquired characteristic of slight webbing passed on to the next generation, which then developed even better webs for swimming. The process continued for many generations. (3) He thought that the long legs of wading birds, such as the heron, were developed in response to a need to keep their bodies out of the water. (The feathers of wading birds are not waterproof the way those of the duck are.)

Lamarck really did not explain much, because he offered no explanation of *how* the organism changed to fit its needs. His important contribution was the concept that evolution is *adaptive;* that is, that evolution must lead the species in such a direction as to make it better suited to live in its given environment. From his time until today, any theory of the mechanism of evolution must explain adaptation.

Weismann and the Continuity of the Germplasm

A German biologist, August Weismann, designed an experiment to test the hypothesis that acquired characteristics are inherited.

Weismann cut off the tails of mice and then mated the mice. When the offspring were born, he cut off the tails of the second generation mice and mated these mice. He repeated this process for 20 generations and found that the tails of the last generation of mice were identical in length with those of the first generation. The tails did not shorten or disappear. Weismann concluded that acquired characteristics are not inherited.

To explain why acquired characteristics are not inherited, Weismann developed a theory called the *Continuity of the Germplasm.* This is illustrated in Fig. 23–1.

According to Weismann's idea, all of the cells of an organism may be divided into two groups, the *germ cells* (reproductive cells) and the *somatic cells*

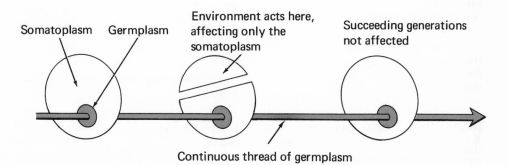

Fig. 23–1. Weismann's theory of the Continuity of the Germplasm.

(body cells). The germplasm consists of the protoplasm of the reproductive cells, which produce the offspring; the somatoplasm consists of the protoplasm of the rest of the cells of the body. The germplasm is set aside by the organism at an early stage of its embryonic development and continues unchanged from generation to generation. Ordinary influences of the outside environment may cause changes in the somatoplasm but not in the germplasm; consequently, these changes are not passed on to succeeding generations. An acquired characteristic would affect only the somatoplasm and could not be a basis for evolutionary change.

While there is a great deal to be learned from this idea, we cannot believe that the distinction between germplasm and somatoplasm is really that sharp. The head of a *Planarian* (somatoplasm) will grow into a whole worm, including gonads. We might clarify the idea that Weismann was groping for by distinguishing between genotype, which remains unchanged and passes to the next generation, and phenotype, which is subject to modification by the environment but does not affect the offspring.

Today, we are concerned because we know that radiations can affect both the somatoplasm and the germplasm. Some examples of the effect of radiations upon the somatoplasm are worth noting. Many people who survived the atomic blast at Hiroshima developed cataract of the eye, a somatic change. X-rays can cause leukemia, which is a disease of the blood. When radiation affects the germplasm, the consequences for mankind may be even more serious. Radiation affecting the germplasm can cause mutations that may weaken future generations of humans.

Darwin's Theory of Natural Selection

In 1858, Charles Darwin received a scientific paper from Alfred Russell Wallace, a fellow naturalist. In this paper, Wallace had presented some of the same ideas on evolution that Darwin had worked a lifetime to formulate and support. The work of both men was presented jointly to the Linnaean Society of London.

Darwin's Theory of Natural Selection may be divided into five distinct ideas:

1. Overproduction
2. Struggle for existence
3. Variation
4. Survival of the fittest, or natural selection
5. Origin of new species by the inheritance of successful variations

Because of the importance of the fourth step in Darwin's chain of reasoning, "natural selection" is used as the name of the theory.

1. Overproduction. One female codfish lays about 9 million eggs a year. If all these eggs were fertilized and the offspring survived, the ocean would be almost solid with codfish in a few years. In other words, the codfish overproduces. *Overproduction* simply means that the number of offspring produced by living organisms is greater than the number that can survive and live to maturity.

Darwin was influenced by Thomas Malthus, an English clergyman who wrote "Essay on Population" in 1798. Malthus explained that organisms increase in number by geometric progression, but their food supply increases only by arithmetic progression. This results in larger and larger gaps between the population of organisms and their food supply as shown in Fig. 23–2.

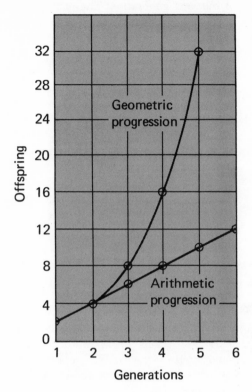

Fig. 23–2. Arithmetic and geometric progressions.

Arithmetic and geometric progressions are shown below:

Arithmetic progression: 1, 2, 3, 4, 5, 6, . . .

Geometric progression: 1, 2, 4, 8, 16, 32, . . .

2. Struggle for existence. The overproduction of living things leads to their *struggle for existence.* In short, they must compete for food and other necessities for survival. The struggle for existence is a competition among members of the same species as well as between members of different species. It need not be a dramatic event such as blood letting with fang and claw. For example, the competition between two seedlings for water, minerals, and light is as much a struggle for survival as is a battle between two wild stallions for leadership of the herd.

3. Variation. No two individuals are exactly alike — even the peas in a pod vary. There may be differences in size, in the ability to produce enzymes, in the efficiency of the cell membrane, and in resistance to disease. Darwin assumed that most variations (including acquired variations) are passed on to the next generation.

4. Survival of the fittest, or natural selection. Some organisms in a species have variations that give them an advantage over other organisms. Because of these favorable variations, they are better able to compete in their environment, to survive, and to reproduce their kind. In nature, the fittest survive; all others die off without leaving offspring. Since nature selects the organisms that survive, this is called *natural selection* as opposed to selection by man, which is known as artificial selection.

5. Origin of new species. New species arise by the accumulation over many generations of inherited variations. When a type is produced that is very different from the original type, it is a new species.

De Vries' Theory of Mutation

Darwin did not explain how variations arise. Many of the variations that he described were acquired characteristics rather than changes in heredity. Darwin did not have the benefit of Gregor Mendel's work to help him with the problem of variations, for Mendel was not accepted until 40 years after Darwin's theory was announced.

Darwin's vagueness about variations proved to be a weakness of his theory.

In 1901, the Dutch botanist Hugo De Vries published the results of his experiments on the evening primrose. De Vries had noticed that markedly different types arose in a single generation. He suggested that the chief agent of evolution was mutation, by which he meant a sudden change that produces a markedly different offspring. He pointed out that the mutant may or may not survive and produce offspring. If it does, it can develop into a new species. Today we know that he was not seeing what we now call mutations. His new forms were not really new species but simply polyploids (page 441) within the same species. Nevertheless, De Vries is often credited with introducing the concept of mutation into evolutionary theory.

De Vries' theory altered Darwin's Theory of Natural Selection by showing that the variations which are effective in evolution are changes in heredity (*mutations*).

Comparison of Lamarck and Darwin

We can compare the different theories of evolution proposed by Lamarck and Darwin. Two examples will be used — the giraffe and the peppered moth.

Evolution of the giraffe. Lamarck would say that the ancestors of our modern giraffes had short necks and were, therefore, obliged to eat the leaves from the lower branches of the trees. This practice soon left the lower branches bare, thereby creating a "need" for longer necks. The animals of one generation stretched their necks in order to reach the higher branches and developed a slightly longer neck. This acquired characteristic was passed on to the next generation. The next generation of animals stretched their necks a little more. As a result, after many generations, the ancestors of the giraffe developed into the modern type with a long neck.

Darwin would say that, among the ancestors of the giraffe, (1) there was an overproduction; that is, more of these animals were born than could be kept alive by the available food supply. (2) These animals competed with each other for the limited amounts of food. (3) There was variation in that some animals had longer necks than others. (4) The animals with the longer necks reached the food and survived but the other animals died off. Nature, in this way, selected the animals most fit to obtain food. (5) The animals which survived, reproduced and passed the character of long necks to succeeding generations. After many generations, the accumulation of variations resulted in a new species, the modern giraffe.

Black pigment in peppered moths. Peppered moths occur in light-colored and dark-colored varieties. They are active during the night, but they rest on trees during the day. One hundred years ago, the region surrounding Manchester, England, had mainly light-colored moths. Since then, the Manchester region has become a smoky, industrial region, and the trees are darkened with soot. Now most of the moths are dark in color.

An outdoor-laboratory type of investigation showed that when equal numbers of light moths and dark moths were placed on *dark* tree trunks, the birds ate more of the light moths. However, when equal numbers of light and dark moths were placed on trees with *light* bark, the birds ate more of the dark-colored moths. Presumably, the contrast between the moths and the bark helped the birds to find the moths.

How would Lamarck and Darwin explain the evolution of these moths in the Manchester region where the dark moths had a competitive advantage?

Lamarck would say that when the environment changed and made the tree trunks darker, the white moths had a "need" to become darker. Each generation became a little darker and passed this characteristic on to succeeding generations.

Darwin would say that there was variation in the pigmentation among the moths; some of them were dark-colored and others were light-colored. As the trees became darker, the dark moths had a competitive advantage. Accordingly, more of them survived to reproduce and pass their heredity on to succeeding generations. After 100 years, the descendants of dark-colored moths become predominant. Note that Darwin did not explain how the variation arose.

Competition and Cooperation

Biologists today point out that cooperation, as well as competition, is an essential feature of the biological world. For example, bees ("social insects") live together in colonies. The cooperation of the animals within a colony is essential to the survival of the species.

It is easier for a wolf to kill an isolated sheep than to kill one that stays with the flock. The herd provides protection for its individual members.

MODERN IDEAS ON EVOLUTION

Modern biologists have added to Darwin's Theory of Natural Selection. They have explained how variations are produced. Recall that the genetic basis for variations within a species is provided by

- gene mutations
- chromosome mutations
- meiosis
- recombination of alleles at fertilization

Chapter 24 considers additional modern ideas concerning the process of evolutionary change.

Resistance of Bacteria to Antibiotics

A physician may find that treating a patient with penicillin causes a dramatic improvement in the patient's health. Sometimes, however, the patient recovers only partially, and he soon becomes as sick as he was before. When the physician again administers penicillin, there is no improvement in the patient's condition.

The explanation is that the bacteria within the patient have evolved a new variety. This variety, or strain, is immune to penicillin. The evolution of immunity by bacteria can be explained in terms that combine Darwinism with the concept of mutation as follows:

We begin by assuming that a *large* population of bacteria existed within the patient. Before penicillin was introduced into this environment, many individual bacteria may have already had a mutation that made them immune to penicillin. However, because the environment did not yet contain penicillin, these immune bacteria had no advantage over the others. The first treatment with penicillin caused the extinction of most of the non-immune bacteria, and the patient's condition improved greatly. The immune bacteria have not been harmed and may have been limited in number.

Now it should be noted that bacteria

reproduce very rapidly. (Three generations may occur within an hour.) Accordingly, the immune bacteria rapidly reproduced and established a large bacterial population that made the patient sick again. Since this new variety of bacteria was immune to penicillin, the second series of penicillin shots did not help the patient.

Physicians are aware of this problem and know how to lessen its effect. The layman should not take antibiotics without the supervision of a physician, for he might do more harm than good.

The pharmaceutical (drug) industry is engaged in a constant race to produce new antibiotics faster than bacteria can evolve immunity to them.

Resistance of Insects to DDT

DDT was formerly very effective in killing insects. However, in many localities DDT must now be supplanted by newer insecticides because of the insects' newly evolved immunity to DDT.

In the cases described above, the bacteria and the insects developed immunity by the evolution of a new inherited trait. The immunity is transmitted from generation to generation. In this respect, it differs from the immunity that a person develops to a disease after receiving an injection of a toxoid. Such immunity is an acquired characteristic that cannot be passed by inheritance through the germ cells.

REASONING EXERCISES

1. What is organic evolution? What are the evidences for evolution?

2. How can we explain the fact that few kinds of marsupials are found outside of Australia even though environmental conditions are favorable to them?

3. What is the most plausible explanation for the homologous forelimbs of the whale, bat, and the horse? (Before trying to answer this question, review Chapter 3.)

4. What objections can you raise to Lamarck's belief that acquired characteristics are inherited?

5. What is overproduction? How does overproduction lead to the struggle for existence?

6. What is natural selection? Apply the idea of natural selection to the evolution of the giraffe.

7. What was the weakness of Darwin's theory? How did the work of De Vries help scientists to overcome this weakness in Darwin's theory?

8. Review the discussion of *Eohippus* (page 55). Combine Darwinism and the concept of mutation to explain the evolution of the modern horse.

B THE ORIGIN OF LIFE ON EARTH

The theory of organic evolution offers an explanation of how simple living things developed into the variety of complex organisms on the earth today. But how did the first living things originate? In this section we will see some of the answers that have been given to this question.

Spontaneous Generation

In order to have the proper perspective on modern theories on the origin of life, it is essential to review the idea of spontaneous generation and the experiments by which it has been disproved.

Spontaneous means "by itself" and *generation* means "the production or creation of." *Spontaneous generation* is the belief that living things can origi-

nate from nonliving matter. Another word for the same idea is *abiogenesis* (*a*, without + *bios*, life + *genesis*, generation), the creation of life without previous life. On the other hand, *biogenesis* is the theory that living things originate only from previous living things.

Belief in spontaneous generation goes back to the Greek philosopher Aristotle. Until the 17th century, "observations" of the rise of living eels, fish, frogs, mice, maggots, and insects from lifeless mud and filth were accepted without question.

Francesco Redi. In the 17th century, the concept of spontaneous generation was challenged by Francesco Redi, an Italian physician. He devised a controlled experiment which showed that maggots do not arise in meat unless living flies have access to the meat. Maggots are a larval stage in the devel-

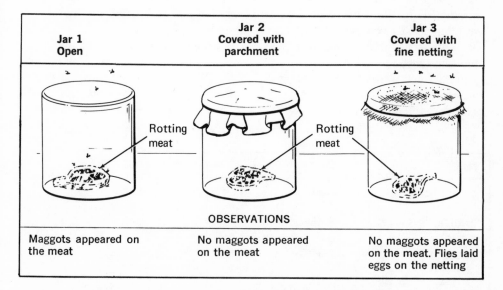

Fig. 23–3. Redi's experiment. This experiment showed that life (maggots) comes only from living things (flies) and not from nonliving things (meat).

opment of eggs into fully developed flies. Redi's experiment is illustrated by Fig. 23–3.

Redi began his experiment with Jar 1 and Jar 2, both of which contained the same kind of rotting fish or meat. Jar 1 was left open but Jar 2 was tightly sealed with parchment. His observations were as follows:

Jar 1: Flies entered and left the open jar. Later, maggots crawled over the decaying contents.

Jar 2: No flies entered the sealed jar and no maggots appeared within.

Redi duplicated the conditions in Jar 1 and Jar 2 with three additional setups of the same kind, and obtained the same results. Redi concluded that new living things (maggots) were produced only when previous living things (live flies) were present.

When opponents objected that there was a lack of "vital" air in Jar 2, Redi introduced Jar 3. This was covered with fine netting which permitted air to reach the meat but kept out living flies. Flies clustered around the netting, and maggots later were found crawling on the netting. However, no maggots appeared on the meat within the jar. Now

Redi had succeeded in proving to his contemporaries that the maggots came from the flies, not the meat.

Redi's Jar 1 and Jar 3 provide an illustration of a controlled experiment.

The conditions of Jars 1 and 3 differed only in one respect — the admission of living things. (See table below.) Scientists call this single difference in the procedure the presence of a *single variable*. Consequently, the difference in results can be attributed to this single variable. If there had been an additional variable, such as a difference of temperature, or a difference in the presence of air, it would be impossible to tell which variable condition caused the different results in the two jars. *A controlled experiment has a single variable.* Controlled experimentation is fundamental to the scientific method. When designing experiments, scientists go to great lengths to make sure that there are no concealed variables in their procedure. The steps in the "experiment" and the "control" (check) must be exact in all respects, except for the condition being tested.

As biologists learned more about the methods of reproduction of plants and animals, the idea of spontaneous generation fell into disfavor. However, the idea was revived when microorganisms

	Jar 1	Jar 3
CONDITIONS IN THE PROCEDURE	meat	same
	jar	same
	temperature	same
	place where kept	same
	length of time	same
	presence of air	same
	life can enter	no life can enter
OBSERVATIONS	living things (maggots) produced	no living things (maggots) produced

were discovered. When 18th-century biologists worked with microorganisms, they tried to sterilize and seal their flasks containing the culture medium (nutrient material). However, no matter how carefully they did this, the flasks often became cloudy with bacteria. They wondered if microscopic organisms arose spontaneously.

Lazzaro Spallanzani. In the 18th century, Lazzaro Spallanzani, an Italian scientist, devised an experiment to disprove the abiogenesis of microorganisms. His experiment may be repeated in the school laboratory, as shown in Fig. 23–4.

Spallanzani used a culture medium consisting of a vegetable infusion; that is, a mixture prepared by boiling plant material (seeds) in water.

Bacteria developed within the sterilized culture medium only when living things had access to it. Accordingly, Spallanzani claimed that spontaneous generation cannot occur. However, his experiment was criticized for lack of

proper controls. Spallanzani's critics said that he had two variables in his experiment. They claimed that the tubes that he used not only varied in their ability to admit life but also in the types of air that they contained. One tube had fresh air and the other had heated air. His critics claimed that the heated air had been deprived of its "vital spirits."

John Needham. Among the supporters of abiogenesis was John Needham, an 18th-century English scientist. He performed experiments quite similar to Spallanzani's, but his technique was not as good.

Needham boiled mutton for his cultures and allowed them to cool in flasks sealed with cork stoppers. He let the flasks stand for a few days and opened them to examine their contents under a microscope. The contents were filled with living microorganisms. Then he performed the same experiment, using vegetable cultures and other meat cultures. He always got the same re-

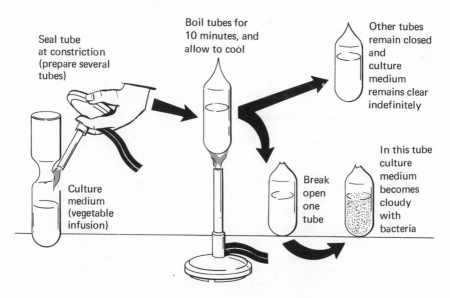

Fig. 23–4. Spallanzani's experiment.

sults. This convinced him of the accuracy of his work.

The error of his approach was that he did not boil his culture medium long enough to kill off all microorganisms at the beginning of his experiments. In addition, microorganisms may have entered through his cork stoppers. The microorganisms originally present reproduced to form a culture teeming with living organisms.

Louis Pasteur. In 1864, Louis Pasteur, a famous French scientist, performed many convincing experiments to disprove abiogenesis. He solved Spallanzani's problem of supplying fresh air to both parts of the experiment by using a flask with a curved tip, as shown at the right in Fig. 23–5. He boiled both flasks shown in the diagram and allowed them to cool.

Since Pasteur's time, scientists have agreed that living things today originate only from previously living things. The names of some of the men who have been involved in the controversy over spontaneous generation can now be summarized.

Supported Biogenesis: Redi, Spallanzani, Pasteur

Supported Abiogenesis (Spontaneous Generation): Aristotle, Needham

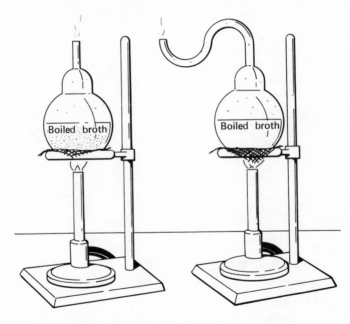

	Flask 1	Flask 2
PARTS OF THE PROCEDURE	Tip straight up. Both bacteria and air can enter.	Tip curved. Bacteria cannot enter, but air can enter.
OBSERVATIONS	Flask became cloudy, showing growth of organisms.	Flask remained clear, showing that no organisms developed.

Fig. 23–5. Pasteur's experiment.

How Did the First Life Originate?

If living things can arise only from previously living things (biogenesis), how did the very first life on earth originate? There are several possibilities:

1. *Life always existed on earth.* According to this idea, the earth and its living inhabitants had no beginning. However, scientists are convinced that the earth itself had a beginning about 5 billion years ago. Therefore the earth and its living inhabitants could not have existed for all time.

2. *Life came from some other world in outer space.* The objection to this hypothesis is that we know no form of life capable of surviving the extremes of temperature and the deadly radiation that are present in space. Meteors striking our atmosphere burn up because of friction. Scientists have found some forms of carbon within meteorites that have reached our planet and have wondered whether these are the remains of living forms. However, even if life did reach earth from another world, this merely pushes back the question: How did life originate in this other region? Nevertheless, we cannot rule out this possible explanation for the origin of life on our planet.

3. *Life originated on earth at some remote time in the past.* This hypothesis does not contradict the concept of biogenesis. According to the idea of biogenesis, life under *today's* conditions can arise only from previously existing life. However, conditions existing on the primitive earth might have been different, and the first life might have gotten its start from nonliving materials under these different conditions.

The autotroph hypothesis and the heterotroph hypothesis are two ideas which attempt to account for the origin of the first life on earth about two billion years ago, under the primitive conditions which existed then.

Autotroph hypothesis. As previously explained, an autotroph is an organism that synthesizes its food from relatively simple substances in its environment. Green plants (including algae) and a few bacteria are autotrophs. Most autotrophs use light as the source of energy for food making (photosynthesis).

According to the autotroph hypothesis, the first living things were autotrophs and could make their own food. This hypothesis seems attractive at first glance; however, it does not take into consideration the complicated biochemistry of photosynthesis. Scientists realize that a living organism that can make its own food is already quite complex and it could not have been the first form of life on earth.

Heterotroph hypothesis. A heterotroph is an organism that relies on outside sources for its food. Heterotrophs include animals, most bacteria, yeasts, molds, and mushrooms. According to the heterotroph hypothesis, the first living things were heterotrophs and took food from their environment instead of making it themselves through the process of photosynthesis.

The main outlines of the heterotroph hypothesis were presented by the English biologist J. B. S. Haldane in 1929; the Russian biologist, A. I. Oparin in 1936; and the American chemist Harold Urey in the 1950's. The series of assumptions of this hypothesis are as follows:

1. *The presence on the primitive earth of an atmosphere that was much different from the atmosphere of the earth today.* In its early stages, the earth cooled from a hot molten mass. The crust and waters came mainly from vol-

canic sources. The ancient atmosphere lacked oxygen, nitrogen, and carbon dioxide. The gases present in the atmosphere were probably the ones shown in the table below.

GASES IN THE PRIMITIVE ATMOSPHERE

Gas	Formula
water vapor	H_2O
hydrogen	H_2
ammonia	NH_3
methane	CH_4

2. *Formation of organic molecules in the atmosphere and in the seas.* Under conditions of intense lightning and unfiltered radiation from the sun, the gases in the primitive atmosphere formed organic molecules. These organic mole-

cules were carried to the seas by rain. Further linkages of molecules resulted in the first proteins and carbohydrates. The accumulation of organic compounds made the ocean resemble a sterile *hot thin soup.* Support for this assumption has been advanced by laboratory studies duplicating the conditions presumed to be on the primitive earth.

Stanley Miller, a pupil of Harold Urey at the University of Chicago, in 1953 circulated water vapor, hydrogen, ammonia, and hot methane past a strong electrical spark for one week. He found that this simulation of "possible primitive earth conditions" resulted in the production of several amino acids. (See Fig. 23–6). He identified the amino acids by the technique of chromatography.

Sidney Fox, of Florida State University, in 1957 heated a dry mixture of

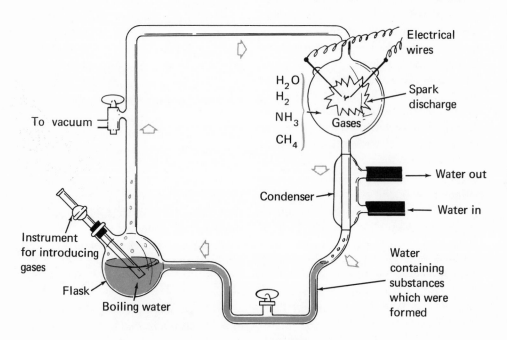

Fig. 23–6. Diagram of Stanley Miller's apparatus.

various amino acids and obtained complicated molecules similar to proteins.

Melvin Calvin, of the University of California, showed that gamma radiation, acting on the gases assumed to be present in the primitive atmosphere, could produce amino acids and sugars. The nitrogen bases, purines and pyrimidines, were also produced. Purines and pyrimidines are components of molecules (such as DNA) that are vital for reproduction and energy exchange in even the simplest of today's cells.

These first complex organic molecules were probably synthesized by energy from the following sources: heat, lightning discharges, X-rays and ultraviolet radiations (solar radiation), and "hard" radiation from radioactive rocks in the earth's crust.

3. *Formation of aggregates.* In laboratories today, biologists study large accumulations of protein-like molecules having a boundary of electrically attracted water molecules. These accumulations are called *aggregates* (also known as "clusters" and as "coacervates"). Aggregates have the capacity to attract other molecules.

It is assumed that "sticky" aggregates formed in the primitive seas and attracted other molecules. In time, the aggregates became larger and more complex and developed a boundary membrane. They developed the ability to take in and use energy-rich molecules from the sea, to grow, and to reproduce with the aid of the purines and pyrimidines. At this stage, these primitive heterotrophic systems could be considered to be "alive." Throughout this process, natural selection sorted out the more efficient systems from the less efficient ones.

4. *Rise of the anaerobes.* Since free molecular oxygen was presumably absent from the earliest atmosphere, the first forms of life must have obtained energy by anaerobic respiration. Anaerobic respiration releases energy by chemical oxidation without using free molecular oxygen. An example of anaerobic respiration (fermentation) is:

$$glucose \xrightarrow{\text{enzymes}} alcohol + CO_2 + energy$$

Thus anaerobic respiration added CO_2 to the atmosphere.

5. *Rise of the autotrophs.* With CO_2 now available, the first primitive autotrophs could arise. These photosynthetic organisms utilized CO_2 and water to produce their own food.

$$CO_2 + H_2O + light \xrightarrow{\text{enzymes}} glucose + O_2$$

This process added oxygen to the atmosphere.

6. *Rise of the aerobes.* The oxygen now available in the atmosphere was used in respiration by both heterotrophs and autotrophs. Aerobic respiration uses molecular oxygen to release energy by chemical oxidation.

$$glucose + O_2 \xrightarrow{\text{enzymes}} CO_2 + H_2O + energy$$

As previously stated, aerobic respiration is more efficient in releasing energy than anaerobic respiration.

These hypothetical steps resulted in early forms of one-celled organisms that could carry on autotrophic or heterotrophic nutrition. These one-celled organisms could obtain energy by aerobic or anaerobic methods. Examples of such unicellular organisms existing today are:

heterotrophs — bacteria, yeasts, molds, protozoa

autotrophs — algae, some bacteria

Although the heterotroph hypothesis provides an explanation for the rise of living things from lifeless matter, this form of abiogenesis could not occur under today's conditions. The first pre-living complex molecules that formed on the primitive earth could not do so on the earth today where the combination of gases in the atmosphere is not suitable. Moreover, if primitive hetero-trophs should be formed by abiogenesis in today's oceans, they could not compete with modern heterotrophs that would use them as food.

It should be pointed out that although most of today's scientists agree with the broad outlines of the hetero-troph hypothesis, there is considerable divergence of opinion concerning some of the details.

COMPLETION QUESTIONS

[A] 1. The title of the book written by Darwin in which his theory of evolution was stated is ....

2. The only living kind of native marsupial in the United States today is the ....

3. The concept of use and disuse was a major assumption of the explanation of evolution given by ....

4. The inheritance of acquired characteristics was part of the theory of ....

5. An experimenter who disproved the inheritance of acquired characteristics was ....

6. All the body cells are known as the ....

7. The constant competition between organisms for the necessities of survival is called the ....

8. Another name for survival of the fittest is ....

9. Darwin's theory is frequently referred to as the Theory of ....

10. De Vries said that variations which are inheritable arise by ....

[B] 11. Another expression for "spontaneous generation" is ....

12. The belief that mice can arise from filth illustrates the idea of ....

13. That maggots do not arise spontaneously from dead meat was shown by the experiments of ....

14. A controlled experiment has a .... variable.

15. Pasteur disproved abiogenesis by using a flask with a .... tip.

16. A famous Greek philosopher who believed in abiogenesis was ....

17. According to the .... hypothesis the first living things could make their own food.

18. According to the .... hypothesis the first living things obtained high-energy organic molecules from their environment.

19. Four gases probably present in the primitive atmosphere were ....

20. The accumulation of organic compounds in the early ocean made it resemble a ....

21. Stanley Miller used a (an) .... as a source of energy in his experiment on the heterotroph hypothesis.

22. Large accumulations of protein and water molecules that attract other molecules are called ....

23. The first forms of life probably obtained energy by .... respiration.

24. The early anaerobes added .... to the atmosphere.

25. The earliest living things probably used the .... type of nutrition.

MULTIPLE-CHOICE QUESTIONS

[A] 1. The fact that digestive enzymes and body hormones are similar in many mammals illustrates (1) physiological similarity, (2) embryological similarity, (3) similar vestigial structures, (4) similar homologous structures.

2. From a comparative viewpoint, man and the ape show many similar but also many unlike characteristics. This fact indicates that (1) man evolved but the ape did not evolve, (2) man has descended from the ape, (3) man and the ape probably had a common ancestor, (4) man has descended from the ape.

3. Natural selection is a theory that attempts to (1) account for the mechanism producing variations, (2) provide the evidence for evolution, (3) confirm the work of Lamarck, (4) explain the process of evolution.

4. To assume that his hypothesis of evolution was correct, Darwin also had to assume that (1) natural forces which made the world as we see it are no longer in operation, (2) it is impossible to determine what natural forces made the world as we see it, (3) the world as we see it resulted from the operation of the same forces that are now in operation, (4) natural forces are determined by the environment in which they are found.

5. Which statement most clearly contradicts Lamarck's hypothesis that acquired characteristics are inherited? (1) The cactus plant, when introduced in Australia, spread very rapidly. (2) Most zebras can run faster than lions. (3) *Staphylococci* that are resistant to penicillin have been discovered. (4) The seeds of a wind-twisted pine grew into tall straight trees in the sheltered valley.

6. Bobbing (cutting) the tails on some breeds of puppies has to be repeated in each generation because (1) there is a physical need for the tail to be long, (2) the tail is a vestigial structure, (3) no gene change is produced, (4) the germplasm has no effect on the somatoplasm.

7. Which idea is *not* basic to Darwin's theory of evolution? (1) Mutations cause changes in living things. (2) Nature selects the most fit organisms to survive. (3) Variations are inherited by organisms. (4) Organisms increase in number by geometric progression.

8. The modern theory of evolution does *not* support the concept that variations within a species are the result of (1) meiosis, (2) gene and chromosomal mutations, (3) use and disuse of an organ, (4) fertilization.

9. That the embryo of the reptile has gill slits is best explained by the fact that (1) some reptiles live in water, (2) reptiles need gills to breathe, (3) embryos can live without lungs, (4) reptiles descended from fishlike ancestors.

10. Which best suggests that evolution is still taking place? (1) fossil discoveries, (2) volcanic activity, (3) mutations in fruitflies, (4) melting of polar icecaps.

B 11. The assumption that life comes from pre-existing life is called (1) the heterotroph hypothesis, (2) biogenesis, (3) spontaneous generation, (4) abiogenesis.

12. The first living organisms probably were able to obtain energy directly from (1) enzymes in the environment, (2) oxygen in the atmosphere, (3) carbon dioxide and water, (4) organic molecules in the water.

13. According to the heterotroph hypothesis, which is the proper sequence? (1) autotroph — heterotroph — organic molecules — aggregates of molecules, (2) organic molecules — aggregates of molecules — heterotroph — autotroph, (3) heterotroph — aggregates of molecules — autotroph — organic molecules, (4) aggregates of molecules — organic molecules — heterotroph — autotroph.

14. A high school boy carrying on a research investigation could produce "a hot, thin soup" by use of (1) an electric spark, methane, and carbon dioxide, (2) gamma rays and pyrimidines, (3) cosmic rays, water, and carbon dioxide, (4) a hot plate, water, and a chicken.

15. The energy available for the formation of organic molecules during primitive earth conditions probably included heat, ultraviolet rays, and (1) enzymes, (2) gamma rays, (3) ATP, (4) chlorophyll.

CHAPTER TEST

1. According to the doctrine of natural selection, the toes of *Eohippus* changed because (1) the environment favored the survival of a swifter one-toed animal, (2) the hard soil caused a gradual wearing down of the lateral toes, (3) radioactive fallout changed the genes for multiple toes, (4) there was hybridization with a new type of one-toed animal.

2. The term that is closest in meaning to spontaneous generation is (1) fermentation, (2) biogenesis, (3) abiogenesis, (4) reincarnation.

3. Many antibiotics which formerly were effective in combating infection have lost their effectiveness. The best explanation is that the (1) bacteria have become accustomed to these antibiotics, (2) environment has no effect on survival of bacteria, (3) need for protection from harmful drugs built up immunity in bacteria, (4) the descendants of drug-resistant types of bacteria survived and multiplied.

4. The weakness in Darwin's theory of how evolution occurs was his inability to explain the (1) reasons for overproduction, (2) role played by natural selection, (3) mechanisms which produce variations, (4) adaptations of living organisms for survival.

5. The mutation theory was important to the understanding of evolution because it explained the (1) value of natural selection of animals, (2) production of variations in animals, (3) use of organs in animals, (4) differentiation of animal embryos.

6. Cottontail rabbits closely resemble the fall and winter vegetation in which they hide. This is most likely because (1) in earlier generations, the rabbits nearest this color were the ones that survived and reproduced, (2) the rabbit has learned to

undergo a seasonal change in fur color, (3) all gene mutations act to fit an organism to meet the problems of survival, (4) this color provides more warmth than other colors.

7. About how many years ago did Darwin propose his theory of evolution? (1) 50, (2) 100, (3) 200, (4) 2000.

8. De Vries observed several new forms that appeared in the evening primrose and that reappeared in the offspring. This observation led to his theory of (1) variations, (2) use and disuse, (3) adaptations, (4) mutations.

9. Malthus' ideas most directly lent support to Darwin's idea of (1) variation, (2) overproduction, (3) adaptation, (4) inheritance of variations.

10. In a population the number of individuals possessing a beneficial mutation may increase as a result of (1) diffusion, (2) contagion, (3) use and disuse, (4) natural selection.

We have already noted that geneticists are concerned with how traits are inherited by individuals and in families. Geneticists are also interested in the total number of different kinds of genes present in an entire population. This is important to us now because modern explanations of evolution are based largely on the genetics of populations. *Population genetics* is the study of the genetics of large groups of sexually reproducing organisms. You will understand this concept better as you study this chapter. Before you begin, study carefully the definitions in the table below.

DEFINITIONS IN POPULATION GENETICS

Term	Definition
Species	An interbreeding unit.
Population	All members of a species inhabiting a given location.
Gene pool	The sum total of all of the genes collectively present within a given population.
Gene frequency	The fraction of all members of the population that have a particular gene, expressed as a decimal. (The sum of all the gene frequencies for the genes of any allelic series is always 1.)
Evolution	Any change in gene frequency.

(A) ## THE POPULATION EQUILIBRIUM

Every population is subject to two opposing trends: (1) the tendency of the population to remain stable, and (2) the tendency of the population to change as a result of forces acting upon it. Evolution results when the forces for change get the upper hand.

The chief stability factor is the nature of the gene itself, which can pass unchanged through many generations. Furthermore, genes are linked together in groups, within chromosomes.

The basic law of population genetics is known as the *Hardy-Weinberg Law*, since it was derived independently (in 1908) by the English mathematician G. H. Hardy and the German physician W. Weinberg. As with many laws, this one was derived by assuming the simplest possible conditions. Since this law is based on probabilities, it was necessary to assume a large population, for small populations often behave in improbable ways. The effect of any possible mutations was ignored, and it was assumed that the population is isolated, so that there is no migration in or out. All individuals survive to breed. Finally, it was assumed that mating is completely random; that is, that any male is just as likely to mate with any one female as with any other. This means that the individuals must travel at random all over the range of the population, and that they have no basis for preferring one mate over another.

In short, the assumptions of Hardy and Weinberg were as follows:

- large populations
- no mutations
- no migration
- random mating

No real population ever meets any of these requirements, but these assumptions made it possible to write the basic mathematical law of populations. The law could then be modified to take account of all the special conditions. The law is simple: Under the stated ideal conditions, the *gene frequencies in a population remain constant from generation to generation.*

An Example of the Hardy-Weinberg Equilibrium

Let us assume that there is a population of hamsters in which a dominant gene (*B*) produces black-coat color and a recessive gene (*b*) produces gray-coat color. Suppose these are the only two alleles, and the dominant gene has a frequency of 80% (or 0.80). Then the frequency of the recessive gene must be 0.20. What are the frequencies of the pure black, hybrid black, and pure gray in the population?

To answer this, we must appeal to a basic law of probability. Let us suppose that a pair of dice has been thrown. What is the probability that both of the dice will come up 4? The probability of *either* die coming up 4 is $\frac{1}{6}$, since the 4 appears on one of the six faces. The probability of two 4's coming up simultaneously is the *product* of the separate probabilities. Thus, the probability of throwing a double 4 is $(\frac{1}{6})(\frac{1}{6}) = \frac{1}{36}$.

Now let us apply this principle to

gene frequencies. As previously stated, we are assuming that the dominant gene (*B*) for black-coat color in hamsters has a frequency of 0.80. Accordingly, we are assuming that 0.80 of all sperm have the gene *B* and 0.80 of all eggs have the gene *B*. Since mating is completely random, any sperm may fertilize any egg. The probability that both the egg and sperm in any mating are *B* is therefore $(0.80)(0.80) = 0.64$. Similarly, the probability that both egg and sperm in any mating are *b* is

$$(0.20)(0.20) = 0.04$$

We can calculate the probabilities of all the possible matings by using a Punnett Square:

	Sperm	
	0.80 B	0.20 b
Eggs 0.80 B	0.64 BB	0.16 Bb
0.20 b	0.16 Bb	0.04 bb

F₁ Generation

genotypes	phenotypes
0.64 BB ⎤ 0.32 Bb ⎦	96% black
0.04 bb	4% gray

As shown in the table above, the gene frequencies are 0.64 *BB*, 0.32 *Bb*, 0.04 *bb*. Now let us suppose that this population of hamsters produces another generation. Since we assume no mutations, no migrations in or out, and no loss of population, there will be no change in these gene frequencies during the lifetime of the hamsters. What are the frequencies of the two kinds of sperm they produce?

Parents	B sperm	b sperm
0.64 BB	0.64 BB	0
0.32 Bb	0.16	0.16
0.04 bb	0	0.04
	0.80	0.20

A similar calculation holds for eggs. We find, then, the frequencies of the gametes coming out of this generation are 0.80 *B* and 0.20 *b*. These are the same as the frequencies of the gametes that produced it! If mating is still random, the same Punnett Square we used above applies to the next generation. Thus, the population remains stable at **96% black** and **4% gray**.

General Statement of the Hardy-Weinberg Law

Previously, we used specific numbers for the gene frequencies. As you will recall, the gene frequencies for the entire population of hamsters were 0.80 for allele *B* and 0.20 for allele *b*. These two frequencies add up to the sum of 1 because they represent 100% of the population. Let us now proceed to a general mathematical statement of the Hardy-Weinberg principle that will be applicable to other situations. We can let *p* represent the frequency for one allele and *q* represent the frequency for the other allele. (The letter *p* is used for the dominant allele in cases where dominance is present.) We may write:

$$p + q = 1$$

Thus, the frequency of the dominant allele plus the frequency of the recessive allele total 100% of the gene pool for this pair of alleles. In the example above, where $p = 0.80$ and $q = 0.20$,

$$p + q = 1$$
$$0.80 + 0.20 = 1$$

The males and the females each have *p* and *q* amounts of each allele. Consequently, we can set up a Punnett Square using *p* and *q* as the frequencies of the alleles.

Sperm

		p	q
Eggs	p	pp	pq
	q	pq	qq

The gene frequencies in the offspring are thus $pp + pq + pq + qq$. By combining similar terms this becomes $p^2 + 2pq + q^2$. Inasmuch as this expression constitutes 100% of the offspring, we can proceed to this general equation:

$$p^2 + 2pq + q^2 = 1$$

After one generation of random mating, the genotypes in a population will tend to distribute according to the relationship shown above. In this relationship, $p = $ the frequency of the dominant allele (A) and $q = $ the frequency of the recessive allele (a). Thus, in the population

$p^2 = $ the frequency of homozygous dominant (AA) individuals

$2pq = $ the frequency of heterozygous (Aa) individuals

$q^2 = $ the frequency of homozygous recessive (aa) individuals

In the example of the hamsters where $p = 0.80$ and $q = 0.20$, substitution in the general equation gives: $0.64 + 0.32 + 0.04 = 1$. Accordingly, the genotypes of the population are in the proportion **64% pure black, 32% hybrid black,** and **4% gray.**

Students of algebra will recognize that the expression $p^2 + 2pq + q^2$ is an expansion of the binomial $(p + q)^2$.

Applications of the Hardy-Weinberg Law

1. In a gene pool where the percentage of gene R is 70%, what is the percentage of gene r?

GIVEN: $p = 0.70$
SINCE: $p + q = 1$
$0.70 + q = 1$
THEN: $q = 0.30$, or **30%**. *Answer*

2. In a population where the frequency of gene D is 0.60, how many individuals are hybrid Dd?

GIVEN: $p = 0.60$
SINCE: $p + q = 1$
$0.60 + q = 1$
THEN: $q = 0.40$
THUS: $2pq = 0.48$, or **48%**. *Answer*

(Note that $2pq$ is the frequency of heterozygous individuals.)

3. The gene for brown eyes B is dominant over the gene for blue eyes b. In a certain population 9% of the people have blue eyes. What percent of the people are hybrid brown eyed?

GIVEN: $q^2 = 0.09$ (Note that q^2 is the frequency of phenotypes possessing the recessive trait.)
THEN: $q = 0.30$
SINCE: $p + q = 1$
$p + 0.30 = 1$

THEN: $p = 0.70$
THUS: $2pq = 2(0.70 \times 0.30) = 0.42$, or **42%**. *Answer*

4. Some persons find that PTC (the chemical phenyl-thiocarbamide) has a very bitter taste. These persons are known as "tasters." Individuals who do not taste this chemical at all are known as "non-tasters." This is a case of simple Mendelian dominance in which allele T (taster) is dominant over t (non-taster). A research team finds that 64% of the people in a certain city are tasters and 36% are non-tasters. Of the tasters, how many are pure tasters and how many are hybrid tasters?

GIVEN: $q^2 = 0.36$ (Note that the pure tasters and hybrid tasters together = 0.64. Therefore, we cannot say that $p^2 = 0.64$.)
THEN: $q = 0.6$
SINCE: $p + q = 1$
$p + 0.6 = 1$
THEN: $p = 0.4$
THUS: $p^2 = 0.16$, or **16%**
THUS: $2pq = 2(0.4 \times 0.6)$
$2pq = 0.48$, or **48%**
16% pure tasters, 48% hybrid tasters
Answer

Note that for a population in equilibrium we can determine the frequency of the alleles in that population if we know the percentage of *one* class of the phenotypes.

PROBLEMS IN POPULATION GENETICS

Assume that the gene for white wool in sheep is dominant over the gene for black wool. If 25% of the sheep in a large population have black wool,

1. What are the gene frequencies for these two types of wool?
2. What percent are pure white and what percent are hybrid white?

Ⓑ EVOLUTIONARY CHANGE

No population stays in Hardy-Weinberg equilibrium for very long. The gene pool of a population is very unstable, precisely because of those factors we had to ignore to arrive at the Hardy-Weinberg equation. The gene pool is constantly under the influence of two kinds of processes, one tending to make it more variable, and the other tending to make it more stable by keeping the amount of variation within limits.

Factors That Increase Variability

There are processes that increase the variability of a population:

1. *Recombination of genes.* At every fertilization, genes are united into new combinations. The number of possible combinations is so great that it is unlikely that, in the whole history of life on earth, there have ever been two individuals with completely identical genes (except identical twins). Genes tend to be transmitted in groups (linkage), but even these groups are frequently recombined by crossing over.

2. *Mutations.* Gene mutations are a rare event for any given gene, of the order of one mutation per million genes per generation. However, there are so many genes in every individual, and so many individuals in the population, that there are a few new mutations every generation. If they produce recessive genes (as they usually do), the genes may remain in the population for many generations without having any effect on the phenotypes. Chromosome changes, involving duplication of sections, translocations, etc., also provide new variation.

3. *Migration.* Influx of individuals from other populations with a slightly different gene pool will provide a constant supply of new genes, unless the population is isolated, for example, on an island or a mountain top.

These processes tend to keep making the population more variable as time goes by. If nothing else were acting, every population would consist of widely different individuals. However, this does not happen. Natural populations are surprisingly uniform, considering the wide variety of genes in them.

Normalizing Selection

The explanation for the uniformity of natural populations lies in the process of *normalizing selection.* As we saw in Chapter 22, every species occupies a particular ecological niche. It is adapted to a particular set of conditions which may include food sources, light, temperature, other organisms, etc. An individual that differs too widely from the norm for its species will not be as well adapted as the normal individuals. Accordingly, it is more likely to die young. Furthermore, if its appearance or behavior does not agree with the norm for the species, it is unlikely to obtain a mate, and its particular combination of genes will not pass on. Thus, the extreme variations are constantly being removed from the breeding population, and the population remains fairly uniform in spite of mutations, recombinations, and immigration.

The Guiding Factors in Evolution

Although mutations are random occurrences, evolution itself is not random. The changes that occur in the gene pool are not chaotic. As plants and animals developed throughout geologic time, they exhibited sets of traits that adapted them to the environment in which they lived. *Adaptive variations* are modifications of structure or function that aid the organism to survive in its environment. The guiding factor in evolution is the *adaptation of organisms to the changing environment by the process of natural selection.* In the struggle for survival among members of the same species, those individuals possessing the most favorable combination of genes for meeting the problems of the environment were the ones that most often survived to reproduce. The succeeding generations thus possessed a higher frequency of genes for adaptive variations.

The environment as a guiding factor in evolution. According to Lamarck, the changing environment and the changing needs of organisms shaped the course of evolution. (See page 416.) Biologists today also emphasize the role of the changing environment in guiding the path of evolution, but their concept of how this happens differs greatly from Lamarck's. A variety of organisms does not obtain a new characteristic just because it is "needed."

Today it is recognized that random genetic changes (such as mutations) which introduce new alleles into the gene pool can cause changes in gene frequencies if the new traits are favored by the environment. Most of these new alleles are associated with traits that are harmful for the organisms. Some, however, are beneficial in the struggle for survival. A changed environment may make some alleles, which ordinarily might be harmful, have an adaptive value — a result which permits success in the new environment.

The percentages of various alleles in a population tend to reflect the adaptive value of the traits controlled by these alleles. Alleles for traits that offer high survival value tend to increase in frequency. Alleles for traits that have low survival value tend to decrease in frequency.

If environmental conditions change, new selective forces act on the gene pool. Gene frequencies, and the traits they control, may change markedly. Traits that formerly had low survival value may have greater survival value in a changed environment; the corresponding gene frequencies may increase accordingly. On the other hand, traits that formerly had a high survival value may diminish considerably and even disappear in the changed environment. In this way, the environment acts as a selective force that interferes with the constancy of the gene pool. Evolution consists of changes in the gene pool, thus building a population that is adapted to the changing environment.

The peppered moths. Let us consider again the peppered moths of Manchester, England, which were described on page 420. As you will recall, the dark moths gained an advantage over the light moths when industrial soot darkened the trunks of trees. Accordingly, birds picked off more of the light moths than the dark ones. Most of the moths in that area today are dark colored, although the light-colored ones predominated in 1850.

We can now understand the change in color of the peppered moths as an example of gene pools that changed under the influence of natural selection in a changing environment. The combination of a changing environment and natural selection had produced an increase in the frequency of genes for melanin (dark pigment) in the local population.

The Hardy-Weinberg Law is still the basic law of population genetics, but modifications have been added to it. Thus, there are mathematical terms indicating the mutation rate, the selective value of genes, the size of the population, and so on, from which the *changes* in gene frequencies can be calculated. *A changing environment stimulates evolution to be adaptive, not by causing new traits to arise, but by changing the frequencies of genes already present in the population.*

In living things, survival is promoted not only by strength, fangs, long claws, and success in direct struggle, but also by many other factors; for example, cooperative behavior and mimicry (page 383). Survival may depend upon the *increasing efficiency* of an organ such as the brain, or upon the increasing efficiency of an enzyme-directed biochemical reaction. It is a mistake to justify aggressive or selfish human behavior on the basis of a one-sided "fang-and-claw" interpretation of "survival of the fittest."

Speciation. A *species,* according to the definition on page 434, is an interbreeding unit. Since genes flow freely from one population of a species to another, the whole species changes together, although there may be minor differences between populations at any given time. Nevertheless, the number of species is much greater now than in the past, in spite of the fact that many species have become extinct. Since a species can only arise from an existing species, there must be some process by which a single species can separate into two or more. As you will recall, this process is called *speciation.*

Once a species has divided into two parts, the two new species may become adapted to different ecological niches.

They may even come to occupy the same region, but they can no longer interbreed. The process of speciation may be repeated, until many species are formed from a single ancestral one.

A newly developing species stands a better chance of survival if it is not in direct competition with existing species. It tends to evolve so as to occupy unused ecological *niches.* Two species of birds may live in the same territory but one lives on weed seeds while the other lives on grain. They have the same *habitat* but occupy a different *niche* in the total community. Newly developing species tend to fill the available niches, rather than compete with existing species for their niches.

Geographic isolation. When Darwin arrived at the Galapagos Islands, he found different species or sub-species of finches (page 415). Probably all the species of these birds had descended from a few original individuals that reached these islands from the mainland. Because distance over water prevented much interbreeding of birds living on different islands, evolution took a slightly different course in each region.

For plants and animals living on land, geographic isolation may be provided by barriers such as mountains, canyons, and large bodies of water. For fish, isolation of populations may be provided by land areas. Any such isolation allows two populations of a species to evolve in different directions.

Genetic isolation. Once separation occurs between two groups, further mutations and natural selection lead to *reproductive* or *genetic isolation.* The chromosomes of the two groups may become so different that they cannot form a zygote that will develop. The sperm of a duck will not survive in the oviduct of a chicken. The pollen grains of one species of flowering plant may die when they start to develop in the flower of another species. Female frogs of one species may lay eggs at a time of the year when males of another species are not fertile. Different courtship patterns may isolate closely related species of birds and closely related species of fish. When two closely related groups of organisms can no longer breed together, they are said to belong to different species.

A man-made species. The evolution of a new species is normally a gradual and lengthy process. However, in some plants a new species may arise suddenly by *polyploidy.* Polyploidy is a doubling, or other increase, in the sets of chromosomes. For example, $2n$ gametes produced by $4n$ parents are not fertile with the normal n gametes. Because the two kinds of parents cannot breed with each other, they constitute different species.

There is convincing evidence that new species of plants (but not animals) have suddenly arisen by polyploidy. In 1927, the Russian geneticist Karpachenko used artificially induced polyploidy to form a new species. He crossed tetraploid ($4n$) radishes with tetraploid ($4n$) cabbages. Normally, the hybrids of a cross between different genera are sterile because the chromosomes of the monoploid gametes are not homologous. However, the presence of gametes containing $2n$ radish chromosomes and $2n$ cabbage chromosomes permitted each chromosome to find a homologue with which it could pair. The hybrids produced in this manner could mate with each other but not with either the radishes or cabbages. They were thus a new species.

Unfortunately, the hybrid plants possessed the undesirable characteristics of each ancestor – the leaves of the radish and the roots of the cabbage. However, this experiment is noteworthy as man's first successful attempt to develop a new species.

The Multiplication of Species

Adaptive radiation is the production of a number of different species from a single ancestral one. "Adaptive" refers to hereditary variations that provide for success in a changed environment. "Radiation" here does not refer solely to the geographical spreading of organisms. It means the divergence of the traits of organisms from the original type.

Further radiation can divide species so that the original species expands

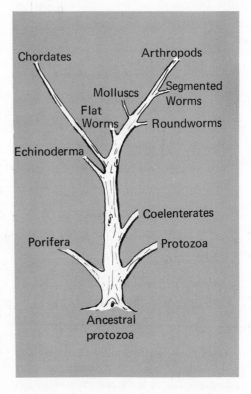

Fig. 24–1. Evolution as a branching tree.

into genera and larger groups. This is a branching type of evolution. Sometimes evolution is incorrectly described as a kind of ladder, with some organisms higher than others. A more accurate picture would represent evolution as a greatly branching tree (Fig. 24–1), with present-day organisms at the tips of the branches.

An example of adaptive radiation is the divergence from the ancestral arthropod. From this type developed the crayfish which swims, the bee which flies, the beetle which walks, and the grasshopper which hops. Another example of adaptive radiation is the divergence from the ancestral mammal to the whale which swims in the water, the bat which flies in the air, and man who walks on earth. Adaptive radiation has been intensively studied in the Galapagos finches. Here several species and varieties of ground-feeding and tree-feeding finches developed from the early inhabitants.

Summary of the Modern Concept of Evolution

Darwin's theory remains as the basic outline for the modern theory of evolution. However, Darwin's theory has been modified by the contributions of modern genetics. The modern concept of evolution may be summarized as follows:

1. *Overproduction.* This is accompanied by stability of the population.

2. *Struggle for existence.* Competition occurs within the species.

3. *Heritable variations.* Mutations, migrations, and recombinations increase heritable variations in the gene pool.

4. *Natural selection, or differential reproduction.* Changes in the environmen select those random mutations that are best adapted to the new environment (*natural selection*). Individuals having these adaptive variations are more likely to survive and produce offspring than those not having them (*differential reproduction*). Accordingly, differential reproduction increases the gene frequency of these adaptive variations and weeds out the less successful combinations.

5. *Origin of new species.* Isolation prevents mixing of the gene pools. When the accumulation of genetic changes is sufficient to prevent the mating of individuals of the two groups, a new species has been formed.

COMPLETION QUESTIONS

(A) 1. The study of the heredity of large numbers of sexually reproducing organisms living in an area is called . . . .

2. All members of a species inhabiting a given location constitute a (an) . . . .

3. The fraction of all members of a population that have a particular gene, expressed as a decimal, is called the . . . .

4. Changes in gene frequency constitute . . . .

5. The basic law of population genetics is known as the . . . . Law.

6. According to the Hardy-Weinberg Law, under ideal conditions the gene frequencies tend to . . . . from generation to generation.

7. For the Hardy-Weinberg Law to apply, the size of the population must be ....

8. When non-random mating occurs, the chance of the gene frequencies remaining constant in the next generation (increases, decreases) ....

9. The Hardy-Weinberg Law applies to the transmission of genes in ... . rather than in individuals.

10. If p is the frequency of one gene and q is the frequency of its allele, the frequency of genotypes is given by the equation ....

Ⓑ 11. New gene combinations are promoted by (asexual, sexual) .... reproduction.

12. Factors that increase variability in succeeding generations are ...., ..., and ....

13. Extreme variations are continually removed from the breeding population by .....

14. Modifications of structure or function that aid the organism to survive in its environment are called ....

15. In order for two diverging groups to form new species they must be prevented from ....

16. The process by which a single species separates into two or more species is called ....

17. A doubling, or other increase, in the sets of chromosomes is known as ....

18. The Russian geneticist, Karpachenko, used tetraploid radishes and cabbages to produce a new ....

19. The divergence of organisms from the ancestral type, permitting them to survive in various environments, is called .... radiation.

20. Differential reproduction is an expression that has essentially the same meaning as Darwin's concept of ....

MULTIPLE-CHOICE QUESTIONS

Ⓐ 1. All of the genes that contribute to the heredity of the next generation of a population constitute the (1) gene frequency, (2) DNA code, (3) species, (4) gene pool.

2. Gene frequencies in a population will remain constant only if the population is large with (1) migration, random mating, no mutations, (2) no mutations, no migrations, nonrandom mating, (3) random mating, no migration, no mutations, (4) random mating, migration, mutations.

3. The observation that vertebrates have certain structural similarities suggests that they probably (1) are the most advanced form of life, (2) are the least complex of all living organisms, (3) have similar habitats, (4) have many similar genes.

4. If the frequency of allele A in a population is 0.80 and the frequency of allele a is 0.20, the frequency of hybrid individuals is (1) 0.04, (2) 0.12, (3) 0.32, (4) 0.16.

5. Individuals who can roll the tongue are known as "rollers" and those who cannot are "nonrollers." The gene R for roller is dominant to the gene r for nonroller. If 64% of the people in a large population are rollers, the frequency of the gene r is (1) 0.20, (2) 0.60, (3) 0.80, (4) 1.00.

(B) 6. Which usually results in variations among offspring? (1) crossing over, (2) budding, (3) fission, (4) mitosis.

7. Genetic and geographic isolation are factors which (1) have no effect on the production of new species, (2) prevent the production of new mutations, (3) encourage the mixing of gene pools, (4) help to maintain any new variations that develop.

8. Which of the following is *not* a factor leading to the origin of a new species? (1) variation, (2) change in existing environment, (3) inheritance of acquired characteristics, (4) differential reproduction.

9. Which is most likely to contribute to the evolution of a new species? (1) Two hybrids are mated. (2) Asexual reproduction occurs. (3) Mitosis occurs in a cell with the diploid chromosome number. (4) The cells of a hybrid plant have twice the normal number of chromosomes.

10. Among members of domesticated species, variability is much more extensive than among members of comparable wild species. One partial explanation for this condition is that (1) domestic species live in a greater variety of habitats, (2) man frequently protects mutants which would be eliminated in nature, (3) wild species must be completely homozygous in order to show variations, (4) the process of domestication increases the rate of gene mutation.

CHAPTER TEST

1. Which would *not* be considered a major force in organic evolution? (1) changes in the DNA of somatic cells, (2) natural selection, (3) mutations, (4) competition among living things.

2. Which process tends to cause genetic changes in populations? (1) mitosis, (2) budding, (3) grafting, (4) mutation.

3. The existence among many other mammals of blood groups similar to our own A, B, AB, and O groups is best explained by the (1) principle of common ancestry, (2) occurrence of chance events in the development of species, (3) existence of random mutations in all organisms, (4) existence of basic similarities in all living things.

4. Most long range adaptations in organisms occur as a result of mutations and (1) protective coloration, (2) environmental change, (3) physical necessity, (4) psychological influences.

5. Organisms that are most likely to adapt rapidly to changing environments are those that reproduce (1) by fertilization, (2) by parthenogenesis, (3) by self fertilization, (4) asexually.

6. Gene frequencies in a population will remain constant only if (1) the population is small and mutations occur, (2) mating is random and mutations do not occur, (3) mating is nonrandom and the population is small, (4) the population is large and the mutation rate is constant.

7. The effect of mutations in evolution is most clearly related to changes in (1) ATP, (2) 2,4-D, (3) DNA, (4) ACTH.

8. If fission were the only means by which a particular species of one-celled organism could reproduce, it may be assumed that (1) mutations could not occur in this species, (2) the rate of evolution of this species would probably be slower than one which carries on sexual reproduction, (3) its numbers would remain constant, (4) this species would belong to the animal kingdom.

Directions (9–11): For *each* statement write the letter of the term, *chosen from the list below,* with which that statement is most closely associated.

Terms

A. Adaptive radiation
B. Inheritance of acquired characteristics
C. Isolation
D. Natural selection
E. Stable gene pool

9. Strains of the same species of squirrels are prevented from breeding because they live on opposite sides of the river.
10. Those members of a species of birds with the genetic code for a long narrow beak have increased in numbers, while those having the genetic code for short beaks have decreased. Both groups live in the same geographical location.
11. Members of a large population mate at random.

Directions (12–15); For *each* statement, write the letter of the term, *chosen from the list below,* with which that statement is most closely associated.

Terms

A. Homologous structures
B. Biochemical similarities
C. Gene mutation
D. Genetic isolation
E. Adaptive radiation
F. Analogous structures
G. Genetic recombinations

12. A white-eyed fruitfly suddenly appears among a population of purebred red-eyed fruitflies.
13. Variations are transmitted through a population by crossing over during meiosis.
14. Different species of sparrows do *not* interbreed even if they live in the same locality.
15. The framework of the wings in birds and bats is basically the same, and the wings of both have a common origin.

APPENDIX

A Classification of Organisms

Kingdom Protista

THE PROTOZOA

PHYLUM SARCODINA.   Move by pseudopodia. Rhizopoda (*Amoeba, Endamoeba*); foraminifera; radiolaria.

PHYLUM SPOROZOA.   Lack power of locomotion; reproduce by spores. *Plasmodium, Monocystis.*

PHYLUM MASTIGOPHORA.   Move by flagella. *Trypanosoma, Astasia.*

PHYLUM CILIOPHORA.   Move by cilia. *Paramecium, Stentor, Vorticella.*

THE ALGAE

PHYLUM CYANOPHYTA.   Blue-green algae. *Nostoc, Oscillatoria.*

PHYLUM EUGLENOPHYTA.   Photosynthetic flagellates not possessing a rigid cell wall. *Euglena.*

PHYLUM CHLOROPHYTA.   Grass-green algae. Possess a rigid cell wall. One-celled, multicellular, and colonial forms. *Spirogyra.*

PHYLUM PYRROPHYTA.   Flagellated unicellular organisms conspicuous in the oceans. Produce poisonous "red tides" which kill fish. Dinoflagellates.

PHYLUM CHRYSOPHYTA.   "Golden algae." Diatoms.

PHYLUM PHAEOPHYTA.   Brown algae. Rockweeds, kelps, *Fucus, Laminaria.*

PHYLUM RHODOPHYTA.   Red algae. *Chondrus crispus* (Irish moss).

THE BACTERIA

PHYLUM SCHIZOMYCOPHYTA.   "True bacteria." *Spirocheta, Streptococcus, Rickettsia.* (Viruses often classified in this phylum.)

THE FUNGI

PHYLUM EUMYCOPHYTA.   Single celled or composed of hyphae. Lacking chlorophyll. Reproduce by spores.

CLASS PHYCOMYCETES.   Produce spores in a sporangium. Lack cross-walls in the hyphae. *Rhizopus.*

CLASS ASCOMYCETES.   The sac fungi. Spores produced in a sac-like cell, the ascus. Hyphae with cross-walls. *Saccharomyces,* yeast, *Penicillium, Neurospora, Aspergillus, Morchella.*

CLASS BASIDIOMYCETES.   The club fungi. Spores borne on club-shaped structures called basidia. Cross-walls present in hyphae. Mushrooms, shelf-fungi, puffballs, truffles, rusts, smuts. *Puccinia, Ustilago, Psalliota.*

CLASS DEUTEROMYCETES.   The fungi imperfecti. Fungi whose method of sexual reproduction is not known. They therefore cannot be properly classified.

THE LICHENS.   Mutualistic combination of an alga and a fungus. The alga is usually a blue-green or a grass-green alga; the fungus is usually an ascomycete. Rock tripe, reindeer "moss," goblet lichen.

PHYLUM MYXOMYCOPHYTA.   Very simple organisms with both plant and animal characteristics. The spreading-mass stage is the plasmodium which contains hundreds of nuclei and no cell boundaries. Slime molds, *Lycogala, Physarum, Stemonitis.*

Kingdom Plantae (Plants)

PHYLUM BRYOPHYTA.   Small plants without true roots, stems, or leaves, but which look like higher plants. Gametophyte the dominant generation; sporophyte reduced in size and often parasitic on the gametophyte. No well-developed conducting tissue.

CLASS HEPATICAE.   Low-lying plants which are often liver-like in shape. Thin, leathery leaves which grow flat on the moist soil or ground. Liverworts: *Marchantia, Riccia.*

CLASS MUSCI.   Plant body is an erect shoot that bears tiny spirally arranged leaflets. No specialized water-conducting vessels. Mosses: *Sphagnum, Polytrichum.*

PHYLUM TRACHEOPHYTA.   The vascular plants, containing specialized water-conducting vessels. Sporophyte is the dominant generation; gametophytes often microscopic and parasitic on the sporophyte.

SUBPHYLUM LYCOPSIDA.   Small spirally arranged leaves which resemble those of mosses. Spores borne in club-shaped strobili. Club mosses: *Lycopodium, Selaginella.*

SUBPHYLUM SPHENOPSIDA.   Leaves arise in whorls from the main branch. Horsetails and scouring rushes: *Equisetum.*

SUBPHYLUM PTEROPSIDA.   Possess relatively large leaves with many veins.

CLASS FILICINEAE.   Underground stem (rhizome) gives rise to leaves (fronds). Numerous sporangia often clustered on lower leaf surface. Ferns: *Marsilea, Azolla, Polypodium.*

CLASS GYMNOSPERMAE.   Seeds not enclosed in a fruit ("naked"). Many are evergreens. Cone-bearing plants: pines, spruces, firs, hemlocks, cedars, redwood, *Ginkgo,* cycads.

CLASS ANGIOSPERMAE.   Seeds enclosed in a fruit ("hidden"). Flowering plants.

SUBCLASS DICOTYLEDONEAE.   Embryo with two cotyledons. Floral parts in groups of fours or fives; net-veined leaves; stems with a cylinder of vascular tissue. Dicots: buttercups, snapdragon, carnation, magnolia, rose, bean, parsley, honeysuckle, oak, maple, dandelion.

SUBCLASS MONOCOTYLEDONEAE.   Embryo with one cotyledon. Floral parts in groups of threes; leaves parallel-veined; stem with scattered vascular bundles. Monocots: lily, onion, palms, orchids, irises, tulips, and grasses such as wheat, rice, and corn.

Kingdom Animalia (Animals)

PHYLUM PORIFERA.   Pore-bearing; two cell layers. Sponges. (*Grantia*, bath sponge).

PHYLUM COELENTERATA.   Hollow-bodied; two cell layers; tentacles with stinging cells. *Hydra*, jellyfish, sea anemone, coral.

PHYLUM PLATYHELMINTHES.   Flattened bodies; three cell layers; mostly parasitic. Flatworms: tapeworm, liverfluke, *Planaria* (free-living).

PHYLUM NEMATODA.   Long, cylindrical, unsegmented bodies; tubular digestive system with mouth and anus; parasitic and free-living. Roundworms: *Ascaris, Trichina,* hookworm.

PHYLUM ANNELIDA.   Body segments very similar; complex organ systems. Segmented worms: earthworm, bloodworm, leech.

PHYLUM MOLLUSCA.   Soft-bodied, many protected with shell. Mollusks: snails, slugs, clams, mussels, squid, octopus.

PHYLUM ARTHROPODA.   Jointed appendages; exoskeleton; segments often fused to form definite regions of body; respiration by gills or trachea; ventral nerve cord.

CLASS CRUSTACEA.   Lobster, crayfish, crabs, shrimp, barnacles, pillbug, *Daphnia*.

CLASS CHILOPODA.   Centipedes.

CLASS DIPLOPODA.   Millipedes.

CLASS ARACHNIDA.   Usually has a fused head and thorax (cephalothorax); four pairs of legs. Spiders, scorpions, ticks, mites, horseshoe crab.

CLASS INSECTA.   Three pairs of legs; body usually divided into head, thorax, and abdomen. Silverfish, grasshoppers, roaches, termites, lice, aphids, bugs, dragonflies, butterflies, moths, flies, mosquitoes, beetles, bees, ants.

PHYLUM ECHINODERMATA.   Skin usually spiny; adults with radial symmetry; water vascular system. Spiny-skinned animals: starfish, sea urchin, sand dollar.

PHYLUM CHORDATA.   Notochord; hollow dorsal tubular nerve cord; gill slits; segmentation. (The first three subphyla may be grouped as the Protochordates.)

SUBPHYLUM HEMICHORDATA.   Acorn worms.

SUBPHYLUM UROCHORDATA.   Tunicates.

SUBPHYLUM CEPHALOCHORDATA.   Lancelets (*Amphioxus*).

SUBPHYLUM VERTEBRATA.   Vertebrae form backbone; enlarged brain; appendages in pairs; olfactory organs, ears, and eyes.

FISH (NOT A TAXONOMIC DIVISION). Aquatic; breathe by gills; cold-blooded; air bladder; two-chambered heart; scales. (The fish consists of the next three classes, the jawless fishes, cartilage fishes, and bony fishes.)

CLASS AGNATHA. Jawless fishes: lamprey eel.

CLASS CHONDRICHTHYES. Cartilage fishes: sharks and rays.

CLASS OSTEICHTHYES. Bony fishes: perch, flounder, trout.

CLASS AMPHIBIA. Moist, smooth, scaleless skin; larvae usually aquatic and adults usually terrestrial; three-chambered heart; cold-blooded. Frog, toad, salamander, mud-puppy.

CLASS REPTILIA. Breathe by lungs from birth; scaly skin; three-chambered heart or four-chambered heart. Snake, lizard, alligator, crocodile, turtle.

CLASS AVES (BIRDS). Feathers; warm-blooded; skeleton with hollow bones, eggs with shell. Robin, penguin, ostrich, *Archaeopteryx*.

CLASS MAMMALIA. Mammary glands; give birth to living young; hair; warm-blooded; four-chambered heart.

ORDER MONOTREMATA. Egg-laying, nonplacental, primitive mammary glands. Duckbill Platypus, spiny anteater.

ORDER MARSUPIALIA. Nonplacental; young born immature and carried in marsupium (brood pouch). Opossum, kangaroo.

ORDER INSECTIVORA. Shrews, moles.

ORDER CHIROPTERA. Bats.

ORDER PRIMATES. Much enlarged brain; usually stand erect; thumbs opposing the other fingers; eyes on front of head; fingers with nails instead of claws. Monkeys, apes, man.

ORDER EDENTATA. Anteaters, armadillos.

ORDER RODENTIA. Rats, mice, beavers, squirrels, guinea pigs, hamsters.

ORDER LAGOMORPHA. Differ from rodents in having an extra pair of incisor teeth. Rabbits and hares.

ORDER CETACEA. Whales, dolphins, porpoises.

ORDER CARNIVORA. Dogs, cats, bears, seals.

ORDER PROBOSCIDEA. Elephants.

ORDER SIRENIA. Manatees.

ORDER PERISSODACTYLA. Odd number of toes. Horse, zebra, hippopotamus.

ORDER ARTIODACTYLA. Even number of toes. Pigs, sheep, deer, cattle.

REFERENCE GLOSSARY

absorption   The diffusion of water and dissolved materials into cells.

accelerator nerve   A nerve which is attached to the heart and which speeds heartbeat.

acetylcholine   A neurohumor which functions in the transmission of an impulse across a synapse.

acids   Compounds that ionize to form hydrogen ions, H^+.

acquired characteristics   Characteristics that are developed by an organism during its lifetime.

active immunity   Immunity to a disease from having had it in a strong or mild form. (The body builds its own antibodies.)

active transport   The process by which molecules move against the concentration gradient. (It is called active because energy must be contributed by the cell in order for it to occur.)

adaptation   A change in a living thing that better fits it for survival in its environment.

adaptive radiation   The production of a number of different species from a single ancestral one.

addition   A change in which a portion of a chromosome is repeated and the corresponding genes are present twice.

adenosine diphosphate (ADP)   A compound containing adenine, a ribose, and two phosphate groups. (ADP is formed when ATP gives up a P and releases energy.)

adenosine triphosphate (ATP)   A compound containing adenine, ribose, and three phosphate groups. (ATP stores energy temporarily and makes it immediately available to the cell.)

adrenaline   A hormone which is secreted by the adrenal medulla.

aerobic   Occurring with the use of oxygen.

agar   An extract from red algae which is used in culture media.

agglutinins   Antibodies that cause clumping of large particles such as bacteria or blood cells.

air sacs   The functional units of the lungs containing alveoli.

algae   A largely photosynthetic group of protists. (The algae include more than one phylum.)

alimentary canal   In the animal body, the tube through which the food mass passes.

alleles   The alternative genes for a trait (for example, in the case of the trait of height, the genes for tallness and shortness). Alleles occupy the same site on a chromosome.

alternation of generations   A life cycle found in plants in which an asexual stage (sporophyte) alternates with a sexual stage (gametophyte).

alveoli   Microscopic structures where the exchange of gases between the lungs and the blood occurs.

amino acids   Compounds that have at least one amino group and one carboxyl group (the acid group) bonded to a central carbon atom. (Amino acids are the fundamental unit in proteins.)

amino group   The $-NH_2$ group in an amino acid.

ammonia   A highly toxic substance having the formula NH_3.

anaerobic   Occurring without the use of oxygen.

anal pore   The pore in the pellicle of *Paramecium* where solids are eliminated.

analogous structures   Structures that have the same function but different basic compositions and different evolutionary origins.

annelid   A segmented worm — for example, the earthworm. (The phylum name is Annelida.)

anterior   Front end or head of an animal.

antibodies   Proteins produced in the blood as a reaction to the presence of antigens in the body. (An antibody neutralizes the effectiveness of the antigen that caused it to be produced.)

antigen   A substance (usually a protein) which induces the body to form antibodies. (Most antigens are foreign substances.)

antitoxin   An antibody that neutralizes a specific toxin.

anus   An opening at the posterior of an alimentary canal.

aorta   The large artery leading from the heart to the body.

aortic arch   In man, an arching curve near the aorta of the heart.   In the earthworm, one of the lateral pumping arteries (sometimes called one of its "hearts").

arteries   Vessels that carry blood from the heart.

arterioles   Small arteries.

arthropods   Joint-legged animals, including crustaceans, insects, etc.

asexual reproduction   Reproduction in which there is no fusion of nuclei.

assimilation   The incorporation of digested molecules into the makeup of an organism.

atom   The smallest particle of an element that can take part in a chemical reaction.

atomic mass   The exact mass of an atom in atomic mass units.

atomic mass unit   The unit used in the measurement of small masses. It is defined by giving the value 12 to the isotope carbon-12. Thus, a single atomic mass unit is by definition exactly 1/12 the mass of a carbon-12 atom.

atomic number   The total number of protons in the nucleus of an atom. (The atomic number is an indication of the electrons in the shells.)

ATP   *See* adenosine triphosphate.

atrium   In the heart, a thin-walled upper chamber that receives blood from veins.

autonomic nervous system   The part of the nervous system which regulates the involuntary activities of the body.

autotroph   An organism that synthesizes its food from relatively simple substances in its environment.

auxins   Chemical substances produced in one part of a plant and transmitted to another part where they promote or inhibit plant growth. (Also called plant hormones.)

axon   The long portion of the neuron.

bacteriophages   Viruses that attack and destroy bacteria.

barrier   A factor which interferes with the radiation of a species.

bilateral symmetry   Having a right side and a left side which are alike. (An animal with bilateral symmetry can be divided into two equal parts by a single plane passing between its symmetrical sides.)

binary fission   A type of asexual reproduction which occurs by mitotic division into two more or less equal parts.

biochemistry   The study of the chemistry of living things.

biogenesis   The theory that living things originate only from previous living things.

biology   The study of living things.

biome   A climax community of plant and animal life that is typical of a broad region with one kind of climate.

biosphere   The portion of this planet in which ecosystems operate.

biotic   Refers to life or living. (For example, the part of the environment which is living is called the biotic environment.)

bone   A living tissue consisting of bone cells and a matrix of calcium phosphate.

bryophyte   A moss, liverwort, or hornwort. (The phylum name is Bryophyta.)

budding   A type of asexual reproduction in which the new individual is produced as an outgrowth of an older one.

capillary   Thin-walled blood vessel where exchange of materials between the blood and the cells takes place.

carbohydrate   Organic compound containing the elements carbon, hydrogen, and oxygen with the hydrogen and oxygen in a 2 : 1 ratio.

carboxyl group   The $-COOH$ group in an amino acid.

cardiac muscle   A special type of involuntary muscle found only in the heart.

catalyst   A substance that affects the speed of a chemical reaction without itself being used up.

cell   A unit of structure and function in organisms.

cell membrane   A thin, selectively permeable membrane enclosing the cell.

cell wall   The fairly rigid nonliving material which surrounds the cell membrane of plant cells.

cellulose   A carbohydrate present in the cell walls of plant cells.

central nervous system   The brain and the spinal cord.

centrioles   Cylindrical particles in the centrosome.

centrosome   A small body found mainly in animal cells. It contains centrioles that seem to function during reproduction of animal cells.

chalk   A soft limestone consisting chiefly of the shells of foraminifers.

chemical bond   Any force of attraction that links atoms together in molecules.

chemosynthesis   The formation of carbohydrates by the use of the energy of chemical reactions (rather than light).

chloroplast   Plastid containing chlorophyll.

chordates   Animals having a dorsal notochord. (The phylum name is Chordata.)

chromatid   One of the strands of a double-stranded chromosome formed by replication during cell division. The two chromatids are joined by a centromere. (After the two chromatids separate, they are each considered to be chromosomes.)

chromatography   Any of various techniques by which related compounds are separated by seeping through an absorbent material.

chromosomes   Structures in the cell nucleus composed of DNA and protein, and bearing genes. (The number of chromosomes in the nucleus is usually constant for a given species.)

cilia   Tiny projections of living cytoplasm which are hairlike in appearance. They are found in *Paramecium.*

circulation   The transport of materials within cells or between parts of a many-celled organism.

cleavage   The early stages in the development of the embryo characterized by a rapid division of cells.

climate   The general atmospheric and weather conditions of a region.

climax community   A stable, self-perpetuating community in which populations exist in balance with each other and with the environment.

closed circulatory system   One in which the system of blood tubes is considered to be closed because the blood does not normally leave the vessels.

coenzyme   A molecule which assists an enzyme in its action.

cold-blooded animal   One that has the same body temperature as the environment.

colony   An association or clumping of organism-like cells which may exhibit varying degrees of specialization. A colony is too primitive to be considered multicellular.

commensalism   A symbiotic relationship in which one organism benefits by consuming the unused food of the other organism.

community   Populations of different species in a given location, interacting with each other.

compound Two or more elements that are combined chemically.

compound microscope A light microscope with an ocular system of lenses and an objective system of lenses.

concentration gradient The difference in concentration of molecules on each side of a membrane. (It is this difference that must be overcome in active transport.)

conditioned reflex A reflex in which a substitute stimulus results in the same response as the original stimulus.

conifers Cone-bearing plants.

conjugation A type of sexual reproduction found in simpler organisms in which isogametes fuse. (In *Spyrogyra,* conjugation occurs as the contents of two cells unite. The term is also applied to the exchange of nuclear materials in *Paramecium.*)

contractile vacuole The vacuole associated with the regulation of water balance in protozoans.

contraction The shortening of a muscle.

control A duplication of every part of the main experimental set up with the exception of one variable factor.

cotyledon Within the seed, an embryonic leaf which supplies food to the developing embryo. (It may serve as the first photosynthetic organ of the seedling as it begins to grow.)

covalent bond A chemical bond formed by a pair of shared electrons.

crossing-over The breaking of linkage groups and the exchange of these linkage groups between homologous chromosomes. (Crossing-over occurs during meiosis.)

cross-pollination Transfer of pollen from an anther of one plant to a stigma of another plant of the same kind.

crustaceans A class of arthropods which are mainly aquatic — for example, lobsters, crabs, etc.

cyclosis Circular flow of cytoplasm within a cell.

cytoplasm The contents of the cell between the nucleus and the cell membrane.

dark reactions The reactions of photosynthesis which do not use light as a source of energy. (This is the synthesis phase of photosynthesis.)

deamination The removal of an amino group from an amino acid. (The amino group ($-NH_2$) is converted to ammonia.)

decomposers Organisms that feed upon dead organic matter, returning inorganic materials to the environment where they are used again by other living things.

dehydration synthesis The synthesis of a large molecule from small ones by the loss of a water molecule.

deletion A change in which a portion of a chromosome is omitted and the corresponding genes are absent.

denitrifying bacteria Those bacteria that convert the nitrogen of ammonia to free nitrogen.

deoxyribonucleic acid (DNA) Nucleic acid having large molecules consisting of alternating nucleotides containing deoxyribose sugar. (DNA controls the metabolism of the cell and stores the hereditary information of the cell.)

deoxyribose A 5-carbon sugar having one oxygen atom less than ribose.

depletion The removal of organic matter and minerals from the soil.

diaphragm A sheetlike layer of muscle separating the chest cavity from the abdominal cavity.

diatomaceous Containing diatoms or their remains.

dicot A plant which produces seeds with two cotyledons.

differentiation The formation of different kinds of cells in a many celled organism.

diffusion A general term for the migration of materials because of random molecular motion. (The net movement is from the region of higher concentration to the region of lower concentration.)

digestion The process of breaking down large organic molecules into smaller, soluble materials.

dihybrid An organism that is hybrid for two traits.

dipeptide Two amino acids linked together by a peptide bond.

diploid number The number of chromosomes in the double set of homologues found in all body cells.

disaccharide A double sugar.

DNA *See* deoxyribonucleic acid.

dominant gene A gene that exerts its full effect no matter what effect its allelic partner may have.

dominant species The species in each stage of ecological succession which exerts control over the other species present.

dorsal Having to do with the upper surface of an animal.

echinoderms Spiny skinned animals such as the starfish. (The phylum name is Echinodermata.)

ecological niche The role that a species plays in the community.

ecological succession The orderly process by which one biotic community is replaced by another.

ecosystem A system containing a living community and its nonliving environment.

ectoderm Outer layer of cells in a simple animal body. Also, the layer of the embryo from which the skin and nervous system develop.

effectors Parts of the body that respond to stimuli.

egestion The removal of undigested food.

egg An ovum. Also, a fertilized egg (or zygote).

electron A negatively charged particle with a mass that is 1/1846 of the mass of a proton. The mass number of an electron is considered to be zero.

electron microscope An instrument that uses electron beams to produce a magnified image of an object.

element   The basic form of matter. (Each element is a single kind of matter. Elements may exist as molecules; some of them may also exist as single atoms.)

elimination   The removal of undigested food.

embryo   An organism in an early stage of development.

empirical formula   A representation showing the kinds of atoms in a molecule and the number of atoms of each kind. (For example, $C_6H_{12}O_6$.)

emulsify   To form an emulsion.

emulsion   A mixture composed of droplets of one liquid suspended in another.

endocrine gland   A ductless gland which secretes hormones into the blood.

endoderm   Inner layer of cells in a simple animal body. Also, the layer of the embryo from which the digestive tract, respiratory system, liver, and pancreas develop.

endoplasmic reticulum   A network of tube-like structures extending throughout the cytoplasm.

environment   All of the substances, forces, and organisms which affect an organism during its life.

enzyme   Organic substance produced by living things that acts as a catalyst in biochemical reactions.

epidermis   The outer layer of the skin, consisting of epithelial tissue. Also, the outer tissue of a young root or stem.

epiglottis   A cartilaginous flap which diverts food and liquid toward the esophagus and prevents it from entering the larynx.

erosion   The action of wind and water to remove the soil.

esophagus   The food tube which connects the pharynx and stomach. Also called the gullet.

evolution   *See* organic evolution.

excretion   The process by which an organism gets rid of the wastes of metabolism.

external fertilization   Fertilization outside the body of the female.

fat   A combination of three fatty acids and one glycerol.

fatty acids   Organic acids in which the R group is a long hydrocarbon chain.

feedback   The automatic reversion of a process which occurs when it exceeds the limits imposed by the system.

fermentation   Anaerobic oxidation of glucose resulting in the formation of alcohol and carbon dioxide or in the formation of lactic acid. (Fermentation is a synonym for anaerobic respiration.)

fertilization   Sexual reproduction by the union of dissimilar gametes (heterogametes) — for example, the fusion of a sperm and ovum.

fibrinogen   A protein in the plasma which functions in the clotting of blood.

filament   A threadlike group of cells found in some algae such as *Spirogyra*. Also, one of the threadlike structures in a gill.

filtrate   The material which passes through a filter.

filtration The separation of materials that occurs as a filtrate passes through a porous substance (filter).

flagellated Having a flagellum or flagella.

flagellum A whiplike protrusion of cytoplasm employed by certain simple organisms and by sperm for movement through a fluid medium.

focus Achieving the proper distance between the objective and the specimen so as to produce a sharp image.

food web Interconnected food chains in an ecosystem.

fossil Any remains, impression, or trace of an organism that lived in the geologic past.

fraternal twins Twins that result when two ova produced simultaneously are fertilized by different sperm cells.

fruit An enlarged ovary with its seeds and any parts of the flower which remain attached.

fungi A mixed group of nongreen plants. (They are classified in more than one phyla of the protista kingdom.)

gamete A cell which unites with another cell in sexual reproduction. (A sex cell.)

gametogenesis The process by which gametes are produced.

gametophyte The gamete-producing stage in an alternation of generations.

gamma globulin A fraction of the blood protein which includes the antibodies.

ganglion Cluster of neurons which may serve as a relay center for directing impulses in several directions.

gastrula In the animal embryo, the stage which marks the beginning of differentiation and which consists of an inpushing of a hollow sphere.

gene That part of the DNA molecule which holds a unit of genetic information.

gene frequency The fraction of all members of the population that have a particular gene, expressed as a decimal. (The sum of all the gene frequencies for the genes of any allelic series is always 1.)

gene pool The sum total of all the genes collectively present within a given population.

genotype The genetic makeup of an individual.

genus A group of closely related species having a common ancestor in the recent past.

germination The growth of a new plant from the seed.

germplasm The reproductive cells, as distinguished from the body cells.

gill openings Openings to the gills found in cartilage fishes.

gills Specialized respiratory organs consisting of platelike, or threadlike structures. (Gills enable some aquatic animals to absorb dissolved oxygen from the water.)

glucagon A hormone secreted by the pancreas which helps to change glycogen to glucose.

glycerol An alcohol with three hydroxide (OH) groups.

glycogen A polysaccharide which is a major storage product in animals.

Golgi apparatus   A stack of saucerlike structures in the cytoplasm, especially prominent in secretory cells.

gonad   Organ that produces sex cells.

green plants   Those that contain chlorophyll.

growth   The increase in size of an organism that results from the change of food molecules into the kinds of molecules which are characteristic of that organism.

gullet   In *Paramecium,* the tract to a food vacuole.   In man, the esophagus.

gymnosperm   A seed plant in which the seed is not enclosed in a fruit. (The evergreen conifers.)

half-life   The time required for the disintegration of half the atoms in a given sample of a radioactive element.

heart   In higher animals, the basic blood pump.   Also applied, but more loosely, to one of the lateral pumping arteries (aortic arches) of the earthworm.

hemoglobin   An iron-containing protein compound which helps to transport oxygen and carbon dioxide. (It gives blood its red color.)

heterogametes   Dissimilar gametes. (The sperm cells and egg cells.)

heterotroph   An organism that takes in and uses preformed organic molecules containing much energy in their bonds.

heterozygous   Pertaining to an organism in which the genes of a pair of alleles are unlike. (Hybrid.)

highway   A factor in the environment which helps a species to radiate.

homeostasis   The maintenance of a stable internal environment.

homologous structures   Those that are fundamentally similar in structure and have the same evolutionary origin. (They may have different functions.)

homologue   A single chromosome in a homologous pair.

homozygous   Pertaining to an organism in which the genes of a pair of alleles are alike. (Pure.)

hormone   A chemical substance produced in one part of the organism and carried to another part of the organism where it has its effect. In vertebrates, hormones are produced by endocrine glands and carried by the blood. Plants produce hormones called auxins.

host   The living organism from which a parasite takes its food.

humus   The organic portion of soil.

hybrid   An organism in which the genes of one pair (or more than one pair) of alleles are not alike.   A cross between species.

hydrocarbon   A compound containing only hydrogen and carbon atoms.

hydrogen acceptors   Coenzymes which undergo reduction by accepting hydrogen atoms.

hydrogen ion   The H^+ of acids.

hydrolysis   The decomposition of large molecules into smaller units by combining them with water.

hydroxide ion   The OH^- of bases.

hypothesis   A tentative assumption to be tested with experiments and observations.

identical twins   Twins that result from the separation of a single fertilized egg.

igneous rocks   Rocks formed by the solidification of molten lava.

immunity   The resistance of an organism to disease.

immunological reaction   The building of antibodies.

impulse   Electrical and chemical changes that travel along a nerve fiber.

index fossil   A fossil of an organism which can be used to determine relative ages of fossils or rocks.

inferior vena cava   A vein from the trunk region which empties into the right atrium.

insulin   A hormone secreted by the pancreas which helps to convert glucose to glycogen. (It is one of the smaller protein molecules, and consists of 15 different kinds of amino acids arranged in two chains.)

internal fertilization   Fertilization inside the body of the female.

intestinal glands   Tiny glands in the walls of the small intestine which secrete enzymes.

inversion   The reversal of the sequence of genes in a portion of a chromosome.

ion   An atom or group of atoms which has gained or lost one or more electrons. (Ions that arise from the loss of electrons are positively charged; those that arise from the gain of electrons are negatively charged.)

ionic bond   A chemical bond between ions, resulting from a transfer of electrons.

ionization   The process of separating into ions.

isogametes   Similar gametes.

isolation   The confinement of a population to a particular region by barriers.

isomers   Compounds with the same empirical formula but different structural formulas.

isotope   One of a group of atoms having the same atomic number but different atomic masses.

kidney   An excretory organ that removes urine.

lacteal   A projection of the lymph system into the villus. The lacteal absorbs the end products of fat digestion (fatty acids and glycerol).

lactic acid   A 3-carbon organic acid which is formed during anaerobic respiration from pyruvic acid. (Lactic acid has the formula $CH_3-CHOH-COOH$.)

larva   An immature stage in the metamorphosis of an animal. (The larval period begins with hatching and ends when the larva changes to another life form.)

larynx   The voice box, situated just above the trachea in man. (Air enters the trachea through the larynx.)

lichen   A combination of an alga and a fungus which exists in a relationship of mutualism.

life cycle   A term used to describe the life history of an organism which has different life forms at different stages of its life.

ligaments   Tough tissues that connect bones at the movable joints.

light reactions   The reactions in photosynthesis which require light as a source of energy. (This is the "photo" phase of photosynthesis.)

limestone   A rock formed chiefly from shells and consisting mainly of calcium carbonate.

limy mud   Soft mud containing calcium carbonate and various remains.

linkage   The "sticking together" of genetic traits that occurs because their genes are present on the same chromosome.

lipids   Fats, oils, and waxes.

liver   A large glandular organ which is associated with digestion, excretion, and glucose metabolism.

locomotion   The ability to move from place to place.

lung   An organ for breathing air found in higher animals.

lymph   A colorless fluid found in tissue spaces and in lymph vessels. (Also called tissue fluid or intercellular fluid.)

lymphatics   Lymph vessels.

lysosomes   Oval bodies in the cytoplasm which serve as centers of cellular digestion.

macronucleus   Large nucleus found in *Paramecium* and some other protozoans.

magnification   The ability of an optical system to enlarge the image of an object.

mammal   A vertebrate having mammary glands, hair on its body, and a birth by separation from the placenta of the mother.

mammary glands   Glands that secrete milk.

mass   A measure of the quantity of matter in a body.

mass number   The mass of the fundamental particles in an atom relative to that of a proton, which is considered to be unity (1). Also, the total number of protons and neutrons in an atom.

mass spectrometer   An instrument used for identifying isotopes on the basis of their differences in mass.

matrix.   Material between cells. (Also called intercellular material.)

matter   Anything that has mass and occupies space.

meiosis   A kind of cell division in which the number of chromosomes is reduced to half.

meristem   Undifferentiated cells in a plant. (For example, cambium.)

meristematic zone   A region at the apex of a root where undifferentiated cells increase rapidly in number.

mesoderm   Middle layer of cells in a simple animal body.   Also, the layer of the embryo from which the skeleton, muscles, circulatory system, and gonads develop.

metabolism   The sum total of all the life processes of an organism.

metamorphic rocks   Rocks produced as a result of metamorphism of igneous or sedimentary rocks.

metamorphism A change in the chemical composition of rock, resulting in a more compact structure.

metamorphosis A series of changes from an egg to an adult.

methane An organic compound with a central carbon atom covalently bonded with four hydrogen atoms. (Its empirical formula is CH_4.)

micromanipulator An instrument that lets the user control the movement of very delicate instruments under a compound microscope.

micron 1/1000 of a millimeter.

micronucleus Small nucleus found in *Paramecium* and some of the other protozoans.

microorganism Any organism so small that it can be seen only under the ordinary microscope or a microscope capable of higher magnifications.

migration Movement of materials or organisms from one system into another.

mitochondria Cytoplasmic organelles serving as sites of cellular respiration.

mitosis The exact duplication of the nucleus of a cell so as to form two identical nuclei. (This involves a doubling and subsequent separation of the nuclear material.)

mixture A substance in which several kinds of molecules are associated physically without being combined chemically.

molecular weight The weight of a molecule in atomic mass units.

molecule The smallest particle of a substance that has a stable and independent existence. (It may be a combination of atoms of the same kind, O_2, or of different kinds, CO_2.)

molting The discarding of an exoskeleton during metamorphosis.

monocot A plant which produces seed with one cotyledon.

monohybrid An organism that is hybrid for only one trait.

monoploid number The number of chromosomes in a single set of homologues. (This is the number found in the sperm or ova.)

monosaccharide A simple sugar that cannot be hydrolyzed into simpler carbohydrates.

mucus A largely protein substance secreted by living cells.

multicellular organism An organism that is composed of many cells which function together as a coordinated unit.

multinucleate Having many nuclei.

mutation A genetic change which can be inherited.

mutualism A relationship in which two organisms live together to the advantage of both.

nematode A roundworm. (The phylum name is Nematoda.)

nerve A bundle of axons covered with connective tissue.

nerve net The simple network of nerve cells found in *Hydra*.

neurohoumor A chemical secreted by the ends of nerve cells which permits impulses to be conducted from neurons to other neurons or to muscle and gland cells. (Neurohumors are also called neurohormones.)

neuron   A nerve cell. The basic unit of which the nervous system is composed.

neutralization   The removal of hydrogen ions and hydroxide ions from a solution to form water.

neutron   A subatomic particle with a mass number of 1 and no charge. (It acts as though it were composed of a proton and electron.)

niche-splitting   Refers to two species taking over different parts of a niche.

nitrifying bacteria   Bacteria that oxidize ammonia to nitrates or oxidize nitrites to nitrates.

nitrogen-fixing bacteria   Bacteria that produce nitrogen compounds by fixing free nitrogen.

nitrogenous wastes   Nitrogen-containing wastes such as ammonia (NH_3), urea, and uric acid.

nondisjunction   Failure of the chromosomes to separate from each other after synapsis.

nongreen plants   Plants that do not contain chlorophyll.

noradrenaline   A neurohumor which resembles adrenaline in its action.

normalizing selection   The stabilizing process by which extreme variations are continuously eliminated from the population.

notochord   A stiff supporting rod in the dorsal part of the body found at some stage in the life of a chordate.

nucleic acid   Compound consisting of nucleotides linked in a chain. (Examples are DNA and RNA.)

nucleoprotein   A complex material containing nucleic acid and protein. (The material of the chromosomes.)

nucleotide   A unit composed of three elements: a nitrogen base, a phosphate, and a 5-carbon sugar. (The nucleotides in RNA have ribose sugar, those in DNA have deoxyribose sugar.)

nucleus   In the cell, the spherical body which contains the chromosomes. In the atom, the central mass which contains the protons and neutrons.

nutrients   The usable portion of foods.

nutrition   Those activities of an organism by which it takes raw materials from its environment and makes them usable.

nymph   A larva that greatly resembles the adult, except it is smaller.

ontogeny   The developmental history of an individual organism.

oogenesis   The production of ova. [By oogenesis, a primary egg cell (diploid) develops into ova (monoploid). The reduction in chromosome number occurs by meiosis.]

oral groove   An indentation near the middle of *Paramecium* which serves for the intake of food.

organ   Different kinds of tissues that function together.

organ system   A group of organs which carry on one of the major functions of the body. (For example, the digestive system.)

organelles   Specialized structures in a cell or unicellular organism. (For example, the food vacuole, which is analogous to an organ, the stomach.)

organic acids   Compounds having the general formula R—COOH.

organic chemistry   The study of carbon compounds in living and nonliving things.

organic evolution   The change of kind with time by which new species develop.

osmosis   A special kind of diffusion which refers to the passage of water molecules through a selectively permeable membrane.

ovaries   The gonads of the female animal. (They produce ova.)

overproduction   The birth of more young than can be kept alive by the available food supply.

ovulation   The discharge of ripe ova from the ovaries.

ovule   Structure in the ovary from which the seed develops.

ovum   A female gamete.

oxidation   The loss of electrons or the loss of hydrogen atoms from an atom or molecule. This definition holds whether the electrons are received by oxygen or any other substance.)

oxyhemoglobin   Substance in the blood, formed by the loose union of hemoglobin and oxygen.

pacemaker   A region near the top of the right atrium which regulates the speed and extent of contraction of the heart.

pancreas   A gland located near the stomach. (It is both an endocrine and an exocrine gland.)

parasite   An organism that lives on or in a host from which it takes its food.

parthenogenesis   The development of an egg into a new individual without its being fertilized by a sperm cell.

passive immunity   Immunity that an individual has to a disease because he has been inoculated with antibodies.

passive transport   Diffusion of molecules from a region of high concentration to a region of low concentration. (It is called passive because it occurs without energy being added to the process.)

pasteurization   The process of killing or slowing the growth of bacteria in a food by heating it at a particular temperature that will preserve its flavor.

pellicle   Thick, flexible outer covering of *Paramecium*. (It surrounds the cell membrane of *Paramecium*.)

peptide bond   The bond between two amino acids.

peptides   Short chain of amino acids. (Two amino acids linked together comprise a dipeptide; three, a tripeptide; and a large number, a polypeptide.)

peripheral nervous system   Includes spinal nerves to and from the spinal cord and the autonomic nervous system.

peristalsis   Wavelike muscular contractions and relaxations in the alimentary canal which force the food mass along.

phagocyte   A cell in the body which is capable of engulfing particles. (White corpuscles are phagocytes.)

phagocytosis   The process by which a cell engulfs undissolved large particles by flowing around them and enclosing them in a vacuole.

pharynx   The throat. (Serves as a passageway for air and food.)

phase   The time relationship of two moving waves. (Two waves are in phase if they reach their crests simultaneously; they are out of phase if their crests and troughs are out of step.)

phenotype   The observable appearance of an individual resulting from its genotype.

phloem   Thin-walled plant cells that primarily transport food downward in the stem.

photolysis   The breakdown of water by light energy during the light reactions of photosynthesis.

photosynthesis   The synthesis of carbohydrates from simple chemicals by using the energy of light.

phylogeny   The evolutionary history of a species.

phylum   A large division of classification below the kingdom.

pinocytic vesicle   A pocket formed in the cell membrane, which breaks off inside the cell to form a vacuole.

pinocytosis   The process by which a cell takes in water and large molecules through pinocytic vesicles. (Pinocytosis requires the expenditure of energy by the cell.)

pioneer organisms   The first organisms to populate a given region.

placenta   The structure by which the mammalian embryo is attached in the uterus of the mother.

plankton   The mass of minute organisms floating near the surface of the sea.

plasma   The liquid part of the blood.

plasmodium   Multinucleate mass of protoplasm capable of locomotion. (One stage in the life cycle of a slime mold is a plasmodium.) *Plasmodium* is also the genus name for the spore-forming protozoan that causes malaria.

plasmolysis   The shrinkage of cell contents due to outward osmosis.

plastids   Small bodies in the cytoplasm of cells of many-celled plants and some one-celled animals. (Many plastids are colorless; chloroplasts contain the green pigment chlorophyll.)

platelets   Living cell fragments which are components of the blood and which help to start blood clotting.

pollination   Transfer of pollen from an anther to a stigma.

polypeptide   A large number of amino acids linked by peptide bonds. (Polypeptides combine to form proteins.)

polyploidy   A condition in which the cells have extra sets of chromosomes, beyond the diploid number. (Polyploid individuals may have $3n$, $4n$, $5n$, $6n$, etc., chromosomes.)

polysaccharides   Carbohydrates with more than two glucose-like units.

population   All members of a species inhabiting a given location.

population genetics   The study of the genetics of large groups of sexually reproducing organisms.

posterior   The rear portion of an animal.

precipitate   An insoluble solid formed by a reaction between liquids.

proteins   Tissue building compounds composed of numerous amino acids linked by peptide bonds. (The main elements present are C, H, O, N, and S.)

protist   One of the protista.

protista   A kingdom that includes the organisms which are not easily classified as plants or animals. (Usually, it includes the protozoa, algae, bacteria, fungi, lichens, and slime molds.)

protochordate   A primitive chordate whose notochord remains throughout its life.

proton   A particle in the nucleus of an atom that has a mass number of 1 and a positive charge.

protozoa   A group of simple animal-like protists. (The protozoa include more than one phylum.)

pseudopod   Outpushing of protoplasm used in locomotion and food getting, as in *Amoeba* and white blood cells.

pulmonary   Pertaining to the lungs.

pulse   A wave of alternate stretchings and contractions which proceed from the aorta along all arteries and arterioles.

Punnett Square   A checkerboard method for showing the gene combinations that result from the possible fertilizations.

pyrenoids   Starch-containing bodies on or near a chloroplast. (These are found in *Spirogyra* and other protists.)

pyruvic acid   A 3-carbon molecule with the empirical formula $C_3H_4O_3$. (It is produced during the anaerobic phase of respiration and oxidized during the aerobic phase.)

R group   In an organic compound, a general expression for any of a number of combinations which can substitute for a hydrogen atom.

radial symmetry   Having parts of the body radiate from a central axis. (An organism having radial symmetry can be divided into two equal parts by any plane passing through the diameter and the central axis.

radioactive   Refers to an isotope that spontaneously emits radiations.

radio-carbon method   A method used for dating fossils not much older than 40,000 years.

receptor   Specialized tissue in contact with nerve cells, and sensitive to a specific stimulus.

recessive   Pertaining to the gene that is masked if the allelic gene is dominant.

recombination   Refers to the recombination of linkage groups during crossing-over.

rectum   The part of the large intestine just before the anus.

red corpuscles   The blood cells which contain hemoglobin.

reduction   A gain of electrons or a gain of hydrogen atoms by an atom or molecule.

reflex   An inborn, automatic act.

regeneration The ability of an organism to grow back a missing part. Also, the ability of one of the parts to develop into an entire animal.

regulation The coordinated response of an organism to a changing environment so as to maintain its stability.

relative age The age of a fossil or layer of rock stated on the basis of a comparison made between it and its immediate surroundings.

repair The growth of new cells to replace lost, dead, or diseased cells.

replication The process by which the DNA molecule makes an exact copy of itself.

reproduction The unique characteristic of living organisms by which they produce more of their own kind.

resolution The ability of an optical system to distinguish clearly and in detail between objects that lie very close to each other.

resorption The process by which materials are absorbed again.

respiration The oxidation of food in an organism, resulting in the release of energy and waste products. (Much of the energy released is used to make ATP which supplies energy for the immediate needs of the organism.)

response The secretion or movement resulting from a stimulus.

rhizoids Rootlike structures that absorb nutrients.

rhizome A thickened, woody underground stem.

ribonucleic acid (RNA) Nucleic acid containing nucleotides in which the sugar is ribose.

ribosomes Cellular organelles that are sites of protein synthesis. (They are attached to the walls of the endoplasmic reticulum or move freely in the cytoplasm.

RNA *See* ribonucleic acid.

saccule Tiny saclike structures.

saprophyte An organism that takes its food from dead or decaying organic matter.

saturated compound A compound in which the carbon atoms are attached to all of the atoms that they are capable of holding.

scion The portion of a stem grafted onto a rooted stock.

secretions Essential chemicals that are synthesized by the organism but which are not incorporated into its basic makeup. Also, the production of secretions.

sedimentary rocks Rocks formed from sediments on the bottom of bodies of water. (Sedimentary rocks may contain fossils.)

seed A ripened ovule containing an embryo.

selectively permeable membrane A membrane that permits some materials to pass through readily but not others.

self-pollination Transferal of pollen from an anther to a stigma of a flower on the same plant.

serum Plasma without fibrinogen and other clotting factors.

sessile Fastened to one spot; incapable of locomotion.

sexual reproduction   Reproduction by the fusion of the nuclei of two cells.

skeletal muscle   Voluntary muscle consisting of long fibers having striations.

small intestine   The tube in the alimentary canal where the end products of digestion are absorbed.

smooth muscle   Involuntary muscle consisting of spindle-shaped cells with no striations. (It is found in the alimentary canal, respiratory passages, arteries, veins, and diaphragm.)

solute   The substance dissolved in a solution.

solution   A uniform mixture of the molecules of a solute and a solvent.

solvent   The medium in which a solute is dissolved.

somatoplasm   The body cells, as distinguished from the reproductive cells.

sori   Sacs containing many sporangia which may be found on the undersurface of certain fern leaves.

speciation   The process by which a single species can separate into two or more.

species   An interbreeding unit. (In nature, a species defines itself by breeding exclusively, or nearly so, within the group.)

spectrophotometer   An instrument that measures the amount of light of various wave lengths which is absorbed by a solution.

sperm   A male gamete.

spermatogenesis   The production of sperm. (By spermatogenesis a primary sperm cell — diploid — develops into sperm — monoploid. The reduction in chromosome number occurs by meiosis.)

spinal column   The spinal cord protected by vertebrae.

spinal cord   In vertebrates, a dorsal tube containing a bundle of nerve fibers.

spontaneous generation   The belief that living things can originate from nonliving matter.

sporangia   Spore cases which contain spores.

spore   An asexual reproductive cell which is capable of developing into an adult directly.

sporophyte   The spore-producing stage in an alternation of generations.

stimulus   A change in the environment that affects the sensitive cytoplasm.

stock   The rooted portion of a plant to which the scion from another plant is grafted.

stomach   An organ in the alimentary canal which functions mainly in storage.

structural formula   A representation which uses dashes to show the bonds between atoms, and which shows the arrangement of the atoms in the molecule.

substrate   The material acted upon by an enzyme.

succession   The orderly process by which one biotic community is replaced by another.

superior vena cava   A vein from the head region which empties blood into the right atrium.

susceptibility An organism's lack of immunity.

symbiosis A nutritive relationship in which organisms of different species "live together." (The relationship may be mutualism, commensalism, or parasitism.)

synapse The gap between the terminal branches of one neuron and the dendrites of a second neuron.

synapsis The pairing of homologous chromosomes during meiosis.

synthesis The process by which simple compounds are united to form more complex materials.

tagged atom An isotope used to trace an element within a series of chemical reactions.

taxonomy The science of classification of presently existing and extinct organisms.

tendons Tough connective tissue that connects bones and muscles.

testes The gonads of the male animal.

tetrad A structure consisting of four chromatids.

theory A general statement that unifies many isolated facts into a broad idea.

thoracic duct A large lymph vessel which empties into a large vein and adds absorbed fats to the circulating blood.

thyroxin A hormone containing large amounts of iodine which regulates the general rate of the body's metabolism.

tissue A group of cells specialized for a particular function.

tissue fluid A colorless fluid found in the spaces around the cells. (Also called lymph when found in lymph vessels.)

toxin Poisonous chemical produced by microorganism. (Toxins cause disease.)

toxin-antitoxin Toxin weakened by the addition of antitoxin.

toxoid Toxin weakened by heating or by the addition of a chemical such as formaldehyde.

trachea One of the respiratory tubes in an insect or spider. Also, the human windpipe.

tracheophyte A vascular plant. (The phylum name is Tracheophyta.)

translocation The transferal of a portion of a chromosome to another chromosome that is not homologous with it.

transpiration The process by which a plant gives off water.

transport The absorption and circulation of materials throughout an organism.

tropism A growth movement of a part of a plant toward or away from a stimulus.

turgor Turgidity and tension in living cells caused by the inward osmosis of water.

unsaturated compound A compound in which the carbon atoms are bonded by double or triple bonds and which can thus take on additional hydrogen atoms.

uranium-lead ratio A method based on the radioactive disintegration of uranium which is used for dating ancient igneous rocks.

vacuoles   Spherical sacs in the cytoplasm which act as reservoirs for water, dissolved materials, and wastes. (They maintain the internal pressure of the cell.)

vagus nerve   A nerve attached to the heart which slows the heart beat.

variable factor   In a controlled experiment, the condition being tested.

variations   The differences that exist in the offspring of a particular species.

vascular bundles   Conducting tubes in plants, consisting of xylem, phloem and cambium.

vegetative propagation   The production of new plants from roots, stems, or leaves.

veins   Vessels that carry blood to the heart.   Also, conducting tubes in leaves.

ventral   Pertaining to the under surface of an animal.

ventricle   Muscular chamber of the heart that pumps blood into arteries.

vertebrates   Chordates which have backbones.

vestiges   Structures in modern organisms which do not appear to have any function but which seem to have been derived from useful structures in ancestral forms of life.

virus   A particle composed of a nucleic acid core and a protein shell. (It is parasitic. Within a host cell, it may reproduce and mutate.)

warm-blooded animal   An animal that maintains the same body temperature despite variations in the temperature of the environment.

waste   The portion of a food that does not serve as a nutrient.

weathering   The action of agents of the atmosphere in altering the earth's surface.

weight   A measure of the force of attraction between an object and the earth. (In practice, weight is often used to mean mass. However, the difference between the two should always be understood. *See* mass.)

white corpuscles   Colorless blood cells that engulf bacteria.

X chromosome   The sex chromosome present doubly in the human female and singly in the human male.

xylem   Thick walled plant cells which provide support and which conduct materials upward through the stem.

Y chromosome   In man, the sex chromosome present only in males.

zygote   The cell which is formed as the result of the union of two gametes.

Selected Readings

Unit One. *THE STUDY OF LIFE*

Asimov, Isaac. *The Chemicals of Life.* New York. New American Library. 1962.
————. *A Short History of Biology.* Garden City, N.Y. Natural History Press (distr. Doubleday). 1964.
Baker, J. J. W. *Cell.* Middletown, Conn. American Education Publications. 1966.
————. and G. E. Allen. *Matter, Energy, and Life.* New York. Addison-Wesley. 1965.
Cain, A. J. *Animal Species and Their Evolution.* New York. Harper and Row. 1960.
Carson, Rachel. *The Edge of the Sea.* Boston. Houghton Mifflin. 1955.
Curtis, Helena. *The Marvelous Animals — an Introduction to the Protozoa.* New York. Natural History Press. 1968.
Giese, Arthur C. *Cell Physiology.* Philadelphia. W. B. Saunders Company. 1968.
Hall, Thomas. *A Source Book of Animal Biology.* New York, McGraw-Hill. 1951.
Haffner, R. E. and J. J. W. Baker. *The Vital Wheel: Metabolism.* Middletown, Conn. American Education Publications. 1963.
Hanson, E. D. *Animal Diversity.* Englewood Cliffs, N.J. Prentice-Hall. 1961.
Hoffman, K. B. *Chemistry of Life.* New York. McGraw-Hill. 1964.
Hurry, S. W. *The Microstructure of Cells.* Boston. Houghton Mifflin. 1964.
Jacker, Corinne. *Window on the Unknown: a History of the Microscope.* Scribner's. 1966.
Lehrman, R. L. *The Long Road to Man.* New York. Basic Books. 1961.
Loewy, A. G. and Phillip Siekevitz. *Cell Structure and Function.* New York, Holt, Rhinehart, and Winston. 1963.
Meyer, J. S. *The Elements, Builders of the Universe.* Cleveland and New York. World. 1957.
Moore, Ruth. *Man, Time, and Fossils.* New York. Knopf. 1961.
Steiner, R. F. and Harold Edelhoch. *Molecules and Life.* Princeton, N.J. Van Nostrand. 1965.
Stoutenberg, Adrien and Laura Baker. *Beloved Botanist.* New York, Scribner's. 1961.
Swanson, C. P. *The Cell.* Englewood Cliffs, N.J., Prentice-Hall. 1964.
Williams, Greer. *Virus Hunters.* New York, Knopf. 1959.
White, E. H. *Chemical Background for the Biological Sciences.* Englewood Cliffs, N.J. Prentice-Hall. 1964.

Unit Two. *ANIMAL MAINTENANCE*

Ames, Gerald and Rose Wyler. *Food and Life.* New York, Creative Education Press. 1966.
Asimov, Isaac. *The Human Body: Its Structure and Operation.* Boston. Houghton Mifflin. 1963.
————. *The Human Brain: Its Capacities and Functions.* Boston. Houghton Mifflin. 1963.

Buchsbaum, R. M. *Animals Without Backbones*. Chicago. University of Chicago Press. 1948.

Burnett, A. L. and Eisder, Thomas. *Animal Adaptation*. New York. Holt, Rhinehart, and Winston. 1964.

Carthy, J. D. *The World of Feeling*. New York. Roy Publishers. 1960.

Hare, D. J. *The Skin*. New York. St. Martin's Press. 1966.

Horrobin, D. F. *The Human Organism — An Introduction to Physiology*. New York. Basic Books. 1966.

Langley, L. L. *Homeostasis*. New York. Reinhold. 1965.

Maisel, A. Q. *The Hormone Quest*. New York. Random House. 1965.

Morrison, T. F. and others. *Human Physiology*. New York. Holt, Rhinehart, and Winston, 1967.

Schneider, Leo. *Lifeline: the Story of Your Circulatory System*. New York. Harcourt Brace and World. 1958.

Storer, T. I. and R. L. Usinger. *Elements of Zoology*. New York. McGraw-Hill. 1961.

Turner, C. D. *General Endocrinology*. Philadelphia. Saunders. 1966.

Wilson, Mitchell. *The Human Body and How It Works*. New York. Golden Press. 1959.

Unit Three.   PLANT MAINTENANCE

Bonner, J. F. and A. W. Galston. *Principles of Plant Physiology*. San Francisco. Freeman. 1952.

Coulter, Merle C. *Story of the Plant Kingdom*. Chicago. University of Chicago Press. 1964.

Esau, Katherine. *Anatomy of Seed Plants*. New York. Wiley. 1960.

Fuller, H. J. and Z. B. Carothers. *The Plant World*. New York. Holt, Rhinehart, and Winston. 1963.

Galston, A. W. *The Life of the Green Plant*. Englewood Cliffs, N.J. Prentice-Hall. 1964.

Hutchins, R. E. *This Is a Leaf*. New York. Dodd, Mead & Co. 1962.

Meyer, B. S. and others. *Introduction to Plant Physiology*. Princeton, N.J. Van Nostrand. 1960.

Sinnott, E. W. and K. S. Wilson, *Botany*. New York. McGraw-Hill. 1963.

Van Overbeek, Johannes and H. K. Wong. *The Lore of Living Plants*. New York. McGraw-Hill. 1964.

Wilson, C. L. and W. E. Loomis. *Botany*. New York. Holt, Rhinehart, and Winston. 1967.

Unit Four.   REPRODUCTION AND DEVELOPMENT.

Anderson, M. D. *Through the Microscope*. Garden City, N.Y. Natural History Press (distr. Doubleday). 1965.

Baker, J. J. W. *In the Beginning*. Middletown, Conn. American Education Publications. 1964.

Ebert, J. D. *Interacting Systems in Development*. New York. Holt, Rhinehart, and Winston. 1965.

Lehrman, R. L. *Reproduction of Life*. New York. Basic Books. 1964.

McEwen, R. S. *Vertebrate Embryology*. New York. Holt, Rhinehart, and Winston. 1957.

McLeish, John and Brian Snoad. *Looking at Chromosomes*. New York. St. Martin's Press. 1962.

Patten, B. M. *Foundations of Embryology*. New York. McGraw-Hill. 1958.

Rosenberg, Jerome. *Photosynthesis*. New York. Holt, Rhinehart, and Winston, 1965.

Sussman, M. B. *Animal Growth and Development*. Englewood Cliffs, N.J. Prentice-Hall. 1960.

Wendt, F. W. and the editors of *Life Magazine*. *The Plants*. New York. Time-Life Books. 1963.

Unit Five. GENETICS

Asimov, Isaac. *The Genetic Code*. New York. New American Library. 1963.

Frankel, Edward. *DNA — Ladder of Life*. New York. McGraw-Hill. 1964.

Haffner, R. E. *Genetics, the Thread of Life*. Middletown, Conn. American Education Publications. 1964.

Sinnott, E. W. and others. *General Genetics*. San Francisco. Freeman. 1965.

Stern, Curt. *Principles of Human Genetics*. San Francisco. Freeman. 1960.

Unit Six. PLANTS AND ANIMALS IN THEIR ENVIRONMENT

Buchsbaum, R. M. and Mildred Buchsbaum. *Basic Ecology*. Pittsburgh. Boxwood Press. 1957.

Farb, Peter. *Ecology*. Morristown, N.J. Silver Burdett. 1963.

Kendeigh, S. C. *Animal Ecology*. Englewood Cliffs, N.J. Prentice-Hall. 1961.

Milne, L. J. and J. G. Milne. *The Balance of Nature*. New York. Knopt. 1960.

Odum, E. P. and H. J. Odum. *Fundamentals of Ecology*. New York. Saunders. 1959.

Rienow, Robert and Leona Rienow. *Moment in the Sun*. New York. Dial Press. 1967.

Unit Seven. EVOLUTION

Curtis, Helena. *Biology*. New York. Worth Publishers. 1968.

Darwin, Charles. *The Origin of Species*. New York. New American Library. 1958.

Dobzhansky, T. G. *Evolution, Genetics, and Man*. New York. Wiley. 1955.

Lehrman, R. L. *Race, Evolution, and Mankind*. New York. Basic Books. 1966.

Life Magazine editors. *The Wonders of Life on Earth*. Englewood Cliffs, N.J. Prentice-Hall. 1960.

Moody, P. A. *Introduction to Evolution*. New York. Harper and Row. 1962.

Moore, Ruth. *Evolution*. Morristown, N.J. Silver Burdett. 1962.

Oparin, A. I. *The Origin of Life*. New York, Dover. 1953.

Simpson, G. G. *The Meaning of Evolution*. New Haven, Conn. Yale University Press. 1960.

INDEX

A

Abdomen (insects), 104
Abiogenesis, 423-426
Absorption, 2
  small intestine and, 108-109
  defined, 119
  root hairs and, 251
Accelerator nerve, 123, 186
Acetic acid, 210
Acetylcholine, 187
Acid(s), 29
  amino, 32
  organic, 38
  nucleic, 39, 320, 360-367
  pyruvic, 145-146, 147, 171
  lactic, 146-147
  uric, 167, 168, 169
  acetic, 210
  hydrochloric, 210, 211
  indole acetic, 260, 261
  naphthaleneacetic, 260
  gibberellic, 261
  See also Amino acids; Deoxyribonucleic
    acid; Ribonucleic acid
Acromegaly, 199
ACTH, 200
Active gamete, 286
Active immunity, 132, 133
Active transport, 117-118, 251, 254
Adam's apple, *see* Larynx
Adaptation, 2
Adaptive radiation, 441-442
Adenoids, 129
Adenosine diphosphate (ADP), 144
Adenosine triphospate (ATP), 142-148,
  256, 257, 272
Adrenal glands, 202
Adrenaline, 127, 202, 203, 204
Adrenocorticotrophic hormone, 200
Aerobes, 145; rise of, 429
Aerobic respiration, 145, 147, 149, 156,
  256, 257
Afterbirth, 303
Agar, 72
Agglutinins, 132, 136
Agglutinogens, 136

Aggregates, 429
Agnatha, 90
Air pollution, 401
Air sacs, 160
Alanine, 32; deamination of, 171
Alcoholic fermentation, 257
Alcohols, 38
Algae, 71-72
  blue-green, 12, 72
  brown, 72
  red, 72
  alternation of generations, 309
Algebraic method, for solving problems in
  genetics, 353-355
Alimentary canal: in animals, 102; in man,
  105, 106, 297
Allantois, 294
Alleles, 325, 332; multiple, 343-345
Allergy, 131
Alligator, 92
Alternation of generations, 74, 309
Alveoli, 160
American Association for the Advancement
  of Science, 17
Amino acids, 32-33, 34, 212; deamination
  of, 171; code triplets for, 367
Ammonia, 167, 168, 171
Amnion, 294, 302
Amniotic fluid, 294, 302
Amoeba, 8, 13, 15, 70
  classification of, 61
  digestion, 101
  food-getting, 101
  nutrition, 101
  ingestion, 101, 118
  circulation in, 119
  cyclosis and, 119
  excretion, 167
  locomotion, 213-214
  binary fission in, 275
Amoeboid movement, 213
Amphibians, 91
Amylases, 42, 43
Anacharis, 16; osmosis and, 116; cyclosis
  and, 119
Anaerobes, 145; rise of, 429